Jim Cocke

OcT 2011

Reborn in America

Atlantic Crossings

Rafe Blaufarb, Series Editor

Reborn in America

French Exiles and Refugees in the
United States and the Vine and Olive
Adventure, 1815–1865

Eric Saugera

Translated by Madeleine Velguth

The University of Alabama Press
Tuscaloosa

Copyright © 2011
The University of Alabama Press
Tuscaloosa, Alabama 35487-0380
All rights reserved
Manufactured in the United States of America

Typeface: Adobe Caslon

∞

The paper on which this book is printed meets the minimum requirements of American
National Standard for Information Sciences—Permanence of Paper for Printed Library
Materials, ANSI Z39.48-1984.

Library of Congress Cataloging-in-Publication Data

Saugera, Eric.
Reborn in America : French exiles and refugees in the United States and the vine and
olive adventure, 1815–1865 / Eric Saugera ; translated by Madeleine Velguth.
p. cm. — (Atlantic crossings)
Based on the author's thesis written in French in 2007, entitled "Renaître en Amérique."
Includes bibliographical references and index.
ISBN 978-0-8173-1723-2 (cloth : alk. paper) — ISBN 978-0-8173-8511-8 (electronic)
1. Vine and Olive Colony. 2. French Americans—Alabama—History—19th century. 3.
French Americans—Land tenure—Alabama—History—19th century. 4. Agricultural
colonies—Alabama—History—19th century. 5. Alabama—History—19th century. I.
Title.
F335.F8S28 2011
976.1′05—dc22

2011002960

Publication of this book was made possible in part through the generous support of
Bradley and Anne Hale.

The maps are based on original cartography by Jean-Pierre Rousseau.

Cover: Detail of a panoramic wallpaper painted in France in the 1820s depicting
scenes from the French settlement of Aigleville, in Marengo County, Alabama. Courtesy
of the Alabama Department of Archives and History, Montgomery, Alabama.
Design by Robin McDonald.

This book is for Marie-Madeleine Guesnon (1916–2005) and Gwyndolyn (Gwyn) Collins Turner, who, an ocean apart, worked so diligently to preserve the memory of their remarkable towns, Sainte-Foy-la-Grande, on the banks of the Dordogne, in Gironde, and Demopolis, Alabama, on the white bluffs of the Tombigbee. My heartfelt thanks to them for permitting me to write of the exploits of these French people exiled to the United States of America after 1815.

Contents

Illustrations

Acknowledgments

My thanks go to Bradley and Anne (d. 2010) Hale for their unfailing financial support and enduring patience; Madeleine C. Velguth for her translation, research assistance, and vigilance; Gwyn Turner for funding the translation; Alice Meriwether Bowsher, William Henry (Harry) Britton (d. 2008), Kent Gardien (d. 2006), Deborah (Debby) Hunt, and the Marengo County Historical Society for their invaluable contributions; Nicole Sauvage, Michel Coustou, and other descendants of Jacques Lajonie Lapeyre (1787–1878), a lieutenant in the Empire's dragoons, for the preservation and loan of their family archives; my daughters Valérie and Albane, living in Bloomington, Indiana, and Brussels, respectively, for greatly facilitating my research in the United States and Belgium; Jean-Pierre Rousseau, teacher of history and geography in Nantes, for creating the maps; Fabrice Caruso (d. 2010), Annick Foucrier, Bernard Desmars, Michel and Monique (d. 2010) Konrat, Bernard Quintin, Marie-Jeanne Rossignol, Jean-Pierre and Catherine Rousseau, and Jacques Weber, for their meticulous reading; Ed Bridges (Alabama Department of Archives and History, Montgomery), Jérôme Cras (Centre des Archives diplomatiques de Nantes), Roy Goodman and Valerie Ann Lutz (American Philosophical Society, Philadelphia), Janet Hilyard (Hagley Museum and Library), and Mark Wetherington and Jim Holmberg (Filson Historical Society, Louisville) for access to collections of archives; Kirk Brooker (Demopolis), Daniel Cluis (Quebec), T. M. Culpepper III (Demopolis), Bucky Delaroderie (Baton Rouge), Stephanie Dupont de Nemours Speakman (Greenville, Delaware), Warren Fournier (New Orleans), Robert Hunt (Mobile), Joseph Turner (Demopolis), Ann Tremoulet-Davidson (New Orleans), Laurette Vallegeas (Nantes), Louise Webb Reynolds (Demopolis), and Suzanne Wolfe (Tuscaloosa) for their aid and support; and to the hundreds of French and American internet users with whom I maintained "genealogical correspondences," indispensable to the creation of a biographical repertory of thousands of French refugees and exiles in America.

Introduction

In the mid-1960s a crisis shook the old Franco-American alliance. France asserted its independence, left NATO, and condemned the actions of "the American war apparatus" in Vietnam.[1] Yet on October 28, 1967, the two nations exchanged tokens of friendship in Demopolis, Marengo County, Alabama. After months of preparation and two weeks of commemorative celebrations, the people and their political and religious leaders welcomed the French ambassador, who had come down from Washington, D.C., for this specific occasion.[2] One hundred fifty years earlier, in the searing heat of the summer of 1817, scouts of a society of French émigrés based in Philadelphia had chosen the place known as White Bluff, at the confluence of the Tombigbee and Black Warrior Rivers, as the site of a future colony. All were now here to honor the memory of the founders of Demopolis, the "city of the people."

The ambassador gave his speech outdoors, under a blazing noonday sun. He, who a year earlier had handed to President Johnson General de Gaulle's letter denouncing the stationing of allied American forces on French soil, had not come to speak of the disagreements between the two nations, but rather what united them. France had sent Lafayette; the United States, Pershing and Eisenhower. He extolled the blood shed by each country for the freedom of the other, and reminded his audience of the unique fact that they had never crossed swords. This reciprocal esteem was destined to last because it was knit together by common ideals, membership in the same family, and respect for the same "traditions of freedom, social progress and free expression."[3] According to historians, he continued, these traditions were seen in Demopolis as a distant heritage of its builders, combining "courage, cheerfulness and a delicate feeling for human relations." As some of them had been officers of Napoleon's Grand Army exiled to the United States after Waterloo, this led him to say that they were here at the heart of the Napoleonic legend.[4]

The imperial epic and southern sensibility thus mingled in Demopolis, "city of hope" in this "beautiful State of Alabama," the ambassador assured his audience. It is true that the mayor and police chief had succeeded in avoiding the racial clashes that had, two years previously, bloodied neighboring Marion and Greensboro,[5] but it was only the diplomat's self-assurance and the fact that his audience was entirely white that enabled him to speak in this way. Alabama had been subjected to Governor George C. Wallace's segregationist policies (1963–67) and Ku Klux Klan activism, with racial killings and police brutality against blacks struggling for their civil rights.

The ambassador's remarks were favorably reported in the press,[6] and that very evening he was the guest of honor at a reception for several hundred people in the historic residence of Bluff Hall, and later at a ball held in the civic center and opened with the "Marche Consulaire" that had been played at the Emperor's coronation. Mrs. David Turner, chairwoman of the 150th anniversary celebration committee, was radiant on the arm of the ambassador, who thanked her for keeping alive the memory of the city's origins. The "Marseillaise" rang out as the French flag was unfurled and attendees danced a quadrille. Marengo had been a dry county since Prohibition. The champagne, purchased in a neighboring county, was not as bubbly as in Paris, but this did not keep Demopolis from resounding with French verve until late into the night.

This pride of lineage was not merely an excuse for festivities. Several days earlier it had been officially inscribed on a historical marker at the site of the original town.[7] Under the title "Vine and Olive Colony," that marker records that a congressional act of March 3, 1817, granted, around this site, four townships —each six miles square, making a total of 92,000 acres—to exiles, Bonapartists, who founded Demopolis, Aigleville, and Arcola but very soon gave up the attempt to cultivate the grapevines and olive trees for which they had come.[8] Near a fountain dedicated to the exiles by schoolchildren, a second marker was inaugurated in 1971, confirming their arrival at the place called White Bluff (because of the white cliffs overlooking the Tombigbee, or Chickasaw Gallery, after the Indians who controlled navigation on the river hemmed in below).[9] The region's founding past, rich in its native population, explored by the French early in the eighteenth century,[10] then colonized by Bonapartist exiles, was recognized as an asset for its image and promotion.

The ambassador could not have been ignorant of the seriousness of recent events in Alabama. However, the point of his visit was rather to recall the role of the Bonapartist officers in founding Demopolis with the same romantic fragrance that people had enjoyed distilling for over a century in recounting this story. After this speech, an assessment of the discrepancy between myth and reality as to people, place, and chronology is in order.

The people who in Philadelphia created a Colonial Society to farm and perhaps manufacture in the southern United States were not in the least all officers outlawed for their political ideas, nor were they all French. Alongside them, on one hand, were foreigners and civilian compatriots who had emigrated after 1815 as they had, chiefly for economic reasons when peace reopened the shipping; on the other hand were colonists from Saint-Domingue who had sought refuge in America a quarter century earlier. Émigrés, refugees, exiles, banished or proscripted men[11]—how could these people with such diverse backgrounds and motivations strive toward the same goal and, for some, live together in a place located "on the Outermost Limits of Civilized America?"[12]

The society shareholders who were to go to Demopolis to farm the lands the American government had granted them did not move to this place in one great surge. Quite a few cofounders of the city and its county were so only on paper; they never went there. Studying their venture is not limited to their journey or possible settlement in Alabama, but must extend to the other states to which they migrated and where they settled, from Pennsylvania to Louisiana via Kentucky, Missouri, and so forth. In spite of the distance separating them, did these new French emigrants, originally united, remain in contact, or did they eventually disperse and melt into the American landscape?

As to chronology, the emigrants of the Colonial Society are generally thought to have arrived in America two years before the birth of Demopolis, beginning in 1815, after the defeat at Waterloo, which was fatal to the Emperor's partisans. But the story that interests us begins before this, in 1814, after Napoleon's first abdication; even earlier, in 1792, after black slaves revolted against white colonists in Saint-Domingue. Instead of concluding in the 1830s, when the French presence in Demopolis was only a memory, the story will be extended to the American Civil War, which saw the descendants of society members, sons and grandsons of slaveholding Domingan planters and imperial officers, defend the South. How can this reality be aligned with the traditions of freedom and social progress and the sense of human relations praised by the ambassador?

These are important questions not asked in 1967, although the subject had already elicited a good number of works offering anyone interested the possibility of a bibliographic survey.[13] In 1851, Albert J. Pickett, relating in his history of Alabama the clumsy attempts of the officers and their wives to plant vines and olive trees,[14] launched a popular American historiography glorifying the exile of a military-aristocratic elite when the Emperor was on St. Helena, and expressed in words a romanticized set of prints contemporary with the events: the soldier-farmer in uniform among his companions in misfortune, abandoning, in a tropical setting, his saber for a plow; the women rigged out as if in Paris, watching over both the warrior's rest and their children in the shade of palm trees.[15] In its concern to depict the antebellum South, the post-Reconstruction era of the last quarter of the nineteenth century saw a succession of publications that continued to favor the picturesque.[16] At the turn of the century, facts finally emerged despite everything, thanks to J. W. Beeson's articles in the *Demopolis Express,* the study by Gaius Whitfield Jr., grandson of a Confederate general, and finally Jesse Reeves's groundbreaking work on the Napoleonic exiles.[17]

World War I, in which French and American soldiers fought side by side, and the centennial celebrations of Demopolis and the state of Alabama aroused increasing interest in the subject, but still not in a scholarly way.[18] Rather, during the interwar period, the Franco-Alabamian episode, escaping the strict framework of research, came to the attention of literature and film, which cemented its mythological foundation. In 1934, Carl Carmer's book *Stars Fell on Alabama* again emphasized the quality of the French exiles, Bonapartists, and aristocrats, among the most influential men of their country, and the contrast between the refinement to which they were accustomed and the rusticity of their transplantation. In 1937 Emma Gelder Sterne published *Some Plant Olive Trees,* a fictionalized account of the story of the Vine and Olive Colony. In 1949 there was a motion picture, pure kitsch in style, *The Fighting Kentuckian,* with John Wayne, back from the Battle of New Orleans in 1815, helping the French threatened by land speculators. Morality wins out, and the leader of the colony, one of Napoleon's generals, gives his daughter Fleurette to her savior who has fallen in love with her.[19]

This motion picture gave the subject publicity and generated new interest,[20] as it confirmed a century of publications that had turned all the French émigrés into important figures of the Empire punished by the Bourbons for rallying to Napoleon during the Hundred Days. It was not until the 150th anniversary of the founding of Demopolis that the standard reference work of the following decades appeared: *Days of Exile.*[21] Its author, Winston Smith, an English professor born in Demopolis, did not really broaden the field of knowledge nor rid it of all the clichés surrounding the colony's history, but he reduced the Bonapartist preeminence in favor of the formerly minimized role of the colonists who were refugees from Saint-Domingue. The latter, cofounders in 1816 of the Colonial Society of French Emigrants in Philadelphia and its most active members, had been overshadowed by the Napoleonic stars.

Smith's book made people want to go further, among other things to know more about the original participants in the founding adventure. From 1974 to 1976 the local daily *Demopolis Times* published a long series of articles by James and Emogene Armistead on the pioneers of Marengo County, and then, in 1983, another series of ten articles, "The Vine and Olive Puzzles," by Kent Gardien. The descendant of a former officer of the royal guard who emigrated to Alabama in 1829,[22] Gardien greatly advanced knowledge of the history of the French colony, its military, Domingan, and Philadelphia roots, through the rigor and breadth of his research on both sides of the Atlantic.[23] This resulted in a document of seven hundred typed pages, a who's who of the Vine and Olive Colony and the Champ d'Asile, its avatar in Texas.[24] Gardien did not draw from this comprehensive survey the work that should have followed, but in 2005 Professor Rafe Blaufarb did, publishing a summary of the biographical data collected by Gardien in the first full-length scholarly study devoted to the role of Bonapartists in the states bordering the Gulf of Mexico: *Bonapartists in the Borderlands.*[25]

While expanding on the work of his predecessors, Blaufarb differs in his presentation of the ins and outs of the French presence in Alabama. He sees this presence as strategic in a South that had shown itself vulnerable in Louisiana when a British force had landed at the mouth of the Mississippi in 1814, as well as now in Alabama and Georgia where Indians and fugitive slaves were conducting guerrilla operations from Spanish Florida. These states had to be made secure, and their defenses reinforced to check the cross-border raids and wipe out the risk of renewed

Royal Navy activity along the coasts of the Gulf of Mexico. Congress, he argues, therefore offered the French colonists 92,000 acres of land in Alabama because some were veterans of the Napoleonic Wars, including a few renowned generals like Grouchy, Lefebvre-Desnoëttes, the younger Lallemand and Clauzel, and because many others, like the elder Lallemand, Rigau, and Vandamme, were yet to come. They were expected, in time of need, to take up arms and defend their lands as well as the nation that had granted them.

Argentine scholar Emilio Ocampo gives his version of the origins of Alabama's French colony in a work published in English in 2009, *The Emperor's Last Campaign*.[26] He sees the Vine and Olive Colony as a mere screen hiding a completely different Bonapartist enterprise: military and financial support of the patriotic independence movements in the Spanish colonies of South America, but also of the Bonaparte family, destined for a new dynastic future on the throne of one or several liberated countries, such as Mexico. This mainly centered on the former King Joseph, then living in New Jersey, or on Napoleon himself, who would be brought from St. Helena by steamship, submarine, or hot-air balloon, as some suggested. By obtaining the concession of these lands in Alabama—which, furthermore, were near the Spanish territory of Florida that could be invaded and used as a base for other conquests—the Bonapartists did not in the least intend to develop them, but to sell them quickly for a good price to purchase, among other things, the provisions, uniforms, arms, and munitions necessary for the liberation of South America and the Emperor. Ocampo has revived this old story of the Bonapartists' plans in America based on his research and interpretation of a vast amount of material in the public and private archives of many countries, some of it unpublished. I am not, however, convinced by the ideas advanced in this book and the preceding one regarding the place of the Vine and Olive Colony in the Bonapartist saga in America, and shall therefore in the present study open the debate by confronting them with my own view of things.

A review of the American bibliography thus shows that interest in the French emigration to Alabama has been continuous, but that, with the exception of the most recent publications—regardless of the difference in assessment—the quality of the historic output has often suffered from its numerous clichés perpetuating the illusory romantic vision of exiles from imperial high society at grips with a hostile natural and human environment.

Whereas other French ventures in America have enjoyed definitive studies, such as Annick Foucrier's *Le rêve californien* or Jocelyne Moreau-Zanelli's *Gallipolis,* the French bibliography on the subject is disappointing. At best, French works deal with exile during the Restoration, the fate of Napoleon's soldiers in France, their participation in various national movements,[27] and the exile of Bonapartists in America in a general way, with special mention of the Champ d'Asile in Texas, whose rapid and tragic end in 1818 had moved contemporaries and resulted in published accounts.[28] There is thus a collection of chapters, articles, and scattered allusions, never very numerous, lacking in depth, and often dated.[29] Republished several times since 1950, Simone de La Souchère-Deléry's *À la poursuite des aigles* is considered a classic, followed by Inès Murat's *Napoléon et le rêve américain* in 1978. Lacking a work of synthesis, the only recourse is the study of the exiles' personal careers in biographical dictionaries and individual biographies.[30]

In 1893 Georges Bertin devoted a book to the leading Bonapartist in exile, Joseph Bonaparte, the Emperor's older brother, who lived for fifteen years near Bordentown, New Jersey. In 1972 Fernand Beaucour defended a dissertation on Joseph's secretary, Jean Mathieu Sari, a Corsican naval midshipman; he also studied two other French émigré officers sometimes confused with one another because of their names, Nicolas Raoul and Jacques Roul.[31] Two generals who played leading roles, Charles Lefebvre-Desnoëttes in the Alabama colony and Simon Bernard at the head of the American Corps of Engineers, were also the subjects of dissertations, defended in 1961 and 1988, respectively.[32] The American adventure was not limited to military men; important civilians of the Revolution and the Empire shared the lot of the officers, individuals like the former minister Regnaud de Saint-Jean d'Angély, the former police prefect Réal, and former Convention members Hentz, Pénières, and Lakanal. Each of them has his biographer.[33]

The large amount of information and many analyses published between 1851 and 2009 both in France and the United States do not preclude further research as, in the end, information and interpretations are often repeated from one publication to the next, to the detriment of a significant progression of knowledge. The present work is an opportunity, in chapter after chapter, to correct errors, fill in gaps, clarify facts, redefine viewpoints, and put forward a detailed version of the successive stages of the Vine and Olive Colony's history. Beyond the efforts to plant vineyards and olive groves in Alabama, the relocation in the United States of French émigrés whose country no longer wanted them or was unable to retain them must be described. This necessitated exploring dozens of archival collections on both continents, often facilitated by digitalization of documents and their availability online, thanks to institutions, individuals, historical societies, and genealogical associations.[34]

French public archives made it possible to follow the individual paths of émigrés for whom there was sometimes only a name. Since they were assumed to be chiefly officers, research began with personnel files in the army and naval archives in Vincennes. The realization that a greater number were Domingan refugees led me to spot-check the Saint-Domingue public records and notarized document collection at the Centre des archives d'outre-mer in Aix-en-Provence. The biographical database compiled from these national archives was supplemented by material gathered in departmental and communal archives in all the regions of France and, in the United States, in county courthouses, principally in Alabama and Louisiana, in search of public records and any document signed by a justice of the peace or notary public concerning the French immigrants. Establishing their identity and following their paths also meant taking an interest in the circumstances of their emigration. For information regarding their departures from Europe: the inexhaustible subseries F[7] of the French National Archives contains police reports, exiles' files, and especially passports for foreign travel issued by the minister of general police, the Paris police chief, and the prefects in their administrative districts; the M series of the departmental archives of Gironde also offers a harvest of passports, which the departmental archives of Loire-Atlantique and Seine-Maritime complete with the names of the passengers on French ships leaving Nantes and Le Havre for America; in Brussels, Anderlecht, and Antwerp the royal archives also give information on the presence and departure of French exiles. For research into their arrivals in the United States: the American National Archives has the names of passengers disembarking from American ships in Philadelphia prior to 1820, and in New Orleans, New York, and Boston after that date.

Research on the French émigrés' identities was conducted concurrently with that concerning their settlement in America, particularly in Alabama. This second aspect was studied in depth in the archives of the Ministry of Foreign Affairs (Paris and Nantes), by far the most important, and in the United States in the National Archives (Washington, D.C., and Philadelphia), the archives of the states of Alabama (Montgomery) and Louisiana (in Baton Rouge), and various public and private archival collections in Wilmington (Delaware), Louisville, St. Louis, New Orleans, Mobile, Birmingham, and Pensacola, among others.

The perusal of public archives furnished the material necessary to construct the body of the subject; the consultation of private documents gave this body the organ it needed to come to life: its heart. On the south bank of the Dordogne River, the walled Huguenot town of Sainte-Foy-la-Grande (Gironde) has been home to the Lajonie family uninterruptedly since the arrival around 1470 of its earliest known ancestor: Antoine Lajonie, a cloth merchant. At the dawn of the twenty-first century, two of his descendants, Nicole Sauvage and her brother Michel Coustou, no longer bear his name, but have kept their centuries-old family papers, among them the correspondence of an ancestor who served as officer in Napoleon's armies. Jacques Lajonie, a victim of the White Terror that struck southern France during the second Restoration, had to flee his country and take refuge in Philadelphia. Arriving in January 1817, he met other outlawed men who urged him to join the Colonial Society of French Emigrants. Until he returned to France in March 1829, he wrote more than one hundred letters to his family back in Gironde, from various places in the United States, but especially Alabama, relating the itinerary of a Frenchman in the New World, along with those of his compatriots, Domingan refugees and exiles from France, founders with him of the Demopolis colony.[35] Together with some nine hundred other letters from émigrés on the same subject, gathered and compiled in France and the United States, they make possible a sincere, intimist approach to this story that official documents cannot offer.

In France, these other letters come from Jacques Lajonie's passive correspondence: his family's answers to his letters have disappeared, but not those of society members who were friends and settled in Alabama, Louisiana, and Arkansas. Many others from the files of the French consulate in New Orleans, housed in the Centre des archives diplomatiques in Nantes, are from French citizens worried at being without news or eager to receive the supposedly fabulous estate of a husband, father, or brother who died in America. Published memoirs and letters from exile complete this group: thus, from 1873 to 1989 those of Marshal Grouchy, Joseph Bonaparte, and Convention members Billaud-Varenne, Lakanal, and Pénières.[36]

In the United States, Philadelphia's Girard College houses banker and ship-owner Stephen Girard's passive correspondence. French exiles asked this extremely wealthy American, born in Bordeaux, to manage their money when they had it and for help or employment when they did not. In Wilmington, Delaware, where their powder plant was prospering, the Duponts were solicited for the same reasons, as shown by letters received and archived in the company's boxes at the Hagley Museum. The library of the University of North Carolina in Chapel Hill owns eleven letters written by General Lefebvre-Desnoëttes in Alabama to General Clauzel (1818–19), while the Filson Historical Society of Louisville has dozens of others exchanged by French immigrants, concerning, among other things, the sale of their Alabama lands. Finally, like Lajonie's descendants in France, Americans have also kept the papers of their exiled French ancestors: in Demopolis, Louise Webb Reynolds has those of John M. Chapron, and in Birmingham, Mary Walker Lamkin holds those of Francis L. Constantine. This corpus of letters and documents drawn from the most diverse sources has made a reexamination of this story possible.

The Franco-American epic unfolds in five parts. The first, placed under the sign of treason and terror, presents the political and military situation in France from Napoleon's abdication in April 1814 to the beginnings of the second Restoration a little over a year later. How did the deterioration of the relationship between a portion of the army and the Bourbons end in a plot in northern France and a rallying to the Emperor in March 1815? Waterloo condemned to death or exile fifty-seven important figures guilty of supporting the Hundred Days and, in a ricochet effect, banished the survivors of the National Convention who had voted for the death of Louis XVI. In the climate of terror created by ultra-royalists in southern France, the case of Gironde, whose capital of Bordeaux had been the first to rally to the Bourbons in March 1814, stands out. Among the Bonapartists harried by the local authorities and forced into exile were Lajonie and the American consul William Lee, two key figures in this study.

An important emigration center, Bordeaux was not the only port from which people left for America; they sailed from all of coastal France, from Le Havre through Nantes to Marseille, and beyond the borders from Belgium and Italy. Part two—across the Atlantic—opens with an account of this exodus and continues with the conditions of welcome in American ports, from Boston to New Orleans. Could these Frenchmen, driven from their country or fleeing a regime they despised, enjoy official hospitality on American soil, without this offending the new ambassador, Hyde de Neuville, whose uncompromising loyalty to the king was feared by all? Alarmed by plans fomented in America to help Napoleon escape St. Helena or to create a Napoleonic Confederation around Joseph, he quite understandably approved of Congress's passing the act of March 3, 1817, that granted the French of so-called recent emigration—were the refugees from Saint-

Domingue truly "recent"?—lands in the Alabama Territory where they were to promote the cultivation of grapevines and olive trees.

How could the French have been entrusted with the task of acclimating these plants if there had not been precedents in the United States? The answer to this question introduces the third part, on Alabama's exotic roots, in an overview of attempts already made in this area, from the Huguenots' vineyards in New Bordeaux, South Carolina, in the second half of the eighteenth century to those of Swiss winegrowers in Kentucky and Indiana at the beginning of the next. Alabama's soil and climate, thought to be comparable to those of southwestern France, were supposedly assets, but skeptics may have been right to criticize the incompetence of the future farmers, city intellectuals unfit for hard work in the fields. A meticulous task of biographical research was indispensable to determine whether city dwellers actually outnumbered rural people in the society and consequently whether the backgrounds of the majority prepared them or not to prune vines and pick olives.[37] But how many émigrés realized this at the time, imbued as they were with the idea of a return to nature promoted by the authors of the day, in the wake of Rousseau's *La nouvelle Héloïse*?

The experience of their inevitable woes, since they were no more adventurers than they were farmers, began with the long, perilous journey to the Tombigbee on the various routes they took to reach the "French Lands" chosen for them in Alabama. This is the fourth part. The apprentice pioneers were rewarded by one difficulty after another, while in Philadelphia Domingan refugees were quarreling with henchmen of the elder General Lallemand who wanted to sell the lands granted by Congress and found a more ambitious military-economic colony in Texas: the Champ d'Asile.[38] This secession could only further compromise the chances of success of those members who persevered regardless of the pitfalls of which Jacques Lajonie, the model French colonist, who took his work the most seriously, gives an edifying inventory in letter after letter. What did the French colonists lack: mastery of the language, American citizenship, money, pragmatism, business sense, enough slaves to work their plantations, friendly nature, or the will to settle permanently in Alabama? Homesickness got the better of a great many émigrés who went back home, some after being pardoned by Louis XVIII, others after the Revolution of 1830 in France.

By this time, Demopolis, the French colonists, and the vineyards, three inseparable elements of this story, were dying. But none of them perished, and those who had persevered could then be truly reborn in America.

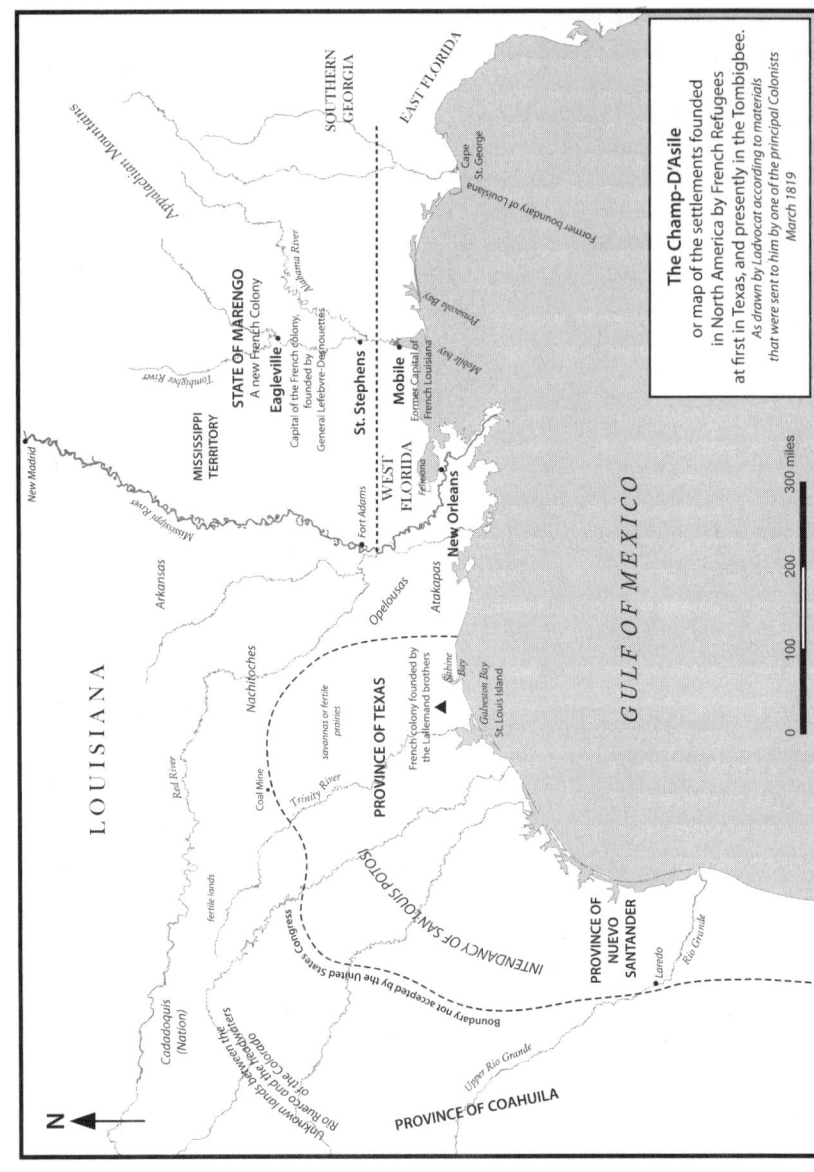

Map 1. The Champ d'Asile, based on a map originally drawn in Paris by Ladvocat in 1819.

I

Treason and Terror

1

A Critical Moment

For each of us there is a critical moment; well or badly chosen, it decides our future.

—Chateaubriand, *Les mémoires d'outre-tombe*

In the early months of 1814, an exhausted French nation watched almost impassively as its territory was invaded, and Paris offered vain resistance to the allied forces that took it on March 31. Napoleon had worked wonders to drive back the invaders, but the enemy's crushing numerical superiority finally overwhelmed what was left of the Grand Army. The Emperor abdicated and left for Elba. The man who legitimately claimed the French throne—vacant since the execution of Louis XVI in 1793—replaced him: Louis XVIII restored the Bourbon monarchy and granted his subjects a constitution that evidenced his desire for conciliation and suggested a gentle reappropriation of the country. The peace-loving king spared the French a civil war, but he lacked tact, firmness, and foresight. His many blunders and errors aroused contempt for his regime and led to the Hundred Days of Napoleon's return.

This return generated quite a bit of anxiety. To the post-revolutionary generation that had come to maturity during the Republic and the Empire, the name Bourbon did not mean much. People scarcely knew the new king; his brother, the Count of Artois, with his two sons, the Dukes of Angoulême and Berry; and finally his niece, Marie-Thérèse, daughter of the martyred king. The majority of the French certainly had no need to fear the anger of an heir frustrated at being so long deprived of his legacy, but many must have been alarmed as to his true intentions toward them. This was particularly so for the surviving members of the National Convention who had voted for Louis XVI's death, the military men who had turned monarchical Europe upside down, and the civil servants who had run the imperial machine. In one proclamation after another, the pretender to the French throne had shown a rather positive and encouraging evolution in his thought—but was he sincere?

Very early, he had promised a return to the ancien régime in all its purity and eternal damnation to those monsters, the regicide members of the Convention, before relaxing his position, notably when Napoleon was crowned in 1804: absolving the crimes of the Revolution and compromising his legitimacy was out of the question, but if he returned to France, he would proclaim a general amnesty, keep the military men at their ranks and the civil servants in their positions. In 1813, near London, when Napoleon's star was beginning to pale after the Russian disaster and the possibility of a restoration of the monarchy was taking shape, he went even further, promising to forget the events of unhappy times, abolish conscription, confirm the reforms in the administration and the army, and reward those soldiers who would join his cause. Finally, in May 1814, in Saint-Ouen, near Paris, the king—who had replaced the pretender—announced a new constitution that included all the basic freedoms but which would under no circumstances question the principle of the divine right of kings. This was the limit of Louis' renunciation: he had merely granted his people a constitution, the Charter, read in the Senate and the Chamber of Deputies on June 4, 1814. Despite appearances, the king remained staunchly conservative, and there would be no lack of uncompromising royalists to support him in his reaction and get him to go back on his declarations.

INTEGRATE OR PURGE

Once the time of promises and concessions had come to an end, the task of the government proved difficult, if not impossible, as it tried to amalgamate the France of its supporters, whose loyalty had to be repaid with rewarding positions, and that of the imperial soldiers and civil servants. The first Restoration could not bring together these two versions of France, divided by what the Duke of Berry called twenty-five years of banditry.[1] It served those who had remained faithful to Louis, before those who had supported the Emperor in his conquests.

Even so, most of the general officers who had fought brilliantly during the Empire accepted the change. The marshals,[2] whom Napoleon had led to the summit, refused to follow him in his fall, and became the liegemen of their new overlord. After voting to depose Napoleon or approving his abdication, they all pledged allegiance to Louis XVIII.

The *Dictionnaire des girouettes* ("Dictionary of Turncoats"; literally, "weathervanes"), published in 1815,[3] had ammunition aplenty, and some contemporaries condemned such a rapid passage to yesterday's enemy. But the military men could justify their declarations politically by pointing to unambiguous official proclamations: on April 2, 1814, the Senate, in the name of the entire French nation, had released them from their oaths; in Fontainebleau, Napoleon himself had asked his Imperial Guard to continue to serve France with honor and to be loyal to their new sovereign.[4] The officers' fickleness is thus only one part of a whole that included this release from their oaths, the wearing effect of years of war, and a major change in their attitude: they were now the servants of the government and not of an individual in power; the army was no longer royal, republican, or imperial, but French.[5] As for the civilians, they did the same, being in the habit, since the Revolution, of taking oaths without keeping them.

Finally, the marshals, with their sudden royalist feelings, were also keeping their eyes on their patiently amassed nest eggs: "great lords of the Empire united to their pensions by sacred and indissoluble bonds, no matter what hand is dispensing them."[6] Napoleon had given them all they could wish for; Louis was able to mollify them by promising always to rely on their support. His intelligence and courteous ways enabled him to carry out an enterprise of seduction that consisted of giving prestige to the new clothes of the imperial aristocracy by sticking onto them the labels of the aristocracy of the ancien régime. He bought their services without waiting for the proofs of trustworthiness that his generosity would have been expected to recognize. Napoleon had created the imperial nobility; the king confirmed it and made ten marshals peers of France. Napoleon had distinguished them in the Legion of Honor; the king promoted them into the royal Order of Saint Louis.[7]

The generals were just as easy to persuade. As soon as Napoleon abdicated, they began to support the government and continued to flock to it.[8] Clauzel is a good example: Knight of Saint Louis on June 1, 1814, inspector-general of infantry on December 30, count the next day, and Grand Cross of the Legion of Honor on February 11, 1815. How many men could display, side by side on their chests, the Legion of Honor and the Cross of Saint Louis? Even those one would not suspect of compromise had accepted investiture into the Royal Order or strove to be admitted.

Because of his role in the battle of Austerlitz and his military talent, Napoleon had come to terms with the touchy General Vandamme's reputation as a plunderer and a ruffian. Corpulent, his neck buried between broad, round shoulders, with a slight stoop and a small forehead, Vandamme's massive physique suited his bad temper perfectly. Returning from captivity in Russia, where he had succeeded in exasperating the czar with his retorts, he asked for an audience with the king, who refused to receive him. On September 24, 1814, the minister of war requested he leave Paris within twenty-four hours for his estate in Cassel in northern France. The general took offense at the calumnies that he thought had motivated the decision. He had never personally slain émigrés: "A brave man like me fights, kills, but does not murder."[9] Vandamme's great fear was that the oblivion in which it seemed he was being enveloped would make him lose the fruits of twenty-five years of glory, including twenty-two as general. The king did not give in. Vandamme was added to the already-long list of officers retired from active duty.

Appointing General Dupont as the first war minister was not a good idea from the point of view of national reconciliation. His name was linked to the defeat of Baylen in Spain, one of the first to shake the pedestal of Napoleon's invincibility. While that might retrospectively earn him the gratitude of the royalists, it was nonetheless true that this disaster was a millstone around his neck, all the more unbearable because he did not deserve all its infamy. The danger was that his decisions would be interpreted as the vengeance of a humiliated man. Dupont began by reducing the size of the army, both the enlisted men and the officer corps, but this was only in response to peace treaty stipulations, budgetary necessities, and a reduced need in times of peace.

A royal edict eliminated infantry and cavalry regiments and sent into retirement the officers with seniority, the injured, and the infirm, condemning to destitution those who had not served long enough to qualify for retirement. It suspended thousands of others from active duty, those resistant to the idea of serving the Bourbons, and, above all, those who were no longer needed. The latter were put on half pay, barely enough to live on. The government had created a new social category, that of half-pays whose image, with cane, top hat, and frock coat, was perpetuated by the engravers and writers of the time. Joining the malcontents, they did their best to influence public opinion and create a poisonous climate in the army, whiling away their time and forgetting their poverty by venting their spleen on a hated regime, in Paris as in the provinces. The army, formerly the terror of families, became friendly and popular, and its past glory became national property.[10]

Soldiers were not the only target of the first Restoration. In all of the senior branches of the civil service there were high-ranking officials associated with the imperial and—even worse—the revolutionary past, such as the members of the Convention who had voted for the death of the king. In 1814 many were still in responsible posts where they could not remain. Some did not need to be booted out, understanding on their own the incongruity of their situation. How could one shout "Long live the king!" after having his brother guillotined, defending the Republic, and then serving the Emperor? Others did not have the same good grace, but the result was the same; the regicides were gotten rid of. And they were not the only ones to disappear. The king refrained from appointing former senators known for their revolutionary opinions to the Chamber of Peers, which replaced the Senate, and the Council of State was purged of pillars of the imperial regime.

The first Restoration, however, burned no one at the stake, condemned no one. It acted like a political force taking back the control of the sensitive sectors of the state with reins long abandoned to other hands.[11] But while tact was used in dismissing the elite, it was forgotten in firing thousands of modest civil servants and reducing anonymous junior military men to half pay.

THE NORTHERN CONSPIRACY

The only thing we are certain of is the alarm this news has occasioned throughout the country. Parties are forming, and if the Bourbons do not arrest this evil in the bud, the most serious consequences may follow. The army, it is feared, are in his favor, while the conduct of the old *Noblesse,* Clergy and Emigrants has disgusted the people and divided their sentiments. The King is growing very popular, but the rest of the Royal family are not liked, except the Duchess of Angoulême, who inspires a general interest in her favor, from her virtues and her sufferings. The English, who are detested on the continent, are suspected of being at the bottom of this affair, with a view to create a civil war in France.[12]

This is the picture of France that William Lee, since 1801 United States consul in Bordeaux, sketched for Secretary of State James Monroe on March 12, 1815. For twelve days Napoleon had been marching toward Paris at the head of a band whose original corps had landed on a Mediterranean beach. No one had seen it coming, particularly not those whose assignment it was to keep their eyes and ears wide open, because the notion of an imperial return seemed so ludicrous to them. Even when Napoleon was again on French soil, the seriousness of the event was played down; in Gironde, the authorities pretended to consider it the last effort of a madman. But the American consul knew by a letter from Lainé, president of the Chamber of Deputies, to his friends in Bordeaux that the situation was worrisome. He himself was not surprised. Unlike courtiers and ministers incapable of making a diagnosis, Lee felt the arrhythmic pulsations that were troubling a large portion of the country and was able, as an attentive observer, to analyze their cause.

THE END OF ILLUSIONS

The hope of a France at peace, free and just, had led people to accept the Bourbons, but this hope dwindled away and discontent spread through all levels of society. The common people, who had everything to gain after being bled dry by twenty years of war, felt completely lost, as did people of note, irritated by the aristocratic pretensions encouraged by the monarchy. An era thought bygone was resurfacing, and the throne, priesthood, and nobility spared no effort to make this clear to one and all and directed their hatred against the friends of the imperial cause. The Count of Artois, heir to the throne who looked back nostalgically to former times, was disappointing, as were his sons. Married to her cousin, Marie-Thérèse, Duchess of Angoulême, who had miraculously escaped the scaffold, was the exception because people felt sorry for her. The Charter was flouted, the press censored; political opposition increased, and satire had a field day. The ministers seemed incapable of controlling the country, which was going to the dogs. England was accused of running it behind the scenes and of savoring the nation's collapse. Even the army was plunged into gloom.

At the beginning of the Restoration, it is true, a great number of general officers had been enthusiastic because the king had successfully flattered them and received them at court to ensure their loyalty. These men, who were well taken care of, intended to enjoy their incomes and pursue in tranquility careers that until now had had their share of wounds and bruises. The situation was quite different for many military men, who could no longer put up with the government's indifference to their service records. Paris police reports abound with their "grousing," their "despicable" comments about the king during meals labeled "scandalous orgies." The reports also deplore the bad frame of mind of the troops and the fact that the officers did nothing to remedy any of this.[13]

In December 1814, the appointment of Marshal Soult as war minister had seemed a wise move. Talented and hardworking, he was supposed to make people forget his predecessor's erring ways, but in his frenzy to please the king he managed to antagonize everyone—both the Bonapartists, who accused him of going back on the imperial religion, and the royalists, who accused him of going too far to be truly sincere.[14] Soult was universally hated.

The king was aware of the increasing discontent, but he thought it had more to do with his entourage than with himself. He also, however, was included in the army's disaffection with his regime, not only because he seemed to go out of his way to reopen wounds that had barely healed, but because his very person displeased them. This was a far cry from the power of fascination that Napoleon had exercised over his men, and their going over to the king had not changed anything. A shrewd mind and a regal way of holding his head imposed respect, but a heavy body, worn out at sixty years of age, earned him from his detractors the nickname "fat pig." "These men deprived of their captain were forced to salute an old King, disabled by time rather than by war," explained Chateaubriand.[15] Born of defeat, he could not be imagined in combat, only at the table, appeasing a compulsive appetite, or confined to his wheelchair by gout. The comparison with Napoleon, a hero on horseback born of victory, was cruel.

The military men, despairing of seeing the end of a regime they could no longer abide, began to plot under the aegis of a triumvirate dominated by the ex-minister of the imperial police Fouché, Duke of Otranto.[16] This defrocked priest and regicide former member of the Convention was willing to work with the Bonapartists as long as they did not insist on reinstating the Emperor, whom he did not want at any price because Napoleon had dismissed him several years earlier. Nor did Fouché want the king, whose government, headed by Talleyrand, had dispensed with his services. He preferred the solution of Napoleon II with Marie-Louise at the head of a regency council, of which he himself would be a member.[17] But the King of Rome was with his grandfather in Vienna, and his mother, the ex-empress, was gradually forgetting her husband and French matters. Swift action was necessary, for the risk of Napoleon's escape from Elba had never been greater. None of the people responsible for his surveillance were doing their job: neither the chief of police, to whom Fouché expressed the worry, in January 1815, that the French seacoasts were not being guarded, nor the postmaster general, an old friend of the king, who was falling asleep at his desk and did not know what was going on in his department.[18]

Fouché persisted in seeing Napoleon as a jaded figure whose return would make all Europe, armed to the teeth, swoop down on France; meanwhile, the Bonapartist party continued to grow stronger. All of Paris was whispering, in meetings secret and public. The police noted that generals were taking turns hosting gatherings at night, that at Vandamme's, ten to twelve of them had reveled in subversive measures, and that a get-together at the home of another general included one of the most fiery critics of the new government.[19] People met in salons, such as those of Hortense de Beauharnais, ex-queen of Holland, or Madame Maret, Duchess of Bassano, whose home the police considered the headquarters of all who had remained devoted to Napoleon. This finally led to the hatching of a plot whose active elements were Generals Lallemand, Drouët d'Erlon, and Lefebvre-Desnoëttes. The elder General Lallemand summarized the general feeling: "This regime is unbearable and we shall break it with the sword: this has been decided."[20]

And so it was decided in mid-February that the time for action had come, even if no agreement had been reached as to who would replace the king; the essential thing was that he be replaced. There was consensus on this last point, as on the timing of the intervention, set for the spring, and its point of departure in the northeastern part of the country: the thirty thousand men of Lieutenant General Drouët, Count of Erlon's Sixteenth Territorial Military Division were stationed there. Wholeheartedly devoted to the Emperor, Drouët had nonetheless agreed to serve the king.[21] But in Paris he soon began to participate in meetings with the generals mentioned above. He was given the lead role in setting the plot in motion. When the time came, he would march on Paris with his troops, joined by other garrisons along his route, swept along in a snowball effect. It was expected that the soldiers, informed en route of the operation's goal, would go along with it, because rejection of the Bourbons would prevail over their military discipline. After this, nothing would prevent the taking of the capital, and the king and his family would be chased to the border.

The conspiracy was waiting to be activated when an event caught its instigators unawares. Known in Paris on March 5, Napoleon's landing in Juan Bay was unexpected but not unforeseeable. According to Mme De Staël, "the simple good sense of the Swiss peasants" led them to predict Napoleon's return.[22] The uncertainty had less to do with this possibility than with the manner: "We had a vague feeling that he would come back, that a life of miracles would not fade out on a rock between Italy and France; but how and in what way?" wondered his devoted former postmaster general.[23]

A flotilla that left Elba with a small troop arrived in southern France on March 1.[24] Having seen the errors committed by the Bourbons and noted the accelerating dissatisfaction with the regime of the French in general and the army in particular, Napoleon had decided to reconquer France.[25] On March 20 he entered the Tuileries palace, empty of its royal resident, who had packed up and left. Napoleon had needed only twenty days to resume his position as reigning emperor, "without a single rifle shot being fired, a drop of blood spilled, and without any conspiracy on the part of the inhabitants of the country," marveled his valet.[26]

When this invasion of France by a single man is seen in parallel with the military bustle of a few generals in the north, it becomes evident that there was perhaps nothing more extraordinary in Napoleon's entire career—aside from the later episode of Waterloo and its dramatic consequences for the country. At very nearly the same time, two attempts to free France from the Bourbons had set out from opposite points of the compass without ever consulting one another. While it is certain that Napoleon knew what was being plotted in Paris, it seems equally true that the conspirators had no inkling that the Emperor was leaving the island of Elba.

AN ASSOCIATION OF PLOTTERS

Napoleon's return did not merely upset Fouché's plans; Fouché simply did not want him back. So, expert tightrope walker that he was, he hedged his bets. He decided to activate the military movement that was to march on Paris from the north to take the city. If the plan succeeded before Napoleon reached his goal, Fouché could form a government of national union, call the Chamber of Deputies and the Senate into session, bar Napoleon's way, and that would be the end of him. If it failed, he would go over to the Emperor and rejoice at his success. Since both options would turn out in his favor, he set things off by lighting the fuse that had been ready for weeks.

On March 5 Fouché summoned the Lallemand brothers to his home in Paris.[27] At forty, François Antoine, known as Charles, was the elder and served in the cavalry. Since August 1814 he had been in command of the subdivision of the department of Aisne in the city of Laon. An enlisted man during the Revolution, he had been promoted in one campaign after another, from Egypt to Saint-Domingue. A colonel after the battle of Iéna, a baron before being made a general in Spain, he was an outstanding officer, speaking several languages, and an elegant horseman, according to his contemporaries.[28] He was said to be "capable and discreet, calmly vigorous and resolute,"[29] but he was also considered proud, impulsive, and extremely hypersensitive. His love for the Emperor was equaled only by his hatred for the king. The police were keeping an eye on him in Paris, where he was spending more time than in Laon: his home there visited by many officers meeting at set times and appearing to have "very bad principles."[30]

Henri Lallemand was his younger brother. Unlike François Antoine, who was promoted from the ranks, Henri had attended the École polytechnique[31] to go into the artillery. Appointed captain of the Foot Artillery in the Imperial Guard in 1806, he became its chief of staff in Russia. He was promoted to general when the allied armies invaded France. In October 1814 the Restoration government put him in charge of settling the accounts of the ex-Guard's Foot Artillery at the arsenal of La Fère. This was near Laon, where his older brother was stationed. The younger Lallemand was not as spirited, but shared his convictions.

Alerted before his co-conspirators, Fouché kept the news of Napoleon's landing in southern France from them, easily misled the Lallemand brothers, and urged the elder to go to Lille and give the signal setting events in motion. On March 7 Drouet ordered the regiments of his military division to set out for Paris. Not knowing they were being committed to a rebellion, they began to comply the next day. The conspiracy was on the move but, quickly informed, Soult ordered Drouet's arrest: unsuited to intrigue, the latter became frightened and told his troops to turn back. Lefebvre-Desnoëttes' Mounted Royal Chasseurs Corps, which had left Cambrai, continued. Drouet notified him of the turn of events, but the messenger was intercepted on the tenth at the La Fère arsenal, where Desnoëttes had arrived the evening before with the intention of seizing it. This general was one of the chief plotters. Eleven months earlier he had escorted the Emperor from Fontainebleau on the road to Elba.

Desnoëttes was a model imperial officer: a captain and Bonaparte's aide-de-camp at the battle of Marengo, a cavalry general at the age of thirty-three, and Jerome Bonaparte's first aide-de-camp; Napoleon could not have found for this position "a man better-liked and more respected by the army."[32] The Emperor thought very highly of him, had married him to a Corsican woman related to the Bonapartes, and had given him the house in Paris where he had lived with

Josephine.[33] In 1808 Desnoëttes was promoted to colonel of the regiment of horse chasseurs of the Imperial Guard, one of the most coveted positions in the army. The title of count, which he received the same year, crowned the first part of his career, but the second began with his December capture in Spain.

In 1812 he escaped from England, where he had been a prisoner on parole, and got back in time for the Russian campaign, from which he returned with the Emperor. After distinguishing himself in Saxony and France, he recommended his men to the new king, assuring him that he could not be better guarded than by such brave, loyal soldiers, and then asked to return to his family. But in April 1814 Desnoëttes gave his support, and that of his troops, to the various actions of the provisional government and proclaimed his devotion to the dynasty of the Bourbons. In Paris he was presented at court, where the new organization of the army, in which he had his place, was explained to him. The king had not shared Napoleon's presentiment that, as only he could lead the Guard, it should be dismissed, with good pensions given to the noncommissioned officers and soldiers, and promotions to those who wanted to continue in service. While the Imperial Guard as such was indeed dissolved, the regiments composing it were preserved intact and dispersed to various provincial garrisons. The men retained the privilege of a higher rank, but their pay was cut by a third and they lost the prestige of being a guard close, no longer to the Emperor, but to the king, in favor of a new royal guard composed of young men of means who had never seen military service. In the end, this ceremonial army could not fail to arouse the sarcasm and rancor of soldiers relegated to the sidelines. Desnoëttes, however, could not yet have been at this stage of disenchantment when he was offered the command of the Royal Chasseurs of France, which amounted to his retaining his position: "I must admit that in accepting this command I gave myself entirely to my new duties. My devotion was unlimited and I did all I could to make people love and respect the government of the Bourbons. I would happily have shed all my blood for them."[34]

He was garrisoned in Saumur, where he arrived in May 1814, and proceeded to adjust his conduct to his words. On an inspection visit, Marshal Ney praised Desnoëttes as "a consummate general officer of the greatest distinction in every way possible," and the war minister confirmed him in his rank of lieutenant general.[35] All of this augured well for the future, but the Bourbon idyll did not last, and the army became increasingly disenchanted with the government. The public's feeling of helplessness, undermined by the half-pays' complaints, rubbed off on the state of mind of military men suffering from the great difference between the present situation where they were nothing and the past where they had been everything. For the first time in years, the army was not treated with the utmost consideration, and it felt discredited. Carried along by the general current of doubt and gloom, Lefebvre-Desnoëttes in turn was grieved by this and lost his head:

But soon we became worried about the future, as public opinion worked on the minds of officers and soldiers alike, and left nothing but bad impressions; the transition had been too abrupt; we, the ex-Imperial Guard, were not always well-treated and we thought that France was going to take away whatever good might be left of the Revolution. After so many wars and up-heavals we needed something to be enthusiastic about and our most ardent hopes were for freedom and the Charter. We became convinced that plans were to take us back to the Government of Louis XV. We got all worked up and distracted.[36]

Despite his initial profession of loyalty, his determination had never been strong and sincere enough to dissolve his natural Bonapartist fervor in a royalism of convenience. The atmosphere in Saumur was rebellious. In Countess Desnoëttes' salon, reports a police spy, people said the most outrageous things about the court; the lady of the house, Bonaparte's cousin, herself made unseemly remarks about the royal family, and the regulars predicted that Napoleon would soon return.[37] On her husband's name day, the Countess gave a dinner for two hundred people, which the local sub-prefect did not want to attend. The fear of being at a banquet at which he could not have offered the regulation toast to the health of the king prompted him to take his leave before the beginning of the festivities—where there was indeed no question of good wishes for Louis, but of fueling insurrection. The general was soon denounced "for not having drunk to the health of the king," and suspicions regarding him were strengthened.[38] In January 1815 he received the order to leave Saumur and be garrisoned in Cambrai.

Led by General Lion, the second-in-command, the regiment set off on February 1 for its new posting under the inquisitive eyes of the police, who were instructed to record the actions of the highly suspect chasseurs. Every stopover on the journey was an opportunity for a drinking bout at which they toasted "the health of the Exile," recalls a witness.[39] At one of them,[40] their uncalled-for remarks were reported to the war minister, who warned Drouët about these troops coming to his military region.[41] Did this warning assure Drouët that the Royal Chasseurs and their leader were ready to rally to the cause? It is known that Desnoëttes came over well before the return from Elba, since from late January to March 7, 1815, he was in Paris, from where he transmitted his own and Lallemand's orders to General Rigau, their accomplice in the department of Marne.

FAILURE AT LA FÈRE

When he was asked to march on Paris with the troops from the northern garrisons, Desnoëttes committed to the Emperor's cause without hesitation: "I thought that general opinion in France was in his favor and I wanted to serve him when it was very dangerous to do so; I did not want to be swept along by his triumphal chariot. I received orders."[42] He went from Paris to Lille, where on March 8 he probably received his orders from Drouët, then on to Cambrai, which he reached after nightfall. There he joined his men, in place since February 22 and waiting for his arrival to receive the white standard of the Bourbons in a very official ceremony. Instead, the next morning, he led them out. On the evening of March 9 they were ready to get to work at La Fère, a strategic fortified town housing the Paris arsenal, the Second Regiment of Foot Artillery, and the artillery school headed by General d'Aboville, who had lost an arm in the battle of Wagram.

The conspirators had thought they would seize La Fère easily. The elder Lallemand could assert his status as commander of the Aisne; additionally, he could count on the aid of his younger brother, then stationed at La Fère to settle the accounts of the Guard's Foot Artillery, whose chief of staff he had been during the French campaign. He had just been returned to active service on March 9, 1815, like all the officers on half pay and the soldiers on leave whom the king had ordered recalled.

Desnoëttes' unexpected arrival puzzled d'Aboville, for it coincided with the announcement of the imperial return, but he agreed to quarter his troops within the fortress. On the morning of March 10, the younger Lallemand, back in La Fère, took advantage of his position to try to convince the leaders of the Second Artillery Battalion to join the conspiracy, but he failed. La Fère then went into defensive mode and decided to arrest the mischief-makers, who had to negotiate to get the gates reopened . . . so that they could get out. Desnoëttes harangued the gunners, and the younger Lallemand enjoined them with drawn saber to follow them: the men, reminded of their duty by their officers, remained unyielding.

The chasseurs, empty-handed, attempted with no better success to entice away the dragoons of a neighboring garrison, nor did they find the ten to fifteen thousand men who, according to Desnoëttes, were to accompany them to Paris. On March 11 they pinned their hopes on a regiment of horse chasseurs stationed in Compiègne, but they turned a deaf ear. In the city where not a soldier remained, the conspirators engaged in a few pathetic demonstrations of force. What was really important was taking place elsewhere. This was the last straw. Most of the officers realized how isolated they were. Not only did they see no one come to swell their ranks, but the various units they came upon either stood fast or fled at their approach.

The decision made at this particular moment by the individual military men—whether to continue or turn back—influenced their destinies. General Lion, major of the Royal Corps of Horse Chasseurs, and the chief officers rejected their leader's proposal to rush toward Lyon to meet the Emperor as partisans: "Changing my plans, I wanted to join Napoleon in the Alps, where we thought he still was; my officers rebelled: they were horrified by my plan and I was forced to abandon it," wrote Desnoëttes.[43]

The expedition had been as badly prepared as it had been conducted. The conspirators had misjudged the spirit of obedience and the conduct of the rank and file, who were more moderate and disciplined than expected, as well as the officers' unanticipated resistance. Apart from Fouché's triggering role, the responsibility for its failure in the field falls on the elder Lallemand, who had not been able to feel the military pulse of a department under his command, as well as on the younger, incapable of bringing over the gunners of La Fère. When the time for action came, neither of them, and Desnoëttes no more than they, had the necessary authority or bearing to assert themselves to men who would have been ready to follow them. Perhaps they could not find the right words or define their goal clearly, since they themselves did not know exactly what they were about or where they were going. In any case, their failure was a windfall for monarchist propaganda. In the Chamber of Deputies, the minister of the interior lost no time praising the resistance of La Fère and denouncing the traitor Lallemand.

Lallemand had fled with his brother, Desnoëttes, and other officers ready to play their personal cards and ride flat out to join Napoleon. This slapstick escapade had an outcome that was ridiculed at their trial, as a police squad had easily intercepted them: "All were armed with sabers, daggers and pistols. But men who act in accord with their duties have such a great influence over guilty ones, that seven soldiers could not resist five policemen, and after a short defense, four of them were seized; the two Lallemands were among them. They were carrying on their persons seven to eight thousand francs in gold and a letter in which Lefebvre-Desnoëttes wrote to the elder Lallemand that it was time to act."[44]

Alexandre Dumas, who was only thirteen years old at the time, saw the Lallemand brothers pass through Villers-Cotterêts on March 14 on their way to prison, each of them sitting in a police cabriolet between two gendarmes: "They were pale, but seemed calm," he wrote.[45] Their failed attempt to buy off a police sergeant led a field officer of the La Fère arsenal who had opposed but not stopped them to comment ruefully: "Their cowardice when they were arrested convinced me that already in La Fère we could have gotten the better of them without resistance."[46] Of the three who had managed to escape the policemen, only one succeeded in joining Napoleon. Desnoëttes, in his distress, took refuge with General Rigau in Châlons-sur-Marne.

In Châlons, informed by a major of the plan to give its Emperor back to France, Rigau had agreed to inform him in secret of the state of public opinion in his department of Marne, and had, in February, contacted Generals Desnoëttes and Lallemand, whose orders he awaited thanks to a well-rewarded liaison officer in Paris. Rigau was taking a serious risk. Promoted to the rank of brigadier general after twenty years of service, he was an old soldier, called the "martyr of glory" because of his countless injuries. The Emperor had decorated him, given him a great deal of money, made him a baron, and planned a battlefield promotion to the rank of major general after Rigau distinguished himself one last time on March 20, 1814. Since Napoleon had not been able to keep his promise, Rigau had three months later resigned himself to accepting the Cross of a Knight of Saint Louis and the command of the Marne region. None of the conspirators would have held it against him if he had waited quietly for retirement. It would seem that his rejection of the Bourbons, his need for action, and his love for the Emperor prevailed over the prudence that should have come with age.

The mere fact of taking in Lefebvre-Desnoëttes was a reckless act, for on March 12, 1814, it was still too early to know whether Napoleon's attempt would succeed. The loyalist attitude of the garrisons visited by the conspirators invalidated the opinion that the army no longer wanted the Bourbons. But Rigau did not stop there. Using departmental tax funds, he printed and distributed a proc-lamation exhorting his troops to disobey the king and gave the soldiers money to spend in Châlons' cabarets.[47] Won over to his cause, the troops sported the tricolor cockade and set off with him for Paris. The northern conspiracy had been a disas-ter, but on the day when the die was already cast in Paris, Rigau had saved its face.

So this muddled conspiracy was of no help whatsoever to Napoleon, who, far from thanking the Bonapartist party for its hue and cry, is said to have criticized Generals Drouët, Lallemand, and Lefebvre-Desnoëttes: "In his view, it was an attack on royal majesty, a revolt against authority, a bad example. It was impor-tant to his honor to owe his return solely to his own audacity and the affection of the people and the army."[48] Paradoxically, no one dared boast of having con-spired, and the generals were freed by the one they had wanted to help. Desnoëttes and Drouët, who had escaped, reappeared in public on March 20. The next day, the prefect of Aisne, at Napoleon's request, ordered the release of the Lallemand brothers.

Chateaubriand scoffed at the General Police, incapable of preventing the con-spiracy and the imperial return: "They saw nothing; they knew nothing. Napo-leon's parcels traveled publicly via the postal service, the couriers were his men: the Lallemand brothers traveled with arms and baggage . . . ; the usurper had just dis-embarked [in Cannes] and the police did not know a thing about it."[49] Politician and historian Thiers was critical of the conspiracy itself: "If we only consider as a conspiracy a well-planned project and that by serious-minded men firmly deter-mined to attain their object even by risk of their lives, and who have arranged their means with prudence and precision, it would be impossible to assert that there was anything of the kind here."[50]

Most important is the fact that the king and his council never forgave Napoleon or the conspiring generals for deceiving them. The second Restoration dealt even more severely with them than with the others—those who rallied to Napoleon be-fore March 23 or signed the "Additional Act to the Constitutions of the Empire," a sort of "improved Charter," in the words of Chateaubriand, that had been pro-mulgated a month later.

After they had gone over to the king, Napoleon succeeded in getting his gen-eral officers to do an about-face, often, admittedly, with bad grace. Among those who were enjoying a peaceful existence, many resented the fact that he forced their hand and engaged them in an adventure whose short-term survival was uncertain. For despite Napoleon's efforts to avoid a confrontation, it was inevitable. The al-lied powers could not let the man who had pushed them around for twenty years come back to power.

Napoleon counted his troops and called back barons and vassals, but he could not do without draftees, despite restive public opinion. Factories and shops worked around the clock to equip them. Napoleon excluded from top positions those whose past attitude permitted no redemption, but he cast a wide net and pardoned the officers who had opposed his return. On the other hand, his steps to regain control of the marshals did not augur well: in the end only seven were available, among them Davout, appointed war minister, and Soult, who was pardoned and made major general of the Army of the North.[51] Had he run them down? Napoleon was not ungrateful toward the apprentice conspirators.

Drouët d'Erlon was given command of the First Observation Corps, created in the departments of northern France, then that of the First Army Corps—placed, along with the Second, under Ney's orders at Waterloo. The Imperial Guard had been reestablished around the Royal Chasseurs: on March 24 in Paris, they were reviewed by the Emperor and presented him with the eagle, which their regiment had kept. On April 14, Napoleon reappointed Desnoëttes colonel of the horse chasseurs of the Imperial Guard, seconded by the elder Lallemand, promoted to lieutenant general. What Napoleon told O'Meara, the Irish doctor on St. Helena, about him invalidates what he is supposed to have said after the failure at La Fère: "When I returned from Elba, he came out in my favor at the most perilous moment. Lallemand has much resolve, is astute and there are few men who could conduct a hazardous enterprise better than he."[52] Given Lallemand's flagrant incompetence at La Fère, it is difficult to believe that Napoleon was referring to this incident to assert that Lallemand could, better than anyone else, conduct a hazardous enterprise. We will see that in the United States in 1818, in the Champ d'Asile affair, he failed once again, after woolly schemes that tarnished his reputation because, beyond his personal failure, he sent into disaster hundreds of men who had followed him into what was precisely a most hazardous venture. This first chapter's detailed account of the circumstances of the northern conspiracy fiasco and the personalities of its plotters is essential to a clear understanding of certain events that will take place in the course of our story.

But on June 2 all was well for the elder Lallemand, who, along with Drouët and Desnoëttes, was made a peer of France. Lallemand's younger brother was not forgotten. He was promoted to lieutenant general and major of the Foot Artillery of the Guard. Old Rigau, who had acquitted himself the best of the four of them, got nothing in return, other than remaining head of the department of Marne. Waterloo swept them all away.

Desnoëttes and the elder Lallemand had been given command of the Old Guard's Division of Light Cavalry in the Army of Belgium. They were always on the front line, charging in repeated waves at Waterloo on June 18, but the two generals saw all their assaults shatter against impregnable enemy formations.

For his failure to lend assistance at Waterloo, where Napoleon was expecting him with his thirty-three thousand men, many threw the humiliation of the defeat onto Grouchy's not-so-broad shoulders.[53] But tens of thousands of men immobilized on other fronts were also missing, and a plethora of errors were committed, which their authors were quite happy to hide behind Grouchy's, masking them all. He survived Waterloo by thirty-two years, which he spent defending and justifying himself against the attacks to which he was subjected.

Emmanuel de Grouchy did not have the luck of Marshal Mortier, whom an attack of sciatica had confined to his bed before Waterloo. One can imagine the consequences such an unexpected difficulty would have had for him: the two days of

June 17 and 18 would disappear from a well-rounded career. After fighting in the Vendée region of western France and in Brittany, and covering himself with glory in Italy, he distinguished himself during the Empire at the battles of Eylau, Friedland, and Wagram. Napoleon gave him command of a cavalry corps in the army of Italy, then the legendary "sacred battalion" on the retreat from Russia. On April 15, 1815, after twenty years as a major general, he became Napoleon's last marshal. Wearing his title for the first time at Waterloo, Grouchy, inhibited by the dogma of the infallibility of the master, was accused of having neither the energy nor the initiative necessary for this crucial moment.[54] On St. Helena, Napoleon crucified him, adding that starting with himself, no one had done his duty at Waterloo. However, Drouët judged him in a far more nuanced way than general opinion of the time: "He was one of our most distinguished officers, but he had not yet been commander-in-chief and found himself, as he began, in a most difficult position. In such an important command, a great responsibility weighed upon him; he probably thought he should not go beyond his instructions and should await new orders."[55]

The Emperor returned to Paris on June 21. Contrary to what had happened the previous year under similar circumstances, some, like his brother Lucien, advised him to make a stand before Paris, supported by a people that had not abandoned him. But Fouché was not about to miss his chance to finally bring down his rival. Pulling the strings in a Chamber where the opposition was in control, he manipulated the deputies, warning of a dictatorship. Lafayette, harking back to his youth with a flight of oratory on liberty and another on the homeland in danger, proposed that Napoleon be deposed if he did not abdicate.

On June 22 he yielded in favor of his son. The Chamber elected a five-member Provisional Commission; Fouché assumed its presidency and accelerated the vanquished Emperor's departure so that there would be no risk of a last-minute upset. Fouché had deceived the father, betrayed the son, and taken advantage of Lafayette; he used Davout to ask the Emperor to leave Paris, which he did on June 25 for Malmaison and then Rochefort on the Atlantic coast.[56] On July 3, Paris capitulated to von Blücher's and Wellington's armies. The Napoleonic epic as such was finished. Its misadventures continued on the roads of exile.

2

The "Chief Culprits"

Men overwhelmed with your gifts, decorated with your medals, kissed in the morning the royal hand they betrayed in the evening. Rebellious subjects, bad Frenchmen, false knights, scarce had the oaths they had sworn to you died on their lips, when they went, with your lily on their chests, to swear perjury as it were to him who so often declared himself to be a treacherous, disloyal traitor.

—Chateaubriand, report to the king, Ghent, 1815

On the day after the battle of Waterloo, Louis XVIII wrote to Wellington expressing his satisfaction with the success of the allies over the French troops. The allies and their English supreme commander had, it is true, made a distinction—in theory—between the French nation and Napoleon, waging war only against the latter. But the royal delight at the imperial chaos, even if it was comprehensible, was apt to exasperate those who saw the victory of Waterloo above all as the defeat of France. In Ghent, the king, never doubting he would return, had fallen under the influence of his brother, the Count of Artois, for whom liberty was the worst of all evils (as it had permitted the return of the monster) and who wanted to see the accomplices of the Hundred Days punished. But this choice of a hard line offended those royalists who favored moderate, constitutional measures, and displeased the allies who had not decided whether or not Louis was going to regain his throne, after showing so little aptitude for keeping it and even less for defending it. So the king took the initiative, crossed the Franco-Belgian border, and in late June 1815 confirmed in two successive statements that the losers were in trouble.[1] Everywhere, harshness prevailed over the wish to forgive and forget.

REPRESSION AFTER DEFEAT

When the peace treaty was signed in Paris in 1814, the victors had shown moderation toward vanquished France, which was, financially and militarily, neither crushed nor humiliated. This time, indulgence was out of the question. Before a new treaty so harsh that it had nothing in common with the first was signed on November 20, 1815, a paragraph of the surrender agreement of July 3 suggested the worst for those who had particularly exposed themselves to attack. The officers of the ex-Guard who had served at Waterloo would be excluded from the new royal army. These men who had fought so many battles could see this as homage, but to the French army it was a warning. The second Restoration's policy with regard to military men was mapped out. In 1815, there were only victors and vanquished.

The king dismissed the old army and organized a new one reduced to a shadow of its former self: from divisions down to battalions, corps were broken up and troops dispersed into small, insubstantial units. Those considered surplus were demobilized and swelled the ranks of the unemployed: the census of the summer of 1817 shows nearly sixteen thousand officers on half pay, among them 301 generals.[2] But already on July 24, 1815, the king and Fouché, the new minister of police, had signed an edict that targeted important figures in the French army.

THE EDICT OF PROSCRIPTION

The king had to hand over to the most fanatical members of his entourage the henchmen of a party against whom hatred knew no bounds, and satisfy the thirst for vengeance of the allies who demanded punishment for the traitors of the Hundred Days, men whose presence was incompatible with law and order. Sentencing the French officers to death followed logically from the thesis that Louis XVIII had never ceased defending, in which he set the treason of the army in opposition to the loyalty of his subjects: the accident of the Hundred Days resulted from a military rebellion without any connection to civil society. This allowed the king to minimize his responsibility and justify a second return to the throne.

Most of the military leaders, however, could not be accused of lacking devotion or loyalty to the Bourbons. Napoleon himself stated on St. Helena that since no one had "carefully weighed the feelings of the masses and the fervor of the nation," they had been unable to do anything to stem the torrent of opinion that had swept them away.[3] The English knew this, but wanting to see their continental rivals under a weak king, had adopted the official line and gotten the Russians and Austrians to agree with them. Now they had to give meaning to their analysis and strike the perpetrators of the military rebellion without delay.

On the French side, it was very much in Fouché's interest that the French state instigate legal proceedings against the most guilty, whose names the king had requested. Drawing up the list himself assured this former minister in the Hundred Days' government, who had opportunely gone over to the Bourbons, that he would not be on it. He seems not to have raised the slightest objection to the proscription measure, and went on to compose the edict of July 24, 1815, which determined the fate of fifty-seven individuals.[4]

Its first article ordered the arrest and court-martial of eighteen general officers and one civilian who betrayed the king before March 23—the date of the edict the king had signed in Lille, dismissing "all the officers and soldiers who have gone over to the command of Napoleon Buonaparte and his followers"—or who seized power by means of violence (Napoleon's royalist opponents often deliberately used the Corsican spelling of his name as a sign of contempt). In the *Bulletin des Lois* (the official posting of laws), the two-column list of names began with those most compromised. In one column: Ney, La Bédoyère, the Lallemand brothers, Drouët d'Erlon, and Lefebvre-Desnoëttes; in the other: Grouchy and Clauzel. Also high on the list was former police minister Savary, the last inspector general of the national police force.

The second article designated thirty-eight other individuals also guilty of rallying to the Emperor during the Hundred Days. Their immediate punishment was to leave Paris for house arrest in a distant part of the country; the Chambers were then to decide between a court trial and exile. Marshal Soult headed a group of fourteen generals that included Vandamme, plus a colonel and a police captain. The rest were a mixed bag of former Convention members, ministers, senior members of the Council of State, prefects, deputies, and various individuals who had, in one way or another, come to attention: people like Garnier de Saintes, Réal, Regnaud de Saint-Jean d'Angély, and Dirat, who will reappear in this narrative.

The choice of names and their division into two categories were arbitrary; although the interested parties had every right to expect to be tried legally, there had been no preliminary trials sorting them out into two groups. The distinction, based on the degree of participation in the "attack" against the king, was debatable. While the majority could be accused of supporting Napoleon's return before March 23, and some, like the plotters of the North, quite a bit earlier, many who joined the cause after this date had good reason to contest their presence on the lists. As for the civilians of the second article, most had, during the Hundred Days, acted like the majority of civil servants: the deputies had been elected to the Chamber in May, thus later, and some were practically unknown, gaining notoriety merely because their names were on the lists. The whole thing seemed more like a gigantic lottery than a legal document. But neither chance nor the law were involved. The edict of proscription was an opportunity to settle accounts with names linked to past events of painful memory to the royalists, such as the beheading of Louis XVI. Grouchy, who had just captured the Duke of Angoulême, Clauzel, who had chased the Duchess out of Bordeaux, or Vandamme, disqualified from the outset by the many grievances held against him,[5] had done too much to be forgiven. Finally, not content with listing Savary, who had succeeded him as Napoleon's minister of police, Fouché took out his rancor on the small fry close to the man, like his former aide-de-camp and private secretary, Cluis, who had been brought into the national police force in April 1815.

Fouché boldly defended himself against criticism for having drawn up the list, arguing that he could not have avoided it because the royalists' animosity after the king's return was too strong. But while the choice of names, particularly those of his friends, had been difficult, he had been able to contain the persecution, which, without him, would have been far more tragic. All things considered, he gave himself good marks for shortening the list and calming the reactionary fever that was setting the country ablaze. But Fouché had trouble convincing those who owed him their exile. The ambiguity was simply too great. He was never able to guarantee them his good faith, nor prevent the crimes of the White Terror. Just when he had arrived at the peak of success through his marriage into the high provincial aristocracy, the election of the ultra-conservative Chambre introuvable (Unobtainable Chamber) in August 1815 proved fatal, so hostile was it to him.[6] The presence in its midst of a renegade, apostate, and regicide, covered with the blood of thousands of innocent torture victims, would be an affront. People conspired to get him dismissed. The king abandoned him. He was forced to hand over his ministry in exchange for a minor embassy to the German court of Saxony.

In the meantime, the determination of Fouché, whose name remained associated with the crimes of the Revolution and the dictatorship of the First Empire, did not soften the attitude of the excessive royalists, who, far from turning the page, went from one terror to the next.[7] Becoming ultra-royalists, they acted illegally in the field or legally on their benches, passing reactionary laws in the "burning Chamber of Deputies," as Madame de Staël called it.[8]

Fouché denounced the atrocities perpetrated in southern France, favored since late June by a certain administrative confusion and military disorganization. On July 28 he sent the prefects a circular stating that the king wished "to draw a veil over the crimes and the faults that had been committed" and to leave to justice "the task of punishing attacks and treasons."[9] But the king's will dissipated in the southern heat. Fouché had to repeat his calls to order, indignant at the disgraceful acts being committed with impunity and insisting that the exercise of public force belonged only to the king. However, the judges who were supposed to punish in his name dragged their feet and found it difficult to sentence the murderers. The laxness that was long shown them was equaled only by the extreme severity of the courts-martial.

The preamble of the July 24 edict was clear. It was necessary to outlaw the accused because the king's credibility, Europe's tranquility, and public safety were at stake. The first Restoration had offered its hand; the second struck with its fist. The edict of August 2, 1815, gave the court-martial of the First Military Division the exclusive responsibility of investigating the crimes with which the military men listed in the edict were charged.[10] La Bédoyère, who had been made a general and peer of France for going over to the Emperor, was condemned to death on August 19, 1815, and executed by firing squad that very day. The Chamber of Peers, meeting as a court of justice, condemned Marshal Ney to death on December 6, 1815. The king refused to pardon him. He was executed by firing squad on December 7.

The Amnesty Law

The crackdown on the "chief culprits" intensified in the last months of 1815, coinciding with the wind of reaction blowing over a legislative Chamber that was more royalist than the king.[11] Between October 1815 and January 1816, its deputies, united in their hatred of the Revolution and the Empire, passed four special laws abolishing fundamental freedoms.[12] The so-called amnesty law was the most important.

The July edict had given them the power to judge those on its lists, but the deputies, fearing the effects of the royal clemency of a king accused of lacking determination, went beyond his intentions and took the offensive. In November 1815 one of their champions drew up an amnesty bill that did not grant amnesty but rather had the curious feature of strengthening the edict's provisions by proposing other categories of culprits liable to the death penalty. "To break up plots, we need fire, executioners, torture," he thundered in the Chamber.[13] On December 20, 1815, on the eve of his execution, Count de Lavalette, postmaster general during the Empire and the Hundred Days, managed an incredible escape from the Conciergerie prison.[14] This feat strengthened the case of those criticizing the government's incompetence.

The country had recently acquired a new prime minister, the Duke of Richelieu, an intelligent man but one who, returning from twenty-four years of emigration, was not familiar with the situation in France.[15] A mediocre orator not given to intrigue, he was in a position of weakness when he introduced a bill granting amnesty to those less compromised and having the Chambers decide the fate of the others. The amnesty, however, was so relative due to numerous restrictions that foreign opposition newspapers waxed ironic over this government granting a general pardon from which everyone was excepted.[16] But the deputies, for whom this was still not enough, added to the list of people to be banished relapsed regicides—that is, members of the Convention who had voted for the death of the king and subscribed to the Hundred Days. The bill was brought to the floor on January 2, 1816.

Backing Richelieu, those in favor of pardon, synonymous with regained serenity and social cohesion, faced their more relentless, numerous, and better-organized adversaries who claimed that the regicides were incapable of change.[17] The vote on the bill sealed their fate: 153 "murderers" of Louis XVI who were still alive were sentenced to banishment. Richelieu then brought the bill thus amended to the floor of the Chamber of Peers.[18] On January 12, 1816, the amnesty law was passed by the deputies with a majority of ten to one, and soon afterward by the peers with an identical majority. The two assemblies thus punctuated three months of inordinately savage legislative activity. For the former culprits and the new ones so designated by the law, this was the beginning of a long period of wandering. The Empire had had its "martyr of glory" in the person of General Rigau; it now had its "martyrs of loyalty."[19] This grandiloquent designation, intended for the Lallemand brothers, counterbalanced that of "chief culprits" and referred to the trials endured by the victims of the great military and civilian purges of 1815 and 1816.

"MARTYRS OF LOYALTY"

The army did not keep its ministers long, for political ups and downs were replacing them with the regularity of a metronome. The latest was General Clarke, Duke of Feltre, rewarded for making the right choice of the Bourbons when Napoleon returned from Elba. Clarke served the king with zeal. Merciless toward those who had taken part in the Hundred Days, he closed the doors of the army to them, or opened them only to drive them out. Two committees were appointed in October 1815: the first was to examine the conduct of officers of all ranks who had served during the usurpation, and classify those to be eliminated; the second was to replace them with supporters of the royalist cause. The task of purging the former officers corps of the imperial army converged with the reactionary laws of the Ultra deputies. As for the allies, then occupying two-thirds of the country, they approved heartily. People were not only dismissed; they continued to be sentenced and executed. Jails were filled to overflowing with officers, shattered by heavy sentences, prison for life or capital punishment. Thirteen of the first nineteen names in the edict had not waited around for their trials to find out what justice would be like.

The hunt for generals followed the publication of the edict, but there was a period of hesitation, more or less tolerated to help people escape the country, as the king was aware of the risk of developing a generation of Bonapartist martyrs. Thus, on August 1, 1815, Marshal Macdonald supposedly warned that orders of arrest had been drawn up for the generals named in the edict and that they should think of their safety without losing a moment.[20] Couriers with the news were sent to the camps concerned, including Sancerre, where Lefebvre-Desnoëttes was stationed.[21]

According to the prefect of the department of Cher, the general had already left, dressed as a traveling salesman, and had crossed the Loire at a place the Prussians had neglected to guard. On August 3, at the bridge in Orléans, three individuals heading for Paris were arrested:[22] his servants, whose interrogation proved fruitless. The countess's lady-in-waiting had been entrusted with a note giving the names on the proscription list that had just been drawn up. Drouët d'Erlon, who was lunching with Desnoëttes, confirmed that the note arrived;[23] Leontine Desportes, the messenger, later boasted that she had saved their lives, since the two generals thought it wise to take cover.

Noncommissioned officer Chevalier claimed that this took place after an initial alert when a platoon of gendarmes had shown up with an arrest order. But the chasseurs, whom Desnoëttes was reviewing, closed ranks around their leader and saved the day. He was sorely missed by his men: "In my entire life I have not known a kinder and braver man than Count Lefebvre-Desnoëttes, a man of war and a man of the world, educated. Good and just towards the soldiers, he was able to gain everyone's love and respect."[24] Sightings were reported throughout France. The Paris home of a reputed mistress was searched, as were the establishments of manufacturer Richard-Lenoir, a staunch Bonapartist and father-in-law of Captain Zénon Lefebvre-Desnoëttes, the general's brother. An all-out police search turned up nothing.

As for Grouchy, he had not crossed the Loire, having resigned earlier. On August 5, in Paris, he obtained a passport in the name of Jean-Charles Gauthier, merchant, to travel in the interior of the country. The police searched for him in Normandy, where he had family ties, and along the coasts of the Cotentin peninsula, from where it was thought he might embark.

The prefect of Gironde issued two blank passports to General Clauzel and his aide-de-camp; they left Bordeaux on July 28.[25] Anonymous letters soon reported Clauzel's presence in various parts of the country, all widely separated, with a predilection for Ariège, in the southwest, where he was born, and neighboring departments. False leads allowed him to remain hidden for a while with friends in Niort, some 120 miles north of Bordeaux, in spite of the fact that the minister of police had, on August 17, alerted the prefect of Deux-Sèvres to his presence. Long after he had left France, people persistently reported seeing him here and there.

No search was necessary for the elder Lallemand, Savary, and Bertrand, the hard core that had chosen to follow the Emperor to the end. In England, Bertrand, as grand marshal of the palace, was permitted to accompany him to St. Helena, but Savary and Lallemand, listed in the July 24 edict, were made prisoners of war. On the frigate to which they were transferred, they found other officers of the imperial entourage, among them Major Schultz, light horse commander on the island of Elba who had been wounded at Waterloo and who was said to want only to return to his homeland of Poland.[26]

The two generals, who thought they would be handed over to France, were taken to Malta and interned in its fortress, where they remained for six months. Freed in early April 1816, they boarded a British schooner that took them to Smyrna. It was reported that Savary was banking on returning to France to clear his name and that Lallemand was thinking of expatriating himself, while the French diplomatic corps in Constantinople was upset at seeing the region "contaminated by the stay of several of France's chief culprits" and did not hesitate to complain about it in terms of "another plague."[27]

Meanwhile, in France, the political situation had radicalized and the Ultras were talking tough. Fouché had to relinquish his position to the chief of the Paris police force, Decazes, the king's favorite who immediately reactivated the manhunts. In response to his dispatch of September 30, 1815, concerning the arrests, among others, of Grouchy, Clauzel, and Lefebvre-Desnoëttes, the prefect of Ariège wrote that he had quickly ordered that travelers be closely watched, in order to discover the "criminals who, under assumed names and with the aid of passports not belonging to them," would try to take refuge in his department.[28] In January 1816, the minister, thinking that General Lefebvre-Desnoëttes had not left the kingdom, sent the prefects another circular: "I ask you to order, throughout your department, the most thorough searches with regard to him. A sum of ten thousand francs will be awarded to the person ensuring the arrest of this accused person, whose description I am sending you once more."[29]

This price on his head highlighted his special status among the fugitives. But the search was fruitless, though not for lack of zeal. The prefects used every means at their disposal, published his description and the promise of reward.

The prefect of Corsica, where the general's mother-in-law had been born, stated flatly: "If one of the chief culprits . . . should dare to show himself in this department, he would not escape my investigations."[30] This was mere bluster. Neither Lefebvre-Desnoëttes nor any other general had gone underground, on the island or on the Continent. But their absence did not stop the wheels of justice from turning.

The second court-martial met on May 11, 1816, to decide on the fate of Lefebvre-Desnoëttes. The compromising documents had disappeared, opportunely burned during the Hundred Days, but the two principal witnesses for the prosecution were there: Lieutenant General d'Aboville and Colonel Laîné did their best to justify their promotions by relating the attempts of the accused to get around their troops, and by showcasing their success in preventing this. The reporter found no mitigating circumstances for the accused:

> Gentlemen, there is no good Frenchman who can help but feel shamed and indignant at the memory of General Lefebvre-Desnoëttes' treachery. In this episode of our recent disasters, we have seen one of our fellow countrymen veil his perfidy with the appearance of loyalty and candor, and sustain this ignoble role up to the moment when he was given the signal for action; we have seen one of the most valiant generals of the army violate solemnly made oaths, tread in the dust the most sacred duties and sacrifice his King and his fatherland to the hope of some new honor.
>
> His Majesty had permitted the accused to retain command of one of the army's finest elite corps. General Lefebvre-Desnoëttes corrupted this regiment with the conspirators, stirred it up against legitimate authority and wanted to use it as an instrument of ruin against the Government and the capital.[31]

The three charges brought against him were indisputable: first of all, without orders and with the intention of overthrowing the legitimate government, moving the corps of Royal Chasseurs he commanded toward Paris; next, attempting to shake the loyalty of the troops stationed at La Fère and Compiègne; finally, wanting to seize the depot of the La Fère arsenal.[32] But beyond this, he was accused of deceiving the king and setting the chasseurs of the ex-Guard against him at Waterloo. Three generals of the Empire added their voices to those of the royalists with whom they sat, unanimously condemning him to death in absentia.

On May 16, Rigau, who had received Desnoëttes in Châlons-sur-Marne, was brought to trial. His case got off to a bad start. For one thing, the crimes of which he was accused were clearly established; for another, his attitude after Waterloo and since the king's return had confirmed all the ill one might think of him.[33] After crossing the border to elude justice, he had persisted in his bad conduct: he was suspected of having dangerous contacts with officers fearing neither God nor man, propagators of vile lampoons and seditious correspondence. At the time of his trial he was said to be in Sarrebruck, where he was making scandalous remarks against the king. Faced with a man so incapable of repentance, the French government had requested his arrest and extradition.[34]

The court, chaired by a former general of the Empire, had more than enough evidence and witnesses for the prosecution establishing bribery and treason.[35] Of all the exhibits, the most irrefutable was a letter from Desnoëttes praising "this fine general" who "had arranged to get information on several points, to make sure of the state of mind of the troops and the people who were indispensable to the execution of the plan."[36] Of all the depositions, the most overwhelming was that of Marshal Victor, the accused's immediate superior at that time. The court-martial unanimously condemned General Rigau to death in absentia.

The fate of the last three plotters remained to be settled. That of Drouët d'Erlon was decided on August 10. D'Erlon's order for his troops of the Sixteenth Military Division to leave their district, along with the letter to Lefebvre-Desnoëttes intercepted by General d'Aboville, proved his treason. A former general of the Empire presided over the court that unanimously sentenced him to death in absentia.[37] The same sentence was pronounced on the elder Lallemand on August 20, and the following day on his younger brother, blamed for their role at La Fère and their participation in the march on Paris. Between May and December 1816, eight other generals, including Clauzel on September 11, received the same sentence. The court-martial convened on October 19, 1816, declared itself incompetent in the case of Grouchy.

These convictions were tainted by bias. Former imperial generals sat on courts alongside émigrés and noblemen whose careers as soldiers outside of France and aversion for Bonapartists stood in the way of dispassionate justice.

Death sentences or long prison terms were the most spectacular manifestations of repression during the second Restoration, but not the only ones. With a host of important civilians, a great many other military men were asked to leave French soil.

VIA BELGIUM

While most of the men liable to court-martial attempted to flee, many others, listed in the second article of the proscription edict or victims of the amnesty law, were in a different position, since their lives were not threatened; it was their right to remain on national territory that was questioned. They therefore tried, sometimes with the utmost energy, to exonerate themselves and delay the moment of departure.

Vandamme had spent the first Restoration in his château in Cassel. Napoleon brought him back, made him a peer of France, and entrusted him with the Third Army Corps and then command of the rear guard in the retreat to Paris on June 28. On the proscription list, Vandamme left the army of the Loire in August to live in the country near Limoges, where he waited until his fate was decided, stating in his defense that he had been constantly French and had done nothing unworthy of this name.[38] Since his home had become the meeting place of too many officers, he had to choose another. He found it in the department of Cher, where the prefect praised his good conduct: the general was busy drawing up a statement imploring that he be struck from the edict.[39] Countess Vandamme, who had joined him, also sent a petition to the king at the time when discussion of the amnesty law was getting under way in the Chamber: "He is not guilty, Sire; his enemies have deceived Your Majesty. My husband had no part, either direct or indirect, in Bonaparte's return. He was peacefully in the midst of his family during the events that turned France upside down, and it was only when Your Majesty had left the kingdom that he went to Paris. General Vandamme does not deserve the harsh fate that threatens him and I dare to beg Your Majesty not to keep him on the exile lists. Your Majesty will always find this general officer to be a loyal, obedient and devoted subject."[40]

Given his reputation, amnesty was unthinkable; the royalists would have received it badly. The general chose neighboring Belgium, like many other Frenchmen considered undesirable in their country.[41]

Newly freed of French occupation, Belgium had been partially joined to Holland in the first Treaty of Paris, according to the wishes of the allies, particularly England, which wanted a strong barrier-state north of France. The Belgians were not as infatuated with this union as were the Dutch, but the victory of Waterloo, to which they both contributed, developed the spirit of concord that had been missing until then and cemented the union of the two countries, proclaimed by William I in March 1815.

A new era was beginning without Napoleon, but not without his friends, whom Restoration France was pushing out willy-nilly. Banished men of all kinds, famous or not, with or without passports, imperialists or liberals, settled in the southern part of the kingdom, particularly in Brussels where they knew they would be able to speak French and obtain the active sympathy of Belgian liberals. Beyond this, they owed their hospitality to the Fundamental Law bestowed upon his country by William, which ensured to all foreign residents protection of their persons and possessions. The king's generosity was due less to his love for the exiles than to his desire to seem liberal—which he was not, in his heart of hearts, remaining an autocrat convinced of the divine essence of royal power. He was ready to accept from refugees what he would not have tolerated in his own subjects, because this gave him a way to harm France, a country he continued to distrust, even though it was now under Bourbon rule.

Before the pressure of the allies and the conduct of some of the banished men finally made William change his mind, the years 1816 and 1817 were an effervescent period for the Bonapartist and republican opposition in Belgium, as well as a period of extreme tension between the king and France. Many Frenchman lived quietly in the chief cities of Belgium, but they were not equal in exile. There was quite a distance between renowned, wealthy individuals and former officers unwanted by the army of Belgium and the Low Countries; most, however, were united in a common desire to combat the Bourbons or overthrow them.[42]

By permitting the opposition to express itself freely in his country, William demonstrated courageous tolerance in a Europe scarce recovered from its Napoleonic trauma. His minister of foreign affairs urged him to be less accommodating toward Bonapartists and republicans who were turning Belgium into an active base against the Bourbons. William was incurring the anger of France and the allied powers who had signed the second Treaty of Paris on November 20, 1815, which excluded his kingdom, Germany, Switzerland, and Italy from the countries that could accept French exiles belonging to the second group of the July edict. Those with passports from the French government were allowed only in Austria, Prussia, and Russia. Keeping them far from the French borders would cut them off from their base of operations and allow for easier and stricter surveillance, thus strongly reducing a possible threat.

The kingdom of Belgium and the Low Countries remained deaf to the protests of the French government as long as it was giving the exiles passports that were in order, as the civil servants responsible for issuing them long remained ignorant of the clauses of the Treaty. In June 1816 the minister of foreign affairs, deploring this situation, forbade issuing passports for Brussels to those banished by the edicts[43] and increased his pressure on Belgium to expel the beneficiaries of hospitality contrary to the resolutions of the Treaty. The court of Brussels was far from satisfying France completely, but it did compel the men most compromised to inquire about another land of exile. General Vandamme and former Convention member Garnier de Saintes were among these men.

In February 1816 Vandamme settled in Ghent, his wife's native city and conveniently near his residence in Cassel, to which he made several illegal round trips that came to the attention of the authorities. So he was forced to move to Amsterdam, where other French generals had preceded him. Since the French consul found nothing to complain of in his conduct, Vandamme rebelled against a notice of extradition to Denmark. Three physicians called to his bedside attested that due to his sedentary life he had gained so much weight and so damaged his health that there was a danger of apoplexy.[44] Their certificate enabled him to buy time and send for his wife, but not to return to Ghent.

Jacques Garnier de Saintes had chosen exile in Brussels upon publication of the July edict. At the time this destination was legitimate, since the Treaty had not yet been signed, but his past gave him no hope of indulgence. Louis XVIII could not forget that as a member of the Convention Garnier had asked that his brother be treated, not as an ordinary defendant, but as an enemy, and then voted for his death without appeal or reprieve. Nor could he forget that after Waterloo, he had advocated resistance and continued combat, assuring those who would be killed that this would be the day of resurrection. If this had not been enough to banish him, Fouché could testify to his servility, for during the Hundred Days, Garnier had had the lack of foresight to appeal to him: "Your Grace, for two days I have been unable to contact you. Persecuted under the Bourbons, am I to be nothing under the Emperor? In thirty years of service and sacrifice, it is indeed cruel to have obtained nothing. May Your Grace deign to tell me that I shall have a position; employ me in Paris, for any duty whatsoever and I shall be useful; the interests of my family hold me here."[45]

Three months later Garnier wrote to Fouché, who had gone over to the Bourbons, in surprise, but accepting his fate: "Your Grace, I was stunned by the unexpected blow that has just befallen me, but my strength is returning with the feeling of a clear conscience. His Majesty has included me on the list that is right now being published. I am neither dangerous nor spiteful; however I am resigning myself to the cruelest sacrifice, and to forestall useless severity I am subjecting myself to expatriation."[46] On August 1, 1815, Garnier's lawyer thanked Fouché for his "kindness" in issuing his client a passport for Brussels, and a week later Garnier himself agreed with Fouché's decision to banish him.[47] But as soon as he arrived in Belgium, his activity caused concern: he was accused of writing and planning to have printed a biography denigrating the royal family of the Bourbons. Despite a denial in the *Surveillant,* a Brussels newspaper for which he stopped working in January 1816, pressure from the French government put an end to his stay in Belgium. Faced with a new, unavoidable exile, Garnier, joined by his son Athanase, managed only to put off his departure for health reasons.[48]

Unlike officers accustomed to physical exercise and in the prime of life, the surviving members of the Convention were considerably older and in poor health.[49] This was taken into account when the amnesty law, expelling the regicides, was put into effect, since a circular from the minister of police to the prefects allowed exceptions for those unable to travel because of a serious health condition duly certified. This consideration was noted by those who saw in it a reprieve from the proscription measure. Tormented for seven months by a chest ailment and for two weeks by rheumatism depriving him of the use of his right leg, former councillor of state and police prefect Réal claimed on January 21, 1816, that he found it impossible to undertake such a long journey in the winter.[50] The minister denied a postponement, and Réal went to Belgium to grow hops on an estate he reportedly bought near Alost. But like Vandamme and Garnier, he could not remain there. He was considered a high-risk individual, a troublemaker, welcomed with open arms by refugees to whom he was suggesting subversive ideas.

Jean-Augustin Pénières had not prepared for exile as had Garnier and Réal, alerted six months earlier. Born in the department of Corrèze, where his father, a judge under the ancien régime but receptive to the new ideas, had encouraged him to go into politics, Pénières was in 1792 one of the youngest deputies of the Convention, before serving in most of the assemblies that succeeded each other until 1815. A sincere republican, he had avoided revolutionary excess; although he voted for the death of the king, he asked that in the future the death penalty be abolished. This "philanthropic regicide," believed only in the virtues of the Constitution and of progress. Expelled from the Tribunate[51] in 1802 and the Legislative Body in 1812, he evidenced a moderate devotion to Bonaparte and was later to admit that he had not liked the Emperor. When the glassworks he owned was near bankruptcy, he returned to politics in 1815 and was elected to the Chamber during the Hundred Days. This final term of office along with public declarations he made after Waterloo were to cost him dearly.

In late December 1815, with no illusions as to the result of the vote on the amnesty law, Pénières knew he had better "quickly wax his boots" and leave the country.[52] On January 22 he was in the capital of his department, where the prefect was to inform him of his exile, when he was knocked senseless by a sack of fodder that accidentally fell from a loft where it was being stored.[53] Seriously injured and taken to his home, he informed the prefect and took the opportunity to state his confidence in royal justice, since he had never accepted an appointment from the usurper or voted for Napoleon's Additional Act. The authorities, pointing out his position as deputy, rejected the political argument but did consider his state of health. The prefect notified the minister of police that Pénières had met with a very serious accident and that there was even fear for his life. The request for a reprieve could thus not be ignored, but the prefect nonetheless thought that "Monsieur Pénières is not the worst, but in my opinion the most dangerous, because he has friends and influence and because his well-known and firm opinions corrupt his entire surroundings. He is the idol of two cantons where, because he is so obliging, leads a good life and is popular, he no doubt inspires sentiments similar to his own. I therefore consider his accident as very regrettable since, in this region, it will make many people feel sorry for what has befallen Monsieur Pénières."[54]

Two medical examinations permitted Pénières to prolong his convalescence until it was decided to place him under round-the-clock police surveillance. On March 1, 1816, barely recovered, he boarded a little schooner and sailed down the Dordogne River, which, with the end of the bad weather, was again navigable. But the "devil blew" so hard "into the sails of the executioner" towing him that Pénière's river journey finished in a shipwreck.[55]

On October 28, 1815, in Dresden, Fouché had presented his credentials to Frederick Augustus, king of Saxony. Waiting to return to favor, he undoubtedly thought he had hit rock bottom, but his fall was only beginning.[56] In Paris, the deputies had sworn his downfall. While the amnesty law was being prepared, his name came up obsessively: the most guilty in the past, he would be the most dangerous in the future.[57] Fouché may have thought that the law against regicides targeted him more than anyone else. The king, who had strong reasons for wanting to part company with him, did not wait for the vote on the law to remove him from office, and never dreamed of rehabilitating him. After five years of exile in Europe, Fouché died on the day after Christmas 1820 in Trieste.

Perhaps Fouché had twice missed his appointment with another destiny, not among the European elite who never really admired him, but in America where people were not snobbish. In 1810, believing himself threatened by the Emperor, Fouché, astonishingly panicked, had wanted to flee and charter a boat in Leghorn to seek in the United States "a refuge inaccessible" to his enemies.[58] Boarding a small ship in the morning, he had disembarked again in the evening, overcome by seasickness and fear of English warships in the Mediterranean. Fouché had turned his back on the ocean on whose shores he was born: he was not cut out for the role of a New York immigrant.

In 1815 Fouché confirmed his lack of enthusiasm in more favorable but humiliating circumstances. When his dismissal became inevitable, Talleyrand, the prime minister, had to find a way to get rid of him. Pasquier, a witness to this comedy,[59] tells how, one September day, he mentioned to his ministers that he had at his disposal "the finest position the king could give," that of his ambassador to the United States, the only country where the person holding it would retain all the advantages of his rank and enjoy definite influence. Furthermore, what place on earth was more desirable than the United States, which he himself had visited twenty years earlier when the Convention had exiled him there? "It is such a beautiful country . . . , I know it, I have traveled there, I have lived there; it is a superb country. There are rivers unlike anything we know: the Potomac, for example; nothing more beautiful than the Potomac! And then those magnificent forests . . ." What more could one wish for one's best friend than the opportunity to observe and study this great country, which already occupied such an important place in the world? The best friend had no difficulty recognizing himself, but this brilliant alchemy of humor and geography was not to his taste. Fouché, who liked nothing better than to mock his adversaries, found the farce sour and preferred Saxony to the offer of a compensatory transatlantic ticket.[60]

It would have been interesting to see Fouché in the United States, confronted with compatriots who had emigrated at the same time as he, and in some cases because of him. For the choice of America was the obvious one for many, often, it must be admitted, by constraint or default, thus very far from the American dream.

3

Political Reaction in Gironde

Pressed by time and the fear of being discovered in my refuge, [I am] obliged at all times to avoid appearing in public, not wanting to be seen by anyone.

—General Count Bertrand Clauzel, August 1815

Chased day and night from refuge to refuge, I no longer have, even in the midst of the remotest woods and wilderness, a place where I can rest for an hour, an instant, without running the greatest risk of being discovered.

—Jacques Lajonie Lapeyre, September 1816

In the Europe of 1816, Belgium was the refuge of the Frenchmen rejected by the Bourbons; it was the closest and most accessible continental destination from Paris and northeastern France. But others, whose legal situation was particularly critical and urgent, had to flee as soon and far as possible. This, we shall see, was so for General Clauzel, a famous outlaw, and Jacques Lajonie, a former dragoon officer and anonymous outlaw, both from southern France, at that time consumed with royalist fanaticism.

In the southern half of the country, hatred, latent during the first Restoration, struck those who had committed to the Revolution and the Empire. Thus the White Terror came to the department of Gironde, whose capital, Bordeaux, had never truly come over to the Empire. Thanks to its colonial and slave trade and the agricultural and viticultural resources of its inland countryside, the port was experiencing its Golden Age in the late eighteenth century, when the Revolution broke out. The war against England which followed hurt the French navy and led to the decline of Bordeaux's maritime trade, despite recourse to attacking the enemy's commercial shipping and chartering neutral vessels.[1] Seriously affected, the city sank into depression:[2] it lost population, and—"This says it all!"[3]—the price of a barrel of red wine tumbled by more than two-thirds. Its inhabitants consequently longed to climb out of the slump that a traveler from Hamburg had noted as early as 1801: "Bordeaux's ancient splendor is no more."[4]

FROM EXAGGERATION TO INJUSTICE

The city's inhabitants were ready to deal with the devil to regain this splendor, warned the prefect of Landes in 1813: "Devoid of any feeling for national glory, they see honor and prosperity for France only in the reestablishment of their trade relations, and to accomplish this, they would consent to become vassals of England."[5] A police report from the following year confirmed that Bordeaux was among the cities whose motto was "*No salvation without commerce,* and which subordinate or sacrifice everything to this most important maxim."[6] While its citizens, on the brink of bankruptcy, wanted peace, their motivation went beyond this: after the war, of which they were as tired as their compatriots, they could recover the tradition of independence and free speech that the Empire had taken from them.

It would, however, be rash to claim that the inhabitants of Bordeaux agreed on the measures necessary to regain peace and renown, such as getting rid of the Emperor, when a royalist had to admit that the majority, and particularly the merchants, were attached to him.[7] The position of the upper middle class, which held the purse strings, remained ambiguous toward a man whom they had initially supported when he restored the order necessary to commerce, then disavowed and detested when the rise of dictatorship, strengthening of the blockade, and slide into chaos in Spain and Russia thwarted their interests. Unlike the working class and nobility, the merchant elite both wanted and feared change, but they were inclined to accept the idea as long as the views expressed were sensible and a credible alternative was offered. It so happened that in Bordeaux men capable of satisfying these two requirements had been at work for a long time.

THE "CITY OF MARCH 12"

In the Revolution's climate of total terror and disorder, a working-class royalist movement, the Philanthropic Institute, was born among elements of Bordeaux's youth who wished to return to power the legitimate king and church, using all means, both civil and military. Its activity came to an end during the Consulate, but did not disappear entirely, picking up every now and then in convulsions that were supposed to shake the Empire. Its leaders[8] were never able to destabilize the regime by their actions, but their chipping away succeeded in fixing monarchial doctrines firmly in people's minds and driving out the revolutionary principles that might have lingered there.

After 1810, the institute was revived in another form and its members got together with the Bordeaux representatives of another royalist organization, the Association of Charity, a secret order of knighthood whose eminent members, the Knights of Faith, were devoted body and soul to the return of the king and the church. They recruited chiefly in the west and south, from the department of Gironde on the Atlantic to Var on the Mediterranean, where disaffection with the Empire remained strongest.

These various currents of royalist opposition were thus ready when, in early 1814, the Anglo-Iberian forces crossed the Pyrenees and set up their headquarters in Saint-Jean-de-Luz, where the Duke of Angoulême joined them. At a time when the return of the Bourbons was still hypothetical for the allies (except the English, who supported it fully), they had to be shown that a royalist France was capable of participating in its own liberation. Its leaders went down to meet Wellington and suggested that he have Louis XVIII recognized as king of France by the city of Bordeaux. The city would be easy to take, since public opinion was growing ever more hostile to the Empire, and better yet, since its mayor, Count Lynch, had made contact with the royalists, despite owing the Emperor his career and title.[9] Wellington let himself be persuaded all the more readily as the French defeat at Orthez on February 27 opened the road to Bordeaux, whose occupation would facilitate communications with England. Under Wellington's second-in-command, an Anglo-Portuguese corps set out and a detachment of eight hundred men arrived at Bordeaux on March 12, 1814. The mayor and his municipal councillors received them at the city gates and presented the English general with the keys to the city. Lynch gave a welcoming address in which he promised that his people would rally to the cause.[10] They applauded their liberators. People sported the white cockade and flew the royal white flag. The Duke of Angoulême had followed the English. Feted that very afternoon, he congratulated Bordeaux on its loyalty and three days later posted an appeal for calm and reason in which he beseeched the French to abstain from all partisan dissention and avoid a misfortune even worse than tyranny.[11] William Lee, the United States consul in Bordeaux, did not believe a word of this. He wrote indignantly to his state department:

> The enemy entered Bordeaux on the 12th under the command of Marshal Beresford. They have committed no excesses and the city enjoys a perfect tranquility or rather stupor. The mayor received them at the gate, threw his decorations of the Legion of Honor under his feet, and took that of St. Louis with a white sash. The Duke of Angoulême, the representative of Louis XVIII, followed the enemy in, and was received with shouts of *Vive le Roi!* A Te Deum was sung at the cathedral by order of the Archbishop who was present. At the theatre, *Vive Henri IV* was played by the orchestra, and *God Save the King* which the populace heard standing with their hats off.[12]

The consul then mentioned his fear that Napoleon, at the time victorious in the North, would make the people of Bordeaux pay dear for their conduct; but his abdication saved them. The king thanked the mayor and the first large French city to have publicly come over to the monarchy.[13] He would never forget the services rendered and would be truly happy to discharge his debt.[14] Invited to Paris, Lynch received the sash of the Legion of Honor, which few Frenchmen had worn to that date. Those who had rallied to the king on the first day received the decoration of the Bordeaux Armband created for the occasion with this inscription: "Bordeaux, March 12, 1814."[15]

One year later, on March 5, 1815, the Duke and Duchess of Angoulême, on an official visit to Bordeaux to commemorate the event, received an indescribable welcome,[16] the prelude to a series of important gatherings, which the duke, recalled by the king, had to miss. In Bordeaux, whether by negligence or excessive confidence, little attention was paid to the rumors concerning the Emperor's return. Madame chose to remain. The anniversary of March 12 was celebrated in jubilation. A week later, Bonaparte entered the Tuileries Palace.

Most of the inhabitants remained faithful to their rallying, ready to defend their city, but the troops stationed in Bordeaux and the neighboring citadel of Blaye were solidly for the Emperor and had already implicitly come under the authority of the new governor of the Eleventh Military Division, Lieutenant General Clauzel. The son of a cloth merchant from Ariège, this forty-year-old soldier had risen fast, favored by his stature, people skills, and exceptional military talent. Under Grouchy in Italy, he had next served in Saint-Domingue and the Iberian Peninsula, where he was commander in chief. Napoleon numbered him among his best generals. Clauzel had the makings of the marshal he was later to become. After fighting to the bitter end against Wellington's advancing forces, he came over to the king, who trusted him. The Emperor sent him to conquer Bordeaux on March 22, 1815.

Setting out with his wife, her maids, and a few servants, Clauzel was reinforced on the way by an escort of policemen, and then on the northern edge of the city by a small infantry corps. Things went very quickly. With the exception of the mayor, Clauzel promised to respect the inhabitants of Bordeaux, of whose royalist convictions he was aware, and to treat the Duchess of Angoulême with the consideration due her birth.[17] As the regular troops had indicated to her that they would not obey, the duchess avoided the bloodbath of the all-out defense advocated by the deputy Lainé to save the honor of the city; she preferred to leave Bordeaux on April 2 and board a corvette belonging to the king of England. The mayor did the same on a merchant vessel bound for Plymouth. The tricolor flag again flew over Bordeaux, until July 20, 1815. Waterloo had long sounded the death knell of imperial hopes, but it was only on the previous evening that Clauzel had signed his submission to the king. A week later he went into hiding, a prelude to exile.

Thus the excitement of a Bordeaux back in the royalist fold and the reasons for its intransigence toward the traitors are understandable. On July 30, Tournon, the new prefect of Gironde, described a city where the execration of Bonaparte was at its height and where "little girls danced in the streets while singing songs against him."[18] In August 1815 the delirious city again welcomed their Highnesses of Angoulême. In Bordeaux, the first Restoration had had its "excessive" royalists; the second had its "Ultras."[19] The same people, along with others, even more determined.

A Wave of Americanophobia

The Ultras, supported by all classes of society, took over the city and bullied the Bonapartists and their friends. The American consul was obliged to complain to the prefect[20] that people were making life difficult for him, that he was being threatened to get him to remove the American flag and the eagle, symbols of his country, from the consulate, that he was being bombarded with anonymous letters and insults shouted beneath his windows.[21] He requested protection. His compatriots were also insulted and molested at the slightest excuse. On August 29, 1815, the crew of the schooner *Midas* was attacked because the ship's private flag was tricolor; on another day, the son of a Jewish naturalized American was attacked.[22] Americans in Bordeaux no longer felt that they were "France's best friends." Lee scathingly denounced an Anglo-royalist faction of "renegade" English, Irish, Russians, Austrians, and Dutch who held sway in Bordeaux and the department's aristocracy who worshiped the Angoulêmes.[23] "What we have here is not love for the king; it is not affection for a monarchic form of government; it is idolatry for Mylord the Duke of Angoulême and above all for Mylady the Duchess of Angoulême," wrote Prefect Tournon.[24] The antagonism between the consul and their highnesses was long-standing. Shortly after March 12, 1814, Lee had asked to be presented to the duke, but the prefect of the time had justified his refusal with the explanation that it was not proper to present to the prince the representative of a nation that had dared to declare war against the brave English.[25]

A year later, Lee's name was struck from the list of consuls invited to a ball given by the duchess for Bordeaux's high society. Why this ostracism? The United States of America was guilty not only of declaring war on England, but of winning it, and its consul in Bordeaux had committed the wrong of publishing his opinion in a book offending the British allies of Bourbon France.[26] William Lee justified his animosity against the English by their disgraceful war against his just government, and claimed that during the Hundred Days he had never uttered words hostile to the king. His country and he himself supported the august family called upon to direct the French nation. But the consul, whose career in Bordeaux (1801–16) coincided with that of the First Consul and Emperor, never managed to convince the royalists that he was telling the truth.[27]

Though victims of xenophobia, Americans were more or less protected by their status as foreigners. But the inhabitants of Bordeaux who had rallied to the usurper had no safeguard against the administrative purge that hastened transfers and retirements, or against the political purge that sentenced accused men to prison, banishment, or death, as in the case of the "Twins of La Réole."[28]

LEGAL MURDER

Born in La Réole, brothers César and Constantin Faucher,[29] politically and militarily active during the Revolution, attended to public affairs in their town, then enlisted as volunteers in 1793. Both came out of the Vendée war[30] with the rank of *adjudant general chef de brigade*[31] but were discharged due to injuries; they held minor departmental offices during the Consulate. After resigning and going into business in Bordeaux, they held no official positions during the Empire, but the Hundred Days made one mayor of La Réole, the other a deputy, and both members of the Legion of Honor and brigadier generals, appointed by Clauzel four days before Waterloo. After the defeat, with Gironde in a state of siege, Constantin was officially the military commander of La Réole and Bazas until, by order of the new war minister, Clauzel relieved him of his duties on July 21, 1815. The Fauchers fell back into line. They had compromised themselves, but the matter might have gone no further had events not worked against them.

On July 22 soldiers passing through La Réole pulled down its white flag, and the next day, in an engagement with royal volunteers, one of the royalist soldiers was killed. This was more than enough to declare the town a center of subversion stirred up by the two brothers. The sub-prefect alerted the king's commissioner in Bordeaux, who informed the war minister that he was sending royal guards there to reestablish order. Feeling threatened by these fanatics, the Fauchers appealed to Clauzel, but their letter, which arrived the day he left, fell into the hands of the newly installed Prefect Tournon. The prefect ordered that their house be searched: arms were found, not very surprising in the home of generals, but enough to convince the crown prosecutor, who had them brought to Bordeaux and imprisoned. A slap-dash investigation followed. The lawyers of the Bordeaux bar recused themselves and left the Fauchers to conduct their own defense before the court-martial, which condemned them to death for encouraging civil war. They were executed by firing squad on September 27, 1815.[32]

During the Third Republic, a schoolteacher raised them to the rank of martyrs and worked to rehabilitate their reputations posthumously.[33] Initially victims of their own past and of deep-rooted local hatreds, the Fauchers had come up against a sort of sacred union: the Ultras made them pay for disowning their class and fighting against the Great Catholic Royal Army in Vendée; the mayor, sub-prefect, crown prosecutor, and presiding officer of the court-martial—the local leader of the Knights of Faith and one of the organizers of March 12, 1814—were all more or less settling personal accounts with them; the prefect wanted people to forget his imperial past and justify the confidence just shown him; finally, the police, the law, and the army, who had lent their unwavering support, were acting in accord with the results of the preceding month's legislative elections. In the *Unobtainable* Chamber, two of the seven men elected from Gironde were moderate royalists, but the other five were Ultras. On September 5, 1816, the dissolution of this Chamber presided over by Bordeaux's Lainé aroused consternation described as extraordinary in his city.

Thus covered by the institutions devoted to the king, violence was used against groups for ideological and religious reasons, but also against individuals guilty of simple verbal outbursts. In December 1815 a sergeant of the Bordeaux city toll bureau was sentenced to four years in prison for shouting "Long live the Emperor, long live Napoleon, s—— for the royalists."[34] And in the summer of 1816, Jacques Lajonie Lapeyre got into trouble with the mayor of his village after a harmless altercation.

FROM ONE CAMPAIGN TO ANOTHER

The Lajonie family had lived for centuries in a region of vineyards and humanism, centered on the town of Sainte-Foy-en-Agenais[35] and linked to Bordeaux, the capital of the province of Aquitaine, by the Dordogne River. A walled town founded in 1255 as a defense against the English, it had attracted merchants, among them the Lajonies' earliest known ancestor, Antoine Lajonie, known as Figeac, who had settled there in the late fifteenth century. His sons were soon solid members of a class midway between the bourgeoisie and the minor rural nobility. A century after the founder's arrival, his great-grandson Simon de la Jonye, a wealthy cloth merchant and municipal magistrate of Sainte-Foy,[36] had converted, along with all the inhabitants, to Protestantism. In a climate of religious wars, followed by intolerance that culminated in the revocation of the Edict of Nantes in 1685,[37] the Lajonies, in spite of harassment and forced abjurations, remained loyal to their faith and their region, where they continued to live and marry within a twelve-mile radius.[38] In 1707, Isaac Lajonie, an infantry lieutenant in the Normandy regiment, married Jeanne François, the daughter of a lawyer and judge in La Roche-Chalais, Dordogne. Owner by marriage of the house of Lapeyre near the village of Gensac, he settled there and took its name, which he passed on to his descendants. One of them, Jacques Lajonie, sieur de Lapeyre (1784–1804), a bourgeois and well-to-do farmer, became the mayor of Gensac during the Revolution. At his death he left a name, personal property, lands, a widow, and five children, the last of whom was born in Gensac on July 24, 1787: Jacques Lajonie Lapeyre Jr., the eleventh generation known to us.

An Imperial Career

A boarding student at the lycée of Bordeaux, young Lajonie was congratulated by the headmaster for his conduct, church attendance, and ability in Latin, history, geography, and mathematics.[39] As there was no other career open to a young man, Lajonie wrote later, he applied to the special imperial military school in Fontainebleau.[40] Admitted on December 21, 1806, he took a four-month accelerated course: the Empire needed relief troops to seal the recently instituted continental blockade, ensure the extension of its conquests toward Poland, and offset losses after Pyrrhic victories like that of Eylau.[41]

Graduating as a second lieutenant on April 30, 1807, Lajonie was assigned to the Seventh Dragoons in Eugène de Beauharnais' army of Italy,[42] recovering in Crema, Lombardy, from a nightmarish campaign in Calabria. When he arrived at his posting he found a regiment that was again looking good, with thirty-four officers, eight hundred men, and almost as many horses. Italy was no longer a theater of operations, and the young man experienced "the tumultuous and licentious life of garrisons and camps." In March 1809 he finally left Milan in General Grouchy's First Division, to do battle with Austrians "quite inferior to the French in courage and military talent," thought Lajonie.[43]

Among all the battles fought as far east as Hungary, the regiment won particular renown at Wagram on July 6, 1809, inflicting heavy losses on the enemy. Given the mission of continuing the pursuit northward, the Seventh Dragoons took the city of Nikolsburg in Moravia, arriving at the Taya River before the Austrians, and on the tenth reached Unterwisterlitz on the river's banks. The French charged onto the bridges to prevent the Austrians from burning them, and that is how Lajonie, leading his platoon of scouts, had his left arm shattered by a bullet.[44] This was the cost of his regiment's lack of experience with such operations. When the bridge was taken and the Austrians driven off, Grouchy complimented his troops; on July 14, the Znaim armistice put an end to the campaign. The Seventh Dragoons inscribed the name "Battle of Wagram" on their standard and on July 20, at headquarters in Schönbrunn, Lajonie was awarded the cross of a Knight of the Legion of Honor.[45]

The waters of Petersbaden did not heal his arm. Because considerable swelling prevented Lajonie from holding his reins in one hand and his saber in the other, the regimental surgeon finally forbade him to ride and sent him to the depot of Lodi in Lombardy for consultation. In June 1810 Colonel Séron gave him a three-month convalescent leave to go reassure his family and enjoy the benefits of thermal treatment at Barèges in the Pyrenees. His request for paid leave was well received at headquarters, since Séron sang the praises of the gallant officer and zealous and brave young man. The colonel had lost an arm at Wagram; he had no trouble finding the right thing to say.

Another course of treatment at Acqui in the Piedmont had no better results. In early March 1812 Lajonie was notified by the war minister that the Emperor, in a decision dated February 12, exempted him from military service due to ill health and authorized him to return home. On April 24 the governing board of the Seventh Dragoons, which had deleted his name from the regimental rolls, placed the notice "resigned" next to his regimental number.

Bitter and disappointed at having to abandon his chosen career, Lajonie was nonetheless happy for his family: "I think that you will be pleased that I am leaving a career in which your friendship does not tell me to continue," he wrote his mother. "I also think that my sisters and brother will be as pleased to see me as a *bourgeois* as they were unhappy to see me as a soldier. Finally, I thought that you would all be glad and that is why I gave in easily."[46] His regiment left for Russia without him. Of an initial strength of more than seven hundred dragoons, only forty came back across the Niémen River on December 31, 1812.

CHÂTEAU LIFE

As with all military men, Lajonie's reentry into civilian life was tricky. After five years in uniform, he did not want to become a farmer, but planned to live on bread and air, assuring his mother that all one needed to be happy was to enjoy the little one had.[47] His portion of his father's inheritance and the hospitality of his older sister Marie, known as Nauzille, and her husband, Jean Pierre Taupier-

Letage, made this happy-go-lucky attitude possible.[48] They lived in Juillac, a village twinned with Gensac, in the manor house of Le Soulat. Situated on the edge of a plateau, this centuries-old dwelling overlooked the admirable Dordogne valley, but Lajonie described it in 1816 as a ruined château that had lost its onetime luster. Taupier had inherited it from his father, a former infantry officer turned farmer, and had settled there with Marie Lajonie, whom he had married twenty years earlier. The couple lived there childless, surrounded by vineyards and fields whose produce gave them affluence and respectability. In the years of the Revolution and Empire, Jean Pierre Taupier was the mayor of Juillac.

On May 14, 1814, Lajonie started his own family. In the nearby village of Les Lèves-et-Thoumeyragues he married Dorothée Noguey-Maransin, a sweet, pleasant woman, says Lajonie, born at the château of la Beauze. She was twenty years old. Her mother, Catherine Obre, was said to have been the servant of Michel Noguey-Maransin, her father, who had died during the Consulate. This former secretary of the Parliament of Bordeaux had attracted attention in the district by making reactionary speeches against the Revolution that could have cost him his head. He had doubtless chosen a bad year to become the new lord of la Beauze, since in 1789 he succeeded the Marquis of Rabar, whose excesses had finally exasperated the population. The château itself had lost much of its proud aspect after a royal punishment had amputated it of its towers and reduced it to an inelegant, massive stone block.[49]

Peace had brought back to their womanless homes men whom war had kept too far away for too long.[50] Marrying was the way for them to reconnect with civilian life and get around their precarious material situation. For most junior officers, the vast majority of whom were of rural origin, country life thus succeeded bloody engagements.[51] So rumor, fed by supplanted rivals, had it that Lajonie had married a dowry. Living at his brother-in-law's, he had neither a residence of his own nor an income. The division of his father's estate into five equal parts had left him fewer than thirty-five acres in fields, vineyards, meadows, and woods surrounding the manor house.[52] The widow Maransin provided her daughter with a pension of two hundred livres[53] and was keeping for her a portion of her father's estate.[54] Dorothée found Lajonie attractive. They lived a quiet love story amidst farm work and pleasures of the hunt, and soon had a family: Marie-Elbine Lajonie was born on March 26, 1815, during the Hundred Days, and Pierre-Léopanno on November 28, 1816. Both babies' names illustrate Lajoine's devotion to the emperor: "Elbine" refers to the site of Napoleon's exile and "Léopanno" is an anagram of the name "Napoleon."

Lajonie bore the mark of his past: raised in the cult of the Emperor, seasoned in the army that he left reluctantly, he had chosen his camp and made no secret of it. With his eternal officer's hat, black sideburns trimmed to a point, long slender build, and red decoration in his buttonhole, Lajonie cultivated his image, affixed *m.d.1.h* (*membre de la légion d'honneur*) to his signature, and ostensibly displayed a loyalty that might well offend a rural world hostile to the Empire that had brought war. In the end, these signs of homage exasperated the conservatives who had champed at the bit hoping for Waterloo, and whose frenzy was unleashed once Napoleon was on his rock in the South Atlantic.

THE LAJONIE AFFAIR

As has been mentioned several times, as of July 23, 1815, Gironde had a new pre-fect.[55] The Count of Tournon was an intelligent and sensible administrator who, back from a posting as prefect in Rome during the Empire, had gotten a posi-tion with the Restoration government after refusing to serve during the Hun-dred Days. He prided himself on a sense of moderation that led him to condemn the spirit of reaction noticeable in the department, and to curb royalist excesses.[56] His attitude regarding the Faucher brothers must have reassured the Ultras, who were sizing him up suspiciously, and he confirmed this good impression by initi-ating a thoroughgoing reorganization of the civil service and town councils: hun-dreds of people were dismissed for their political opinions and replaced by right-thinking individuals.[57]

In Sainte-Foy-la-Grande this was not necessary, as the mayor and his town council had no difficulty turning with the wind, in April reaffirming their "love, admiration and respect" for Napoleon, and in July, on the opposite page of the same register, cherishing Louis XVIII as a good father and offering "to shed their blood for the defense of [his] sacred person."[58] With this excessive royalism, the town councillors were expressing the same sentiments that their colleagues in Bor-deaux had shown earlier with even greater elation.[59] In Juillac things were differ-ent. Pierre Jacques de Lachaud had taken the "de"—a sign of nobility—out of his name to better establish his legitimacy during the Empire, and had taken ad-vantage of public feeling to push aside his rival in the town hall, Taupier-Letage, whose deputy he had been since July 1812; Jacques Lajonie, back from the army, had been appointed to replace him as town councillor. Although they were re-lated, old enmities separated the families, so that Lachaud would not pass up the opportunity to settle a dispute that until then had been going against him. When the time came to reappoint mayors and deputy mayors, the new sub-prefect of Li-bourne confirmed Lachaud as mayor.[60] Now attacking Taupier publicly, he forced him to leave Le Soulat, which Lajonie was to see to in his absence, but he then turned his hatred against the new occupant. He spread rumors to the effect that Lajonie was wrongfully wearing the cross of the Legion of Honor and had in-sulted him in connection with a citation for damage caused by his dogs chasing a hare.[61] Lachaud manipulated the authorities so that they soon saw Lajonie as an enemy of law and order to be apprehended. Fearing incarceration, Lajonie ignored three injunctions from the sub-prefect to come to Libourne. On July 20 police ser-geant Descas of Sainte-Foy was ordered to bring him in by force.

At 6 A.M. on July 23, four gendarmes appeared to arrest Lajonie, who refused to open the doors of Le Soulat. So they left to get reinforcements on the pretext that the place was a fortress where fifty men, armed to the teeth, had entrenched themselves. According to Lajonie, gendarmes came from all the neighboring police stations and rummaged through the dwelling that he had deserted, turning everything upside down on two successive visits. There are numerous accounts of similar situations elsewhere, involving a general mobilization against supposed conspirators. The affair of the Faucher brothers had begun in just this way.

Descriptions of the wanted person were issued. Lajonie went into hiding and considered fleeing the country: "Will I be forced to expatriate myself and go to a foreign land in search of the safety and protection that would be refused me in my own country?"[62] In early August 1816, 120 mayors, deputy mayors, town councillors, notables, and most important taxpayers of Gensac and the surrounding towns signed a petition attesting that they "knew Jacques Lajonie Lapeyre, knight of the Legion of Honor personally, and that in the nearly five years since he retired from military service and has been living at Le Soulat in the village of Juillac, his conduct has been calm and peaceful, occupied with agricultural tasks and occasionally with the pleasures of the hunt, and that to our knowledge there has been only one altercation with another person in the market of Gensac last July."[63]

Lajonie's hopes as to the impact of the petition were all the more disappointed in that Lachaud's hostility was backed by that of the sub-prefect in collusion with him.[64] Shortly thereafter, Lajonie sent him the manuscript of a carefully worded first justificatory statement, in which he asked to be heard impartially and not harassed over minor offenses blown out of proportion by an informer. The sub-prefect turned a deaf ear. The tone of the second statement, printed in Paris in September 1816, was completely different. Lajonie had nothing more to lose against an "implacable avenger of partisan hatreds and resentments": he denounced the unprecedented fury and incompetence of a civil servant arrogating to himself power that the law had not given him.[65]

Tension had reached its height with the murder that was blamed on him. During the night of August 29, 1816, four members of the Sainte-Foy police force met, at some distance from the main road, two or three travelers or hunters—it was not determined which—who refused to give up their arms. A policeman was mortally wounded in the struggle that ensued.[66] Someone was said to have uttered Lajonie's name in the melee. The mayor and sub-prefect pounced on this bit of hearsay and alerted the prefect in Bordeaux, who, on September 2, ordered posted and circulated the description "of an officer not on active duty charged with the murder of a policeman trying to arrest him in accordance with an arrest warrant from the court of Libourne."[67] Nothing was accurate except the description of the fugitive: light, striped waistcoat, five feet ten inches tall, very dark brown hair and eyebrows flecked with gray, black beard, thick sideburns coming to a point on his face, brown eyes, nose a bit long, average mouth, full lips, long thin face, tanned skin slightly marked by smallpox. A search was launched to bring him to justice. Less than a year earlier, in Trévoux in eastern France, the Bacheville brothers, two captains of the ex-Guard who had survived Waterloo and been accused of plotting, had experienced an identical situation before crossing into Switzerland and wandering throughout Europe and the Near East in search of a peaceful haven.[68]

Lajonie, however, had grounds to complain of an illegal and arbitrary order because of the lack of evidence and the absence of an arrest warrant in due form. Only an investigating magistrate, and not a sub-prefect, had the right to issue one,[69] and the magistrate of Livourne did not sign the arrest warrant until December 26. On January 14, 1817, the bailiff of the Pujols justice of the peace, at the request of the prosecutor, handed a copy to a young woman at Le Soulat: Lajonie was not there, she stated; she did not know where he could be found. As it happened, he was no longer in France.

His family under surveillance, his description posted everywhere, liable to be shot without warning, Lajonie was done for. Rather than turn himself in to die on the scaffold, he fled to Geneva, from where, on October 12, 1816, he sent his wife a document giving her power of attorney. He did not remain in Switzerland long. Before the end of the month, in the Gironde estuary, he boarded the American brig *James Murdock* bound for Philadelphia with a cargo of varied merchandise and nine registered passengers.[70] On November 4, 1816, he wrote his first letter as an outlaw to his family in Le Soulat: "I am going to venture upon a career that is new to me. How many times did I not, in my anxiety, turn my eyes to a quiet life of obscurity? But how many times also, did I curse the fate that forced me to leave my fatherland. Never have I seen myself weaker, or if you prefer more vulnerable. The strongest idea that I can give you of this weakness is to admit it to you, at the very moment when you need to be strengthened by my example."[71]

This first letter was followed, from January 1817 to March 1829, by a long series of others mailed from Pennsylvania, Delaware, Maryland, Louisiana, and particularly Alabama, addressed to the château of Le Soulat, in Juillac, in Gironde.[72] The 102 letters that have come down to us constitute an exceptional account of the stay in the New World of an unknown but remarkable letter writer. In his trials, partly because he wanted to be worthy of his family's respect, he demonstrated a curiosity, integrity, and courage, at times worn down by the harshness of his exile and the suffering of separation, that are not to be found in the letters to the wife of someone like Marshal de Grouchy, bearing up badly under adversity and the injustice of his fate.[73] And so, Jacques Lajonie's cat's-paw handwriting[74] will, from here to its end, lead us as we pursue our study.

II

Across the Atlantic

4

Maritime Exodus

America was in every respect our true refuge . . .

—Napoleon, St. Helena, 1816

I have taken passage on a ship that is leaving for America, where I shall dig my grave . . .

—General Savary, Smyrna, August 1816

Jacques Lajonie's stay in Switzerland was brief. After years under Napoleon's rule, the new Confederation had just returned to the control of the old elite, from whom the fleeing French could expect no help: the right to stay was denied to many, and others were taken back to the border and handed over to waiting French gendarmes.[1] Accused of murder, Lajonie had to fear the zeal of the Swiss constabulary and consider Switzerland a place of transit, not a permanent refuge. He thought it better to cross the seas: "In the United States there is liberty."[2] Beyond an ocean far more difficult to traverse than the Alpine barrier, it was the ideal destination, reached chiefly from the ports of Antwerp and Le Havre, Marseille and Leghorn, Nantes and Bordeaux.

After returning home, Lajonie could easily go down the Dordogne to the Gironde estuary, gateway to the New World, open in the eighteenth century to tens of thousands of people from the Aquitaine region of southwestern France, driven out by poverty and religion, or dreaming of a better life where they would—why not?—be fabulously rich "Ameriquains."[3] People were then thinking of Louisiana and the islands, with Saint-Domingue, "the pearl of the Antilles," as the destination of the vast majority. But in the 1790s, the slave revolt, the maritime war with England, and then with the United States slowed the movement of men and merchandise. Bordeaux's function as the port of welcome and redistribution for the Aquitaine population declined until peace was restored. Then transatlantic shipping resumed its regular rhythm and the United States of America became the recipient of a new wave of immigration, following that of the refugees from Saint-Domingue, its finest colony, which France had lost for good.

TWO BROTHERS ON THE ATLANTIC

Freedom and fortune were all the more cherished by the emigrants because they had been won with difficulty, at a time when sailing was still a heroic act for sailors and passengers, even those accustomed to risks, like military men. Artillerymen, infantrymen, and cavalrymen were most often strangers to the world of the sea, since France, from the Revolution to the Empire, had not been known for naval exploits, foundering at Aboukir and Trafalgar. After Waterloo, the British squadrons relaxed their hold on French shipping, except for one traveler whose escape was unthinkable.

Before he conceived the idea of an American exile after his second abdication, Napoleon had already received encouragement in this direction . . . from Fouché, wanting to get rid of him. The Emperor ought to flee Elba, which slanderers intent on his ruin would quickly label a dangerous base of operations for stirring up a revolt. For his own safety he should put half of the earth, that is, the Atlantic, between himself and his enemies, and begin life anew in the United States, among new people who could admire him without fearing him: "You will prove [to them] that if you had been born in their midst, you would have felt, thought, and voted as they do, that you would have preferred their liberties to all the dominations of the earth."[4]

Fouché would have liked to see Napoleon leave the Mediterranean and permanently renounce his ambitions so as not to compromise his own. Napoleon on Elba was for Europe Vesuvius next to Naples, he prophesied to the Count of Artois. Finding the comparison persuasive, Artois sent his protégé, Hyde de Neuville, back from a long American exile, to investigate the possibilities of the Emperor's escape and then to speak highly to him of America: "In this virgin land of liberty, the name of the great conqueror would have impressed Europe; the setting matched his stature and renouncing his ambition would be a heroic end."[5] Fouché was thus not alone in wanting the Emperor far from France, and Count Anglès, trained by Fouché, went so far as to extend the need for an oceanic quarantine zone to the entire family in a police report of July 1, 1814. The presence of the Bonapartes on the Continent constituted a threat to the Bourbons and to peace: "It would truly be desirable to get the [allied] powers to recognize how useful it would be for the public peace to put the ocean between this family and Europe."[6] In the autumn of 1814 the subject was discussed behind the scenes at the Congress of Vienna, but nothing was decided.

After his abdication, left with no other choice but England or the United States, Napoleon told the scholar Monge of his desire for a new career in which he would explore America to study those phenomena of physics and the earth on which the scientific world had not yet expressed an opinion.[7] Scientific instruments were purchased. Malmaison was emptied. The banker Laffitte was to send three million francs in gold across the Atlantic. Napoleon counted on the entire Bonaparte family making the journey.

His departure depended upon the goodwill of Fouché, who headed the Provisional Commission. On June 26, 1815, it was decided that the frigates *Saale* and *Méduse* would be put at the Emperor's disposal at Rochefort-sur-Mer and could get under way after safe-conducts were delivered—except that Fouché had not the least intention of issuing them. Helping the Emperor flee would reduce his chances of becoming a minister in the second Restoration, which was being planned with Wellington's backing. Fouché knew that Wellington, who had gotten burned once, would not be satisfied with Napoleon's abdication but would demand he be handed over to him. On July 8, Napoleon was on board the *Saale*. The winds were contrary and the port blockaded by the English. The situation was certainly doubtful, but there were feasible escape plans, and one of them could have succeeded.

Frigate Captain Charles Baudin was, according to a retired vice-admiral who had proposed his name to the Rochefort police port authority, the only man capable of taking the Emperor to America.[8] Baudin had two corvettes, the *Bayadère* and the *Infatigable,* at anchor in the Gironde estuary. He agreed to help, wanting to spare France the humiliation of seeing its sovereign fall into the hands "of its most implacable enemy."[9] He knew of two extraordinarily fast American privateers waiting in the port of Bordeaux that would have no difficulty running the blockade. Getting their captains to agree to take the Emperor aboard would be a mere formality. But the Emperor's agreement was long in coming.

When the elder General Lallemand went to give Captain Baudin Napoleon's assent to his plan, precious time had been lost, but the captain promised, despite everything, to keep his word and went to Bordeaux to see General Clauzel, who still headed the Eleventh Military Division. All that Clauzel could offer was to go together to discuss the matter with the American consul, William Lee, who had for so long made a show of his attachment to the Emperor. This cooperation resulted in finding an American ship that agreed to take him aboard, but when Baudin wanted to return to the *Bayadère,* he saw in its place two English frigates that had managed to slip in with the complicity of the commander of the coastal forts whose cannons had remained silent. As five war cruisers closed off the mouth of the estuary, any attempt was now hazardous. Tired of resisting, Napoleon gave himself up. Baudin wrote to Lee in dismay. Their efforts to save the Emperor from the affront of finding himself a prisoner of Albion were fruitless, for on July 15 he had surrendered to the English squadron off Rochefort, with his retinue and baggage. In the margin of the letter he had just received from Baudin, Lee noted the escape plan devised with Clauzel and added the last word: "We arrived one hour too late."[10]

Napoleon had not listened to those in his entourage who, like the elder Lallemand, categorically rejected British asylum, preferring American hospitality. He doubtless chose the riskiest solution, the most worthy, whereas the others would have diminished him.[11] This would explain his refusal of Joseph's offer to take his place, taking advantage of their resemblance. In the end, more flexible than his brother on the subject of his late royal majesty, Joseph tried his luck and became the first American exile after Waterloo.

Joseph had arrived in Rochefort under a pseudonym and the identity of a merchant, but he quickly made himself known to the local notables whose membership in Freemasonry could be useful to him, the Grand Master of the Grand Lodge of France. A shipowner and Worshipful Master of a lodge in Rochefort said that he was ready to help him, and had his son search in Bordeaux for an American ship that would agree to take him across the ocean. Captain Misservey's brig was chartered for eighteen thousand francs and a rendezvous arranged in Royan. The *Commerce* cast off on July 25 at 3 A.M. The weather was superb and the moon and tides favorable, relates James Caret, a young American whom Joseph had engaged as interpreter.[12]

Two warships inspected the ship at the mouth of the estuary, but the officers who searched it were not the shrewdest of people. As they were no longer looking for Napoleon, who was now aboard the *Bellerophon,* they paid no attention to the passenger whose passport bore the name Surviglieri—Captain Misservey himself supposedly did not know with whom he was dealing. After thirty-two days at sea, the *Commerce* came in sight of Long Island, then escaped two English frigates ready to stop Napoleon, who was said to have left France for the United States: the news of his surrender was not yet known. The *New York Evening Post* announced on August 29, 1815, that a brig, arriving the previous day from Bordeaux, had set ashore near Brooklyn five French passengers; another article added that Generals Clauzel and Carnot were among them.[13] Wishing to remain incognito, Joseph did not refute this claim; therefore, when the mayor came to call on him at Mistress Powell's boardinghouse on Park Place, he believed he was speaking to Carnot. It was only an unexpected encounter with a former officer of his guard on Broadway that revealed the secret of his identity. So Joseph Bonaparte became as such an exile in the United States—the first of any importance, preceding many others.

EXILED FROM EXILE

In 1815 one could be happy to cross the North Atlantic in less than five weeks. Ships had become faster thanks to slimmer hulls and better trained captains, but this progress was not widespread, and the unscientific navigation of ancient times had not disappeared. Those setting out on a journey were sure of nothing: neither the date of departure nor the duration of the crossing nor, consequently, the date of arrival, the port of arrival, nor even arrival itself. Setting out was a relief, sailing a trial, arriving a deliverance.

DEPARTURES FROM FRANCE

During the entire period of Napoleon's Empire, British cruisers prevented French oceangoing ships from sailing. When the return of peace removed this obstacle, there remained the struggle against natural elements, even in the summer: the northern latitudes could still unpleasantly surprise the traveler with wandering icebergs in June.

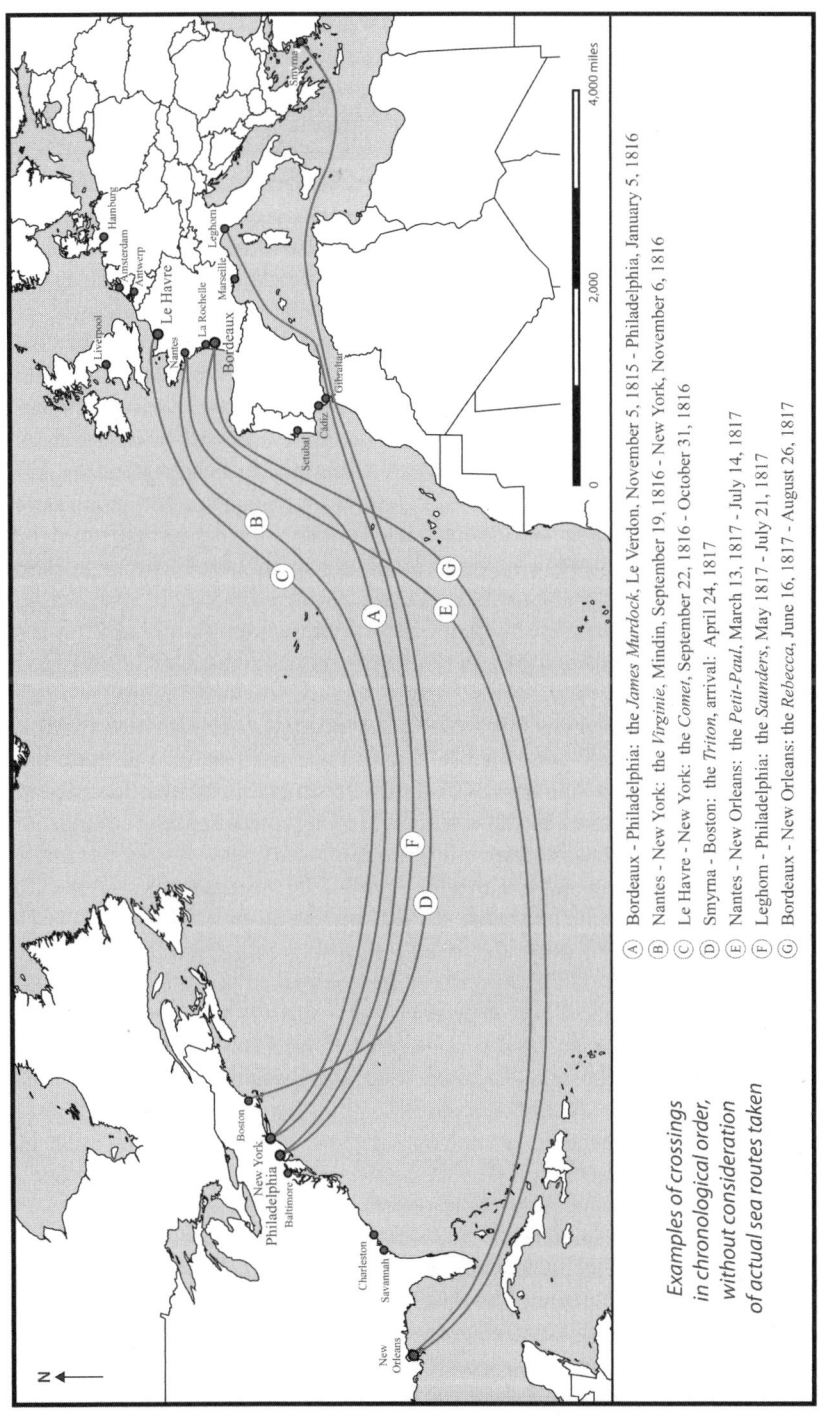

Examples of crossings
in chronological order,
without consideration
of actual sea routes taken

(A) Bordeaux - Philadelphia: the *James Murdock*, Le Verdon, November 5, 1815 - Philadelphia, January 5, 1816

(B) Nantes - New York: the *Virginie*, Mindin, September 19, 1816 - New York, November 6, 1816

(C) Le Havre - New York: the *Comet*, September 22, 1816 - October 31, 1816

(D) Smyrna - Boston: the *Triton*, arrival: April 24, 1817

(E) Nantes - New Orleans: the *Petit-Paul*, March 13, 1817 - July 14, 1817

(F) Leghorn - Philadelphia: the *Saunders*, May 1817 - July 21, 1817

(G) Bordeaux - New Orleans: the *Rebecca*, June 16, 1817 - August 26, 1817

Map 2. The French emigrants' ports of departure and arrival (1815–1818).

Seasickness. The ship that Lajonie boarded was to leave Gironde on November 1; the weather was bad, and he immediately realized that he did not have sea legs. Nor did Guillaume Tell Poussin.[14] Poussin had preferred exile to the consequences of the capitulation of Paris, against which he had fought in 1814. In his opinion, with no future possible for young men in France, their only hope was to go far abroad to seek what their homeland could no longer offer.[15] Like Percier, the Emperor's architect who had hired him in 1813 for his name and drawing talent, the United States would value him. Leaving Le Havre on an American ship on October 3, 1815, he reached New York after forty days at sea: "I arrived completely exhausted from having had to remain immobile in my bed, due to seasickness."[16]

In September 1815, Édouard de Montulé, a lieutenant, had been obliged, like Lajonie three years earlier, to resign because of an injury, and like Poussin, he could expect nothing from the Bourbons.[17] But wealthier than they, he was able to set out on his journey simply to satisfy his curiosity "to observe the capacious irregularities of nature and to inquire into the traces that succeeding generations have left on earth."[18] Montulé shared with them a taste for writing and a lack of experience with the sea. On September 19, 1816, the French brig *Virginie* left the Loire estuary for New York.[19] The sky changed color, the wind began to blow, the sea rose and became, in his oceangoing neophyte's eyes, monstrous. Montulé looked back nostalgically on the French coast before it had even disappeared from sight.

Lajonie was in no better shape when his ship left "after the most horribly bad weather and something of a health problem" that made him look longingly toward his "dear home."[20] Contrary winds forced a second departure. Crossing the Bay of Biscay was torture until they reached the region of the Azores, Canaries, and Madeira. Lajonie said that he vomited 140 times during two storms, coming out of them exhausted and sorry he had not brought along a personal stock of healthy food; their fresh provisions used up and fishing impossible due to storms, all that was left was salt meat and sea biscuits. The *James Murdock* passed Bermuda and, after another hurricane, finally reached Philadelphia on January 5, 1817, after eight trying weeks. It was preferable to leave between March and September in order to avoid these frightening experiences, but aside from the fact that an outlaw had no choice, this was not a guarantee either.

In March 1816 the prefect of Gironde ordered the Customs Administration to search two American ships to make sure that three banished men, two of them regicides, were on board. After being checked, the *Magnet* and the *Harriet* were escorted until they reached the sea on March 14.[21] Fortune was to smile on them differently.

Louis Marie Dirat, dismissed from his position as sub-prefect of Nérac (Lot-et-Garonne) by the first Restoration, had joined the opposition on the editorial team of *Le Nain Jaune* (The Yellow Dwarf), an illustrated satiric newspaper founded in December 1814 to denounce the Bourbons.[22] Its writers, whom the authorities associated with Napoleon's return from Elba, were asked to leave the country. The paper reappeared in Brussels as *Le Nain Jaune Réfugié* without Dirat, who received a passport for the United States on February 14, 1816, in Agen. To pay the passage for himself, his wife, and their two sons, he had to fall back on the captain's generosity, since he did not have the money the French government is said to have promised him. On May 4, 1816, the *Magnet* arrived in Philadelphia,[23] where Dirat was welcomed with open arms by Simon Chaudron, editor of the French-language weekly *L'Abeille Américaine*: "We have seen the Yellow Dwarf; he is not ugly, is above all very witty, and we cannot imagine why he has been persecuted by the OTHER DWARFS, unless it is precisely because of this."[24]

The two regicides were Pénières, whose trials we have mentioned, and Bernard, known as de Saintes, a lawyer infamous for his bloody excesses during the Revolution. He had just attracted public notice again as a deputy during the Hundred Days and with his remarks boasting of his regicide vote and expressing delight at Napoleon's return.[25] He denied in vain that he had ever held a public position "under this man" whose defeat he had welcomed, and did not manage to convince anyone of his most sincere wishes for the good fortune and long reign of King Louis XVIII.[26]

The bad luck that dogged Pénières soon contaminated his companion in escape and the other passengers of the *Harriet,* bound for Louisiana.[27] Two days after sailing, the ship, tossed about by extraordinarily violent winds, sprang a leak.[28] The crew had to take turns constantly pumping out the water flooding into the ship, but after four weeks of effort there were seven feet of water inside the hull, which was sinking so badly that the order was given to abandon ship. The passengers and crew boarded two lifeboats and were miraculously picked up after five days adrift by a Philadelphia sloop sailing from Leghorn, Italy.[29] Put ashore in late April 1816 in Madeira, Bernard de Saintes remained there, in the hope that his unhappy lot would move the king,[30] but his pleas were fruitless and he died alone and destitute in 1818.[31]

Pénières had been fortunate to remain aboard the ship that saved them, which took him and his maid to Philadelphia, reaching that port on July 9.[32] The circle of French émigré refugees was delighted: "Another outlaw has just, after an awful shipwreck, disembarked onto the free and hospitable lands watered by the Delaware River. Mr. Pénières, former member of the Tribunate, ex-member and *questeur* of the last Chamber of representatives . . ."[33]

Besides the need for a hasty departure for some and a secret one for others, fear of the crossing often explains the absence of wives and children (who were sometimes very young), their delay in going to the United States, or their decision to remain in France. People thought that crossing the sea was a masculine adventure, and statistics of transatlantic migrations show that this was a reality.[34]

Traveling Solo. The Countess of Montholon managed to convince her husband to take her along to America with the Emperor when this was still the plan in Rochefort: "The journey was long. It was a bit harsh. Only men could go," she wrote.[35] They went instead to St. Helena. Neither Julie Clary, Joseph Bonaparte's wife, nor the Countess Lefebvre-Desnoëttes had been able to accompany their husbands, nor could they join them. In May 1817 the former had arranged to board, in the Netherlands, the American ship *Montesquieu* bound for Philadelphia, but the captain had to refuse her passage on the orders of the ship's owner, Stephen Girard, who feared that his ship would be stopped by the Royal Navy.[36] She eventually chose to go to Italy, letting her two daughters join their exiled father later. The general expected the countess to arrive in New York at about the same time, but was merely told of her failure to sail by the French consul in Philadelphia: her badly damaged ship had to put into port in England; the countess had been so frightened that she had abandoned all ideas of a sea journey and had returned to Paris, letting her companion, Léontine Desportes, go on without her.[37] Dorothée Lajonie also wanted to join her husband. Very early, she stated this as her firm intention, but Lajonie exhorted her to do nothing of the sort until the conditions were right for their reunion. His greatest fear for her was the crossing. She carried out her plan regardless.

Although the vast majority of men traveled solo, there were some group crossings with a relative, a mistress, one or more friends, or a servant when embarkation was not hurried by a forced departure. General Simon Bernard's is a case in point. He had been Napoleon's aide-de-camp at Waterloo and wanted to accompany him to America when plans for this were made at Malmaison in June 1815, but the Emperor had refused on the grounds that Bernard could not leave his wife, Joséphine de Lerchenfeld Siesbach, and their two daughters, Pauline and Sophie, four years and eighteen months old, respectively, alone in France. The *Comet* deserved its name: leaving Le Havre on September 22, 1816, they arrived in New York after five weeks at sea.

Uneventful Crossings. During his Atlantic crossing in 1816, Merle d'Aubigné, from Geneva, Switzerland, remarked that a people whose history is boring is happy, and applied this adage to the sea traveler, boredom being by far preferable to storms or the authority of an irascible captain.[38] Many expeditions went well, with American captains considerate of their French passengers, despite their administrative or monetary situations. A spy in the pay of the French diplomatic corps in Washington introduced his observations with the remark that "there are few American ships coming from France that do not have on board individuals who may be wanted by the police. They always manage to embark without passports, for money and sometimes that consideration is very small."[39]

General Clauzel left Bordeaux on July 28, 1815. His last public act was to forward to the minister of police a justificatory statement of his conduct during the Hundred Days, for the king's attention. In the preamble to an edition of this statement, he said he intended to go to Louisiana, but this went unnoticed. The prefect of Clauzel's native Ariège did not believe the rumors circulating about him, particularly one from his wife saying he was in America. And yet, the prefect of Gironde had reported the statement in a New York paper of January 17, 1816, that General Clauzel, "one of Buonaparte's favorites," had arrived in that port on the ship *Medora,* sailing from La Rochelle.[40] On March 1, 1816, Clauzel officially confirmed his arrival in a long letter addressed to the French authorities from Philadelphia: "I am one of the nineteen generals on the first list of the edict of July 24 and my name is Clauzel."[41]

The police sleuths had also gone astray in search of Grouchy, whom they rightly thought to be in Normandy. In the days following Waterloo, the marshal had, in an exemplary retreat, brought back all his troops intact and then, on June 29, 1815, resigned from his position as commander in chief. In Paris he is said to have attempted to resume contact with the Bourbons to assure them of his good intentions and ask for an opportunity to prove his zeal.[42] Grouchy then went to his château of Laferrière in the department of Calvados. When he saw his name heading the edict, he took fright and went into hiding in a secluded cottage.[43] When the intensified search threatened his safety, he decided to leave France with the help of his family and neighbors.

Under cover of darkness, Grouchy was taken to the shore. A boat was to transport him to the island of Guernsey, but the rounds of coastal patrols interfered with his getting aboard. He later told how he dug himself a lair in the sand and huddled there until he was able to swim to the boat, and then reach Guernsey where he lived for a month under a pseudonym.[44] In December 1815 the *Two Brothers* took him to Baltimore in a difficult two-month crossing. On February 3, 1816, the French consul announced the news to the minister of foreign affairs.[45]

Grouchy did not go to the United States with the idea of remaining there. Before leaving, he had told his older son that he preferred the southern Netherlands, where he would acquire property and send for his hunting dogs.[46] In a last letter to his wife, he confided that if he managed to get out of France, he would want "nothing more than to return," perhaps at the time of the coronation when the king might declare an amnesty.[47] Grouchy claimed to be the victim of an injustice that would eventually be exposed, and he counted on his family to have his name cleared as quickly as possible. His trial took place on October 19, 1816. His wife, two sons, and sister, the Marquise of Condorcet, were united behind the man accused in absentia. Colonel Alphonse de Grouchy spoke at length in his defense, certain of the sanctity of his cause. This filial plea was found moving in the faraway United States: "Touching piety worthy of ancient times. Sublime effort of love and honor!"[48] But the judges' verdict did not permit the return of his father, whose fear that the doors of France would long remain closed to him was now justified.

Alphonse de Grouchy and his brother Victor, a lieutenant, were on half pay. In April 1817 they were granted a year's leave to spend with their father in the United States.[49] Accompanied by their first cousin, artillery Lieutenant Le Doulcet de Pontécoulant,[50] they left Le Havre on April 20 on the American brig *Ocean* and arrived in New York in record time on May 16.[51] Many other exiles whom the police did not have to hunt down left through the port of Le Havre. The best-known of these were former minister and member of the Council of State Regnaud de Saint-Jean d'Angély and former regicide Convention members Quinette, Lakanal, and Hentz.

Destination New York. Regnaud did not delay in conforming to the proscription edict of July 24, informing Fouché a month later that he was going to Le Havre, from where he had reserved passage to America.[52] The prefect of Seine-Inférieure confirmed, on the eve of his departure, that he had issued him a passport, as well as to the people accompanying him, a Madame Frossard and his son, Auguste Regnaud, aide-de-camp to Napoleon during the Hundred Days and a captain at Waterloo. The prefect asked the police to facilitate their boarding of the *United States* on August 30; the *Evening Post* announced their arrival in New York on October 25, 1815.

Quinette de Rochemont's haste regarding the amnesty law was equally great. On January 31, 1816, he was in Le Havre with a passport for the United States, the next day his baggage was taken onto the *Abeona*, and on February 8 he was on his way to New York. The regicide vote of this former finance manager of the communes and hospitals of France, and his promotion during the Hundred Days to the Council of State and the Chamber of Peers, had earned him this one-way ticket that he would rather have done without. When he returned to Le Havre, the ship's captain reported that Quinette had been very sorry to leave his country and had said several times during the crossing that he liked neither the Bourbons nor Bonaparte, preferring the republic to both.[53] Regnaud and other refugees boarded the ship in New York to welcome him and three days later invited him to a gala dinner with Joseph Bonaparte.

Of the twenty-two French passengers aboard the American ship *Eugene,* Lakanal and Hentz were the only ones affected by the amnesty law. Had it not been for this exile, nothing would ever have brought together these two men, whose destinies had diverged after their days in the Convention. In Le Havre, before leaving, Lakanal had expressed his sadness: "I have not been entirely useless to my country which is rejecting me and which I miss."[54] During the Revolution he had played a crucial role in developing France's public education system. Given his republican principles, he refused important positions during the Consulate and Empire, working as a professor of classical languages, then as bursar of the lycée Bonaparte in Paris, and finally, from 1809 on, as inspector of weights and measures in Rouen. The second Restoration dismissed him, but had difficulty finding anything of which to accuse him. Granted, as deputy from Ariège, Lakanal had voted for the king's death and had demonstrated both hatred for the monarchy and an anticlericalism of which there was written proof, but since then he had kept his distance from the imperial regime and had not voted for the Additional Act during the Hundred Days. In the end, it was by going abroad that he proved the amnesty law could be applied to him.[55] Unlike Grouchy, he did not leave France with the intention of returning soon, but with his wife, their two daughters, the money from the sale of his possessions, his books, correspondence, scientific equipment, and three letters of introduction: the first from the American consul in Paris for General George Mason; the second from Thouin, superintendent of the Jardin du Roi (King's Garden) in Paris for Thomas Jefferson; and the third, also for Jefferson, from Lafayette.

Although Nicolas Joseph Hentz, Lakanal's former fellow deputy, also left with his wife and children and a loan of three thousand francs to finance his exile, his situation was totally different. While history has recorded Lakanal's name, that of

Jacobin Hentz has faded into obscurity, despite his sensational beginnings during the Revolution. Deputy from Moselle in the Convention, he had been, at the age of twenty-four, its youngest member. As a delegate to the army, he had played the organizer, appointing and transferring generals, before being sent to Vendée to figure out how to eliminate its inhabitants, members, he said, of a bad race.[56] The end of the Reign of Terror, during which he had shown what he was capable of in the east of the country, turned him into an "old-fashioned piece of furniture" that no longer fit in anywhere.[57] Forgotten by the government for the next twenty years, he was directing a girls' boarding school after various other jobs when he was reminded of the martyr-king's death. The *Eugene* left Le Havre on January 23 and reached New York in six weeks.[58]

Jean Jérôme Cluis needed coaxing to purchase his ticket to exile. Aide-de-camp and then secretary to General Savary, the minister of police, he was appointed a police captain by Savary, who had become the inspector general of the national police force (the gendarmerie) during the Hundred Days.[59] Cluis considered his listing in the proscription edict to be an error, having, he claimed, nothing to feel guilty about: his offense was that his name was linked with that of his mentor, whom Fouché had never forgiven for succeeding him as minister of police.[60] First placed under surveillance in Versailles, where he was being treated for joint pains, Cluis left in February for his hometown of La Châtre, where he had a wife and three daughters, before being asked by the police prefect to hasten his departure from France according to the terms of the amnesty law. Cluis obtained a passport for the United States. On March 17, 1816, he sailed from Le Havre on the *Shakespeare* with his mistress Emilie Louise Mézières, a former companion to Josephine at Malmaison, and their natural son Victor.[61]

Departures from Abroad

The journey taken by Savary and Lallemand had been more tortuous, from Rochefort to Plymouth with the Emperor, then to Malta and Smyrna, where they placed themselves under the protection of the Ottoman authorities who were delighted to welcome such important individuals. The French consul himself did not complain of their presence at first, but during a dinner at the consulate, Lallemand, hypersensitive about the Emperor, slapped a naval officer for making insulting remarks. The consul shipped him off to Constantinople, where the presence of the French exiles disturbed the new ambassador of France, the Marquis of Rivière, the last person from whom they might expect compassion.[62] A zealous royalist, imprisoned and then deported for conspiring against the First Consul, he considered the Bonapartists the most dangerous of sects, spreading like wildfire if left alone.[63] He demanded they be deported. The Ottoman government gave the generals notice to leave.

Savary's answer was theatrical: "I shall obey, and have already reserved passage on a ship that is leaving for America where I shall dig my grave and arrive in the difficult season of the equinox. I am now at the disposal of the ship's owner and the wind."[64] But Savary was not ready to cross the seas, stating in his *Mémoires* that he had felt no desire to sail for America.[65] Letting opportunities for Salem and Philadelphia go by,[66] he finally went to Trieste, on the Adriatic, where the Austrian police kidnapped him. Reduced to boredom in Gratz, he harassed the Austrian minister with requests for a one-way ticket to North America, or for a ticket to return to Smyrna, which was finally given him.

The elder Lallemand had long abandoned his vague commercial ventures in Smyrna, preferring to make money from his military skills. He offered his services as instructor to the Turkish army, but France reminded the sultan that Lallemand was a rebel against the king. The same offer to the shah of Persia in Teheran had the same result, as did his offer to the pasha in Cairo, where they were recruiting Frenchmen to reorganize the army. Lallemand had several reasons for thinking of America after failing in his eastern overtures.

After the failure of the French expedition to Saint-Domingue and the collapse of the peace treaty of Amiens with the English, French officers had passed through the United States, where some found wives: on December 24, 1803, in Baltimore, the First Consul's young brother, Jerome Bonaparte, married Elizabeth Patterson, the daughter of one of the city's wealthiest shipowners;[67] on April 30, 1804, in New York, Charles Lallemand married Henriette Roberjot-Lartigue, a young Creole daughter of a wealthy Domingan merchant refugee.[68] Lallemand's choice was additionally motivated by the presence in the United States of Joseph, a great many comrades, and his younger brother Henri, who had arrived in Philadelphia on May 19, 1816.[69] A year later, on April 24, 1817, Charles arrived in Boston on the schooner *Triton,* sailing from Liverpool.[70] According to Captain Rea of Salem, Massachusetts, whom he met in Smyrna, the elder Lallemand went to England—probably via France—for the vague reason of wanting to see to some property. The British, who had detected neither his stay nor his departure, could not have clarified this; in fact, a London newspaper stated that he had arrived in Boston directly from Smyrna, "under the name General Cotting."[71] From there he went to New York, where, on May 8, 1817, he was warmly welcomed by his exiled friends; a week later, with his old confederate Lefebvre-Desnoëttes, he went to Philadelphia to visit Joseph Bonaparte.[72] Few exiles traveled as long a road as they had from La Fère. Most leaving from abroad sailed from the ports of Antwerp or Amsterdam on the North Sea or from Leghorn in Italy.

Prefect Réal could not get an extension of his stay in Belgium, ostensibly to grow hops, and returned to his initial idea of going to America.[73] On May 7, 1816, he wrote from Antwerp to the minister of police that he was about to embark for New York: "Three days before leaving France I sold the largest of my properties to procure funds with the help of which I can hope not to die of hunger and cold on the 1,760 acres of land that I am going to clear . . . at the age of 60."[74] The large property was the château of Ennery near Pontoise, acquired at the end of the Directory period. The seventeen hundred acres of land awaited him at Cape Vincent, where Lake Ontario flows into the St. Lawrence River. After Joseph Bonaparte, who purchased twenty-six thousand acres, Réal had just bought these lands in northern New York State from Jacques Donatien Le Ray de Chaumont, who had been born in France, but was an American citizen and large land speculator in the United States where he was returning that year after a long stay in France.[75] On June 9 the maritime bailiff in Antwerp stamped Réal's passport for New York on the *Swift*, with his nephew Jean François Roland de Bussy, a police captain and secretary-general of the police prefecture during the Hundred Days, his wife, and their children. The ship arrived around August 10, 1816.[76]

The idea of moving to a rural life did not come naturally to someone like Garnier de Saintes, whose pugnacity did not weaken with age. The French government accused him of participating in the "plan of the revolutionaries" teaming up in Belgium to swamp France with despicable writings, and managed to have him deported. Garnier published his *Adieux à messieurs les habitants de Bruxelles* (Farewell to the Gentlemen Living in Brussels) with this epigraph: "Today an exile from exile is being imposed on me."[77] In 1798 he had refused the Directory's offer of the position of vice-consul in Wilmington, Delaware: the American adventure did not attract him. On July 2 and 3, 1816, Garnier and his son's passports for America were stamped in Antwerp; Jean Guillaume Taillefer, a former Convention member and physician, accompanied them. On September 10 they disembarked from the *Prince of Orange* in Philadelphia.[78]

Several generals left from Antwerp. Desnoëttes, whose presence in Aix-la-Chapelle was reported in December 1815 by the prefect of Nord, had arrived there with a false passport stamped by a Prussian major. Once there, he claimed he had been offered a commission in Prussia, but the *Zoé* took him to Philadelphia, where he arrived on April 6, 1816.[79] General Fressinet, living in Brussels, left without protest. In February 1816 he announced that he intended to embark for America on May 15 at the latest, simply asking for permission to remain in the kingdom until that date. General Rigau needed persuading. He left Antwerp much later, with his daughter Antonia and his son Narcisse Rigau of the ex-Guard. The *Tybee* put them ashore in New York on November 5, 1817.[80] He was the last banished general officer to reach the North American continent.

Other generals left from Amsterdam. Jordan, Napoleon's Polish aide-de-camp on Elba who had been made prisoner at Waterloo, was in New York on July 9, 1817,[81] and Vandamme arrived shortly thereafter. Driven farther and farther north, he refused to go to Denmark, a dreaded stopover on the way to Russia, of which he had bad memories. Grumbling about an American exile, he spoke of injustice and to show his good faith announced that his first visit would be to the French ambassador in Washington.[82] In late May 1817 he left Amsterdam under the name of Louis Baert (his mother's birth name) with four hundred Swiss emigrants; on July 27, after sixty-four days at sea, he disembarked from the *John* of Baltimore at the health control building in Philadelphia.[83] "This fine soldier, one of the most generous defenders of his country, is fortunately safe from the attacks of legitimacy," announced *L'Abeille Américaine* with obvious pleasure.[84]

Some also left from the South. Two American ships brought a great many émigrés to Philadelphia from the port of Leghorn, in Tuscany: Italians, Corsicans, former imperial officers and civil servants, some with ties to the Bonapartes exiled in Italy (Louis, Lucien, and their mother, Letizia) and given the mission of seeing Joseph in Bordentown, New Jersey. On July 21, 1817, the *Saunders* put ashore Barthélémy Colonna d'Ornano, a former judge of the magistrates' court in Ajaccio, related to the Bonapartes, whose participation in Murat's failed 1815 expedition to reconquer the Kingdom of Naples had earned him exile; his servant Toussaint Peraldi; four officers, Louis Gruchet, Pierre Guillot, Louis Taillade, Jean Baptiste Vitalba; and Charles Haraneder, a young clerk whose father from Rouen was a merchant banker in Florence.[85] On the following October 13, the *General Jackson* brought men named Gatti, Astolphi, Ilari, and Scasso.[86]

What balance sheet can be drawn up for French emigration to the United States, from the first Restoration through the year 1818? It is, first of all, difficult to distinguish those who left because they were thrown out by the new regime from those whose choice to leave was part of the natural cycle of migratory movement after a quarter century of maritime disruption. Guaranteeing numbers is tricky as well.

In July 1816 on St. Helena, counting up his partisans whom he hoped to join one day in America, "the country that best suits the French," he said,[87] Napoleon thought that there were already more in the United States than in any other foreign country, but did not go beyond the number of three hundred French families ready to populate the city that his brother Joseph was planning to found there. He added, "Not to mention those coming from all over, announced in the papers, we can count among the individuals we know: Regnaud, Savary, Merlin, Chaptal, Grouchy, Lefebvre[-Desnoëttes], etc." The newspapers in question were no more certain of the names they gave their readers (Savary, Merlin, and Chaptal remained in Europe) than of the quantities of émigrés served up in tens of thousands. Before the end of the year 1817, Parisian newspapers had gotten to a total of thirty thousand—military men, merchants, and artisans, for the most part. On September 6, 1817, the editor of Baltimore's *Niles' Weekly Register* denied the reality of this tidal wave, postulating that, as a transatlantic ship transported an average of twenty passengers, since Louis' first return fifteen hundred ships would have had to sail to bring thirty thousand, which was far from being the case. Had even three thousand arrived?[88] A French naval officer present in the United States during the same period suggested the intermediate number of eight thousand.[89] It is still difficult to quantify with any precision the French influx into America, principally because of the lack, first on the French side, of the two indispensable documentary sources for passenger count: an individual passport received from the authorities before embarkation, and the ship's manning and equipment list on which the names of the people aboard are entered. Due to the uncertainties of archive preservation, these documents have not survived everywhere and totally. The lists for French ships sailing from Saint-Malo, Lorient, Bordeaux, and Marseille have disappeared; there is no longer a record of passports issued in Nantes and Rouen, and the series of passport registers for Bordeaux before 1825 is damaged and incomplete.[90] Then, on the American side, lists of passengers disembarking from American ships are available beginning in 1800 only for Philadelphia;[91] for other ports these statistics begin in 1820.[92] One must therefore take advantage of what each port, in France and abroad, has to offer,[93] but this goes beyond the scope of our study of a certain number, rather than all, of the French citizens who emigrated to America, whether temporarily or permanently. Nonetheless, two sets of representative statistics can be singled out.

From 1815 through 1818, the port of Nantes[94] sent sixty-four French ships to the United States, carrying four hundred persons, that is, an average of 6.3 passengers per journey and 100 passengers per year.[95] Nantes, less important than Le Havre and Bordeaux, but far more important than La Rochelle and Saint-Malo,[96] for instance, is an interesting point of reference. In this same interval, dozens of American ships sailing from the ports of Bordeaux, Le Havre, La Rochelle, Marseille, and Nantes took more than five hundred passengers to Philadelphia.[97] These modest figures corroborate the Baltimore paper's analysis and so give significance to the maximal number of two to three thousand French émigrés, for the most part from France, Belgium, and Italy, on French and American ships. The wave of emigration of the first years of the second Restoration was thus not as large as the French of the time claimed, and it was not to increase during the last five years of Louis XVIII's reign;[98] on the other hand, its emotional resonance on both sides of the ocean was considerable, and its echo was as strong as that of the preceding emigrations.

A TRADITION OF WELCOME

Unlike their predecessors in emigration, the Napoleonic exiles did not systematically consider the United States as the Promised Land. The circumstances of their emigration were different; most had resorted to this choice by default. America was better than the execution squad not only for those condemned to death but also for those banished, subjected to French pressure on Belgium and the Ottoman Empire, the countries taking them in. The United States offered the advantage of total freedom that took effect as soon as they boarded, when American captains closed their eyes to the identity of their passengers. On the other side of the Atlantic those arriving were not really checked, so the passports that would have been necessary in the ports of the Old World were no longer needed in the New.[99] The same was true of personal baggage and tools, quickly inspected at customs and exempted from duty. And no passport was needed to move about in the country, nor was it necessary to present their papers as the law demanded in the authoritarian European monarchies.[100] This freedom, along with freedom of expression, was most attractive. The quality of welcome in the United States was not a pipe dream, particularly for the French, toward whom the debt of gratitude from the War of Independence endured. The French Revolution had given rise to the first wave of exiles to America; the Restoration was the cause of the fourth.

The first wave had brought to America the blueblood Frenchmen who had preferred the ocean's perils to the guillotine. Renowned names of the French aristocracy such as Chateaubriand and Talleyrand very quickly showed the way to an anonymous cohort of former royal officers or churchmen who had refused to swear loyalty to the Civil Constitution of the Clergy. Most stayed only briefly, unlike the refugees of the second wave who began to arrive in 1791, chased from Saint-Domingue by the slave revolt. Outnumbering the whites by ten to one, the slaves had put the island to fire and sword in order to seize their freedom. The proprietors of the large plantations were the rebels' primary targets and had no alternative but to flee. Some of the refugees returned to France; others remained in the Caribbean, going chiefly to Cuba, which was becoming to Spain what Saint-Domingue had been to France; and still others chose to settle in the United States. The third wave left Cuba in 1809 when, in retaliation for the French occupation of Spain, its French refugees were expelled: twenty-five hundred whites and five thousand blacks, both slaves and freed, reached New Orleans.

They joined the twenty-five thousand other refugees who had settled in the United States during the preceding twenty years. Despite intense economic and cultural activity, especially in Philadelphia, this micro-society suffered from its heterogeneity. Ideological divergences complicating cohabitation between groups like royalists and republicans or proslavery people and abolitionists were aggravated by differences in standard of living. Those who had been able to flee France or the islands with securities and servants or slaves had acquired land; those with the necessary skills became silversmiths, blacksmiths, saddlers, tapestry-makers, shopkeepers, seamstresses, or laundresses;[101] ruined aristocrats taught fencing, dancing, and music. But others, arriving destitute, remained mired in poverty, living by their wits or on charity, unable to return to France. The émigrés of the Revolution had continued to hope that their exile would one day end. Bonaparte relieved them in 1802, but on the principle of communicating vessels: while some returned, his opponents left.

It is not the number of those leaving that is important here—it cannot be called a wave—but the personalities of the proscribed men, among them two generals who for a time served Bonaparte, Moreau and Humbert, and two obdurate opponents, royalists Limoëlan and Hyde de Neuville. Chateaubriand wrote concerning Limoëlan and Moreau, his former schoolmates in Rennes, that it was "rare to find at the same epoch, in the same province, in the same little town, in one school, such singular destinies."[102] Their destinies first intersected in America, then encountered those of the Restoration exiles.

The chevalier de Limoëlan had arrived in the United States in the summer of 1802 to flee the guillotine that had claimed his father's head during the Revolution and was awaiting his own after the failed assassination attempt against the First Consul in the rue Saint-Nicaise a year and a half earlier. The police investigation had identified the instigators and the perpetrators, but had not arrested them all: among others, not him and not Hyde de Neuville, another future exile in America. Limoëlan, alias Clorivière, worked as a miniaturist painter[103] before entering Saint Mary's seminary in Baltimore, where he was ordained in 1812. When he briefly returned to France three years later, the king awarded him a medal for his good and loyal service to the counter-revolution. He died in 1826 and was buried in the chapel of the Georgetown Visitation Monastery.

The victory over the Austrians at Hohenlinden on December 3, 1800, was the last glorious feat of General Moreau, followed by difficult years. He was suspected of relations with royalist agents who came to France from England in 1804 to assassinate Bonaparte. Arrested with the plotters but acquitted for lack of proof, he was authorized by the Emperor to go into exile in the United States, where he arrived with his wife in August 1805.[104] The welcome accorded him by local officials displeased the French diplomatic corps, which did not want people to be interested in him.[105] The general learned English, acquired property near Philadelphia and a house in New York, and met important Americans and French émigrés, chiefly through Freemasonry; he farmed and traveled. Moreau had come to the United States unwillingly, but as he came to know the country he quickly grew to like it: "It is the land of tranquillity and liberty," he wrote to his lawyer.[106] But the desire to do battle with the man who had exiled him won out over the pursuit of calm in America; he accepted an offer from the czar, who sought his wisdom to enlighten a Russian general staff woefully lacking it. In 1813, in the battle of Dresden, he was cut down by a French cannonade: "This general went back to Europe only to find the cannonball on which his name had been inscribed by God's finger," concluded Chateaubriand.[107] Moreau was interred with great pomp in the church of Sainte-Catherine des Français in Saint Petersburg, where his body lies to this day.

The solemnity of General Humbert's New Orleans funeral was perhaps more sincere, an homage to a man whose life was not exemplary but who had not betrayed his country. Humbert met Moreau in Philadelphia and rejected his proposal to accompany him: "I shall never turn the point of my saber against Frenchmen, unless I meet them in the ranks of the enemy."[108] A native of the Vosges Mountains, nearly illiterate, a general at the age of twenty-seven after castrating pigs and peddling rabbit pelts, he had in 1798 managed the feat of landing in Ireland and defeating the English on their own soil, which no one had succeeded in doing since the Battle of Hastings. But during the Saint-Domingue military expedition, Leclerc sent him back to France for cowardice in the face of the enemy and corrupt business dealings on the island: "He is an ignoramus, a shady schemer; this man does not even go to the firing line."[109] The First Consul relieved him of his command in 1803.

In 1809 Humbert was recalled from retirement in Brittany to counter an English landing attempt on the coasts of the North Sea. Despite meritorious conduct, the Emperor did not restore him to his former position. After selling his possessions to pay his debts, Humbert turned toward the United States, then at war with England, and proposed recruiting a foreign legion around a core of French Canadians.[110] Although he received no response, he did not give up. In July 1812 Napoleon granted his request for a passport, and Humbert had left France for good by March 1813. Exiled but patriotic, Humbert continued, from America, to want to serve his government and submitted a plan to seize the English colonies in the Caribbean and support the independence movements in the Spanish colonies.[111] Failing again, he thought of Louisiana, where he could buy land and farm it.

In New Orleans dives, he was drinking up his French general's pension in the company of disreputable pirates when the war for which he had come gave him a chance to shine in defending the city against the invading British forces. General Andrew Jackson, associating Humbert with the victory of January 8, 1815, wrote that he had continually exposed himself to the greatest dangers with characteristic bravery.[112]

The only thing the republican Humbert and the royalist Hyde de Neuville[113] had in common was the man to whom they owed their exile. Whereas Humbert began his chaotic life in poverty and ended it in rum, Hyde was viscerally conservative and loved order above all, in the interest of royalty, social well-being, and civilization. One presents the unflattering picture of an uneducated, debauched soldier; the other, that of a man for whom loyalty and dedication were the guiding precepts of existence. Hyde de Neuville owed the first part of his name to his Scotch origins and the second to the property that his father, a manufacturer of hardware and edge tools, had bought for him in the department of Nièvre in central France, where he was born in 1776. He owed to his mother his faith in God and his king.

In 1790 he went to Paris to save the monarchy and define his lifelong goal: to prove his commitment to the royal family. His childhood had taught him to be fanatically devoted to the sacred persons of the king and queen; their martyrdom introduced him to clandestine activities, resistance, and terrorism with the rue Saint-Nicaise assassination attempt. His complicity was not clearly established and Hyde always denied it, but the strong suspicion that remained attached to his name was the blot on his life. Hiding on his lands, he felt that he no longer belonged in France, so unshakable did Napoleon's position appear to be.

Quite a bit older than her husband, Madame de Neuville was considered an admirable woman by all who knew her.[114] Pursuing Napoleon across Germany, she caught up with him at the Schönbrunn Palace in Vienna at Christmas 1805, pleaded her cause, and was heard. The couple took an American ship in Cádiz and arrived in New York in June 1807: "While the first sight of this foreign land aroused in us a pleasant sensation, we felt a very painful one in thinking that in this immense city where we were going to go ashore, no relative, no friend was awaiting, was desiring our arrival," Madame de Neuville was to remember.[115]

Hyde had chosen exile as his last recourse because he considered exile a political error. France was his homeland, and it was there, not abroad, that he felt he should serve the cause of the king. But as the popular support given the Empire weakened the justification of his struggle, all he could do was leave. In the United States Hyde hoped at first that he would be allowed to return to live in France "docilely, far from commotion and a stranger to politics," he wrote to Jefferson, but his hope was fruitless.[116] Having means and contacts made it easier for him and his wife to accept this. Thanks to a letter of recommendation from Nathalie de Noailles, whom they had run into in Spain, Hyde met General Moreau, and these two men, whose convictions and temperaments were so contrary, became friends: "He is cold, not very forthcoming. With him, intimacy does not come quickly," Hyde said of Moreau.[117] Hyde bought a house in New York, on the street where Moreau was living, which facilitated their exchanges. They influenced each other: the republican took on a tinge of royalism and the royalist of liberalism. Hyde was much moved by Moreau's death despite the uneasiness he felt when he went over to the czar—which did not prevent him from publishing his eulogy.[118]

The Hyde de Neuvilles took winter quarters in New York near the Moreaus and in the summer settled in a farm in New Brunswick on the Raritan River. While his wife painted in watercolors, Hyde devoted himself to sheep raising, geography, philanthropy, journalism, publishing, and teaching, all with efficiency and selflessness, earning him the gratitude of the Americans.[119] He had successfully transplanted himself. His love for America was not challenged by his return to Paris on July 20, 1814: "My arrival in France had for me the sweetness of awakening after a bad dream; the years of exile, hopes so often disappointed; time that passes and brings nothing gave way to this intoxicating reality of the return of the King."[120]

For twenty-four years he had been waiting to serve the legitimate throne and be entrusted with a mission that would ensure its continuity. Asked to keep an eye on Napoleon's doings on Elba, Hyde set down his thoughts in a premonitory report in which, like Fouché, he suggested putting "the seas between Europe and him."[121] His report was approved but no action was taken, and he was sent to Italy to investigate further. The Hundred Days proved him right. Then, as devoted as ever to the king, Hyde made himself useful and was rewarded by a Legion of Honor and later a position as deputy on the benches of the Ultras, where his extremist positions (motivated by the need to remove, once and for all, the threat of revolution) made people say that he was "the most violent of the royalist party, the most ruthless promoter of reaction, intoxicated with his own declamations, ridiculous in his pretensions to fidelity, to loyalty."[122]

In his *Mémoires* Hyde did not deny his participation in the ultra-right-wing Unobtainable Chamber, and defended the exceptional measures, but he criticized the amnesty law, believing that it did not fall within the competence of the deputies to judge the authors of crimes from which they themselves had suffered. Those who had committed them were confused men who should be treated leniently after living so long in immorality and ignorance of sound principles. But Hyde's compassion was that of a man made wise by time; he had voted for the law.

When, in January 1816, the Duke of Richelieu suggested Hyde as ambassador of France to the United States of America, he was proposing a man whose reactionary reputation was well established. A pathological legitimist, conspirator, imperial exile, and merciless legislator, Hyde de Neuville would not get along well with the outlawed French on the other side of the Atlantic.

5

A Conflictual Friendship

The Americans surround us with attentions. But do they honor us only as unfortunate proscribed men, or because we are the French who once helped them. I do not know if I should wholly like the Americans, as I do not know either whether they truly like us, their interests or their commerce coming before everything.

—Jacques Lajonie Lapeyre, Baltimore, March 1817

America was for a time to become an area of turbulence between two antagonistic forces of a foreign nation: official France, well established in Washington and in its consulates, and the émigré France of renegades driven out by the newly restored Bourbons. Situated between the good and the bad Frenchmen, the host country strove to be pleasant to one group without displeasing the other, but diverging emotions and interests exacerbated passions and conflicts. This resulted in a difficult cohabitation, made worse by a tense international background, with major regional stakes (the independence of the Spanish-American republics and the territorial expansion of the United States toward the Floridas) for which the crumbling Spanish empire and the British commercial rival, the big loser in the War of 1812, were the prime contenders. For France to join them, it first needed to reestablish itself on a continent with which its relationship had progressively deteriorated during the past quarter century. Its role in the independence of the United States had sealed its friendship with the new republic and initiated a special relationship, but by 1815 all that remained of this was a nostalgic memory, kept alive by the aging glory of Gouverneur Morris in one country and Lafayette in the other. The slackened ties between the two nations had to be strengthened.

THE DECLINE OF FRENCH AMERICA

French presence in the Western Hemisphere had diminished steadily after the Seven Years' War.[1] In 1762 the secret Treaty of Fontainebleau had transferred the possession of Louisiana to Spain, and the following year the Treaty of Paris gave Acadia, Cape Breton, and Canada to England. The 1791 slave revolt in Saint-Domingue had resulted in the independence of Haiti thirteen years later. Relinquishing its North American colonies was for France a huge territorial loss, and giving up the "Grand Isle" with its slave population of five hundred thousand laborers was a major economic one. Diverted from his oriental dreams by his failure in Egypt and by English naval supremacy in the Mediterranean, Napoleon took a fresh look at a continent and country he had until then ignored. Reestablishing himself in America would harm the English and check French colonial decay.

In 1800 the Treaty of San Ildefonso, confirmed by the Convention of Aranjuez, ceded Louisiana back to France as discreetly as it had been given away in 1762, and in 1801 an expeditionary corps was sent to retake Saint-Domingue, but this double attempt to restore French sovereignty came to naught. The Spanish had scarcely handed Louisiana back to the French when the First Consul changed his mind and decided to sell it to the Americans for fifteen million dollars on April 30, 1803. While the decision to make this purchase was one of the most important, in terms of its momentous consequences, that a president of the United States could possibly make, its implementation was to replenish the coffers of the French treasury that were being emptied into Saint-Domingue, where the French army was decimated by the resistance of the rebels and by yellow fever.

France's territorial presence was limited to the confetti of its West Indian islands and a forest in Guyana whose geostrategic uselessness was obvious to all. From this point of view, France was opening new horizons for the United States by permanently evacuating its continent and coastal waters. But Franco-American relations were not merely geographic; much remained to be done to shore them up in the political, commercial, and cultural arenas.

France had represented for the young American nation "an extraordinary opportunity to cease being a cultural province of Great Britain," writes Ronald Creagh.[2] France was not so infatuated with the United States that it rushed in, but exchanges did develop with the accelerated rhythm of more frequent sailings. Already in the 1780s, ships brought merchants come to try their luck and travelers to see the country. French-language newspapers such as Philadelphia's *Courier de l'Amérique* appeared and flourished during the following decade. A royalist émigré founded the *Courier de Boston* in the hope of freeing Americans from the moral slavery in which they were kept by the English papers, "the only source from which they draw all their opinions."[3] His was not an isolated voice. Within cultured American society there was a Francophile or even Francomaniac group, and French, a prestige language taught at Harvard, had its crusaders.

The elite who had founded the nation were naturally partial to France, and some went there to represent their country: Thomas Jefferson before the Revolution,[4] and James Monroe after the Reign of Terror.[5] The French Revolution, which reminded them of their own, was at first well received by Americans. They hailed its precepts, which seemed destined to bind together these two modern republics. The French thought so too; in August 1794, in Bordeaux, a great festival celebrated the union of the French Republic with the United States of America, and the destruction of London was toasted.[6]

But in the United States, cracks very quickly fissured this fine edifice, with

Americans divided between the Anglophile and monarchist Federalists of Treasury Secretary Hamilton and the pro-French Republicans (whom their enemies called Democrats) of Secretary of State Jefferson. President Washington was won over to Hamilton's views. The Federalists despised the French Jacobins, whose terrorist action scorned the ideals common to the two revolutions. Their annoyance increased when the French Republic abolished slavery in 1794, fanning their fear of contagion, and came to a head with the intrusion of Franco-French conflicts onto American soil. The immigrants from France and Saint-Domingue engaged in micro-civil wars concerning events that were occurring thousands of miles from their new home. Rival groups, each with its club and newspaper, hounded each other: in Philadelphia, the republicans responded to a royalist mass in memory of the martyred king with cannon fire.

All the political agitation of which the French were capable was nothing compared to that of citizen Genêt, the new French ambassador to the United States. Entrusted by the Convention with the task of strengthening relations with revolutionary France, at war with the English, Genêt thought, when he arrived in April 1793, that he could lord it over everyone, convinced that, on the authority of the alliance treaty of 1778, the American government would support him.[7] Even before presenting his credentials, he solicited the aid of American citizens and enrolled volunteers against England. Committing one blunder after another, he exasperated the French royalists as well as the American Federalists, and his zeal ended up costing him his job, on the insistence of the American government that he be recalled.[8] Ten years after the Treaty of Paris, Genêt and his replacement, Fauchet, were bitterly "disillusioned with American gratitude,"[9] but quite aside from its own interests, the United States could very well disapprove of the violence done to Louis XVI, to whom it owed much.

In short, the Franco-American honeymoon was coming to an end, and divorce was just around the corner. After proclaiming American neutrality, Washington, in 1794, had John Jay conclude a trade treaty that was most advantageous to England. Outraging the Jeffersonian opposition and a large part of public opinion, it tolled, like betrayal to French ears, "the end of the dream of Franco-American friendship" and the "return to the ties of the past."[10] This reversal of alliances resulted in an almost total suspension of trade between France and the United States and in an undeclared maritime war in 1798 and 1799.[11] Their aversion to revolutionary ideas and the intrigues of the French on their territory had revealed the fundamental Anglophilia of the American political leaders and placed the American elite back under British cultural influence for what was to be a long period of time.

Bonaparte's accession on November 9, 1799, Jefferson's election to the American presidency one year later, and the signing of the Treaty of Mortefontaine (October 1800)[12] between France and the United States and the Treaty of Amiens (March 1802) between France and England were all chances for the three nations to improve their relations. But England's ambition to rule the seas and Napoleon's to rule the Continent precluded any lasting reconciliation and led to a double blockade. The United States was once again caught between the two "superpowers" of the time.[13] Not content with having to pay for a war that did not concern them, the Americans engaged in a sort of self-mutilation by voting, in 1807, to impose an embargo that forbade American ships from sailing to foreign ports and foreign ships from coming to American ports. The measure gave impetus to domestic development but was unpopular in business circles. Largely ignored, it was replaced by the Non-Intercourse Act, forbidding trade only with the belligerents, but even this was too high a price to pay. In 1809, James Madison's presidency offered a chance to reestablish trade relations with France, but not with England, which refused to lift its maritime blockade. The conflict that ensued in 1812 could have reversed alliances, but the United States was angered by the continued harassment of its merchant marine by France and did not want to displease Russia, with which relations were cordial, by making an official pact with the French. The long list of American ships and merchandise seized by France since the Revolution finally became a serious irritation. This resulted in a long-running dispute and requests for indemnities that would be the stumbling blocks of all diplomatic negotiations for nearly thirty years.

In a political game that had been constantly shifting for a quarter century, the round engaged in by the Bourbons and Napoleon between 1814 and 1815 helped muddle an international situation that was just beginning to clear up. Great Britain had signed the peace treaty of Ghent with the United States on December 24, 1814, and was working toward normalizing relations with France, when Napoleon's return threw everything into confusion. After Waterloo, the English remained on the alert, and the Americans remained cautious at the beginning of the second Restoration. Peace freed navigation with the blessing of the Royal Navy, which was not an easy matter. Accustomed to mastering the seas in times of war, British ships tried their hand at peacetime meddling. But the United States, confident in its recent success, had no intention of letting itself be dominated by its former mother country now that the umbilical cord had finally been cut, or of letting its progress be disrupted by European convulsions with which it had nothing to do. The country was aware of its growing strength in all areas—territorial, demographic, and economic: "In this sense, the year 1815 was perhaps more important than the year 1783. It symbolized the end of the beginning and inaugurated a new period in the history of the United States," writes André Kaspi.[14]

France also had to free itself from British pressure and arrogance. This offered France and America an opportunity to get together again, but a bone of contention threatened to ruin everything: the showdown on American soil between the political refugees flooding into the east coast and Louisiana and the new French diplomatic corps. How could the United States position itself with respect to the two groups? What should it expect of each camp?

STUMBLING BLOCKS TO RECONCILIATION

On May 18, 1816, the frigate *Eurydice* set sail from the Brest harbor. Its mission was to take to the United States a partially renewed diplomatic corps. Unfailingly at his side, Henriette Hyde de Neuville was able, for the first time in a life together sacrificed on the altar of legitimacy, to taste the joy of accompanying her husband to Washington, which awaited the arrival of the "Envoy Extraordinary and Minister Plenipotentiary of His Most Christian Majesty to the United States." The title alone spoke volumes about a promotion punctuating twenty-five years of unimpeachable loyalty. The king had appointed him to this position on January 21, 1816, the anniversary of the execution of Louis XVI.

THE CHOICE OF AN AMBASSADOR

This appointment was not without its risks. At the age of forty, all the fiery Hyde de Neuville had to show for himself was his unproductive past as a plotter and oppositionist, his only concrete political experience being a few months of agitation on the benches of the Ultras, and his diplomatic experience limited to belated snatches gleaned in England and Italy. His poor command of the language of a country in which he had lived for seven years was an awkward deficiency. Finally, as already noted, his reputation as a hard-line royalist stained with the innocent blood of the assassination attempt of December 24, 1800, preceded him. Yet the role of the French ambassador to the United States was to be far-reaching in a nation with a front-row seat in a gigantic theater where Spanish and British actors were center stage, and which all observers were praising to the skies for its astounding progress. Furthermore, the distance between France and the United States placed a two-month round-trip sea voyage between the former and its representative, one of whose major qualities consequently had to be initiative. The king said the mission was of exceptional importance and his prime minister considered the American legation one of the most honorable and important. The reasons behind his appointment must have been very strong.

The principal reason was that the king really wanted the exile, faithful to his cause, to become his representative in the very country that had witnessed his hardships, and the most pragmatic reason is that given by Richelieu in his instructions to the future ambassador: "Monsieur Hyde de Neuville is returning to America with the knowledge that he acquired there during a long residence. This previous experience gives him numerous advantages in his mission."[15] During his years of exile, his curiosity and his altruism had led him to expand his knowledge. Hyde was deeply moved by his appointment: "When I was offered [the legation] of the United States, I felt stirred by a violent desire to see again this country where I had left friends, where years of exile had had their hours of calm happiness, where many prejudices had disappeared for me, making way for a mental maturity for which I thanked Providence. The school of misfortune has important lessons, and is there not a certain charm in seeing again the places where they were learned?"[16]

His eagerness to see the United States again was counterbalanced by an impatience to get into politics, of which he had been deprived all his life, so it was agreed that he would serve as deputy until the end of the first parliamentary session. This is why Hyde's appointment and departure for the United States took place long after the decision had been made to choose him as French ambassador.[17]

On April 6, 1816, the king granted him a farewell audience, during which he reminded him of what was expected of him in the United States and created for him this made-to-measure motto: "All for God and the legitimate King." He added: "This is the story of your life."[18]

The *Eurydice* reached New York on June 15, 1816, after a twenty-nine-day crossing, and worked its way through the little boats meeting it. The ambassador's wife was already able to testify to the appropriateness of the choice of her husband: "When, once ashore, we found almost everyone we were acquainted with and a crowd of people most eager to see us and congratulate us, we were moved to tears by such a kind welcome."[19] Welcomed by a seventeen-cannon salvo, they saw how far they had come from the day they had arrived at this same place as exiles, but also how much they had to accomplish in a country bewildered by the second Restoration.

Hyde had not come alone to face the American government, public opinion, and the refugees, but rather with young attachés, novice consuls, and two experienced consuls[20] appointed by the king to replace untrustworthy republicans or Bonapartists whom the new ambassador did not want, as he was anxious to be on good working terms with reliable colleagues "at a time when the United States [was to] become the refuge of France's worst enemies," he wrote.[21] These diplomats joined others who could be counted on. The Count of Espinville, a staunch royalist who knew both the country and the language well, replaced the Bonapartist Cazeaux in New York, a recall that seemed "to denote some material change in the political thermometer of that strange people," according to the local press.[22] Waiting for Hyde in Washington was the legation's first secretary, Charles Roth, appointed in September 1814, who had carried out his duties impeccably pending the ambassador's arrival, and was very pleased to see him: "Monsieur de Neuville's stay in this country left the most distinguished impression of him, but an even greater reputation precedes his return. The character and talent he displayed in the Chamber of Deputies are the qualities most apt to command consideration and confidence in this country, and even the Democrats who may not share some of his opinions and may fear their implementation will eventually understand that one can be the defender of monarchic institutions in a monarchy without being the enemy of republican institutions in a republic."[23]

Even if Roth was laying it on thick, he was speaking words of wisdom. The secretary's assessment of the situation was accurate, his prediction of the relationship he expected between the new French ambassador and the Americans perceptive. Nothing, however, was settled as to Roth's hopes, nor was the United States won over to the cause of France and its representative. There was still much to be done before the two countries had tamed one another.

RECIPROCAL SUSPICION

The latest political commotion had left its mark on the American understanding of the French situation. Its leaders were guided by a kindly caution. They wanted to wait and give the country a chance to find the stability it was seeking. The Democrats, as Roth suggested, were reticent about the reestablishment of a monarchy that the French had not really wanted, but that had been imposed on them by defeat. The ease with which Napoleon had turned out the king on March 20 was proof that the Bourbons were not in the nation's favor and that "Louis the Desired" was not all that desired. It had taken a coalition of foreign powers to restore him again, but what confidence could one have in a government behind which loomed Wellington's shadow? As an indirect consequence, France was enduring some of the distrust and hostility the United States still harbored toward England. American newspapers had a field day outdoing each other in denouncing a king subjected to the slightest whim of a country that had given him everything, from his own failings to the Capetian throne.[24]

In a more sober style, statesmen like John Adams, the American ambassador in London, had similar thoughts.[25] As long as the foreign occupation lasted, it was impossible to consider France as autonomous enough to maintain with it the normal relations of one state to another, that is, to sign bilateral agreements or simply show it friendship. With which France would one be dealing—a France directed from abroad, or a France responding to the wishes of its people? Considering the second option, it could not, in any case, be the same sort of political base that gave its democratic meaning to the Republic of the United States. The French monarchy had decked itself out in liberal lace with a constitution and representative system, but that was not enough to make one want to associate with it. The absolutist temptation showed through the finery. A nostalgia for the "obsolete and impious doctrine of the divine right of Kings" confirmed people's fears, as did crimes of the White Terror.[26]

Following the lead of the openly Bonapartist *Niles' Weekly Register,* the press magnified the news reaching it from France of outbreaks of violence and the state's weakness in putting them down. In the name of France, Hyde de Neuville could only deplore this: "Most of the lies circulating in this country do not originate in America, but arrive fabricated by the European press, sometimes even in French papers. How can we complain about the falsehoods spread concerning Protestants? How can we convince minds already prejudiced by speeches given on the floor of the Chambers that Protestants are not massacred in France?"[27]

Additionally, American citizens who had been victims of Americanophobic acts, chiefly merchants and sailors, were coming home "with the worst impressions, taking the opinion of this domineering party as that of the whole French nation, and no doubt we shall see their opinions with great additions in our gazette," lamented the American consul in Bordeaux.[28] In November 1816 Philadelphia's *Aurora* published the letter of an American in Bordeaux depicting the apocalypse: "Terror is total. Everything is controlled by government agents, people are arrested and searched by the police without prior investigation. Our compatriots have not escaped these measures."[29]

Last but not least, the announcement of Marshal Ney's execution made a disastrous impression: "Everyone is talking about Ney's death: unfortunately Mr. Monroe's daughter was brought up with this general's wife at Madame Campan's[30] and the whole family was very close to her," Roth reported to the minister, more concerned about an unfortunate family coincidence than about the death of a brave man whose treason almost caused the ruin of his country.[31] The result of all this was a rather general disgust with the regime, and the press, in the vanguard, attacked the Bourbons more vigorously than ever, calling them bigots and imbeciles. No politicians or well-known individuals openly proclaimed that they favored the Restoration, nor did public opinion. The support of a few isolated Federalists and the emotional fervor of Gouverneur Morris, one of the founding fathers, were rare exceptions: as his country's ambassador to France at the height of the Revolution,[32] the witty, wooden-legged ambassador had frequented aristocratic salons and the royal court before experiencing Paris during the Terror. In a grandiloquent New York speech in June 1814, Morris gave free rein to his gratitude toward the august family of the Bourbons and his love for Louis XVI (a "virtuous monarch," a "just and merciful prince," a "protector of the rights of mankind") before inveighing against his assassins and against the monsters who, following them, had ruined France. In the forefront of these monsters was "Buonaparte," "the son of an obscure family, on a small island of the Mediterranean," who had become "the terror, the wonder, and the scourge of the nations." But order had been restored: "'Tis done. The long agony is over. The Bourbons are restored. France reposes in the arms of her legitimate prince. We may now express our attachment to *her*, consistently with the respect we owe to *ourselves.*"[33]

Apart from this oration (of limited influence, as Morris had retired from public life), the French refugees who came to the United States could expect the best of welcomes from their American hosts. The task lying before the ambassador and consuls, who did not much care for this fact, thus promised to be difficult: win over the American government and convince public opinion that the regime of the Bourbons was not the abominable thing depicted by the press. To do this, they had to move carefully in all the delicate circumstances that would arise, and follow the instructions of their ministry.

OFFICIAL INSTRUCTIONS

Hyde de Neuville received his instructions on January 26, 1816, shortly after his appointment, but well before he took over his legation, which gave him time to reflect on the subjects submitted to him by the minister of foreign affairs.[34]

Richelieu was deeply concerned that relations had so long been bad between the two countries and blamed it on the system of Napoleonic decrees against American trade interests. While he admitted that there had been embargoes, impoundings, and confiscations of ships and their cargoes, there was no question of conceding indemnities, since the fault was in no way that of the current regime. On the contrary: "All the measures taken by His Majesty toward the Americans were fair and kind." It would be useful on occasion to remind them of the role played by France in their origin and "to take the interest in their prosperity that one puts into finishing a piece of work." On the one hand, the minister was forgetting that Louis XVIII, then dauphin (heir to the throne), had not supported the War of Independence, comparing it to a rebellion; on the other, his words displayed a certain smug superiority with regard to the country that France had helped come into being and a total lack of knowledge as to its development: the Americans, who had cut their umbilical cord early and grown up quickly, did not need France to help them become adults. The ambassador was to remember France kindly to the United States.

Richelieu did realize, however, that the United States was slipping away from Europe, that it assented to the independence of Spain's American colonies, seeing how doing so would benefit its trade and the expansion of its political and territorial influence in the region—beginning with the annexation of the Floridas. Hyde was to assess to what extent the Americans were encouraging these rebellious undertakings and was to maintain the greatest reserve out of regard for Spain, an ally, but also in order to safeguard French chances of profiting from these new nations, if they were one day to come into being through "an inevitable progression." France had no intention of ceding its place on a continent in full transformation. For that matter, Hyde was also to look over at Saint-Domingue, take the pulse of the political situation, and see whether the divisions between the two dominant parties might present an opportunity. There was still hope of returning. This international review concluded with Great Britain, the common enemy. Hyde was to discover secret causes of discontent with the United States, seeds of dissension, whether commercial since the Treaty of Ghent, or territorial, as the determination of the American-Canadian border was creating problems.

Finally, what of the ban, announced by the British chargé d'affaires on November 24, 1815, on American ships putting into port at St. Helena and on communicating with its inhabitants since a new tenant had been settled there? Was it being respected? Were plans being made in American ports, possibly at brother Joseph's instigation, to help him escape? From here he quite naturally passed to the Bonapartist refugees, whom the government divided into two distinct classes: those banished by the edict of July 24 and those "who were not affected by any French law, but had left the kingdom, in the vague hope of making their fortune, or due to the feeling of anxiety that arises amidst war and revolutions, and that persists long after they have ended." As to the former, their return to France was not to be facilitated as long as their exile lasted, it being understood that their sentence was revocable and could thus come to an end; as to the latter, they were to be encouraged not to forget that they could return when, in time, they might wish to do so or if they were unable to realize their American dream. In any case, both groups, as long as they had not been naturalized, remained French subjects and were to be considered as such.

While it was not neglected, the question of the refugees was not a priority in the minister's instructions and was limited to the document's closing paragraphs. The matter became important with the successive arrivals of French émigrés, including the victims of the amnesty law of January 12, 1816, and of the strict enforcement of the Treaty of November 20, 1815.

In February 1816, Richelieu's first instructions to the consul in New York, Count d'Espinville, bore witness to this development and made it the central concern, as indicated in the preamble:

> The United States seems destined, Sir, to serve as refuge to the majority of those excepted from the amnesty granted by the King to Bonaparte's adherents or of those who, included in this amnesty, will not be inclined to take advantage of this kindness of royal clemency. Both groups will bring with them perverse intentions against which we must be on guard, and our surveillance of them must be all the more active since this country, by its democratic institutions and its partisan spirit stirring up unrest, offers a center of intrigue ready to foment plans which, while they cannot be dangerous for the State, would nevertheless spread regrettable anxiety.[35]

Consequently, and so as to better carry out their surveillance, the minister asked the consuls for a systematic census listing of persons who had arrived from France since July without a valid reason, and of the French subjects living in the United States during the "events" who may have shown "sentiments contrary to the King's Government." Each was to send his list to the Ministry of Foreign Affairs in Paris, the legation in Washington, and all the other consulates, so that all might have complete information and coordinate their actions. Thus they could follow the refugees' movements within the country; work out their relationships with one another and with the Americans; discover their communications with France, its colonies, Saint-Domingue, and St. Helena; and learn the names of their French or foreign agents and emissaries. Finally, the consuls were to assess the unfavorable influence the refugees might have on the press and seek ways to counterbalance it. The conclusion echoed the preamble, hardening France's official position toward those most compromised: "As for your relationship with the refugees, those who are excepted from the amnesty can no longer be considered French subjects: they will therefore not enjoy the King's protection and you will have nothing to do with them." All the others had a right to the aid and compassion of the consuls, who were, despite everything, to be cautious and adapt their actions to individuals and circumstances.

On July 18, 1816, one month after his arrival, Hyde de Neuville relayed Richelieu's instructions to the consuls, adding his own recommendations. They did not actually contradict the preceding circular, but gave it a touch of humanity. The three paragraphs relative to the refugees are a fine illustration of Hyde's sensitivity and explain the conduct that was to remain his during his six years as ambassador:

> But while it is important to enlighten foreigners, you will appreciate how much more important it is to His Majesty to bring those of his subjects who have merely strayed back to right thinking. I like to believe that for most of them, their error is more a matter of ignorance than of bad faith. So let us seek to calm these minds still tormented by revolutionary frenzy. The king,

in his inexhaustible goodness, wants us to reach out to all his children. We should abandon only the incurable, those whose perversity would leave no hope of return.

As to the refugees, Sir, this should be your conduct toward them. You cannot reassure them with promises; but you should not do or say anything either that could deprive them of hope. No one has the right to set limits to royal clemency, and besides, who could measure the breadth of the Bourbons' mercy? This mercy is infinite.

You will avoid meeting the refugees; however, if by an incident quite independent of your will, such a meeting should take place, you will then act with caution and dignity; there is always a way for a well-brought-up man to get out of this sort of predicament, without affectation and without making a scene. Scenes must most particularly be avoided, for they are less a proof of loyalty than of anger. Moreover, you must always bear in mind that foreigners are not obliged to share our impressions. In America especially a refugee is often seen as nothing more than an unfortunate man; this feeling is based upon such a kindly principle that you must not be too upset by it. Do not, therefore, tell Americans that they should not receive such and such an individual; merely let them know that you cannot meet him at their home. This frank conduct will be both the simplest and the safest.[36]

With these three circulars of January, February, and July 1816, all was ready on paper to receive the Napoleonic refugees and endeavor to establish a modus vivendi with them. If the image of France, tarnished by the insults and slander of its bad subjects and Americans hostile to the Bourbons, was to regain its brilliance, the conduct of the French diplomatic corps toward its subjects in exile had to be exemplary. But the encounter on neutral ground promised to be difficult. By taking sides against the exiles, the consuls would reinforce the Americans' image of the tyrannical government that had appointed them; but if they were accommodating toward them, they would not be following the instructions they had been given or the law that had banished the exiles. The best course of action was to avoid confrontations and slip away when one might take place, in return for which, with time, tensions would finally ease and the exiles would return home. This was not about to happen soon.

THE SUSCEPTIBILITIES OF COHABITATION

The French diplomatic corps in the United States was certainly expecting difficult beginnings, but it underestimated the amount of inflexibility and the number of confrontations, whether with the American authorities or the refugees.

MÉNAGE À TROIS

Despite some reservations, the American authorities were favorably inclined toward Hyde de Neuville since many had come to know and like him at the time of his exile. Thus the new ambassador had already formed promising friendships with Jefferson and important men like Rufus King and William Crawford. Arriving in New York on June 15, 1816, Hyde came to Washington on July 1 and two days later paid an informal visit to James Monroe, the secretary of state. The two men got along immediately, a far cry from the Genêt era: "I found Mr. Monroe to be a very witty, pleasant man, speaking marvelous French, and so knowledgeable about the events that had taken place in France that, talking with him, I felt I was still in Paris, where he was the United States ambassador for several years."[37]

Hyde next went to Montpelier, Virginia, to pay his respects to President Madison. Madison remembered that when he was in London, Hyde had defended American interests against England, and thanked him for it. Both sides were very favorably impressed: "I was perfectly received by Mr. Madison, but [with] the natural reserve that the supreme leader of a State must have when he sees before him the man appointed to discuss and defend his nation's interests . . . [He] applauded Louis XVIII's idea of sending back to America a man of whom people had such pleasant memories."[38]

The cordiality of these first meetings did not prevent the immediate discussion of sensitive matters. The French ambassador brought up the case of the American consul in Bordeaux, William Lee, whose slander was responsible for the bad attitude pervading the United States. He thought he had convinced his hosts that they should place only salaried employees in their consulates in France in order to make them more respectable. Both men wrote to express their pleasure with the visit. On July 12, Hyde apologized for his lack of English fluency, thanked the president for his welcome, and assured him that the United States was like a second homeland to him.[39] The president answered warmly. When Hyde sent his circular to the consuls, he was able to present himself as a sincere friend of the Americans.[40] These courteous exchanges foreshadowed the rebirth of Franco-American relations. And yet the rot had already set in.

The five-month delay between his appointment and his arrival in the United States had given Hyde the leisure to assimilate what he had to know in order to carry out his mission successfully and deal with the authorities on the matter of the French exiles. It was incumbent upon him to express his government's disapproval at not seeing a single, unified policy regarding the welcome accorded them. But the situation was not clear in a country where people could express the entire gamut of feelings, from indifference to solidarity, including curiosity and rejection in actions, words, or writing, without anyone criticizing them for this. The welcome given the first important exile was a revealing case in point.

Joseph Bonaparte. Joseph Bonaparte, alias Monsieur Bouchard, had chosen to take his first American steps anonymously, and had not revealed his true identity when the mayor and other important New Yorkers had called on him. But in the hotel where he was staying, he was recognized by an American naval officer to whom he had been introduced in Paris. Commodore Lewis offered him several days of discreet hospitality across the bay in a small New Jersey town. Back in New York, an encounter on Broadway with a former officer of his guard put a definite end to his incognito. Held hostage to a notoriety impossible to conceal from journalists avid for news, Joseph, now calling himself the "Count of Survilliers," found his every move reported for all to read and thus learned that he had enemies as well as friends: the *United States Gazette* was annoyed "to learn that an American naval officer was keeping company with this *Corsican adventurer.*"[41] Fearing he would be extradited to Russia as wished by the allies, he went to see the mayor, who advised him to appeal to the president. On September 14 he was nearing the capital when Attorney General Richard Rush met him to inform him officially that the president had decided not to receive him. Joseph turned back. The content of the message is not known, but given Rush's animosity toward Joseph,[42] he was probably no less direct than the president, who wrote to his secretary of the treasury, Alexander James Dallas, on September 15, 1815:

> I was informed, through *confidential* channels several days ago, that Jos. Bonaparte was about to visit me incog. to make a personal report of himself to this government. I immediately wrote to Mr. Rush to have him diverted from his purpose on his arrival at Washington. Protection and hospitality do not depend on such a formality; and whatever sympathy may be due to fallen fortunes, there is no claim of merit in that family on the American nation; nor any reason why its government should be embarrassed in any way on their account. In fulfilling what we owe to our own rights, we shall do all that any of them ought to expect. I was the more surprised at the intended visit as it was calculated to make me a party to the concealment, which the exile was said to study as necessary to prevent a more vigilant pursuit by British cruisers of his friends and property following him. Commodore Lewis consulted his benevolence more than his discretion in the course he took, without, as I presume, any sanction from any superior quarters.[43]

The American president's position was defensible. The United States was carrying out its duty of hospitality and assuring Joseph of its sympathy; more should not be asked—above all, not the favorable treatment that he might have claimed as ex-king of Spain and member of the ex-imperial family. The republic had no use for these vanities. It had not forgotten that one of its citizens from Baltimore, Elizabeth Patterson, had seen her marriage to Jerome Bonaparte, the youngest brother, annulled by Napoleon for dynastic reasons, and had been forbidden to enter France.[44] Above all, the government had not forgotten the losses that Napoleon's policies had inflicted on its maritime trade. It was consequently understandable that the American president referred to the Bonaparte family's former conduct and that he did not want to be embarrassed by any of its members. Furthermore, establishing new relations with the Bourbons dictated a prudence suggested to him by Monroe, who hoped to succeed him as president and would not have wanted his term to begin with a conflict with France. A sarcastic press often relayed the information that the federal authorities were keeping Joseph at a distance and made fun of the ex-king of Spain, "nothing but a *man,*" before whom there was absolutely no reason to go into raptures.[45] Joseph would have been wrong to take the matter personally: Madison's position could be justified by the example of President Washington who, for considerations of public order, had refused to receive Talleyrand. But he could also take offense at the recollection that General Moreau, exiled by Napoleon, had been received by Madison, and Jefferson before him.[46]

Grouchy and Sons. For the moment, France could not fail to be pleased by the way in which the United States had received Joseph Bonaparte. But in September 1815 he was only the vanguard of the contingent that was to follow in the two years to come. It was easy to deal with an isolated case; it was more difficult to contain the others. Grouchy's arrival offered the French diplomatic corps its first taste of trouble. Enforcing the edict of August 1, 1815, which annulled the promotions and awards given during the Hundred Days, the legation refused to use the title of marshal: "the so-called Marshal Grouchy," Hyde de Neuville said mockingly.[47] Charles Roth, the Washington legation's first secretary, expressed his annoyance to his minister on February 13, 1816:

> General Grouchy is, as of now, the first who has dared to come to Washington, where he has been for two days. Knowing how self-assured he is, this does not surprise me. He appeared in Congress yesterday and the *National Intelligencer* was so indecent as to announce his arrival at the same time as that of the Spanish ambassador, which gives an official character to this notice. As there is here no etiquette whatsoever, and foreigners are received without the consent of their legations, I should not be surprised if some members of Congress thought they could take the liberty of bringing Mr. Grouchy to the President's gathering, as his doors are open to all, and this individual is the sort who would show up there.[48]

It was impossible to lodge any sort of complaint, because nothing prevented Grouchy from moving about freely, nor the American politicians from inviting him, nor the press from reporting this, nor the government from taking advantage of "all the commonplaces of republican hospitality, liberty and independence that do not allow the President to be master in his own house." Roth, following the letter of his official instructions, had no alternative but passive resistance. He had to ignore the exile and boycott his hosts if it happened that circumstances risked putting them in one another's presence. "I would refuse," he wrote, "to appear in the circle during the entire time that it would please General Grouchy to remain there; and without making a complaint, without giving a reason, I would not come to subsequent get-togethers." Several days later, as Roth had feared, General Grouchy was taken to the home of the president, who, "without saying a word to him, let him walk by, lose himself in the crowd, and appear in one room after another, on one floor after another for the rest of the evening."[49] Roth, who complained so bitterly about the American press, was as derisive toward Grouchy as the *United States Gazette* was toward Joseph: "After people had satisfied their initial curiosity, they found that General Grouchy was a very ordinary man. These gentlemen lose a great deal when seen close up, and to cure the country of its enthusiasm I should like to see all of them make a pilgrimage to Washington."

While it was important to turn public sympathies away from the exiles, it was essential that the authorities set the example; this gathering was distressing proof that they were not doing so. Roth called Treasury Secretary Dallas's attention to Grouchy's presence—because Dallas was applying for the post of United States ambassador to Paris—and expressed his surprise with feigned indifference "out of sheer politeness, for this sort of American irregularity does not count in Europe," but loudly enough to be heard by the president. Madison got the message and made a point "of talking to [Roth] much more than is customary in these gatherings," without ever mentioning the burdensome presence. At the end of his second term, he could allow himself a certain easing of diplomatic proprieties, a thing impossible for the secretary of state whose ambition was to succeed him. After dissuading the president from receiving Joseph, Monroe, hewing to his line of conduct, disappeared "like a flash, to avoid having to greet him," when Grouchy arrived. Roth would have liked all the members of the government to take this line, but Secretary of War William Crawford, a recent United States ambassador to France (1813–15), honored Grouchy with a dinner because he had known him during his recent mission in Paris. On the other hand, Albert Gallatin, who was in the end preferred to Dallas, waited for Grouchy to leave the federal capital before going there to receive his instructions. Roth was not taken in: he saw this as skill in avoiding difficulties rather than as a sincere approach to the problem.[50] As far as he himself was concerned, Roth had decided no longer to go anywhere where he would risk meeting the undesirable marshal.

As if the father were not enough, an agreement had to be reached on how to behave toward Grouchy's sons, whom the French government had authorized to join him. They turned up at the Philadelphia consulate on May 22, 1817, to have their passports stamped: "Can I and should I return their calls?" a worried Consul Pétry asked Hyde de Neuville.[51] The latter minimized a situation having no relationship with their father's, since they were not on any list,[52] and asked Pétry to keep an eye on their relations and inform him of their conduct: if it was decent and proper, he would be delighted to send word of this to France.[53]

The problem of the welcome Americans gave the exiles and the related matter of what attitude to adopt in the case of a chance meeting were long a headache for the French diplomatic corps, which Grouchy seems to have taken malicious pleasure in provoking. The consul in Philadelphia found it hard to come to terms with this, as the following example illustrates: "Yesterday I was invited to dine with a family. When we were at the gaming tables, General Grouchy arrived. I continued to play. Supper was served and we remained at our game. When it was finished, I went into the dining room and as I do not eat supper, I remained standing behind a lady. Without my noticing, Mr. Grouchy came up to me and asked me three or four questions about Mr. Dupont and the success of my game. I answered them all coldly and he left as soon as the first guest did; I did the same."[54]

Pétry was bitter at having to leave a gathering whose private nature should not have interfered with his public life, but, though he was not on duty, he felt the weight of his position as an agent of the king in a confrontation he found too difficult because it was lost in advance. Unable to do anything or to protest, he preferred to hibernate, give up his old acquaintances, and avoid gatherings where he might meet exiles, rather than suffer a new affront. At the end of this year of 1817, the French consul in New York met with a similar situation: "A family I have known for 38 years, that was receiving exiles before my arrival here, had told me on which days I would not meet any in their home. Faithful to this condition, they did not expect General Grouchy to invade their gathering. In the past three weeks this has happened three times. The first two, I withdrew immediately, although the family apologized for such behavior. Yesterday I experienced a third annoyance. Quite resolved not to be exposed to it again, I put on a bold front and waited at a gaming table where, of course, the general was not, for an opportunity to withdraw."[55]

Grouchy turned up at every event that was even remotely official. On June 19, 1818, ship's Lieutenant Maud'huy, ashore in New York, met him at a tea at the mayor's home.[56] Since Grouchy was well aware that the French consuls and ambassador to the United States had been forbidden to meet the exiles outside the framework of their official duties, d'Espinville wondered unsuccessfully what might motivate him to act this way: "What is this general's aim? I cannot figure it out." A renegade to the nobility into which he was born, the Marquis de Grouchy had nonetheless remained a man of the world, a gentleman at ease in society, where he enjoyed the success his upbringing and good manners brought him. He certainly also wanted to exhibit irreproachable conduct, which the French agents would report to their minister and consequently to the king. In this he was wrong, since the consul in New York was as resolute as his colleagues in warding off what they considered provocation.

In May 1817 the consul in Boston refused a dinner invitation from the former United States ambassador to Sweden because the newly arrived elder General Lallemand had been invited.[57] Consul Pétry congratulated him on his decision, adding that he was astonished that "the people of this country, forgetting proprieties, have so little tact as to invite us along with those who wanted and perhaps still want to tear open their mother's breast."[58] This zeal bordered on the ridiculous. The previous month Hyde de Neuville had thought it his duty to denounce the order that the Swedish ambassador to the United States was said to have received from Bernadotte, the king of Sweden, to treat Joseph Bonaparte with consideration and give a banquet in his honor.[59] He was forgetting that the two men were brothers-in-law and the indisputable fact that the king could hardly be suspected of Bonapartism.

The French diplomatic corps and the Americans continued to misunderstand each other, whether acting as private individuals or officially. Beyond their own case, the French consuls and ambassador were defending the service and honor of the king, whose person was sacred. To the Americans, however, republicans by conviction, these notions of respect and legitimacy were completely foreign. Their newspapers and public opinion were totally free to take whatever tone they wished when discussing those governing them, so they did not think it necessary to compromise that freedom by preserving the dignity of a king of France whose reactionary character they condemned. Lack of understanding was coupled with an exasperation all the greater in His Majesty's appointed subjects in the United States because their margin of protest was slim. Hyde, inflexible regarding the grandeur of the Bourbons, was on the look-out for the slightest incident he could exploit. He agreed that the primary goal of his mission was to bring together two nations that had for too long been disunited, but he would not let this take place on just any condition; his country's credibility was at stake. Two affairs, in which the Americans Skinner and Lee played the lead parts, came to Hyde's attention simultaneously. He did not let them pass.

A Troublesome Toast

During the patriotic festivities on July 4, 1816, there was an outburst against the restored monarchy: "The most revolutionary spirit, the most opposed to legitimate governments, in particular that of France, appears to have presided at all the banquets given for local authorities in Washington, Philadelphia, New York and Baltimore, by the navy, the militia and all the associations," wrote an enraged French consular official.[60] In Baltimore, during a public dinner for about one hundred people presided over by General Samuel Smith, a Jeffersonian congressman from Maryland, and at which General Lefebvre-Desnoëttes was to be the guest of honor, John Stuart Skinner, a twenty-nine-year-old lawyer who had just been appointed city postmaster,[61] gave a toast in which all joined: "The generals of France in exile; the glory of their native land—not to be dishonored by the denunciations of an imbecile tyrant."[62]

After only three weeks at his post in Washington, Hyde had to handle a situation that it was not in his nature to avoid. Since freedom of the press was an inviolable right guaranteed by the Constitution, complaining about the invectives picked up by the American newspapers was an exercise in futility, but a public insult by a federal employee against the king of France was quite a different story. That is what an outraged Hyde made a point of telling Monroe, before demanding official redress and the dismissal of the man who had caused the scandal.[63] The secretary of state, whose moral responsibility was implicitly called into question, took his time writing a blunt refusal that delighted Hyde de Neuville's detractors.[64] "Your past stay in the United States must have taught you that the government has not the slightest influence on the press, that all have the greatest freedom to judge and criticize the government itself," Monroe wrote. "We cannot be responsible for the outpourings of feeling that might be expressed at a public celebration with regard to foreign governments, an incident in which the identity of the officer, especially a person of lower rank, disappears into that of the citizens."[65]

Monroe was attempting to defuse the dispute by taking refuge behind freedom of the press and playing down an event whose perpetrator was supposedly a mere citizen. The government did not feel obligated by an individual's statement of opinion and washed its hands of the affair. There was no question of sanctions. But Skinner was not really the "average American" portrayed by Monroe: he had fulfilled various missions as a government agent during the War of 1812, and his new position in the postal service made him an important part of the Madison administration.[66] What is more, as an influential member of the Society of Red Men, a patriotic organization that defended the principles of the American Revolution, Skinner supported the cause of independence for the Spanish colonies, and admired Napoleon as much as he detested the English and consequently Louis XVIII, who owed them his throne and had proscribed these soldiers whose glory was that of the Emperor. The official response did not therefore calm the French ambassador, who fanned the flames by following his instructions, and turned a toast into a crisis. From New Brunswick, he sent an emissary to Washington bearing three letters for the secretary of state.

Arriving in the capital on August 24, the emissary learned that Monroe had left to see President Madison in Montpelier, from where he was to go on to Monticello to meet with Jefferson. So, whether he liked it or not, he had to play catchup: "I still have three presidents to see, the past, the present and the future; this is a fine prospect, I admit; but all I ask of them is to let me return to Your Excellency as quickly as possible."[67] He missed Monroe, who had already left Montpelier, but he did not give up hope: "This is a regular game of tag; I shall catch him tomorrow for certain; I have only thirty miles to go."[68] Meanwhile he gave the dispatches to the president, who read them and said that "this important affair had been the subject of a conference between himself and the secretary of state and that it was essential that he and Mr. Monroe have a discussion of this matter, which could be lengthy." Nevertheless, Madison held out hope for a solution in keeping with the wishes of the French ambassador, which he understood perfectly—but this diplomatic version was not exactly what Madison told his secretary of state. The president was annoyed by Hyde's arrogance in wanting to dictate to the government what it should do and his efforts to turn a simple incident into a "provoking enormity."[69] Faced with the French ambassador's insistence, Madison thought of asking for his recall. Hyde, not at all intimidated, recalled the consul in Baltimore, and in Paris, Richelieu asked Gallatin for an explanation. The Skinner affair was taking a turn that interfered with the resumption of good Franco-American relations. This minor dispute, beginning as a simple pretext, was most revealing. France was bruised by the repeated assaults of its denigrators, and the American state was oblivious to this. It was important to clear the air and put an end to a deteriorating situation that was bad for all concerned. The United States killed two birds with one stone by granting the French request to relieve William Lee of the office of American consul in Bordeaux, which he had held for fifteen years. His Bonapartist bias, France complained angrily, was incompatible with his mission's duty to be neutral. It had also come to be a problem for his own nation.

THE CONSUL'S "RECALL"

The ease with which this matter was settled was partly due to the fact that it was not necessary to recall him and thereby act in a way that would be humiliating both for the disowned consul and for the country obliged to grant a foreign demand. On December 21, 1815, Madison had written to Lee to assure him of his continued confidence and confirm his appointment as consul in Bordeaux, thus, with a stroke of the pen, consoling him for all the humiliations of the past months, both personal and in the name of the country he represented. But the deluge of intimidations had so worn him down that in his answer of February 16 he had expressed a great desire to return home.[70] He needed to return to the United States to forget, to attend to his business affairs, and to consider the future of his children that a ruined France could no longer secure.

Lee took passage on the *Laguira*[71] with his family and some sixty French passengers whom he had helped emigrate because they exercised the trades with which Lee liked to supply the United States.[72] He received a hero's welcome in New York in August 1816: "The regrets of the good French people accompany him and follow him to the new world. Honor, glory and gratitude to the worthy representative of the American nation . . . Generous protector of the banished, accept here their unanimous gratitude."[73]

Jacques Roul was one of these "good French people," except that he had not followed Lee to America, but preceded him.[74] He had been the Emperor's first aide-de-camp on Elba, had been promoted to colonel during the Hundred Days, and had been taken prisoner at Waterloo. Even before he was freed, the second Restoration had discharged him without a pension for joining Napoleon on his island and following him back to Paris. Returning to France, he went to Bordeaux, where Lee got him a passport under the name Brown and put him on the *Magnet,* which reached Philadelphia on April 28, 1816. Two weeks later, in Baltimore, he proclaimed via the press what he owed the consul: "Colonel Roul impressed with the deepest sense of heart-felt gratitude, offers this public acknowledgement, to the highly respectable William Lee, esq., consul for the United States, in Bordeaux; to his humanity, he is indebted for the preservation of his life, from the bloody hands of the executioners of Robespierre the 18th."[75]

Despite his service record, Roul, a genuine colonel but a false baron, was a braggart.[76] Even if his public homage was sincere, he gave it a publicity that he hoped would gain him the compassion of the Americans and the respect of his French compatriots. He exaggerated his own importance and misunderstood Lee's personal situation.

This demonstrative prose did him more harm than the outrageous poetry published at the expense of Bordeaux, guilty of selling itself to the English.[77] The United States was all the less the hoped-for haven for Lee because Hyde de Neuville, righter of the wrongs done to the king of France, was now there. Since his request that the consul be recalled had been transmitted to the American government, Hyde quite logically thought that Lee's departure from Bordeaux was its consequence. But he learned from Roth and Pétry, whom he had sent to Monroe for news, that Lee had simply returned for a visit. On July 21, 1816, he was obliged to remind the secretary of state of his government's order to request the immediate recall of Lee, who, given his partisan conduct in Bordeaux, could not continue to carry out his consular duties there.[78]

As in the preceding case, Monroe took his time in answering, but his tone was different. It indicated the desire of the American government to settle the affair amicably, without another official request from the king, a possibility that Hyde had suggested: "Mr. Lee having requested some time past permission to make a visit to the United States on account of his private affairs, his return which had been authorized, was entirely owing to that cause. I am instructed by the President to assure you, that the desire of your Sovereign in regard to Mr. Lee's removal from that office will be immediately complied with."[79]

Madison and Monroe, who were continuing to stand up to Hyde de Neuville in the Skinner affair, quickly gave in where Lee was concerned. Perhaps they were acting in compensation, to bring down a soufflé that had risen too high; probably they already had a substitute post to propose to Lee, but that was not all. In the meantime, between the French request of July 21 and the American answer of August 15, a compromising document had surfaced that most imprudently, given the spiteful environment and his absolute duty of discretion, contained Lee's signature.

This brings us back to the Faucher brothers. They alone had been condemned to death, but other people, involved in the same affair, had not been spared in a second trial, that of the "*Fédérés* of la Réole," before the crown court of Bordeaux on December 19, 1815: the republican Pierre Viaud, called "Bombance," a carpenter of La Réole, escaped forced labor, but not ten years of banishment for "conspiracy against the internal security of the state."[80] He chose to spend those years in the United States. On June 4, 1816, in Bordeaux, he had twenty-eight important local people sign an initial certificate attesting that he had been "condemned to exportation [*sic*] merely for having shown too much affection for the imperial government."[81] This support was probably used to obtain a second certificate, on June 6, from Lee, with whom he was going to sail on the *Laguira*. The consul agreed to stand surety in his native language and gave him a document destined to circulate nearly four thousand miles from the place where it was issued:

> The within certificate has been given by the *proscribed patriots* of and about La Reole, a town on this river thirty miles from Bordeaux. It is honorable to the bearer Peter Viaud, particularly so as I find the Curé and Mr. Babire, former attorney general for this district, have signed the same, they are both known to me, and are men of the first respectability. I have procured Viaud a passage to the United States from the representing made to me by his neighbors and on account of his wife and children. He is a house carpenter, industrious, frugal, and of good morals. Given under my hand at Bx this 6st [*sic*] of June 1816. W. Lee, consul of the United States.[82]

Lee was committing himself to a good cause, that of a man who was paying dearly for an outmoded political inclination, but he went further by daring "to call the enemies of the government which he represented *proscribed patriots*," Hyde wrote angrily sometime later,[83] and by facilitating his passage and his family's besides on the same ship that was taking him to New York. In league with American captains, the consul often engaged in this sort of transfer because, beyond the favor done the emigrants, he saw that in the long term it would be an advantage to his country, which lacked Europe's skilled labor and technological know-how. That is why he even suggested to Monroe "to defray the expenses of the passages of some of these manufacturers to America."[84] The carpenter Viaud would find work. While this favored the economic development of the United States, France had a double reason to be annoyed: economic, on one hand, because even though the French authorities did not really become interested in the question of emigration until the end of the 1830s,[85] hiring that harmed national industry was forbidden by article 417 of the penal code;[86] political, on the other hand, because in this individual case, it benefited an avowed enemy of the Restoration. Hyde de Neuville did not have to wait long before denouncing the schemes of the American consul, thanks to policeman Pierre Vauversin, whom he had introduced into the circle of exiles.[87]

Vauversin met Viaud in New York.[88] Barely ashore, the carpenter boasted publicly of his crime and flourished his certificate: it was the attestation of its signatories' hatred for the king and their affection for the Emperor, reported Vauversin, adding, "what is outrageous, what will undoubtedly infuriate Your Excellency, is that on the back of this revolting document there is written in English in Mr. Lee's own hand" an expression of America's gratitude.[89] "If I did not have this document here," he went on, "I would have had difficulty believing in its existence, not being able to imagine that the agent of an allied government is in direct contact with individuals who openly admit, who glory in conspiring against the legitimate government." Hyde's spy furthermore told him that in Bordeaux Lee played the part of a "vile employment contractor," receiving at his home and socializing only with sworn enemies of the king. Hyde had proofs of this: he could cite several cases of inhabitants of Bordeaux who claimed to be victims of arrangements between Lee and people from the Gironde region emigrating opportunely after declaring fraudulent bankruptcies. As to Viaud's seditious opinions, they were well known, since they had earned him his exile: "For France to be happy, she [needs] blood up to her knees for 24 hours," he assured Vauversin.

In any event, recognizing its importance, Vauversin sacrificed his cover and stole the document. This provoked the anger of the group of exiles with whom he associated, including Regnaud de Saint-Jean d'Angély. The subject of the certificate and its principal signer were obviously the most frustrated, and thus the angriest. Viaud said, "It is not so much for myself that I care about the f——ing paper, but what worries me is that among those who signed it, there are some who are currently in important positions."[90] Lee was one of these. He is said to have admitted that for two thousand gourdes[91] he would not have wanted this paper to go astray, and he requested an audience with Hyde de Neuville if the document was not returned to him the next day.[92] This admission arrived just when needed to unmask his "demagogic zeal" and his "hypocritical protests" and to justify his recall. A former officer named Moreau challenged Vauversin to a duel. Frightened, he fled from New York to Philadelphia, but the original or a copy thereof made the rounds of French and probably American offices.

In fact, the decision had by this time probably already been made to relieve Lee of his appointment as consul in Bordeaux. Little more than a week had gone by since the seizure of the document. In early September 1816, from Philadelphia, he sent his resignation to President Madison, who accepted it. Lee never saw France again. The affair concluded to Hyde's advantage, but he continued to wonder. How could a consul of the United States have "given such a certificate, when it would be presented to his compatriots, to the agents of his government? What conclusions should be drawn from this? That his government thinks as he does?"[93] In the end, both sides got something out of the affair: Lee, for whom the government quickly found a suitable position and whom it permitted to look after the French exiles; and Hyde de Neuville, with whom relations could now get off to a better start.

6

Settling in America

It must be acknowledged that this country is favored by everything that could be desired, and that it has before it an immense future of happiness, tranquility and prosperity. If this spectacle alone were not sufficient to make these regions pleasant for me, the welcome and the consideration I received in all my travels would themselves be enough to make up for the sadness felt by a virtuous heart far from a Homeland that witnessed its birth. Nevertheless, while admiring the beautiful country that has accorded me its hospitality, while serving it with thankfulness and a most devoted zeal, my wishes are and will always be for the happiness, peace and glory of France.

—General Simon Bernard, New York, April 15, 1818

After eighteen months in America, General Bernard expressed his gratitude for a welcoming country, blessed by the gods and destined for a bright future. Even if his heart remained in France, like that of his compatriots in exile, his abilities were so exceptional that they could not model their integration on his. The French arriving in the United States, thought Hyde de Neuville, were a second-rate population in which there were relatively few decent people: royalists, the carefree, merchants, artists and artisans, more concerned with their own interests than with politics. On the other hand, there were countless fools, as well as dangerous individuals, ill, incurable, and fanatical, whom the ambassador divided into three main groups: one-third were unmasked Bonapartists or Jacobins, one-third had declared fraudulent bankruptcy or were thieves, and the final third were peddlers carrying their commercial firms in their bags, "adventurers without means of existence, without fortune, young men without education, most of them draft dodgers under Bonaparte."[1]

With few exceptions, Hyde lumped the émigrés with the dregs of France, crooks fearing neither God nor man, street peddlers without assets or prospects, deserters defying society. As his convictions coincided with equally negative official instructions, he hoped to reduce personal contacts to an absolute minimum, and the consuls were to do the same. This did not mean they were not to pay attention to them or watch their every move: spying on them was a daily task for the French agents. A good many émigrés, who owed their fall from favor to the Bourbons, were no more eager to spend time with their representatives in the United States—all the less so because the ambassador's reputation left little hope for compassion. It was not here that they sought aid and comfort, but from their American hosts and from the Franco-Americans who had been living in the country for some time.

HOSPITABLE AMERICA

The emigrants were not equal before American hospitality. The degree of fame and fortune of an ex-king of Spain or an ex-marshal of France was a huge advantage unavailable to the mass of anonymous men banking on their courage, the exercise of a trade, or the aid of contacts made with foresight before their departure. Here too, there was a great disparity between those who benefited from important connections and the ordinary people who were helped in the best of cases by simple family and geographic ties.

Though furnished with addresses, Jacques Lajonie nonetheless experienced difficult beginnings. A first Philadelphia merchant did indeed receive him, but being a secondhand correspondent, he admitted he scarcely knew the French family that had directed Lajonie to him. A visit to shipowner Andrew Curcier, a native of Bordeaux naturalized years earlier in Philadelphia, was scarcely more promising. Lajonie was lamenting those letters of recommendation that led to expense without compensation from their addressees and supposed protectors, when he finally met John Latour.[2] A commission merchant in Baltimore since at least 1795, this Domingan refugee belonging to a family with roots in Gensac and Juillac was the ideal compatriot for Lajonie. Very wealthy, he showered Lajonie with "kindnesses,"[3] offered him bed and board, and entrusted to him, for his first steps in business, jewelry and watches to sell on his way to New Orleans, where he had decided to go. Regional solidarity made up for lack of assets or a ready trade. Based on the confidence that came from belonging to the same soil—"our dear homeland," wrote Latour—and having relatives or friends in common, such solidarity was natural, often effective, and widely used from Pennsylvania to Louisiana, where the Francophone community was strongest.[4]

Appealing to key French figures admired in the United States was a prestigious resource. Lafayette answered these requests all the more readily because he sympathized with the troubles of victims of the Bourbons. Far from politics on his estate in the department of Yonne where he busied himself with farming, he had come over to Louis XVIII in the hope for a constitutional monarchy, but his true return to the political scene occurred when he was elected to the Chamber during the Hundred Days. As vice-president of the Assembly, he opposed Napoleon and then advocated his abdication before attacking the king whom he associated first with the "counter-revolutionary confederation"[5] whose invasion his country was enduring, and then with the measures of repression and banishment contrary to his ideal of liberty. His château became a center of opposition. Hero of two worlds[6] and symbolic ambassador of France to America, the Marquis of Lafayette supported repressed and banished Frenchmen and recommended them to his American friends.

AMERICAN SYMPATHY

In 1818, Secretary of State John Quincy Adams defined his country's immigration policy as follows: encourage no foreigner to come to the United States, but send back and discriminate against no one. Such a broad conception of neutrality in the matter could not help but be favorable to the émigrés. President Madison's refusal to receive Joseph Bonaparte did not set the tone of future relations with the host country. For the American administration, Joseph was a bothersome figure at a time when the second Restoration was getting under way and when the United States was wondering what course to follow, internally as well as externally, just after the War of 1812, and before the Spanish-American republics achieved independence and recognition, a subject to which we shall return in chapter 9.

Joseph and Grouchy were the first famous exiles who created a stir, basically because nothing was known about them or their intentions. In July 1816, Joseph, who had been renting quarters in the United States, purchased the farm of Point Breeze on the Delaware River near the small city of Bordentown, New Jersey, northwest of Philadelphia. As he was not an American citizen, he used a front man to acquire it, and then petitioned the New Jersey legislature for the right to possess his lands in his own name. The fact that he got what he wanted is less important than the way he got it. In January 1817 the governor sent him the deed authorizing foreigners to own land in his state and thanked him for choosing it as his place of residence, adding: "The members of our legislature join me in assuring you of their kind sentiments toward you. The deed they have just authorized is the most convincing proof of this."[7] This was a far cry from President Madison's reserve in his regard, and the obliging attitude of local officials was soon shared by the local inhabitants, called "Spaniards" because of their sympathy for the ex-king of Spain. Joseph was happy at Point Breeze and his relations with his neighbors were always excellent, thanks to a simplicity and affability that his past glory had not led people to expect at first glance. Playing the part of a gentleman farmer busy with landscaping and overseeing a workforce of agricultural laborers was all the easier in that, in his rural exile, Joseph lacked nothing—neither money, nor servants, nor important friends,[8] nor, very soon, a young mistress who bore him two daughters.[9] While Joseph never stopped improving his home—which was said to be surpassed only by the White House—or enlarging his estate (eighteen hundred acres in 1835), he did not remain shut up there. He also resided at Dunlop House in Philadelphia, in July visited the twenty-six thousand acres of land between the Black River and the Adirondacks in northern New York that Le Ray de Chaumont had sold him in 1814, and in August "took the waters" in Saratoga or Ballston Spa.

Grouchy's welcome was no different. Disembarking in Baltimore in late January 1816, he was pleased with the interest surrounding his arrival in this hospitable country. The French consul confirmed that he had indeed been eagerly received by some of the most distinguished figures of the state of Maryland.[10] We have seen that Grouchy later appeared in the company of important politicians and moved in all the right circles, to the displeasure of official France. All that remained was to meet with Jefferson. In October 1817, when he intended to go to Monticello, the illness of one of his sons prevented this. A very cordial exchange of letters has survived to the present day. Grouchy wrote:

> How much I congratulate myself on dwelling in your interesting country; how proud I am, and how thankful for the honorable hospitality which has been bestowed upon me here, and . . . if anything can lessen the bitterness with which a distant exile overwhelms me, and the state of servitude and degradation of my native land, it is to see yours, happy, powerful, free and respected, and all through institutions founded upon the very same principles for the establishment of which I have so often needlessly shed my blood.[11]

Jefferson replied:

> Your name has been too well known in the history of the times, and your merit too much acknowledged by all, not to promise me great pleasure in making your personal acquaintance. If, too, the trouble of such a journey could be compensated by anything which the country between us could offer to your curiosity, it would save me the regret which I could not fail to feel were I to suppose myself the whole object of the journey. In this last case I would certainly think myself sufficiently honored by the written expressions of respect just now received, and should postpone the pleasure of receiving them personally to the unreasonable trouble which such an object would impose on you. As you flatter me with taking the journey in the spring, I am in hopes the face of our country at that season will still better reward the labor of the undertaking.[12]

Other renowned French exiles sought meetings with Jefferson, such as Lakanal, who had arrived in the United States with letters of introduction from Lafayette and Thouin, the administrator of the King's Garden, with whom Jefferson corresponded concerning his agricultural undertakings. In June 1816, Lakanal regretted that he could not deliver them personally because he had changed his plans and left the east coast for Kentucky, where he had purchased land on the banks of the Ohio.[13] Touched by Lafayette's recommendation, Jefferson answered by return mail that he could well conceive what the reality of this new life might be, adding: "But the wise man is at home everywhere, and the mind of the philosopher never wants occupation. I weep indeed for your country, because, altho' it has sinned much . . . yet its sufferings are beyond its sins and their excesses are now become crimes in those committing them. We revolt against them the more too, when we see a nation equally guilty wielding the scourge, instead of writhing under its affliction at the same stake. But this cannot last. There is a date of judgment for that nation, and of resurrection for yours."[14]

The Americans' sympathy for the refugees increased as time went on and the press reported favorably on their arrival in the country, their comings and goings, and the receptions held in their honor. On July 28, 1817, Philadelphia's *Aurora* stood up for the newly arrived General Vandamme, denouncing the malicious rumors that had circulated against him, and clearing the name of this patriotic soldier, devoted only to the defense of his homeland.[15] In late 1818 the Philadelphia *Gazette* in turn was pleased to learn that Generals Clauzel and Lefebvre-Desnoëttes had been very kindly received in Washington, that they had seen the president and dined with the secretary of war, and finally that most of the enlightened people of the city had called on them: "We do not want to think that the French government can continue the exile of such men; they honor not only France but their century."[16] The press reported the welcome that American public figures accorded exiles whose trouble-free stay was reassuring, and this helped them find favor with the public.

FRENCH SOLIDARITY

Many, though at times fragile, ties of solidarity were forged, both with the older French émigrés in the United States and among the newcomers themselves, whose number swelled with every arrival of a ship from Europe. The first welcomed the second in a spirit of mutual aid, remembering their own exile, most of them refugee victims of the revolutions in France and Saint-Domingue. They formed a network of hosts ready to serve those being absorbed by an unknown or hostile world. Just as among the exiles some were more famous than others, with titles and celebrity as their passports, so among the American French there were those whose success encouraged others to appeal to them. We have seen that Jacques Donatien Le Ray de Chaumont had, in France, already sold thousands of acres of land in New York State to Joseph Bonaparte and Pierre-François Réal.[17] Back in the United States in 1816, he also sold land to Grouchy, Clauzel, Garnier de Saintes, Regnaud de Saint-Jean d'Angély, and Quinette.[18] Stephen Girard, Jean Simon Chaudron, and the Duponts de Nemours, father and sons, were also ready to give support.

Stephen Girard: The American Uncle. One-eyed, red-headed, short and stocky, an American citizen at odds with the English language, Girard lacked the social ease that could have mitigated the unfortunate effect of the moral turpitudes of which he was almost always wrongly accused in explanation of his exceptional career. He may well have been a cold, calculating loner, a man of reason rather than passion, but his success as a versatile entrepreneur was due above all to his obsession with work, his clear-sightedness, and his thrift. A sailor who had grown wealthy from trade, arms, and banking,[19] Girard was, on the threshold of old age, a monument—not only in Philadelphia, where he had lived for forty years, but in the entire United States, where his role in financing the War of 1812 had won him national gratitude. Born in Bordeaux in 1750, Étienne Girard had trained on ships sailing between France and the Caribbean before obtaining his captain's license. In 1774, letting his ship return to France without him, he went to New York and then Philadelphia, where his career underwent a meteoric ascent that continued until his death. For half a century Girard sent merchandise across the seven seas, maintained close relations with the major international trading firms, and increased his wealth by diversifying into real estate, maritime insurance, and finance—in 1812, he pulled off a feat of daring by buying the Bank of the United States whose capital stock he subsequently brought up to $1.5 million.[20] This operation gave rise to the same rumors that had dogged him since the beginning of his career. But however rich and egotistical he may have been, Girard could also be of service to others and was not insensitive to misfortune when he considered it worthy, declaring that he took an interest in all unfortunate persons.[21] He helped many Domingan refugees whose sons served apprenticeships with him; he also did much against the great yellow fever epidemic that ravaged Philadelphia in 1793. His support of the expiatory victims "of a government that was the enemy of Republicans"[22] gave him the opportunity to contradict the recurring accusations of misanthropy and avarice, but it is true that he favored prominent individuals with money behind them, whose standard-bearer was Joseph, to the detriment of numerous beggars without money or serious plans.

The alliance between a deposed king and a staunch republican was unexpected but genuine and sincere.[23] Girard was his banker, agent, supplier, and finally a friend soon moving in the same social circles: Grouchy, Clauzel, the Lallemand brothers, and Lefebvre-Desnoëttes entrusted to him their financial interests, particularly with Parisian banker Laffitte (whose younger brother was Desnoëttes's brother-in-law)[24] and incidentally also their love lives.

Girard had been associated with a younger brother in a trading company in Cap Français, Saint-Domingue, a man as expansive and impulsive as Stephen was cold and impassive.[25] John did not enjoy the confidence of his older brother, who considered him a dilettante. Like most other Domingan colonists, John had neither seen the smoldering discontent among the slaves nor gone along with his brother's idea that they should start over elsewhere. It was John who broke off relations. When he died in 1803, he left, in Connecticut, four children[26] and a widow, Eleanor O'Neill McMullin, whom Stephen Girard considered a schemer and a bad mother. After her death, which soon followed, he entrusted the upbringing of his nieces to a Moravian pastor with instructions, given their ancestry, to keep them on a tight rein. Only Harriet, the youngest, found favor in his eyes, for she lived the studious kind of life of which he approved. He placed her with Victoire George, née Le Grand de Boislandry, an energetic woman who ran a boarding school for daughters of Philadelphia's good families. Her husband, Édouard George, with whom she had fled Saint-Domingue, sailed for Girard as supercargo,[27] and their oldest son, Edward Jr., as captain.

Henry Lallemand was twenty-three years older than she. He succumbed to her accent when she spoke to him in French; he wanted to marry her immediately.

Harriet did not force things; respecting proprieties, she opened her heart to her uncle,[28] who granted her hand to her suitor[29] and gave her a dowry, defending himself against accusations that he should have given more. If he gave more to his niece than to others, he feared that people would think he was flattered by a union with one of Bonaparte's generals, when he would have preferred a merchant like himself, devoted to work, the only thing that counted.[30]

And yet, Harriet, daughter of an Irish virago and half sister of a mulatto whom her father had had with a slave, was becoming a baroness by her marriage, even though the baron, as Joseph Bonaparte is reported to have said to Girard, was no more noble than they. As for Lallemand, condemned to death in his country, he was not making a bad match with a well-brought-up young woman whose uncle was, with the possible exception of Girard's friend John Jacob Astor, the wealthiest man in the United States. For that matter, the American uncle was not all that tight-fisted and laid on a lavish reception. On October 28, 1817, the marriage ceremony took place at St. Augustine Church. The cream of Philadelphia society and of the business world, the Francophone community and the Bonapartist microcosm, were invited: Joseph Bonaparte, ex-king of Spain; Marshal de Grouchy; his son, Colonel Alphonse de Grouchy; General Baron Charles Lallemand, the groom's brother; and General Baron Vandamme.

While this marriage consecrated the success of a banished man on American soil and the harmony around him of the older and recent French émigrés, there was one sour note. Grouchy loved society life, but particularly the American variety, for he was afraid of compromising his chances of an imminent pardon if he consorted openly with people from whom he was dissociating himself. According to the New York consul, he had left Philadelphia to avoid the wedding, but Lallemand had sent him a letter urging him to return and offering to postpone the celebration if business prevented him from being back in time. The marshal was not rewarded for letting himself be persuaded, since for "the first time and contrary to all the customs of this country about publishing the names of people attending weddings in newspapers," added the consul, the press had carried an account of the celebration and noted his presence and that of his peers, whose standing with the king of France was at rock bottom. Grouchy was "very upset and provoked" by this, to the point of reproaching Lallemand, whom he suspected "of wanting to put him in line with himself and harm his reputation in France."[31] If the exiles' solidarity was in a bad way, this was in any case not due to Stephen Girard, who had for two years perfectly played his role as supporter of the Bonapartists. In this he was not alone.

Jean Simon Chaudron: L'Abeille Américaine. In the domain of the arts in Philadelphia, Chaudron, a watchmaker, silversmith, orator, poet, editor, and journalist, was a bit what Girard was in the business world: a talented, self-taught hard worker who dabbled in everything. Born in 1758 in Champagne, he had served an apprenticeship in Switzerland and then left to ply his trade in Saint-Domingue, where he linked his professional and personal destiny to those of Pierre Martin Stollenwerck, a watchmaker and jeweler, and his brother Pierre Hubert Stollenwerck, a planter and merchant, whose daughter Mélanie Girard married. In July 1793 the Stollenwerck and Chaudron families disembarked in Philadelphia from the *Charming Betsy,* which had left Cap Français in a mad rush.[32]

Six hundred other Domingan refugees arrived in Philadelphia that year. Pierre Hubert remained there, attracted attention by joining the handful of volunteers who worked with Girard against the yellow fever epidemic, and later moved to New York, where his brother Pierre Martin had set up a watchmaking business. Chaudron prospered in Philadelphia, added eleven more children to his two, and made his home a center of elegance and culture. After a brief association with Charles Billon, a Swiss watchmaker who became his brother-in-law,[33] Chaudron began to produce beautifully made silverware, coffee and tea sets, place settings, and serving dishes in the neoclassical and empire style typical of American craftsmanship of the period.[34] He later cosigned pieces with Rasch, a Bavarian silversmith,[35] but financial problems led them to again work independently.

Chaudron was a born joiner, not only professionally, but of a great many organizations whose common denominator was solidarity among French émigrés. The French made the most of every opportunity to get together, such as political or Masonic meetings, concerts, dinners, or balls and banquets at Oeller's Hotel on Chestnut Street.[36] Besides the Freemasons (to be discussed below), Chaudron belonged to the Grivois club, an outlet for "the natural and irrepressible exuberance of the Gallic nature,"[37] and the French Benevolent Society of Philadelphia, where verve gave way to mutual aid.[38] Founded in 1793 to help refugees arriving in Philadelphia without resources, the society continued, twenty-five years later, to care for those in dire straits whom Hyde de Neuville called "veterans of poverty."[39] In 1810 Simon Chaudron was treasurer of the society and elegist of the tragedy of Saint-Domingue, the origin of these woes. Six years earlier he had expressed in verse his sorrow at the loss of this fruitful isle, and implored Bonaparte to share the mourning of its immolated people.[40] In 1815 the colonists' cause was more than ever his: "People wish in vain to wipe out all memory of you but I shall try to defend you, to avenge you."[41]

Appearing in the first issue, this solemn vow seemed to foreshadow the editorial stance of the newspaper Chaudron launched on April 15, 1815, *L'Abeille Américaine,* the latest representative of the French press in Philadelphia, continuing the line of periodicals published after 1791 by the Saint-Domingue refugees in America. Chaudron more particularly picked up the torch left by his friend Tanguy de la Boissière, a fellow refugee and Freemason whose acerbic style he had admired in the various newspapers Tanguy had edited before his death in 1799.

The paper was a sixteen-page weekly intended not only for the French but also for Americans wanting to improve their mastery of the French language and broaden their knowledge of history, politics, and literature, chiefly those of France and Europe. The matter of Saint-Domingue that was to be the ideological centerpiece of the articles written by Chaudron and his collaborators was sidelined by the effects of the political and military upheavals in Europe.

The charge of Bonapartism was soon leveled against the paper, but Chaudron just as quickly stated that he scorned it.[42] Napoleon's greatness had no equal, nor did his place in the hearts of the French, but he remained a despot given to reprehensible excesses. The man was extraordinary, but his pride and thirst for vainglory had caused misfortune. If he could, he should come to the United States where the atmosphere was conducive to curing his mind of its errors, but it was to be feared that his natural passion for war would regain the upper hand and soon lead him to the turbulent shores of Spanish America, where there was much to be done.[43]

But if Napoleonic France did not possess all possible virtues, that of the Bourbons did have all the vices: letting barbarians pillage the land and allowing England to control the country to avenge itself of the support France had given America[44]—English Bordeaux, where the joy of the inhabitants was at its peak, had its share of disgrace in this sell-off of national honor;[45] the monstrosity of the king whose proscription edict and amnesty law had cut down the victors of so many battles;[46] the complicit ignominy of the assassin Fouché, his associate Talleyrand, and Hyde de Neuville, the ambassador with hands bloodied by the attack of the rue Saint-Nicaise;[47] the perpetuation of crimes in southern France, where terror was feeding on religious intolerance, Americanophobia, and hatred for Bonapartists.

The low-key judgments of Napoleon gave way in the columns of the *Abeille* to savage criticism of those who were hounding the Bonapartists: reactionary France, arrogant England, and hermetically sealed Switzerland. The newspaper unreservedly sided with the outlawed men, happy to see their health toasted in America, grieved by the death sentences and executions of French generals, proud to proclaim itself above all parties and all influence save that of misfortune and honor. It reported the arrival in the United States of Grouchy and Clauzel, "two eminently distinguished outlaws," of Réal, "one of the finest minds of France," of Vandamme, "a brave soldier," deplored the fact that Lucien Bonaparte was forbidden to join them, and erroneously announced the coming of Soult, one of the "foremost generals of Europe."[48]

In contrast to England, Belgium, urged by the allies to expel the exiles, and Switzerland, whose servile nobility had sold out honor and independence, the United States, carried along by republican virtues, remained the sole sanctuary of liberty and the last resort carrying out the august duties of nature with regard to her unfortunate children: "The soil of this hospitable land must therefore offer easy access to men fleeing the tyranny of hereditary monarchies. A nation that, by her divine constitution, judges only the natural rights of man must extend her kindly protection to all those whom involuntary errors or events independent of her government's interests bring to her bosom . . . Admitting these ephemeral demigods to share in the blessings enjoyed by a united America is a duty conferred on her by her political superiority to all the nations of the world."[49]

Finally, what men, what children of the United States could be surprised at the reception accorded these outlawed men, who had been received without being reduced to the humiliating necessity of saying who they were: in June 1817—that is, two years after refuting any Bonapartist collusion—Chaudron begged the indulgence of his readers for his inevitable partiality toward these virtuous citizens who "had had the courage to cross the ocean to seek refuge among free men."[50] On July 9, 1818, in the last issue of the paper, he published an editorial testament that removed whatever ambiguity may have remained: "We have given all our loyalty to the same banners that won renown at York, Fleurus, Marengo and Hohenlinden."

Within the French diaspora in the United States, Chaudron was the most forceful defender of the exiles, who, as remembered by his nephew Fred Billon, frequently visited his home.[51] The sincere and at times excessive fervor of his journalistic and literary writings is striking. With his means, constancy, and opinion—which he said carried no weight whatsoever—Chaudron gave his best in the service of what he considered a just cause. The Duponts de Nemours felt as he did.

Dupont de Nemours: Father, Sons and Company. Witty, talented, and aware of it, ambitious and endowed with a phenomenal capacity for work, Pierre Samuel Dupont de Nemours[52] delighted in the fact that his lot in life was to be happy—not an egotistical happiness, but one freely lived and shared. Freedom was central to his morality in the economic realm, where free trade was the key word, as well as in the political realm: the bases and principles of a prosperous government are "in the obvious development of the rights of man," he asserted in 1768 to Benjamin Franklin, whose struggle he supported.[53] Faced twenty-three years later with the upheaval of his own country, Dupont reaffirmed his faith in "the authority of divine and human law, which cries out in my heart that I owe help to all men illegally oppressed by other men."[54] His disappointment in the French Revolution reinforced his attraction to America, its freedom, institutions, leaders, and finally its people, the hope of the human race.

A philosopher, economist, naturalist, and politician, Dupont de Nemours came from a privileged background. Before moving in the circles of power under the protection of Turgot and then Vergennes, he was known in Versailles as a disciple of François Quesnay, the political economist who held that land was the basic source of wealth, and agriculture the foundation of the economy; Dupont called it physiocracy. But France was not receptive to the physiocrats' ideas, so Dupont, who was, moreover, disenchanted with the political evolution of his country, thought of the United States, an immense new territory where they could be applied.

On January 1, 1800, he arrived with his family in Newport, Rhode Island, as a traveling scholar for the Institut de France but, above all, with the plan of creating in Virginia a Pontian colony where he could fulfill his rustic aspirations and make his fortune. Giving up this plan, Dupont, the pacifist farmer, turned to commerce and industry where his son Irénée, with the help of his older brother Victor, succeeded in the manufacture of gunpowder for hunting and war.

The year 1801 saw the birth of E. I. Dupont de Nemours and Company, a subsidiary of Dupont de Nemours Père & Fils & Cie de New York, which was divided the following year with the creation of a firm in Paris, where the senior Dupont had returned. The Parisian company suffered from the effects of Napoleon's continental blockade, whereas the American enterprise benefited from the effects of the Embargo Act of 1807 and later from orders from the American government at war with England. The gunpowder plant of Eleutherian Mills, located on the Brandywine River north of Wilmington, Delaware, grew very quickly thanks to the talented Irénée, who had studied under the chemist Lavoisier.[55] In 1815 the Dupont brothers, American citizens and suppliers of the United States army and navy, were thanked by their new homeland for their contributions to its defense.

During the Empire, their father had remained in France, where he said he was awaiting the opportunity to return to the holy land, to be buried in freedom: "All my close friends knew that as soon as I became free it would be to America that I would offer my liberty and the work of my last days," he wrote.[56] This did not happen with the return of Louis, who appointed him to the Council of State, but with that of Bonaparte, whose hypocrisy he condemned.

He arrived in America shortly before the first Bonapartist exiles, whom the Duponts welcomed. A great many officers were their guests at Eleutherian Mills and could count on their help. This very much displeased Hyde de Neuville, newly installed in his position of ambassador. In July 1816 the elder Dupont passed on to him a justificatory note, probably from Grouchy, for the king. The answer was polite, but left no hope as to the result of this step and even threatened Dupont that he would be relieved of his position on the Council of State if he continued to associate with the wrong kind of people.[57] Kept informed by his agents of the comings and goings of the officers to Wilmington, Hyde was annoyed by the obviously guilty relations that the Duponts had with them.[58]

Pierre Samuel Dupont set the record straight. The return from Elba had in no way changed his loyalty or his devotion to the king. He had been obliged to leave France to escape Bonaparte, with whom he had refused to compromise himself. This said, in a foreign land, political divisions and animosities no longer affected him; he felt completely free in his choices, such as sympathizing with the misfortune of companions in exile who were responsible neither for the return from Elba, nor the defeat by a superior army, nor the ensuing foreign oppression. In closing, Dupont admitted having "pleasure in finding men who speak and understand our language, so harmonious to my ear, so dear to my heart; who have seen and loved our blessed sun."[59] His attitude toward the refugees is explained first of all by his concern for the rights of man and the respect for freedom, and then by the existence of earlier ties with some of them. Lakanal and Regnaud de Saint-Jean d'Angély had been his colleagues at the Institut de France, and he owed a debt of gratitude to Regnaud, who had prevented the seizure of his property ordered by Napoleon. His sons too had old ties to certain refugees: in 1804 in New York, Victor had been a witness when Charles Lallemand, a cavalry major and assistant staff officer of the Saint-Domingue army, married the daughter of a refugee from the island, Henriette Roberjot-Lartigue.[60]

According to hearsay that reached their father's ears, the French ambassador was said to have called his sons "véritables sans-culottes," which Dupont said could be applied only to his wife, his daughters, and his granddaughters, who were all the nicer for it.[61] The interested parties were not to be outdone in giving the ambassador a piece of their minds. Victor defended himself against criticism leveled at him for consorting with refugees. Since 1799, his oath of loyalty to the American constitution had exempted him from all duties of citizenship with regard to France, which was no longer his country. In these conditions, nothing prevented him from having good dinners with foreigners distinguished by their knowledge, their military and civilian talents, their misfortunes, or their reputation, without worrying "about what part of the world they are coming from or what party they belong to or have belonged to."[62] He also made clear that he had been invited to dinners with "several other Americans of all parties, in all likelihood because Marshal or General or if you prefer Marquis de Grouchy had used [his] house as his headquarters."

Hyde de Neuville did not accept his explanations and continued to reproach him and get worked up over the title of marshal wrongly given to Grouchy. He received a weary response from Victor, indicating his lack of interest in European concerns and pointing out the exaggerated importance he was attaching to the matter "in a country where almost none is attached to calling or not calling Monsieur de Grouchy Marshal."[63]

With time, the tension between the French diplomatic corps and the Duponts subsided, even if, at the death of the dynasty's founder, Hyde de Neuville again flew into a rage. Its cause was a laudatory article in the *National Intelligencer* saying that at Napoleon's first fall Dupont had "accepted the house of the Bourbons."[64] Was there anything more ridiculous and more insolent, sputtered Hyde, "than to have the house of the Bourbons _accepted_ by Monsieur Dupont and Company?"[65] Their father's death did nothing to change the two brothers' friendly feelings toward the refugees, particularly the Grouchys, who had become close friends. They were in a way rewarded when the gunpowder factory exploded, causing deaths and considerable damage. The marshal and his son the colonel, whom Victor had invited to a hunt at Eleutherian Mills, turned their sense of military organization for such circumstances to good use by containing the fire and saving some buildings. Hyde did not wax indignant at the newspapers' tribute to Marshal Grouchy.[66]

The help given the exiles by the Duponts de Nemours differed from that of Girard or Chaudron. Each had his own personality, his history, a particular relationship with France, but they all mingled with the refugees in the milieu of Freemasonry to which they all belonged.

Masonic Fraternity

Like many of his peers, ship captains and merchants, Stephen Girard was a Mason. Raised to the degree of master in 1788 during a stay in Charleston, he later had this ratified by the respectable Loge française de l'aménité no. 73 of the Orient of Philadelphia.[67] His subsequent activity in the French lodge is uncertain; on the other hand, he was a member of the Grand Lodge of Pennsylvania, since he is said to have given twenty thousand dollars for the reconstruction of Masonic Hall on

Chestnut Street after it was destroyed by fire in 1819. Did his status as "Brother" influence his attitude toward the exiles?

There is ample evidence of the solidarity of the Francophone Masonic community in Philadelphia, developed in the 1790s with the arrival of the refugees from Saint-Domingue who brought with them the island's Masonic traditions[68] and formed three lodges: La parfaite-union, La reconnaissance, and La Loge française de l'aménité no. 73. The third was the last to be created, after receiving the consent of the Grand Lodge of Pennsylvania in 1797. A great many of the founders and others on the membership list for the lodge's second year were Domingan refugees: Édouard George was the deputy master of ceremonies and Simon Chaudron the deputy orator.

Having become the official orator, Chaudron composed and delivered to his Masonic brothers the *Funeral Oration of Brother George Washington.*[69] By paying Masonic honors to the memory of the president, who had died on December 14, 1799, the French lodge hoped to dissipate people's prejudice against the Domingan refugees, victims of a wave of Francophobia due to the undeclared maritime war between France and the United States. The long, grandiloquent text was printed and addressed to President Adams, who, in his words of thanks, called it an exquisite piece of elegance. Fifteen years later, Chaudron took advantage of the commemoration of the anniversary of Washington's death to present to the unhappy French people, coming here to weep over their misfortunes, the monument that his admiration and gratitude had dared to consecrate to the memory of the benefactor of humanity, and to make of this funeral oration, published in the newspaper, the exiles' handbook. The eulogy began by urging the proscribed men to walk with respect upon the soil where Washington's ashes reposed, and closed by exhorting them to bless their refuge and to merit, by their virtues, to be numbered among his children.[70]

This literary introduction was followed by a fraternal banquet in the Philadelphia Freemasons' new hall. The citizens of the Franco-American Freemason community wanted to show their respect and esteem for the illustrious leaders whom the political revolutions of their country had forced into exile: Generals Grouchy, Desnoëttes, Clauzel, and senior member of the Council of State Regnaud, grand orator of honor of the Grand Orient of France,[71] and his son.

Though absent, Joseph Bonaparte had his place as former grand master of the Grand Orient of France. His father, Charles, the first Bonaparte to become a Mason, had been imitated by most of his family, including Napoleon.[72] Almost all of the imperial regime's civilian and military dignitaries were involved: at the beginning of the Empire, thirty-two military units had lodges, and half the marshals, hundreds of generals and admirals, and thousands of officers were Masons.[73] Grouchy, the exile with the highest Masonic degree, was also the Mason of longest standing,[74] having joined well before the elder General Lallemand.[75] The Philadelphia Masons' enthusiasm with regard to the outlawed men is therefore understandable. The Duponts received their invitation on April 27.[76] Organizers Monges, Ravesies, and Laussat asked them to come to the Masonic hall the following Thursday to honor the presence of Marshal Count de Grouchy and Generals Clauzel and Desnoëttes.

On May 2, 1816, more than eighty people sat down at the banquet table, with Peter Stephen Du Ponceau presiding.[77] They were served "with as much refinement as lavishness" through to dessert, when toasts were drunk to the health of France and the United States, their perpetual union, the independence of nations, brave men, liberal ideas, the memory of Washington, and the homeland. On the American side, the toast offered by Charles Ingersoll, attorney general of Pennsylvania, was particularly noted, and on the French side, the toasts of the heroes of the celebration, imbued with love for their homeland and the pleasure of finding themselves among fellow Frenchmen. Regnaud de Saint-Jean d'Angély's is representative: "To liberty, to prosperity, to the glory of hospitable America. May all French refugees, of all epochs, remember forever the generous refuge they were given here."[78] Between toasts there was a moving rendition of the "Marseillaise"; songs were sung in both languages; poetry was recited, celebrating the battles of Fleurus and Waterloo, the heroism of the Guard, the glory of Grouchy and Desnoëttes; and the generals were promised radiant and prosperous days on their soil finally rid of dark iniquity.[79] The evening was a success and broke up late. If a witness cited in the *Abeille* is to be believed, at no time was there even a whiff of partisanship; it was simply a republican celebration in a republican country. It is not certain, however, that his report would have convinced even the least fervent royalist.

Far from France, the French émigrés were thus indebted to their American and French hosts for a welcome that helped them withstand the shock of exile. Their hosts assured them of their solidarity and guided or sheltered them, but there were limits to this assistance. To succeed in their settlement on foreign soil, the émigrés had to get involved personally; that is, learn the country's language and find work that would give them financial autonomy.

INDIVIDUAL INTEGRATION

In May 1817 in Philadelphia, Du Ponceau presented an essay on English phonology to the American Philosophical Society.[80] This American of French origin had since childhood demonstrated astonishing facility in understanding and assimilating languages, most remarkably the English language, something rare enough in the France of the time to have earned him the ironic nickname "the Englishman" from his less gifted comrades.[81] From the universe of Shakespeare, he moved to practical use of the language in America, where the Prussian Baron von Steuben had taken him at the beginning of the War of Independence as his interpreter. He remained in the United States, where he acquired great renown as a jurist, author, and philologist. His remarkable integration never severed him from his roots, and he was often the natural adviser of merchants, commercial agents, and French diplomatic personnel. He took the same interest in the French exiles in whose honor he presided over the Masonic celebration in Philadelphia.

TOTAL IMMERSION

Du Ponceau was the French exception in the mastery of English; Hyde de Neuville, the ambassador, was the slow learner. A descendant of Scotch Jacobite refugees, Hyde inherited from his origins neither a gift nor a liking for English. Despite his earlier seven-year stay in the United States, he spoke it very poorly and was incapable of giving a written translation of a text or of answering speeches addressed to him in this language in the course of his official duties.[82] This did not mean that he was unaware of the problem or of the necessity of knowing English well in order to integrate better into American life. During his years of exile, he had found that the educational system left much to be desired. For lack of public schools, instruction was given by private institutions, often expensive, and by various religious groups for their own children. The sons of refugees from Saint-Domingue and Cuba had no access to these schools, and there were few French schools. In 1809 Hyde opened the avant-gardist Economical School in New York: it was coeducational, secular, bilingual, free to poor families, and open to children and adults. A great number of French people in the United States benefited from it.

Many resisted, however, favoring French despite their Americanization. John M. Chapron, who became an American citizen at the age of twenty, had been sent from Cap Français to a boarding school in Germantown, Pennsylvania, in 1800. The distance and the difficulties in Saint-Domingue did not prevent his father from judging the letters of a son whom he encouraged to improve his French.[83] Paternal advice continued through generations. Sixty years later, Chapron vaunted the merits of his native language to his granddaughters, not knowing whether they would be able to answer him in French: "It seems to me that if an American wanting to learn French thought of the fact that he uses a third of all French words when he speaks his own language, he would be more eager to express himself in this language. As I have said before, the French language is the language of the fair sex. It is a language with so many resources of expression and is so pleasant on the lips of a young lady; so try to have a good command of it and remember that the only way to succeed is by speaking."[84]

The French, even those who had emigrated and been naturalized young, remained strongly attached to the language and culture of their origins. Girard, who lived in Philadelphia for three-quarters of his existence, felt more at home on the Bordeaux street where he was born than on Water Street. He spoke French daily with his compatriots, Domingan refugee merchants, and people in his immediate entourage. At odds with written English, he continued to correspond in French and, in christening his ships the *Voltaire, Helvétius, Rousseau,* and *Montesquieu,* made a point of the fact that he came from the century of the Enlightenment when the French language had spread the ideals of his adopted country. The importance of French in the culture of the day gave some a feeling of superiority and even of condescension with regard to English. Thus Dupont, who in 1800 wrote a *Plan d'éducation nationale dans les États-Unis* (Plan for National Education in the United States), mentioned to Jefferson that he might translate the work into English himself, "greatly regretting that this *vigorous* but incorrect and not very philosophic *dialect* is the language of [his] country."[85] Dupont was therefore probably not surprised when, in 1816, he received the following letter in which Regnaud de Saint-Jean d'Angély, his former colleague in the Council of State, told of his difficulties integrating into life in the United States: "If I were to settle here, I would either acquire an inexpensive little property not far from a city, or, because of my son, start a business, or practice law, but this bitch of a language and the hodgepodge of English legislation followed in this country are an obstacle to the latter choice."[86]

Ignorance of English was a considerable hindrance for the French emigrants: "Whatever can they undertake in America when they must first spend a number of years learning the language of the country?" wondered the *Abeille*.[87] In the course of their continental wars, French officers had been able to familiarize themselves with the languages of the countries they occupied, except the only one that had eluded them—as had the country itself—English. Their detention in England gave some an opportunity to make progress in the language of their jailer. Ensign Louis Taillade had been captured on the *Cygne* in 1796 and taken to England, where he stayed more than three years. Applying for promotion when he returned, he successfully requested that his "slight advantage" in speaking and writing English "almost as well as French" be taken into consideration.[88] General Lefebvre-Desnoëttes spent a long time on forced vacation in Cheltenham, a fashionable little spa in southern England, where he was a society darling; he is said to have become a close friend of the Countess of Buckingham.[89] He knew English when he arrived in the United States, as did his second-in-command at Waterloo, the elder Lallemand, famous for his knowledge of several languages. Lallemand had wanted to be part of the last formation on St. Helena. By excluding him, the English had deprived the Emperor of the interpreter he lacked after Las Cases was deported, according to Marchand, who was no more fluent in English than his remaining companions.[90] Joseph Bonaparte, like his brother, knew Italian from birth, but he had gotten around the problem by engaging an English language interpreter in France.[91] However, by 1826, according to an American who visited Point Breeze, Joseph spoke English well enough to make himself understood.[92]

Neither the officers' training nor their careers had prepared them to speak English, but this was also true of civilians. Exile made them realize the importance of knowing foreign languages, and therefore of teaching them correctly. In his *Plan nouveau d'éducation nationale* (New Plan for National Education), written in Brussels, Garnier de Saintes stressed the necessity of developing quality in European instruction in the four major living languages, thanks to teachers who would have spent at least five years in the country whose language they wanted to teach.[93]

In America, this was for many all the greater a problem because their motivation was as poor as the means put at their disposal to learn the basics. Why learn English, rationalized those whose American stay was to be a mere parenthesis or whose age slowed the assimilation of a new syntax and vocabulary. Marshal Grouchy wrote to his wife that their young son, Lieutenant Victor de Grouchy, lacked "the facility and genuine will to learn English."[94] And Prefect Réal, in his sixties, wrote to Joseph that his "overly old ears" would never understand the language of the little group among whom he was living in Cape Vincent.[95]

The Philadelphia papers nonetheless advertised private lessons, by, among others, two Domingan refugees. Peter Stephen Chazotte,[96] an experienced teacher with a perfect mastery "of the principles and philosophy of this language," offered his services "to those of his compatriots newly arrived from Europe, who wish to learn the English Language."[97] Charles Carré offered the same services to "French gentlemen newly arrived from France" at his home on Mulberry Street after five o'clock in the evening, and invited parents to bring their children to his school.[98] For years he had, with his brother Jean Thomas Carré, a highly cultured language professor, and John Sanderson, the latter's son-in-law, run a "beautiful, spacious school, located at the head of the main street, in the healthiest and best ventilated part of the city."[99]

The system of private lessons had drawbacks: prohibitive fees and a prolonged stay in the city, where there were constant opportunities to speak French. Jacques Lajonie opted for complete immersion in a "little town just beginning to grow," Burlington, New Jersey, five leagues from Philadelphia.[100] According to Andrew Curcier, who recommended it to him, this town was "the most apt to give foreigners a taste for the English language," because people there spoke only English. But word must have gotten out, and an encounter with two Frenchmen who had come there for the same reason discouraged Lajonie. In July 1817, when he was in New Orleans, he preferred to resort to the services of a tutor who charged him fifty francs for ten lessons. But while he may have hesitated over the method, Lajonie could not insist enough in his letters on the absolute necessity of learning English, both for himself and those who would come to join him. His expatriate situation convinced him of the usefulness of foreign languages. After five years in the United States, he wrote concerning his son: "I shall wring his neck if he knows one word of Latin. English and Spanish are fine."[101] In any case, how could one do otherwise in the country where one has chosen to live and work? "Without English one cannot go into business."[102]

EMPLOYMENT

The foreign passengers disembarking in America fell into three classes depending upon their occupations.[103] The most numerous, labeled useful and productive, was that of the manual trades: agriculture shared the statistics with foods (grocers, butchers, bakers, and confectioners), construction (carpenters and masons), metals (smiths and jewelers), woods (joiners and coopers), textiles and leather (weavers, tanners, tailors, tawers, and shoemakers), and so forth. The second, useful but unproductive, was dominated, in a heterogeneous spectrum ranging from servants to the ambassador, by merchants and sailors, far outnumbering men of law, engineers, physicians, teachers, architects, pastors, and military men, The third, finally, entertaining and unproductive, grouped Ladies and Gentlemen with hairdressers, musicians, and dancing masters.

A perusal of the jobs listed on passports and passenger lists after 1815 shows that the immigrants' occupations did indeed fall into these three classes and the trades mentioned. In this they were, like their predecessors, destined for a country whose manpower needs were continuous, multiple, and varied. In spite of real difficulties of integration, it was possible to make a place for oneself in the United States.

An important feature of this French emigration, coming after a long period of restriction, was the presence of new elements: the military men and civil servants of the former Empire. Classified in the category of useful and unproductive people, their lack of preparation for a world so different from their own and that had no real need for them was total. Based on practical realities and the experience of the past twenty years of French emigration, the *Abeille* was doubtful of their chances of success and drew up a list of their handicaps.[104] The first stemmed from a birth and upbringing that were essentially urban, preparing them to wield the pen or the sword but not to earn their living manually. The second stemmed from age: the older men could not aspire to the legal professions, and the younger ones, even after a long period of study, would find themselves at a disadvantage compared with native-born Americans. The third stemmed from local particularities that often worked against the French: the army, in which they could have found employment, had no positions to fill; industry, in which they could have engaged, demanded a huge amount of capital that they did not have; commerce, finally, offered to a majority of honest Frenchmen only the short-term prospects of ruin or prison, poverty or shame, in a country where money and credit were king, the banks all-powerful, and immorality the order of the day in business. The newspaper wondered in what sort of endeavor "the unfortunate Frenchman could support himself." Its only answer was to suggest he flee the large American cities, where he could only fail. The picture was dark, and no doubt exaggerated, but in the particular case of the military men, it told the truth, with a few rare exceptions.

Frenchmen in the American Army. The French contribution to the War of Independence did not open the ranks of the American army to Napoleonic officers. The heirs of the "Insurgents," annoyed at having to share their founding feats of arms with adulated foreigners like Lafayette, were less ready than ever to welcome the French. Since their victory in the latest Anglo-American conflict and the subsequent general demobilization,[105] these thrifty nationalists were in no mood to be reinforced by officers who would overshadow them. And yet, the war against the English had emphasized deficiencies, such as border defenses incapable of preventing enemy intrusion and the burning of Washington. The solution for the government was to turn to foreign engineers with the competence and experience that their American counterparts lacked, though the latter denied this.

In 1816, planning to level the terrain for a canal linking the Hudson River with Lake Champlain, the state of New York had just set an example by hiring four Frenchmen, among them a lieutenant colonel in the engineers and a traveling naturalist.[106] The services of the former were so much appreciated that the French authorities were asked to allow him to prolong his stay.[107] When the federal government in turn sought a foreign military engineer capable of fortifying the Atlantic seaboard, far greater personal and strategic stakes made it difficult to get the American military to accept the idea.

While the American authorities were becoming convinced of this need, against the advice of officers in the engineers jealously guarding their prerogatives, General Bernard was developing his plan to emigrate in the Parisian salon of Destut de Tracy, whose son had been his classmate in the École polytechnique and whose daughter was married to George Washington Lafayette, the general's son. This family's empathy with American democracy and its heroes was total. Bernard, imbued with this pervading Americomania, wrote to Lafayette: "I have been looking toward this land that you have made famous by your exploits, and from which have come the ideas in which our generation was reared."[108]

The Restoration permitted the departure of Bernard, an officer whom Napoleon had considered one of his best engineers and had held in high esteem because he came from humble origins and was a self-made man.[109] Courageous and upright, he was one of the most remarkable exiles of the time. Born in 1779 in a prison in the Jura Mountains where his roofer father and servant mother were languishing because of debts, he was a brilliant student, continuing to the army's applied engineering school in Metz and then coming into his own in fortification work. In Antwerp his technical prowess so impressed the Emperor that, in 1813, he made him his aide-de-camp for topographical services and then a general and baron. Coming over to the king without enthusiasm, Bernard was again at the Emperor's side at Waterloo. Under house arrest after the defeat, he had no hope of resuming his career in the near future.

Writing to President Madison and Secretary of War William Crawford, Lafayette praised "the superiority of his military talents, the amiableness of his disposition, and the Patriotic frankness of his opinions."[110] Bernard gave a copy of the letter to the American military attaché in Paris and waited, without suspecting what a stir his candidacy would cause on the other side of the Atlantic. While the president was convinced that he was the ideal man to make the best use of congressional credits for a vast program of fortifications extending to the Gulf of Mexico, he had to begin with compromise before imposing his decision. The military establishment did not realize how far ahead the French were in this field, and in 1815 the success of the battle of New Orleans against the victors of Napoleon had convinced them that "they were the best soldiers of the universe."[111] The corps of engineers and West Point Academy were under the command of a young general in the engineers, brilliant and sure of himself, who did not want to be under foreign guardianship. Born at the close of the Revolutionary War, Joseph Gardiner Swift belonged to the new generation that had just forged for itself a nationalism of steel against England.[112]

In May 1816 William Crawford informed the American ambassador in Paris of Bernard's appointment at the rank of brigadier general if he agreed to be assistant to the head engineer. This condition was a pure formality; all work in progress was suspended until his arrival. Swift said he was mortified and his men rallied round him, finding this distrust insulting: it shamed the country, dishonored the corps, and disgraced its commander. Even Crawford's first secretary saw this recruit as an "exotic talent" and expected no better from his adventurer friends, raised in the School of Ambition and used to changing sides.[113] Weary of such chauvinistic resistance and pride that completely disregarded the national interest, Crawford sent Swift back to his school in West Point and awaited the arrival of the French general who, meanwhile, had accepted the offer extended to him. He did not speak a word of English.

On November 6, President Madison created the "Board of Engineers for Fortifications" and signed Brigadier General Bernard's appointment in his presence. The admission of a foreign officer into the military service of the United States, a first, aroused contrary reactions. The press reported it favorably,[114] but Swift disapproved of using a foreign officer in a key position and eventually resigned. He thus determined an American career that was exceptional in its duration, the scope of its task, and its unique character. No other French officer was ever given such responsibilities, only assignments of secondary importance, like those Guillaume Tell Poussin carried out in his wake.[115]

Bernard had made a name for himself. Poussin inherited his from the painter Nicolas Poussin, whose renown was a passport for integration equal to his letters of recommendation from Lafayette, Madame Dupont de Nemours, and the banker Lafitte, with a letter of credit of one thousand francs for his counterpart Girard. The Duponts received him for a week in Wilmington, and in Philadelphia Girard gave him advice and letters of introduction for a position as architect in Washington. After devoting himself to the study of English,[116] he began working for the architect Latrobe,[117] in charge of restoring the Capitol, but the rapid deterioration of their relationship prompted him to ask Girard to intercede for him with his friends in Congress,[118] and at the same time the president was petitioned for an army appointment. While touring the construction work at the Capitol, Madison met Poussin, noted his illustrious name, congratulated him on the quality of his English, and remarked that for a Frenchman, he spoke with a very good accent.[119] On March 6, 1817, Poussin was appointed captain in the corps of topographic engineers, as assistant and aide-de-camp to General Bernard.[120] They met at the Grand Hotel on Pennsylvania Avenue in Washington: "I found the general who at that moment was talking to our respectable compatriot Monsieur Parmantier. Scarce had my name been pronounced when he affectionately grasped my hands."[121] Bernard had found his interpreter, a collaborator, and a friend.

Paradoxically, the success of this duo brought home to the French the extreme difficulty of entering into the service of the United States. The two men were, in their respective areas of competence, very valuable recruits, additionally possessing a considerable advantage over the other émigrés of their level of expertise: recruiting them did not risk exacerbating the dispute between the French diplomatic corps and the United States, as we saw happen with Grouchy. They were there because they wanted to be, and in their short careers had no liabilities to live down. In spite of this, neither of them had been unanimously welcomed in the United States; Bernard had been targeted by the hostility of his American colleagues and Poussin by that of the architect Latrobe.

Preceded by the dissonant echo of their revolutionary exploits and handicapped by age, the former Convention members who came to America had not offered the services of their talents or attempted to exploit them for their own benefit. This vast country quite naturally offered other possibilities to those who wanted to seize them.

LAND AS RECOURSE

The principal possibility was to follow the advice of the *Abeille Américaine* and flee the corruption of cities where all could be lost, money and honor, for the country's interior where there was everything to be gained, regeneration of body and soul, the purity of new beginnings, happiness in agriculture. This was not a new idea. It had long been spreading from old, decadent Europe to the New World, where imagination stimulated by literature and physiocracy had situated the Garden of Eden and the center of the Golden Age.[122] In the second half of the eighteenth century in France, notably at a time when a special relationship was being formed between France and the infant United States, a great number of private and collective settlement projects appeared, extolling both the return to the land and the myth of a primal refuge where man could blossom in his natural state. The choice made by many soldiers of the French expeditionary force to remain after independence, along with the accounts of those who returned, created an irrational attraction to America where mirage was substituted for philosophic utopia, despite Benjamin Franklin's early warning "to those who would remove to America."[123]

GALLIPOLIS

In 1778, Bernardin de Saint-Pierre submitted to the French government a proposal for a refuge for its poorest subjects to the west of the English colonies where they would be given land to be shared out.[124] Of this plan, which was never executed, there remained the idea of a promised land in the West whose abundance and rich soil stirred desires. In 1784, the French publication of J. Hector Saint-John de Crèvecoeur's *Letters from an American Farmer*, situating it quite precisely on the banks of the Ohio, created a frenzy.[125] Before taking the plunge, people hurried to see the journalist Brissot, a big promoter of European emigration to the United States;[126] Doctor Guillotin, whose machine had not yet made him famous, planned to go settle with his family near Louisville or French outposts in the Illinois country; Benjamin Constant dreamed of farming in Virginia, and, ten years before his father, Victor Dupont made his first trip as part of a diplomatic mission.[127] These were signs heralding a major eruption: sciotomania.

The Scioto Company was born in Paris in November 1789, an offshoot of the Ohio Company of Associates founded three years earlier in the United States to engage in land speculation under cover of a colonization enterprise. Three million acres located between the Ohio and Scioto Rivers were to be sold on behalf of American wheeler-dealers who had no title to the lands, but only an option to purchase from the Board of Treasury. French buyers, knowing nothing of this, flocked from all parts of the country and all social classes, wanting to flee the Revolution and find freedom and exoticism in the American countryside. Some thousand apprentice adventurer shareholders began to arrive, beginning in 1790, in a country where their reception, and then their journey into the West, took place under appalling conditions that augured no good. These young men who had scouted the wilderness only in books were confronted by a reality that discouraged half of them before they even reached the town that had been built for them on the north bank of the Ohio, across from the mouth of its tributary, the Great Kanawha River.

Gallipolis, the city of the French, thus named in honor of its inhabitants, was a clearing in the heart of unbroken forest, planted with four rows of hastily put-together cabins to receive nearly five hundred colonists. Although they had survived multiple pitfalls since leaving France, most of them did not have the energy to cope with any more in a natural environment whose hostility was in stark contrast to the idyllic vision they had concocted in their daydreams, and which the plan's American instigators had carefully done nothing to correct. Indian aggressiveness, harsh farmwork, and the obligation to buy back their lands accelerated the mass exodus. Twenty years later, a census in Gallia County recorded that only 6.6 percent of its taxpayers were French. Born of an unprecedented real estate scam, the French colony had been short-lived, as had its earlier colleagues Asylum, Pennsylvania, Greene, New York, and New Bourbon, Missouri, other refuges from the Revolution.

Despite their fascination with the American mirage, most of the émigrés would gladly have been content to put Rousseau's and Quesnay's bucolic ideas into practice in their own gardens rather than in the depths of dark forests. Chateaubriand did not find, in his exploration of the New World, all the romanticism he was seeking and that his writings would suggest,[128] and Dupont de Nemours dropped his plan for an agricultural colony in Virginia corresponding to his physiocratic views. The émigrés who succeeded them during the Restoration, leaving their country for opposite reasons, were no more eager to answer the now more muted call of American nature. Recourse to the land, rather than a return to the land, met a vital necessity, freed of all sentimental contingencies. The banks of the Ohio, on the other hand, had lost none of their appeal.

THE BEAUTIFUL COUNTRY

The Ohio valley, a strategic link between Louisiana and French Canada in the days of New France, had retained its power of attraction for many émigrés who chose to settle near the Belle Rivière. The northern part of the state of Kentucky was thus singled out: in its capital of Lexington, Joseph Neef, a discharged soldier who had fought under Bonaparte, had opened a school centered on the principles of the Swiss pedagogue Pestalozzi;[129] in 1816, Joseph Alphonse,[130] newly arrived from France, had joined him, while waiting to get together with the naturalist Lesueur and the American philanthropic geologist William Maclure, his "patron," and the partner-patron of Neef and Lesueur, with whom he had signed a contract. The same year, Guillaume Cirode, twenty-four, a tawer from Nantes, was preparing for his family's arrival in Lexington,[131] where he met Antoine Dumesnil, settled there since at least 1812, a silversmith engaged in business and real estate transactions.

Louisville, located above the falls of the Ohio, had attracted people from Marseille and the surrounding region. Jean Antoine Tarascon, who had first gone to Philadelphia, then in 1801 to Pittsburgh, had, with his associate Berthoud, founded a shipbuilding concern and a large watermill in Shippingport, below the falls. He had been joined by relatives, Joseph Martin-Picquet, a merchant and former city worker in Marseille,[132] and Joseph Barbaroux, his nephew, a ship's chandler in Shippingport and in 1818 co-owner with Francis Honoré, a ship's captain, of a steamboat built in Cincinnati. As in Lexington, the teaching vocation of the French found expression in Louisville. In 1814, François Dusouchet, a veteran of General Leclerc's expeditionary force in Saint-Domingue, opened a dancing school and a military academy.[133] He was imitated in 1818 by Henry Guibert, whose dancing school was soon associated with the music school of his wife, Susan Arcambal, daughter of an emigrant French merchant. All invested in land and city lots, but this was only part of their activities. By contrast, André Lecoq Dumarselay had no other business; he was a landowner of independent means, ready to go into trade, but especially interested in speculation. One must wonder what prompted this man, who like his wife, née Bonamy, came from a very old bourgeois family in Nantes, to settle so far from home with his family, particularly since he was over sixty years old.[134] Three letters written to Stephen Girard between 1817 and 1819 show Lecoq brimming with real estate dreams, keeping an eye on the movement of properties and their constant appreciation, dealing in thousands of acres, and conjuring up the founding of the cities of "Stephentown" or "Girardville" on the Ohio or the Mississippi in order to associate Girard with his projects, which the latter certainly never took seriously.[135]

Joseph Lakanal's ambitions were more realistic and in line with the thirty thousand francs in gold that the sale of the priory of Villarceaux northwest of Paris had brought him.[136] Arriving with his family in early March 1816, he had remained in New York long enough to create the Institute of America, a veiled reference to the Institut de France from which he had been excluded along with Joseph Bonaparte, rightfully a member of the new institute, of which Daniel Lescallier, the Emperor's former consul general to the United States, and Regnaud de Saint-Jean d'Angély were also members: "We acted, far from our old homeland, as we had on the Champs-Élysées," wrote Lakanal, who had, however, not emigrated to

America to sit around and talk.[137] According to the *Abeille,* which reported that he was in Philadelphia in April, his aim was to acquire, at a modest price, some one hundred acres in Pennsylvania, at one or two days' journey from the capital, and to devote himself to the sciences that he had enriched "with his inestimable productions."[138] Unlike most of the other exiles, Lakanal had arrived in the United States with a definite project, albeit one which he had not planned to carry out on the east coast, as his letter of introduction from David B. Warden, the United States' consul general in Paris, indicates: "Mr. Joseph Lakanal . . . who enamored of American freedom, bids adieu to his native country, and proposes to spend the remainder of his days as a peaceful cultivator, on the banks of the Ohio. Favored by fortune, he has collected a considerable sum for the purchase of lands."[139]

Lakanal's choice to leave the east coast was due to Warden's encouragement, singing the praises of the Ohio valley—"Beautiful country! Beautiful country!" he exclaimed—and to a work by French botanist and traveler François André Michaux. In 1804 Michaux had published the account of his canoe trip down the Ohio from Wheeling to Louisville via Marietta and Gallipolis and predicted that within twenty years its banks would be the most densely populated and busiest region of the United States.[140] After traveling through Pittsburgh and Cincinnati, collecting plants and examining fossilized bones of mastodons, Lakanal stopped, in 1816, on the north bank of the Ohio, at Vevay. The local newspaper took pride in the arrival of this prestigious person and reported that he had acquired a property two miles above Vevay, but on the opposite bank, in the state of Kentucky.[141]

His four-hundred-acre estate, half of which was already under cultivation, was located in Gallatin County above Port William, at the mouth of the Kentucky River: "The Ohio bathes the feet of my rather pretty chalet which I am working to improve. I have real French people as neighbors," he wrote to Victor Dupont on July 5.[142] Six weeks earlier he had informed Jefferson that he had settled in a pleasant retreat where he was going to divide his time between the cultivation of his lands and the implementation of his great project: writing the history of the United States based on the many documents he had been collecting for the past ten years.[143]

Agriculture was not his priority. Although the breadth of his knowledge enabled him to face this new challenge,[144] after an extremely intense quarter century in France, he had to adapt to the peace and quiet of his lot as an American farmer and perhaps endeavor to become a second Crèvecoeur: "Beautiful nature, which he courted for thirty years in old Europe, is going to reward his loving constancy by showing herself to him, in America, in her virginal magnificence," exclaimed the *Abeille* in delight.[145] Certainly he had courted nature, but it was chiefly that of the Royal Botanical Garden in Paris, which he had been instrumental in converting into a National Museum of Natural History in 1793.[146] Jefferson was conscious of this when he answered on July 30: "The affliction of such a change of scene as that of Paris for the banks of the Ohio, I can well conceive."[147]

Nevertheless, Lakanal liked the pleasant setting and the fact that on the opposite bank, in southern Indiana, stood the village of Vevay, founded two years earlier by Francophone, republican Swiss winegrowers. On July 4, 1816, the fortieth anniversary of the Declaration of Independence, he had gone there to attend a commemorative meeting of three thousand patriots who drank a toast to the exiled Generals Clauzel and Lefebvre-Desnoëttes. The future looked good: "Everything has changed there in the last 10 years, and towns prosper where formerly the boundless forest reigned. It is truly a romantic country. Vevay, whose existence is not known beyond the Alleghenies, is nonetheless an important city. See the note about it in the newspaper printed there. It is universally thought in France that the American soil is not congenial to the vine, and there exists in Vevay a 600-acre vineyard that produces abundantly wine as good as that of Perpignan. The current price is 40 cents a bottle."[148]

Dated July 1816, this is the first known reference of an exile to the vine and winegrowing in America. Wine was not yet being considered as an economic resource for the French, but the editors of the *Abeille* had no doubts as to its place in the development of the United States. On May 9 the newspaper had published a report on the country's progress in a great many areas and foresaw that government encouragement would open prospects in the joint expansion of commerce, agriculture, transport, and demography.[149] The report cited, among other examples, the fact that the new state of Indiana had joined the Union, commenting: "The crops of Europe, of the East Indies and even the vine are already prospering in this fruitful climate that will attract new emigrations, and for which educated, hardworking Frenchmen will leave Europe and come to cultivate a hospitable land that will not be ungrateful like their homeland." The products of the interior would be shipped, according to the landowners' wishes, either to New Orleans or New York, to be sold there or sent on to South America and Europe. Rather than wait impatiently for an improbable return to France, the exiles should be encouraged by all this to settle permanently.

III
Alabama's Exotic Roots

7

Grape Harvests in America

The mulberry tree grows very well there. Cotton fibers are thick and silk very strong. In the two Carolinas, Georgia and Florida there are rice fields. They formerly traded in cotton. Fog and rain prevent wine-growing.

—Napoleon Buonaparte, ca. 1786–89

We hear of the conversion of water at the marriage, in Cana, as a miracle. But this conversion is, through the goodness of God, made every day before our eyes. Behold the rain, which descends from heaven upon our vineyards, and which enters into the vine-roots to be changed into wine; a constant proof that God loves us.

—Benjamin Franklin

Lieutenant Buonaparte's reading notes prove that his interest in the United States, a new but harsh country, began very early. In France, he wrote, four acres of land are enough for a living; there over forty are needed, and fishing besides. There is plenty of wood, but it would not pay to export it. The fur trade is declining. Tobacco grows well in the central part of the country, but it depletes the soil. On the other hand, rice grows well, cotton fibers are thick, and silk is strong. As for grapevines, only the climate prevents growing them. "This summary of America's products comes from a letter from Monsieur Kerguelen and seems quite inaccurate," concluded Buonaparte, without trying to ascertain what was true and what was not. Thirty years later, Napoleon would have been able to see for himself, surrounded by his banished soldiers. "If I had gone to America, I would have farmed, I would have cared for my garden, I would have taken in some old remnants of my army who would have come to be with me and we would have lived together," he is supposed to have said to Dr. O'Meara on St. Helena.[1] The veterans were there all right, but without him. They did not read his notes, but listened to the advice of those who were encouraging them; otherwise they would not have grown grapevines and olive trees in places where this had never before been tried.

Exiled military men who wanted to become soldier-farmers considered the growing of olives and wine grapes a noble task with symbolic values going back to antiquity. Associated with wheat farming in the Mediterranean area, olives and wine grapes had been grown by veterans of the Roman army in their retirement. By simply replacing wheat with cotton in this trilogy, warriors who had followed the eagle standards of two great empires came together at a distance of two millennia. This is how General Lefebvre-Desnoëttes had seen it. In September 1816, preparing to go to the banks of the Ohio and Mississippi to find suitable land, he wrote to Stephen Girard: "We want to be independent of events and if we can no longer serve our country, we still want to be useful to society as well as to ourselves and our families by working in the fields as did the former Roman warriors and as we find models in this very country."[2]

Aside from the link with this glorious imperial past, the vine and the olive tree were associated with essential virtues that Simon Chaudron expressed in an apologue, dedicated to émigrés of all classes, in which the vine (representing abundance) and the olive (representing peace), by their example and their counsel brought an end to a dispute between the oak (strength) and the laurel (valor):

"And in a harsh exile, we share
[say the olive and vine to the laurel and oak]
Both our weal and our woe.
Of the work of proud error beware.
Out of love and need, become equals, so
Your strength and courage you can pool with care.
Imitate our ties, and our work also,
Make a pact as a rival—not as a foe,
And may esteem be its guarantee fair."
The Oak and the Laurel heard the words of this pair:
The counsel of a sage is a treasure rare.
Eternal aid and love they swore;
They took as companions the inseparable ones;
And thus were united, by durable bonds,
Abundance and Peace, Strength and Valor.[3]

The appeal to unity and solidarity was clear: Napoleon's officers knit to the plow in the face of adversity. But the significance of the vine and the olive was not merely symbolic; along with the mulberry tree, they had their place in the United States—in its soil as in its economy.

THE FIRST WINEGROWERS

Contrary to its importance in Europe, on the other side of the Atlantic the vine was not really part of the agricultural scene and even less of the customs of its inhabitants, whether of long standing or the latest arrivals. The American continent did, of course, have grapevines and was consequently suited to their growth and cultivation: one of the very first place-names designating a North American region was "Vinland" or "Wineland," a name given it by Lief Erickson in the year

1000 because of the wild grapevines he found there and the grapes he brought back to Greenland, from where he had sailed. Despite its ability to acclimate in places that would seem hostile to its growth, the vine was much more comfortable in the south, and from the sixteenth century onward the Spaniards continued to develop its cultivation, first in their South American colonies, and later in North America: the Rio Grande valley, New Mexico, and California.

Between these northern and southern latitudes, the north Atlantic shore of America was in an intermediary position. The Florentine navigator Verrazano noticed wild grapevines on the North Carolina coast—the same variety, later called "scuppernong" or "muscadine," as that from which French Huguenots made the first American wine in 1562 in Fort Carolina, Florida. French and Spanish vines arrived in the British colonies a century and a half later, but attempts in Rhode Island, Massachusetts, and New York to grow them and make wine from them most often failed. Success was limited, producing only small quantities, and their efforts were emulated only by Huguenot communities. Other attempts followed, marking stages in the history of the vine in the United States,[4] preceding the moment when the French émigrés came to want to plant vines and olive trees, and contribute to the validity of the statement that "each vineyard region in America has its own romantic story."[5]

THE HUGUENOTS IN NEW BORDEAUX

An essential stage unfolded in the second half of the eighteenth century between southwestern France and South Carolina under the leadership of Pastor Jean Louis Gibert, whose Calvinist intransigence and zeal in preaching the sacred word in Saintonge and neighboring areas managed to bring him to the attention of the constabulary, which wanted to see him hanged, and also to that of his fellow Huguenots, who reproached him with harming their cause at a time when local authorities were beginning to ease up on Protestants.[6] He chose to go live his militant faith in America, an Anglican land of English colonies in the process of being settled, with other Huguenots whom he would guide according to his principles. In order to carry out his plan, he needed the permission of the colonial power and the adherence of emigrants, still ignorant of their fate. Gibert petitioned the government in London for land on which he and his disciples proposed to live together and apply themselves principally to the cultivation of vines and silkworms. The king granted thirty thousand acres in South Carolina to the mere 132 people who accompanied Gibert from Plymouth to Charleston in 1764.

Each colonist who swore allegiance to the king received one hundred acres, plus twenty acres for each dependent, a small allotment in the town to be built, four acres for planting vines and olive trees, all tax- and rent-free for ten years. The town was christened New Bordeaux because most of the colonists came from the province of Guyenne and its capital was both the emigration port for the New World and the benchmark for all things viticultural. Many came from Protestant villages in the area of Sainte-Foy on the Dordogne where Lajonie was born some time later; they furnished twelve winegrowers to fulfill the clauses of the contract stipulating the colony's vocation.

The site selected for them, a buffer zone between a plantation[7] and Cherokee territory, was in an unhealthy and uncleared area. After difficult beginnings, marked by alternating fevers and food shortages, the colony purchased slaves and began to see light at the end of the tunnel, but without Gibert, who had moved into silk production with a Charleston merchant, the son of an émigré from La Rochelle. Confident in the success of their Silk Hope plantation and in financial support from London, the associates opened two spinning mills,[8] whose reputation for excellence did not survive the War of Independence.

Winegrowing had no better luck. Gibert had been replaced as leader of the colony by Dumesnil de Saint-Pierre, who had come there by accident with forty French and German Protestant emigrants. He had remained, impressed by the qualities of a region "perfectly adapted to the cultivation of the vine, silk, and indigo" and by a climate "about the same as in Marseille, [but] with infinitely richer soil."[9] Under his leadership, winegrowing picked up again: he planted vines the very first year and, following his example, the colonists imported twenty thousand French plants. Convinced that he was succeeding in America, Saint-Pierre returned to Europe to announce this. In England he published, for those who might want to imitate him, *The Art of Planting and Cultivating the Vine:* "To promote this important object shall be the grand aim and business of my life," he wrote;[10] in France he procured the best vine plants and hired skilled workers. He returned to New Bordeaux with several dozen new French and German colonists, who received land as soon as they arrived and gave a boost to the viticultural industry. In 1776 the botanist Bartram visited Saint-Pierre's Orange Hill plantation on the Savannah River, and found it flourishing. But the Indians killed Saint-Pierre and the orphaned vines collapsed all the more quickly because the settlers enlisted in the revolutionary army, abandoning the colony, which then ceased to develop. Soon only a memory of the settlement remained on maps: "the old French Town."[11] New Bordeaux's adventure had lasted twenty years. That of the Huguenots continued, without grapevines and silk, but with an economy turned toward cattle raising and more advantageous crops such as wheat, tobacco, hemp, indigo, and then cotton.

The birth of the United States signed the death warrant of still another viticultural venture, an Italian one that opened long-term prospects thanks to the perseverance of Jefferson, who had in a sense sponsored it.

ITALIANS AT MONTICELLO

The Florentine Philip Mazzei was every bit as enterprising as Gibert and Saint-Pierre. Born into a family of merchants and distillers, he spent twenty years in London as an importer of wines, lemon tree shoots, olive oil, candied fruits, and silks. He became acquainted with Benjamin Franklin and businessman Thomas Adams, who suggested he settle in Virginia, where they thought the soil was suited to the flora of the Mediterranean region. Mazzei conceived the ambitious plan of transplanting thousands of vines, olive trees, and other fruit trees. Unable to attract investors, he sailed from Leghorn, Italy, to Chesapeake Bay in 1773 with a reduced number of vines and only ten vine growers.

The plan was for Mazzei to go to the lands that Adams had set aside for him near his own in Augusta County, but en route he stopped at Monticello, where for several years Jefferson had been devoting himself to his passion for agriculture. Jefferson showed him a fifty-acre cleared field for sale near his own lands and two thousand more adjoining it that had not yet been cleared. The vine growers got to work. Mazzei was delighted with the climate and the soil, which he found superior to those of Italy, both for the European vines and, more particularly, for the indigenous grapes of which he and his men had found an infinite variety: "When the country is populated in proportion to its extent, the best wine in the world will be made here. . . . I do not believe that nature is so favorable to growing vines in any country as this."[12] He was far more reserved, however, when it came to acclimating olive and lemon trees. But this could be considered a new Garden of Eden where everything would blossom, provided one had the means. This meant the creation of an association "for the purpose of raising and making wine, oil, agruminous plants and silk," with shareholders as prestigious as George Washington, Thomas Jefferson, George Mason, and Peyton Randolph.[13]

Revolutionary fever carried off the vineyard. While Mazzei was in Europe as agent for the state of Virginia, his estate was rented to a prisoner of war, a Prussian general whose horses, grazing in the vineyard, ruined five years of effort in one week. Finally, it was sold. Jefferson remarked: "and thus ended an experiment which, from every appearance, would in a year or two more have established the practicability of that branch of culture in America."[14] His use of the word "experiment" reveals at what stage winegrowing was in the United States, but short though it was, Mazzei's venture was decisive: it whetted Jefferson's interest in viticulture and wines, oriented his taste, and reinforced his idea that the country could rely on its own resources. The architect Latrobe was also convinced of this, praising Mazzei: "The time is already approaching when our vines . . . will spread your name and gratitude over a great portion of our country."[15]

THE WHITE HOUSE CELLARS

Jefferson, a man with a wide range of interests and an admirer of the French physiocrats, was trying his hand at winemaking at Monticello when Mazzei moved in next door. Mazzei did not succeed in turning him into an accomplished winegrower, but did guide his first steps toward oenology, in which Jefferson became the first great American expert.

The Ambassador of French Wines in the United States. Jefferson began with a palate revolution, rejecting the favorite wines of English colonial power at the same time as the power itself.[16] Mazzei's influence led him to prefer the more delicate French and Italian wines to those of Madeira and Oporto. His appointment as ambassador to the court of Versailles in 1784 completed the evolution of his taste toward these wines as he dined at elite society tables and traveled throughout Europe on the eve of the French Revolution.[17] No important viticultural region escaped Jefferson's exploration. His interest was threefold: gastronomic, scientific, and economic. The notes he left attest, in technically appropriate language, to the pertinence of his observations and the breadth of his knowledge.[18] Later, he took advantage of his two terms as president to extol the culinary excellence of France and the quality of its wines. Guests who wrote of dining with him all mention the profusion of the dishes and the magnificence of the champagne.

For Jefferson, wine was the superior beverage, not only for its flavor and bouquet, but for its medicinal properties and social virtues. Drunkenness was a mortal vice in countries accustomed to hard liquor, whereas it was unknown in those where wine was the everyday drink. An informed connoisseur and moderate drinker who diluted his wine with water, Jefferson saw this as such an important question of public health as well as economics that he deluded himself as to his compatriots' ability to imitate him. Lighter wines, competitively priced, at first imported from Europe and then produced nationally, should be substituted for the sale of the cheap grain alcohol that was ravaging American livers. In any case, integrating wine into Americans' dietary habits would unquestionably be an upset leading to progress, by accelerating the decrease in whiskey consumption, a national scourge. To succeed, it was necessary to set an example, convince the politicians, enact a tariff policy favoring the importation of foreign wines, and, in the long term, aim at viticultural independence for the country by harvesting its own grapes.

Given the eclecticism of his taste, ranging from table wine to Château d'Yquem, it was not difficult for Jefferson to set an example. With the amount of money he was able to devote to his presidential cellar, it was no more difficult to dazzle the congressmen who passed through his home three times a week. Thanks to his oenological erudition, he was a highly regarded counselor to all the presidents, both those who preceded and those who succeeded him.[19]

Washington, who with Jefferson was an associate in Mazzei's company and possessed a small vineyard at Mount Vernon, had always been interested in wine and considered its use a rule of hospitality. Drawing on knowledge acquired in France, Jefferson introduced him to less-known wines like the muscats that had impressed him at Frontignan in Languedoc. At about the same time, Jefferson advised John Adams, the American ambassador in London, and Gouverneur Morris, who joined him in Paris and was often his guest despite their political differences. When he was president, Adams served wine at his table with the conviction that it was "a more convivial offering at State functions than hard spirits."[20] Other American leaders were interested. Rush, Hamilton, and Burr were shareholders in the Vine Company of Pennsylvania, founded by Pierre Legaux, a French émigré who grew grapevines along the Schuylkill River northwest of Philadelphia.[21]

Jefferson had repeated exchanges on agricultural, horticultural, and viticultural questions with his successors: first James Madison, then James Monroe, whose property of Ash Lawn near Monticello was also planted with vines. A month after moving into the White House, Monroe received a long letter giving, among the French and Italian wines that Jefferson had just received and tasted, the list of those worth ordering, with comments as varied as they were precise concerning the price and the best years. Jefferson particularly recommended wines that were little known and produced in small quantities, like the Bellet, from the Nice region, "the most elegant every-day wine."[22] Jefferson knew that the vogue for a wine could be a matter of fad and fashion and that a great name on a label was not everything. Anonymous small proprietors in regions other than Burgundy and Bordeaux were producing simple, inexpensive, but excellent wines that could be imported into the United States. This shows how convinced Jefferson was that wine had its place, not only in the White House, but in the country as a whole.

Import, but Produce. As president, Jefferson had endeavored to give concrete expression to his ideas by adopting favorable customs measures adapted to the wines that he wanted on the American market. In 1807 he confided to Albert Gallatin, his treasury secretary: "I am persuaded that were the duty on cheap wines put on the same ration with the dear, it would wonderfully enlarge the field of those who use wine, to the expulsion of whisky. The introduction of a very cheap wine into my neighborhood, within two years past, has quadrupled in that time the number of those who keep wine, and will ere long increase them tenfold. This would be a great gain to the treasury, and to the sobriety of our country."[23]

Just as his tastes had evolved, so did his ideas on whether it was better to import wine than to produce a national wine with indigenous grapes. Jefferson's viticultural work at Monticello had taught him that, never safe from bad weather and disease, the art of cultivating grapevines was difficult and required the greatest vigilance. His travels through the French vineyards in the spring of 1787 had made him apprehensive about the problems of underproduction and overproduction that regularly agitated the viticultural world. That very year, France had suffered a serious loss of sales due to lowered consumption, from which Jefferson had drawn lessons: "The culture of the vine is not desirable in lands capable of producing anything else. It is a species of gambling, and desperate gambling, too, wherein, whether you make much or nothing, you are equally ruined. The middling crop alone is the saving point, and that the seasons seldom hit. Accordingly, we see much wretchedness among this class of cultivators. Wine too, is so cheap in these countries [of Europe], that a laborer with us, employed in the culture of any other article, may exchange it for wine, more and better than he could raise himself."[24]

In the same letter, Jefferson did not rule out the idea that one day wine could be produced in the United States, particularly when population growth resulted in a labor surplus. "[Wine] is a resource for a country the whole of whose good soil is otherwise employed," he wrote, "and which still has some barren spots, and surplus of population to employ on them. There the vine is good, because it is something in the place of nothing. It may become a resource to us at a still later period; when the increase of population shall increase our productions beyond the demand for them, both at home and abroad. Instead of going on to make an [*sic*] useless surplus of them, we may employ our supernumerary hands on the vine."

If he concluded then that this time had not yet come, twenty years later, in 1808, he thought the same. The demographic situation had not changed enough to furnish the manpower needed for winegrowing without creating a shortage elsewhere. It is true that the population, from a very small beginning, had quadrupled, but at the same time the Louisiana Purchase had doubled the country's area. Under these conditions, Jefferson held to his arguments. The country should not risk losing everything on a crop whose outcome was uncertain, but should produce dependable food crops that it could profitably exchange for foreign wines: "We could in the United States, make as great a variety of wines as are made in Europe, not exactly of the same kinds, but doubtless as good. Yet I have ever observed to my countrymen, who think its introduction important, that a laborer cultivating wheat, rice, tobacco, or cotton here, will be able with the proceeds, to purchase double the quantity of the wine he could make."[25]

Despite his preference for importation, Jefferson had never ruled out the idea of national production from vines native to American soil. Shortly after Jefferson's first inauguration, Pierre Legaux sent his congratulations and offered him thousands of vines to be planted in Virginia. Legaux sold wine made from his French grapes, but also wine obtained, he said, from the variety called "Cape of Good Hope" because, thought he, they were imported from there. In fact, this was the indigenous variety discovered by John Alexander, John Penn's gardener, near the Schuylkill River in Philadelphia and which bore his name.[26] In 1802 Legaux sent specimens to Monticello and Jefferson had them planted.

Beginning in 1809 and for fifteen years thereafter, Jefferson corresponded with New York horticulturist John Adlum, who shared with him the desire to promote the progress of national agriculture, particularly its viticultural branch, in which he too was a pioneer.[27] Adlum thought that grapes capable of producing good wine could be found in every state of the Union,[28] and Jefferson believed that the culture of indigenous varieties should be encouraged, instead of wasting time with foreign vines whose acclimation would take centuries.[29] In 1811 Jefferson reiterated this confidence in American nature to French agriculturist John Dortie: "Wine being among the earliest luxuries in which we indulge ourselves, it is desirable it should be made here and we have every soil, aspect and climate of the best wine countries, and I have myself drank [sic] wines made in this state & in Maryland, of the quality of the best Burgundy."[30]

Despite some unexpected successes in the northern part of the country,[31] Jefferson's hopes were pinned more on the south, particularly North Carolina, a pioneer state in the production of an "exquisite wine," the Scuppernong, which the best tables of Europe would distinguish for its aroma and crystalline transparence.[32] North Carolina's viticultural feats had been reported in 1811 in an article published in the *Raleigh Star*. The author stated that it was his duty to give the results of his inquiry into "this small but very interesting branch of our infant manufacturers."[33] He specified that Washington County alone had produced thirteen hundred gallons of wine from indigenous grapes of the variety he called "large White Grape," and to which the newspaper editors gave the name of the Scuppernong River, on whose banks this variety had flourished from time immemorial.

In addition to his interest in producing indigenous grapes on their soil, Jefferson felt that conditions in North and South Carolina were ideal for receiving new plants with commercial value. During his travels in Europe, he had selected all those he thought could acclimate there, and had seed rice and olive saplings from Aix-en-Provence sent to South Carolina, and muscat vines to the chairman of the state's Agricultural Society. The soil and climate assimilated the rice and the vines, but the earth, being too rich, rejected the olive trees, which Jefferson considered one of the most precious gifts of heaven and the worthiest plant to be introduced into America. If there were still olive trees in the Carolinas, he later observed rue-fully, they were merely garden curiosities.[34]

After struggling so hard to enrich his country with plants profitable to the health of its inhabitants, its agriculture, and its commerce, Jefferson grew weary, in his old age, of results that disappointed his expectations: "In the earlier part of my life I have been ardent for the introduction of new objects of culture suited to our climate, but at the age of 72, it is too late," he wrote in 1815 concerning vines and wine.[35] Two years before his death, although he repeated that America, thanks to its range of climates and soils, could produce wines as good as those of Europe, Jefferson moderated his enthusiasm: "The culture [of the vine], however, is more desirable for domestic use than profitable as an occupation for market."[36] In other words, after completely burying the olive's chances of success, his verdict on the vine was disillusioned.

Now, it was at precisely this moment in his waning life that the French refugees in the United States appealed to Jefferson, asking for his support and wise counsel. Jefferson gave his blessing without becoming further involved, and especially without warning them against an enterprise whose pitfalls he knew better than anyone—as much from a personal point of view with his fruitless attempts at Monticello as from an official point of view with a Swiss attempt he had supported, while president, on the banks of the Ohio.

NEW SWITZERLAND

The examples of Gibert, Saint-Pierre, and Mazzei have shown that the English colonies in America had had their pioneers of viticulture. Motivated by a will to succeed in acclimating grapevines and other foreign plants like the olive and mulberry, they had applied their expertise and benefited from advantageous land and fiscal policies. The results were disappointing, but they were due to extraneous events, so the United States supported new attempts. Early in the nineteenth century, Switzerland took over from its European neighbors and predecessors and gave new momentum to American winegrowing.

THE FIRST VINEYARD

Though he was the son of a pastor and a Calvinist like the founders of New Bordeaux, it was not religion that motivated Jean Jacques Dufour to go to the United States, but his reading of Swiss newspapers.[37] These reported that the French who had gone to the aid of the revolutionaries in America complained of the lack of wine in a land of abundance. The idea of cultivating the vine on a large scale and making his fortune there began to form in the mind of this young boy, born in 1763 in the midst of the vineyards of Vevey on Lake Geneva. He served his apprenticeship, developed his own vineyard, and waited until he was thirty-three years old before sailing from Le Havre to Philadelphia in 1796.[38] He spent the next two years traveling throughout the country, applying himself to studying the climate and the places where the vine had been or might be grown.

Dufour journeyed into the great West, traveled through Illinois, and crossed uninhabited areas, going as far as Kaskaskia, Cape Girardeau, and Saint Louis on the Mississippi. Of the sites that seemed favorable, he preferred the banks of the Ohio, below Cincinnati, but circumstances temporarily diverted him and he turned to Kentucky, where Lexington merchants, backed by John Brown, apostle of western development and the state's first senator,[39] had proposed he come to create with them a society promoting winegrowing.[40] In 1798, Dufour and his partners purchased 630 acres of land above the place where Hickman's Creek flows into the Kentucky River, at the extreme southern end of Jessamine County. The society's headquarters were located in Lexington, and Dufour claimed for it the honor of being the first to get winemaking off the ground in the United States.[41] Henry Clay, James Brown's lawyer brother-in-law, who was just beginning his career as statesman, was one of the subscribers whose national career was to influence the course of this story.

Dufour returned to the east coast and came back with fruit trees and ten thousand vine cuttings from at least thirty-five different varieties—especially Madeira and Alexander—bought from Legaux in Philadelphia. He planted them on the first five cleared acres of his land: The First Vineyard. He now sent for his brother Jean Daniel and five half brothers and half sisters born of his father's second marriage.[42] On January 1, 1801, they got under way, along with ten other people from the canton of Vaud, among them Morerod, a former soldier said to have crossed the Alps with Bonaparte and fought at Marengo; François Louis Siebenthal, a widower; and his son Jean François (who soon married Jeanne Marie Dufour, and was joined much later by his younger brother Jean Louis).[43] They sailed from La Rochelle, reaching Norfolk in April. In July they arrived at the vineyard of which they were joint owners. They settled onto their allotments, planted the vines they had brought, and were filled with hope when they harvested their older brother's grapes in 1803. The yield and quality were so impressive that they sent a messenger to Washington with the good news. The wine was presented for sampling to a congressional committee by the president of the United States in person, in the name of the Kentucky Wine Industry Association and of Dufour, whom Jefferson supported. The previous year, he had signed a law granting the lands discovered by Dufour on the Ohio in southern Indiana Territory, with better terms than those generally given colonists, even since the adoption of more liberal land legislation.

Congressional Aid

The acquisition of land, unregulated by royal legislation in colonial days, had made of speculation a "national commerce":[44] all American capitalists engaged in it, above all the founding fathers, who were great land-grabbers. The federal government wanted to put things to rights and subjected management of the public domain to the Land Ordinance of 1785, which was supposed to favor colonization over speculation. In fact, colonial practices resumed even more to the advantage of eastern wheeler-dealers. After military success against the Indians in the Northwest Territory and the subsequent influx of pioneers, new land legislation was necessary. In 1796 an initial law was designed to stabilize and clarify the conditions of transfer of public lands, at auction, for two dollars an acre, all payable within the year and geometrically divided into townships of six square miles, half of them to be composed of thirty-six indivisible sections of 640 acres each. The law was strict but poorly enforced.

The westward expansion of the United States made imperative the adoption of a land policy better adapted to the interests of the colonists who wanted to settle there, and William Henry Harrison played a decisive role in this respect by promoting the creation and development of Indiana, of which he became the first governor. In July 1800, sixteen years before becoming a state, the Indiana Territory was born, along with the neighboring Ohio Territory, in a division of the former Northwest Territory. This immense space, bounded on the south by the Ohio River and on the west by the Mississippi, had Vincennes as its capital and a white population of six thousand inhabitants under the authority of Harrison, a former military officer who had become secretary of the territory, and then its first delegate to Congress where he chaired the public lands committee. Therefore the land law passed by Congress on May 10, 1800, bears his name: the Harrison Land Act.

The law was a compromise between the needs of the treasury to increase its revenue by the sale of land and the necessity of encouraging large-scale colonization. The price of two dollars an acre was maintained, but the size of acquisitions was lowered to 320 acres and credit went from one to four years. Results were so spectacular that the law, though subsequently considerably modified, remained the model for land legislation until 1820. As for Governor Harrison, he continued to work to open his territory to white settlement, and signed many treaties with the Indians to obtain their lands. The Swiss presence in southern Indiana owed him a great deal, even though the conditions imposed on the colonists, however advantageous they may have been, were still beyond the means of Dufour, who received a bit of a boost from Congress.

Dufour's Petitions. Having invested in his Kentucky vineyard, Dufour did not have sufficient reserves to acquire the coveted lands in Indiana. He then had the idea of addressing a petition to Congress and having it presented by Albert Gallatin, his compatriot. Born into an old Geneva family that provided Swiss guards who, in the words of Voltaire, got themselves killed for France "from father to son since Henry IV,"[45] Gallatin had broken with tradition and preferred America, where he had emigrated at an early age to indulge his Rousseauist reveries in the wildest region of Pennsylvania on the Monongahela River. After failing in his utopias, he had entered politics in Philadelphia and had been elected to the House of Representatives in 1795 in the Jeffersonian Party, where his rise was rapid.

His interest in finance allied with his passion for nature had sensitized him to questions of organizing and extending public lands. Much had been done during the Adams administration, but much remained to be done in the Northwest Territory where Dufour wanted to go. Gallatin shared the preoccupations of Jefferson, John Adams's vice-president and potential successor. On February 1, 1801, Dufour described his difficulties to Jefferson and asked him to use his influence with Congress so that it would agree to sell him, on credit payable in ten or twelve years, the site he wanted and believed to be so favorable to the vine: its cultivation "demands such a great advance of funds and only pays so much later, that it is impossible for me to establish this culture on congressional lands, if I am obliged to buy within the terms of the law," he argued.[46] But he clung to the idea of obtaining this land because he had found no other that was better adapted to his enterprise in this latitude and because no other crops seemed possible to him there. Except for a strip stretching along the river, all the rest for several miles into the backcountry was nothing but steep valleys and hills useless for farmers other than winegrowers, for whom it was a natural habitat.

In his petition to Congress, Dufour repeated all the points of his letter to Jefferson: twenty years' winegrowing experience in Switzerland, the desire to have the United States profit from this, the expected reinforcement of many expert families from the canton of Vaud, and the impossibility for them all to acquire these lands on the Ohio without the help of Congress. He wished to acquire allotments in eighteen sections in two townships for two dollars an acre, payable in the year 1812, without interest. He promised to settle the families as soon as they arrived in the country, to plant ten acres of vines in the two years following their settlement, and in every subsequent year as many acres as possible, and finally, to familiarize the people of the United States with the culture of the vine.[47]

As the petition arrived too late, Dufour had to wait until the beginning of the following year to submit a second one to Jefferson, who received it along with a letter of January 15, 1802, from First Vineyard. Several elements distinguished it from the first: less land was requested, for many Swiss people had been obliged to defer their emigration; Gallatin was replaced by Kentucky senator John Brown in presenting the petition to Congress; Jefferson had become president of the United States on March 4, 1801, and Gallatin had been appointed to the Treasury. Dufour began his argumentation with a reminder of the benefits for the country of a production that everyone was certain would succeed as well as in Europe. The repeated failures were due to the lack of skilled winegrowers or to circumstances. This time things were different. Winegrowing had taken off, and soon American ports would be exporters rather than importers of wine. The Swiss were ready to help the United States provided the country reciprocated:

> It is on the shores of Lake Geneva that the vine is cultivated with the greatest care and thousands of winegrowers from that area can be counted on if the means of settling close to one another is facilitated for them, and this is what the banks of the Ohio are very much suited to, being so hilly that they are good for no other farming, excepting an edge of low terrain right along

the river that will furnish just enough prairie to nourish the cattle necessary to make the manure for the vines that will occupy the hillsides, and I foresee the time when the Ohio will rival the Rhine or the Rhone in quantity of vines and quality of wine.[48]

On February 15, 1802, the Senate received Dufour's petition requesting for himself and his associates the land for their vineyard; on May 1, 1802, the House of Representatives recorded the law: "An act to empower John James Dufour and his associates to purchase certain lands." Congress accorded them the privilege of selecting four sections of land in the zone defined by the petition, payable on credit, without interest, in twelve years, at two dollars an acre.

Vevey-Vevay

The thirteen shareholders divided the land into as many portions of 193 acres each, for the most part stretching diagonally to the north bank of the Ohio. On January 20, 1803, they signed a Covenant of Association repeating the obligation to make the cultivation of the vine their principal economic activity and forbidding them to sell their individual portions before all the land had been paid for, as a guarantee against the risk of land speculation.[49] The Covenant's birth preceded the death of The First Vineyard.

In spite of the care taken in the selection of stock from the Dufours' own vineyards in Switzerland, the vines withered and died. The colonists pinned their hopes on the indigenous varieties, but while waiting for them to bear fruit they put an end to their community endeavor of First Vineyard in 1804 and shared out their property. The announced decline did not benefit the lands in Indiana. Four years after the congressional grant, Dufour and his associates had occupied and cleared only parts of them, and the credit term was getting closer. In view of the preparatory work and late and uncertain wine production, they had to present a new petition requesting a credit extension. The context was not favorable. In early February 1806, the committee of public lands had denied the petition of a Frenchman, a certain Francis Ménissier, who had been experimenting with winegrowing in Cincinnati for six years using vines from France. The committee went beyond this particular case to raise the level of the discussion and criticize the principle of land or financial privileges agreed to by the government on lands that were the common property of the nation and whose sale at a low price would reduce the country's revenue. Congress took the committee's advice and also denied the Swiss petition presented on February 24, 1806.[50]

To meet his financial obligations, Dufour was obliged to return to Switzerland and sell part of his property. Hoping to bring back his wife and his son, Daniel Vincent, whom he had left ten years earlier, he also took this opportunity to tour European vineyards to strengthen his viticultural knowledge. Because of the Anglo-American conflict, he did not return until 1816, alone. His absence did not harm the colony, since before leaving the United States he had settled some of his family on their lands in New Switzerland. In 1803 he had accompanied the Morerod couple, and started them on clearing the land. The others followed progressively until the last crossed the Ohio in 1809. The Kentucky experiment had enabled the Swiss colony to get its bearings in the United States before settling in Indiana, where they permanently founded a town and a county.

The Dufours, their associates, and the other Swiss émigrés who joined them had no immediate urban and landscape development plans other than that of a road giving access to the houses built along either side. Their priority was to clear the lands of their thick forest covering in order to prepare them for their first plowing and the planting of food crops. Wheat fields, orchards, and vineyards began to stretch out. Toward 1807, the first wine was put into a cask. It was good—its quality is even said to have been judged superior to that of Bordeaux claret.[51] Three years later, production reached twenty-four hundred gallons, and it continued to increase as wine benefited from the influx of pioneers, from the War of 1812, which had stimulated patriotism for a local product, and finally from access via steamboat to the Cincinnati, Louisville, and Saint Louis markets. "The site is said to be beautiful and it promises to become the center of the wine business of the West, the grapes raised here being of a very superior quality," wrote the *Niles' Weekly Register* of the day.[52] Founder Jean Jacques Dufour was more farsighted and predicted the vine's long-term success in America: "That precious culture will be tried in different parts of the union and will undoubtedly multiply rapidly. The Swiss will encourage it with all their power. They give slips gratis to whoever will plant them, with directions and instructions as to their cultivation."[53]

The success of the wine, which its Swiss character made unique, was important in the success of the colonization of people from Vevey, who quite naturally came to found another Vevay—this one spelled with an "a." Brothers Jean Daniel and Jean François Dufour were the promoters and planners of the town born in 1814, the first of the future Switzerland County of which it became the county seat. Well situated on the Ohio northeast of Madison, rich in fertile soil with ideal hillside exposure, Vevay grew quickly and soon offered a range of urban services: a mayor, sheriff, justice of the peace, militia, tax collector, post office, newspaper, bank, Literary Society, and so forth.

The viticultural enterprise became secondary. One year before the credit term, looming on January 1, 1814, the colonists were insolvent. Yet another petition to Congress deplored that the terms could not be met, as well as "the length of time which must elapse before vine-dressers can receive a reward for their labor, together with some misfortunes peculiar to themselves."[54] On August 2, 1813, the law accorded them five more years, but in 1818 the Cincinnati tax collector refused the French and Swiss banknotes Jean Jacques Dufour had brought back from Europe. This occasioned a final petition for an extra year's extension, which was granted and respected, otherwise the lands would have been confiscated for nonpayment on time, as stipulated by the Law of 1802. The seventeen years between the law's passage and the final payment attest to the fact that viticulture had not really been a paying proposition. Jean Jacques Dufour died in 1827 near the present town of Markland, east of Vevay, without harvesting the fruits of his labors.

Despite their success in terms of productivity and wine quality, the Swiss winegrowers could not establish wine as part of American dietary habits, nor win out over the competition: it was easier and more profitable to turn corn into whiskey. For the same price, this beverage, with a much higher alcohol content than wine, was preferred by consumers, while its production, preservation, and storage facilitated the task of producers. The Dufours themselves agreed that planting corn or potatoes to distill alcohol was easier than caring for vines: Jean Daniel and two American partners built a whiskey distillery, and Jean Jacques made alcohol from the peaches of his orchard. With this reorientation, the Swiss colony inexorably turned from its viticultural origins and finally destroyed its vineyards, keeping only the number of rows necessary to produce the wine that the oldest colonists could not do without. Daniel Vincent Dufour, the founder's son, could have taken over when he came to Vevay around 1820, but to agriculture he preferred organizing the Swiss Guards, whose captain he was, and whom he presented to Lafayette during his visit to Cincinnati in 1824. The next year, he married a girl from Lexington who bore him nine children.

Dufour had the time, before he died, to publish the first manual for American viticulturists, *The American Vine-Dresser's Guide,* which remained the standard reference for a century: "I want to do everything in my power to help the people of this vast continent to secure for themselves and their children Almighty God's blessing of wine and grapes."[55] Dufour's pioneering work has not been forgotten: his guide has been republished; his Kentucky vineyard was replanted two centuries later under the name of Chrisman Mill Vineyards; and in 1971, in the state of Indiana, liberalization of legislation on winemaking establishments and wine production permitted the revival of a centuries-old tradition. The local historical society honored his memory with a plaque: "Jean Jacques Dufour, loyal and courageous leader of the Swiss family who founded Vevay. Friend of Jefferson. Buried on his old land entry." Though Dufour certainly went to Monticello and corresponded with Jefferson, the Swiss had more exchanges with Henry Clay, who visited the colony and tasted their wine. Both Jefferson and Clay supported the Swiss viticulturists; they would also support the French.

In choosing to cultivate the vine and the olive to secure their existence in the United States, the French refugees did not think they were taking inconsiderate risks: they were, after all, not the first. Two and a half centuries of attempts, experimentation, and efforts preceded them on the North American continent, and the most recent results, in Pennsylvania and Indiana, testifying to genuine possibilities, if not genuine progress, proved that commercial viticulture was possible. Given the diversity of soils and variety of climates in an immense, expanding territory, there was no doubt whatsoever that virgin lands awaited them somewhere, on which the vine and the olive could grow and produce fruits to nourish an equally expanding population. This economic prospect was all the finer in that it could be accomplished, if the American government were interested, off the beaten track, far from inhabited towns and countrysides, in lands still untamed, where everything was yet to be cleared, built, and undertaken, where people could rediscover the happiness of living together. Jean Jacques Dufour and the Swiss families of the canton of Vaud had set the example to follow, no doubt indirectly inspired by another Jean Jacques—Rousseau, their philosophic contemporary and compatriot.

8

The New Thebaid

Will we not prove that we are worthy of tilling the soil together, of making our trees bear fruits that are the most exquisite food . . . ? Why do you not take each other by the hand to go together to a pleasant region and found a new Thebaid?

—*L'Abeille Américaine*, Philadelphia, August 22, 1816

In August 1816, the *Abeille Américaine* published a letter from a subscriber who preferred to remain anonymous so as not to influence anyone and to welcome all initiatives, since the common interest should come before his personal ambition. After a glowing tribute to the superiority of American republican institutions over those "dictated by the governments of old, rundown Europe," the author wondered how the exiles, these "new children of Israel," could take advantage of them, if they had no lands of their own, no homeland to cherish, no place of rest where they could ease their homesickness and weep together. He proposed a project centered on winegrowing in a region they would turn into a second Pennsylvania, the refuge of exiles, or a new Thebaid, in reference to the ancient Thebaid of upper Egypt, where people had lived far from the world in secluded solitude.[1]

Often looking over the map of the United States, the author saw places that were still uninhabited, particularly along the Mississippi, where nature seemed to have prepared everything, combining a temperate climate with healthy air, clean water, and fertile soil, promises of peaceful enjoyment. But this location where the exiles would implant their mores, habits, and customs could only be located far from large cities, "these political Sodoms, where needs create vices, and where corruption forges the chains of servitude." In the meantime, it was imperative to have a subscription in order to know how many families were interested in the project and how much land each would like to acquire, and to elect the three individuals in charge of exploring the country and choosing the appropriate location.

It has been suggested that Garnier de Saintes was the author of this letter to the French living in the United States,[2] but in August 1816 he was at sea, only disembarking in Philadelphia with his son and Jean Guillaume Taillefer, his ex-colleague from the Convention, on September 10, 1816—the *Abeille* reported this two days later.[3] Of the two other former Convention members who might have written the letter, Pénières can be eliminated and, although quite plausible, it is doubtful that it was Lakanal. The author seems to have been Jean Simon

Chaudron, the *Abeille*'s editor himself. As one argument among many, there is the following passage from a letter to Lakanal, expressing the same theme of rejection of cities that we find in the anonymous text: "I want to get out of the mire of large cities where nothing is fashionable but prejudice and pride, and where liberty and civic equality would soon die if republican colonies were not founded in the country's interior. I want to go ask nature's forgiveness for having robbed her of forty years of my existence, and expiate on her maternal breast this error which I shall always regret."[4]

Regardless of who wrote it, the letter touched many émigrés, both older and recent ones. Lakanal had led the way by settling on the banks of the Ohio, in Kentucky, where, as we have seen, many French people had already headed; others were thinking of going on toward the South. Desnoëttes is said to have discussed with Stephen Girard the possibility of founding a colony in Louisiana,[5] where Clauzel had written, even before leaving France, that he wanted to go. The response to the call was therefore as strong as its author's convictions, or as those that others had already formed. So a general partnership was created in Philadelphia that had various names as it evolved: Colonial Society of French Emigrants; Tombigbee Agricultural and Manufacturing Company; Society for the Cultivation of the Vine and the Olive.[6]

THE COLONIAL SOCIETY OF FRENCH EMIGRANTS

As the subscription call was issued in August, preparatory meetings took place that month and in September, but no trace remains of their agendas. It is known that the officers of the committee of the whole were elected and that the venture was advertised among French and foreign exiles present or to come.

The first meeting for which there are minutes took place on October 22, 1816.[7] The president[8] opened it with the minutes of the preceding meeting, thanked the members for their faithful attendance and agreement on issues, and announced that many letters had arrived from Europe from people who planned to come as soon as a site had been chosen. Given the three months it took for letters to cross the Atlantic and reach their destinations in France or abroad, these exchanges must have begun in early August 1816. Since the society was conceived that same August, after a short gestation, September was the month of its birth in Philadelphia. It did not come into the world orphaned. Chaudron, whose paper published its minutes and transmitted information, was its natural father; Jacques Garnier de Saintes, its legal father; William Lee, the Bordeaux consul who arrived in New York on August 2, its tutor; its godfathers were Jean Augustin Pénières and Joseph Lakanal.

THE FOUNDING FATHERS

The men behind the society came from two different groups, united in adversity: on one hand, the exiles of the Restoration, sentenced by the proscription edict of July 24, 1815, and the amnesty law of January 12, 1816; on the other, the

French colonists chased from Saint-Domingue by the slave revolt and settled in the United States for at least the past twenty years. These two subsets are completed by isolated individuals whose professional activities had led to an interest in the lot of the French émigrés, people like the last American consul in Bordeaux. This is borne out by the composition of the society's first executive committee, consisting of a president, a vice-president, and members of the board.

The unanimity vaunted by the president during the meeting of October 1816 gave the impression that the composition of the executive committee had been arrived at without dispute. In reality, it was preceded by dissensions regarding the type of association and by withdrawals, according to Pénières, who had been approached with "generous, kindly offers" relative to the émigrés' settlement plan: "The banks of the Mississippi and even the Missouri seem to be the place where they would like to locate it; but they are already disagreeing on the type of association. The rich and important of bygone days would like to have pre-eminence and lord it over the less wealthy. His own opinion is that those purchasing shares in this association should be independent as soon as they have paid the agreed-upon price."[9]

This system of every man for himself advocated by Pénières to escape a dictatorship of the powerful contrasted with the community system initially proposed. Did these different approaches come into play in the election of the president? Joseph Bonaparte and Réal, both asked to serve, might have gotten everyone to agree, but, adds the consul, they declined the offer "so as not to give rise to suspicions that they had political designs." Furthermore, both were busy with their own land investments, the former in Bordentown and the latter in Cape Vincent, in northern New York State, at the junction of Lake Ontario and the Saint Lawrence River. Generals Desnoëttes and Lallemand could have represented military interests, but they were just as politically suspect. As for the Saint-Domingue refugees, it was preferable to keep their American seniority in the background, so that the forefront could be occupied by men recently banished from their country, whom the United States would be more inclined to pity and help. Had not Chaudron chosen to remain anonymous?

The Presidency. Former Convention member and sixty-year-old magistrate Garnier de Saintes almost naturally became the first president. From his glory days in the Revolution and Empire he had retained the sturdy character, eloquence, and habit of presiding successively over the Jacobin Club, the Saintes criminal court, the La Rochelle customs court, and, as vice-president, the Agricultural Society of Saintes. Basically, the presidency of the Colonial Society of French Emigrants was his by right, although this was ironic, given his past taste for terror and exclusion: after his regicide vote, he had proposed punishing by death all émigrés returning to France regardless of age or sex, and had urged shutting down Bordeaux's Masonic lodges and deporting "bad" journalists—that is, writers with the wrong political ideas. An enemy of religion, a fanatical patriot, muddled and fickle, the republican Garnier received the Legion of Honor and the title of chevalier from the Emperor. After Waterloo, which opened France to its invaders, he called for all-out resistance and mass conscription as in 1792. The king banished him.

The vice-president was William Lee. Writings and actions contrary to his duty of reserve as consul—this is the reproach that the Restoration had formulated against Lee to demand his recall from Bordeaux. The American government had calmed the diplomatic waters and accepted his resignation in September 1816,[10] but without taking disciplinary action against him, immediately appointing him to the War Department. Back in the United States, Lee was to continue assisting the French whom he considered an asset for his country. The migratory influx resulting from the French Revolution had been harmful, he deplored, because it had brought in people of the worst sort: the aristocrats of the ancien régime. This new wave of the Napoleonic elite was completely different, promoted by merit rather than heredity: "Having received their education in the walks of private life, we find them here returning with ease to the source from which they came, and to the dignity of useful citizens."[11]

Lee was to play a pivotal role in the Colonial Society, with his American friendships and French passions. Since his first stay in France in 1796 he had maintained strong ties with Monroe, then the United States ambassador. In 1811 Monroe had become secretary of state and probable successor to Madison, which gave Lee great expectations for career advancement.[12] Appointed to Paris to assist poet-diplomat Joel Barlow in concluding a trade treaty with France, he met important dignitaries, the imperial family, and Napoleon himself[13] before the apotheosis. On February 10, 1812, Lee was invited to a ball where the Empress Marie-Louise and her royal sisters-in-law outdid each other in elegance: "The dresses were splendid, and the *tout ensemble* appeared more like magic or the tales of enchantment than everything I ever saw."[14]

In his own country he could be of great help to Clauzel, with whom he had, in Bordeaux, planned Napoleon's escape; to his close friend Grouchy, whom he planned to introduce to Jefferson; and to Pénières, the man best informed on the most secret latest French doings, he said, and to whom he had given a letter of introduction to the patriarch, Jefferson. Although some vanity and the hope of furthering his own ambitions were involved, he truly cared about the arrival of these illustrious figures who were the honor of France, and of their compatriots, representatives of the most virtuous and civilized nation of Europe. They were sober, honest, and educated, set apart from other immigrants by their good manners: had one ever, in the United States, seen the French interfering in the affairs of the people who had had the indulgence to receive them? Lee liked to think that American public opinion was becoming more and more favorable to them.[15]

After turning it down[16] to launch two manufacturing ventures and give work to hundreds of poor people, Lee changed his mind about the offer of a position in the War Department and informed Madison of this.[17] Setbacks in his industrial projects added to other difficulties had led him, in the end, to prefer the regular salary of a civil service employee. This new position would not divert him from the society; on the contrary, his daily contact with people in power would serve their common plan.

On the way from New York to Washington, Lee stopped off in Philadelphia and, on January 2, 1817, introduced himself to the society members, who had not met since October. In the absence of the president, he chaired the meeting at which he was told of their joy at finally having in their midst the man "to whom they had always wished to express their great esteem," and whose "name could not be pronounced without at the same time calling to mind the numerous and eminent services" he had not ceased to render the unfortunate exiles, who had become the victims of their civic virtues.[18] Lee affirmed his devotion to their cause and assured the members that his heart was with all who were coming here to enjoy a democratic political system whose functioning they had understood, and to declare themselves supporters of the American government. This, coming from a man who now had an official position, was, if not a warning, then at least a reminder to the exiles of their duties to their hosts. He closed the meeting by announcing that in Washington he would see to the society's interests, and that he would soon inform them of its chances of success.[19]

Members of the Board. Neither a refugee nor an exile, the secretary, Colonel Nicolas Simon Parmantier, represented no particular group.[20] In the light of events and subsequent opinions of his conduct, it would seem that he had initially looked out for his own interests in using the society to revive a declining local career. According to his wife, née Françoise-Hélène Le Tallec, who had remained in France with their children,[21] this forty-year-old Breton had arrived in Philadelphia in 1808, where he had operated various businesses including a brandy distillery.[22] In 1812 the American citizen "Parmentew [*sic*]"[23] enlisted in his new country's fight against England as a private in the Second Regiment of the Pennsylvania Volunteers, and came out with the inflated rank of colonel, which made up for his commercial collapse during the war.[24] The creation of the Colonial Society came at just the right time for Parmantier. Although he was not, during the second War for Independence, what Lafayette had been during the first, his participation in the conflict could be an asset for the society in its dealings with a grateful American administration.

Onetime coeditor of the *Nain Jaune* and then of the *Abeille Américaine,* former officer and sub-prefect Louis Marie Dirat became the censor of the Colonial Society. With a humor and bite already employed against the Bourbons, he put his writing talents and the experience of eight years as prefect to use in its service. Banished by the edict of July 24, this relative of Marshal Pérignon's had, thanks to Chaudron, become the link between the Napoleonic exiles and the refugees from Saint-Domingue in Philadelphia. Chaudron, who was playing an essential role by making his newspaper the émigrés' forum, left his place on the board to another refugee, representing the interests of the former colonists of Saint-Domingue.

Treasurer Joseph Martin came from a middle-class family from northern France that had been elevated to a vague nobility with the name du Colombier[25] as a result of making a fortune growing coffee in Saint-Domingue, where he was born in 1760. Educated in Paris in an establishment for sons of the nobility (that a royal edict had opened to children of planters whose economic activity contributed to the national wealth), he recrossed the Atlantic and went to America to fight alongside the revolutionaries. Imprisoned by the English on the worst of the prison ships, the *Jersey,* he regained his freedom and then prosperity by marrying the daughter of the lieutenant-governor of Saint-Domingue. A planter and slave-holder,[26] he defended his acquisitions against the rebelling blacks at the head of the cavalry, which earned him the Domingans' admiration and gratitude.[27] But in the winter of 1792–93 he chose the safety of the United States, settled near Wilmington, Delaware, where he made up his losses in business, and then moved to Philadelphia after his wife's death in 1805.[28] In 1816, at the age of fifty-six (having spent almost half of his life in the United States), he gave up business for medicine, a calling he is said to have followed early in life, and practiced in Nicetown, a Philadelphia suburb, devoting his last twelve years to caring for the poor.

Charles Villar held the title of general agent before becoming the society's second president. This fifty-year-old man was a former naval officer from Lyon[29] who had lived in Philadelphia without interruption from 1798 to 1802 before being naturalized in 1806 under the sponsorship of Charles Badaraque, an émigré from Marseille and future director of the society.[30] Villar was in business in Philadelphia,[31] where Stephen Girard, who held him "in the highest esteem,"[32] recommended him to Hourquebie frères, his consignees in Bordeaux.[33] In 1802 he became their agent and authorized representative in Bordeaux before returning to Philadelphia two years later: due to the continental blockade, Girard's commercial interests had been diverted from Bordeaux and he no longer shipped there. Established as a merchant on the corner of Chestnut and Twelfth Streets,[34] Villar was experiencing economic difficulties when the French colonization project opened new horizons for him.

This sextet was representative of the members of the Colonial Society, but it was only the visible part of a group of strong personalities, including, among others, the exploring commissioners in charge of selecting the colony's settlement site, which had been named before even being found: Demopolis or Proscripolis.

The Town of the People and the "Proscripts"

For the Swiss colonists from the district of Vevey, canton of Vaud, the choice of the name of their town, the seat of New Switzerland, had been obvious. For the members of the Colonial Society, who were from a variety of geographic, social, and professional origins, the situation was very different. What did a Domingan refugee and a Bonapartist exile—one Creole, the other born in France; one already Americanized, the other scarcely arrived—have in common that could be used in a unifying name? Their status as expatriates and the desire, at least in theory, to share a community experience centered on the ideals of a democracy which they recalled had been born in Athens.

The consular and imperial era that had just ended in France had been that of a triumphant neoclassicism and of recourse to the words, values, institutions, and sages of antiquity to guide the new political and administrative orientations and structures.[35] Reference to the Roman Empire had eventually prevailed, his admirers making of Napoleon a new Caesar and of his denigrators a new Brutus, but reference to ancient Greece had also survived: Napoleon was following in the path of great legislators, Solon and Lycurgus, and the heroes of Homer. The French upper classes in exile, imbued with classical artistic, literary, and political models, had brought them to America, where, in architecture, the Greek Revival style flourished.[36] Furthermore, the society members could look to their predecessors in American emigration, who in 1790 had chosen Gallipolis as the name of their town on the Ohio.

The name Demopolis, which some attribute to Pierre François Réal, the former member of the Council of State and prefect of police, is contemporary with the very idea of founding a colony. Lakanal used it in a letter, written at the very latest in September 1816 on the banks of the Ohio, and read by the society's secretary at its October meeting: "I ask you, my honorable friend, to put me on the list of future citizens of Demopolis."[37] Lakanal was answering a letter soliciting subscriptions probably mailed from Philadelphia in August 1816, if one makes allowances for the time necessary for its travel. For several months the name Demopolis was used concurrently with Proscripolis, an ironic name first used by Louis Marie Dirat in the October 10, 1816, issue of the *Abeille*. He launched the subscription campaign for a farming settlement in the American West with humor: "These new colonists, selected among the French proscripts [outlawed men] and émigrés, have conceived of the criminal plan to live as brothers and without kings in this new Beotia, and without fearing there the lot of the companions or soldiers of Cadmus. But it seems certain that these unfortunate souls do not know that the posterity of the good king of Thebes and his wife Hermione awaits them there, in the form of rattlesnakes, and that it will get the better of them. We already have the names of the twenty future inhabitants of Proscripolis."[38]

On January 31, 1817, Jacques Lajonie wrote of his encounter in Burlington, Delaware, with an émigré who had a letter from his brother that "told him of the real advantages that there were in subscribing for the colony of Proscriptspolis (or Demopolis), the name of the colony's town which means, as you see: city of the proscripts."[39] Both names seem to have been used for a while, in any case until March 1817, when Demopolis overrode Proscripolis once and for all: the town of the people, already more universal than the town of the Gauls or the French, also wanted to be less limiting than that of the outlawed.

THE EXPLORING COMMISSIONERS

The founding document, published in August 1816 in the *Abeille,* had foreseen the election, by the subscribers meeting in Philadelphia, of three individuals to be entrusted with exploring the American territory in search of a place, at that time situated between New Madrid and Natchez, along the Mississippi, "one of the most beautiful rivers in the world." But, due to the fact that their route reached them before coming to the banks of the Mississippi, the shores of the Ohio below Pittsburgh were also to come into consideration as a place of settlement for the colony. Given the difficulty of the task and the expanse of the territory, the exploratory mission was carried out by a rather large number of people.

Generals Desnoëttes and the younger Lallemand were the first to set out successively. According to the Philadelphia consul, Desnoëttes, accompanied by an officer whose name he did not know but who had just finished a course of treatment at a spa in Virginia for injuries sustained at Waterloo,[40] had headed for the Ohio in early September. Passing through Lexington, Kentucky, they were in St. Louis on October 25 and in New Orleans on December 21, 1816.[41] Lallemand, who left a month later, is also said to have passed through Lexington, in November,[42] reaching New Orleans in January 1817.

Following them, a group of some twenty society members was reported in Pittsburgh in early December 1816, setting out in search of lands suitable for growing sugarcane, cotton, and grapevines: "Among them were a number of gentlemen of high rank and distinction, both civil and military; and it is with pleasure that we announce that the greatest friendship and harmony existed among them," noted the local newspaper.[43]

The group that included Garnier de Saintes, the society's president, and Pénières sailed down the Ohio to where Lakanal, their former Convention colleague, had settled. On December 19, 1816, Garnier was in Louisville, from where he wrote to the society. Instead of continuing his journey, he settled on the north bank of the river, across from Louisville, in southern Indiana. In June 1817, Montulé, going up the Mississippi and Ohio by steamboat, met him there, through the French owners of Louisville's best inn, where he was staying.[44]

Montulé first made the acquaintance of Basile Meslier, another of the society's exploring commissioners. A native of Barbezieux, Charente, Meslier had arrived in Philadelphia from the Bahamas in May 1816 and set himself up in business as a jeweler specializing in clocks and watches, before joining the society and setting out on its behalf. Meslier interested Montulé—"In his company, time did not drag, and anecdotes followed one another with rapidity"[45]—but not as much as did the young daughter of the Lecoq Dumarselays (who were mentioned toward the end of chapter 6), with whom he became infatuated.[46] She had a younger brother. Already quite good at English and local geography, he took Montulé across the Ohio by ferry to New Albany, where Garnier lived, on higher ground some distance from the bank, to escape flooding: "A log house, in fairly poor repair, was the dwelling place of the former representative of the people. We knocked at the door of a sort of shop, for he sells whiskey, rum, and cigars for a living. A small boy opened the door and led us into the kitchen; it is the bedroom, the living room— in short, the only room of M. Garnier, who came forward to greet us with a spoon in one hand and a notebook in the other."[47]

As one of Napoleon's former soldiers, Montulé was well received by Garnier, who invited him for a dinner of chicken fricassee, beans, and a salad, which he prepared himself. When he was not cooking, Garnier busied himself, in his retreat, by writing a philosophic book, following on his earlier work published in Brussels before his exile, to be entitled *Emérides, ou soirées de Socrate*.[48]

Garnier was on neighborly terms with a woman from the Vendée region, Angélique Audibert, the daughter of a controller-general of taxes in Brittany, Thibaudeau du Fief. The widow of a civil servant in the imperial navy from Nantes, she had left the Loire in 1816 with her four children, among them Aimé, fourteen, employed as Garnier's factotum.[49] Claiming to know about agriculture, she planned to join him and go live in the colony that the French had by this time chosen. In Louisville, Basile Meslier urged Montulé to join the society as well. Obliged to continue his journey south, he had given Garnier the authority to receive Montulé's signature for a quarter share. To confirm his commitment, he paid the seventy-five francs that each shareholder was supposed to advance the society for its initial expenses. But, though delighted at the idea of owning land in the United States, Montulé preferred the role of tourist in Italy and Egypt to that of husband and American landowner.

The link between Desnoëttes, Lallemand, Garnier, and Meslier was Joseph Lakanal: all went to see him at his plantation on the banks of the Ohio. His revolutionary past, his intellectual aura, a level of education unequaled among the French émigrés, and his agricultural experience in Kentucky made him a model to follow,[50] the "salutary sentinel,"[51] the "guiding spirit" and the "Regulator of the Society";[52] he was one of those who had called attention to the Swiss colony of Vevay and the success of its wine production. The society sent him its emissaries, and it was with one of them that he, in turn, left his home to explore.

During the general meeting of October 1816, the committee of the whole had appointed Jean Augustin Pénières to the position of exploring commissioner, to go "into the southwestern region of the United States, to select there land fit to become the refuge of the many French émigrés" who were expected, and of émigrés from other nations who would want to join them. In a letter to his brother dated October 22, Pénières explained: "I am to go see the banks of the Oyio, the Houabache [Wabash], the Missoury, the White River, the Arkansas, the Yassou [Yazoo] River, and I already have all the information I want on all points."[53]

His other instructions advised him to remain in as close touch as possible with the society's vice-president in Washington, the members of the board in Philadelphia as well as Generals Desnoëttes and Lallemand, who had already set out, and finally with Lakanal in Vevay. On October 24, 1816, he informed Dupont de Nemours that he was leaving the next day, and of the intentions of the society members, whose orders he was following: "We would also like to found a colony and imitate you, if possible; as much wisdom as courage is needed for this enterprise; choosing fertile, healthy land next to a navigable river; imitating and initially doing only what our neighbors have done the best; not undertaking too much and not being put off by the first difficulties; I should really have liked to get a mentor's advice, but I do not have the time. We shall definitely adopt your ideas, on which our fathers were in agreement."[54]

From Philadelphia he went via Baltimore to Washington, where he obtained the maps and other indispensable material, regretting, because of the public coach schedule, that he could not accept the invitation to meet President Madison that he had received the day he left. Accompanied by three Frenchmen who had waited for him at Bedford, on the Juniata River, he crossed the Alleghenies on foot and reached Pittsburgh, on the Ohio, by mid-November 1816.[55] To this point he had been warmly welcomed, he said, thanks to the many recommendations with which he had left. The rest of the journey would probably be more difficult, as they would be traveling through less hospitable regions. Pénières embarked with a group of some twenty compatriots who had joined him in Pittsburgh and that probably included Garnier de Saintes. They planned to look in on Lakanal.

Lakanal was the French travelers' rallying point, but he was not to remain long where he was. The society members, meeting in Philadelphia, had commissioned him at the same time as Pénières, his ex-colleague in the Convention and the Council of Five Hundred, to travel the length and breadth of the United States in order to determine the best place to establish the colony. After several months in Kentucky, he did not seem disposed to take to the road, as this letter of October 25, 1816, to a friend in Philadelphia suggests: "I am happy for the first time in 30 years; I am happy because my happiness consists of a few simple elements, domestic harmony, the most ample security, farm work and reading."[56]

However, according to Mignet, his biographer and friend, Lakanal was rather quickly disappointed by first harvests and natural conditions that contrasted with Michaux and Warden's optimism: no orange trees, but corn, a very harsh winter, the great distance to large markets for his products.[57] So in late 1816 he joined Pénières to go explore the American West beyond longitude 90° west. Each of them has left a record of this expedition, which lasted several months, but only Lakanal mentions the presence of Pénières at his side, whereas the latter mentions no traveler but himself in his letters to his brother: "Full of strength, courage and enthusiasm, we pushed, my good comrade and I, our investigations almost to the headwaters of the Missouri, towards the Rocky Mountains," wrote Lakanal[58]—but actually, it would seem, as far as the present-day city of Sedalia, Missouri. They crossed the country of the Osage Indians, who surprised them by shouting "Falanche babichile (Frenchman friend)" and repeated with admiration the name of Bonaparte.[59]

The passages of Pénière's correspondence with his brother selected by Victor Faure give an idea of his exploration in search of a site combining healthy air, fertile soil, and navigable streams on which factories could be built: in January and February 1817, he said, he traveled through territories along both banks of the Ohio and Mississippi as well as those occupied by the Chickasaws and Delawares; in March he went up the Arkansas; in April he reached Cherokee territory, 390 miles from the river's mouth, and discovered nearby an ideal place for the outlawed men's town. The friendly Indians received him, said Pénières, like a father, and smoked the peace pipe with him. From the cabin he had already built on the banks of the Arkansas, he planned to attempt "the finest act of a wandering mortal, that of civilizing a great savage nation"; his mission was accomplished: "I called out loudly to the Israelites who had entrusted me with the responsibility of finding the Promised Land," he wrote on July 1, 1817, from Crystal Rock on the Arkansas,[60] near the hot springs located sixty or seventy leagues from Osage country—

near the headwaters of the Missouri's largest tributary, the Osage River. Everything was perfectly fine. Pénières boasted of being fit as a fiddle. Lakanal confirms this, saying that he was full of energy, constantly busy with botany and mineralogy. After enduring the sanction of exile, narrowly escaping death in Tulle, suffering shipwreck on the Dordogne, escaping drowning near the Azores, and exploring vast regions of the American continent, Pénières was a miraculous survivor.

But eight months had gone by since his departure, during which the society had not remained idle. They were aware in Philadelphia that his journey would unavoidably be too long for its results to be known before the absolute deadline of March 4, 1817, the end of the congressional session and the inauguration of a new president. As a guarantee of the legality of its intentions, the society had asked its president and board to confer with their vice-president in Washington in order to draw up and present to the Congress "the petition in the collective name of the society, that it grant the concession of a tract of land large enough for a settlement as extensive as the one they planned, and whose success depended entirely on the terms that the government of this happy country would be so kind as to accord to unfortunate people seeking a homeland and able to find one only where law, justice, and liberty reigned."[61] The first condition for the petition's success was that it be presented in time to be discussed, amended, and voted on during the session that was to end on March 3; otherwise, the petition would have to be resubmitted the following year, which had been the case sixteen years earlier with that of Jean Jacques Dufour.

The second condition was to be able to give the congressmen at least some indication of the geographic location of the desired land. It was thus necessary not to rely solely on one man but to have several possibilities. A vice-president like William Lee, with an official position in the capital, was a major advantage: he was asked to check out the documents available "at the land office in Washington City, in order to designate the location whose grant the Society could ask for."[62]

On the Tombigbee River

Lee had not yet assumed his position when in December 1816 the society received letters from its members on outposts along the Ohio, such as Garnier's mentioned above, and most particularly "a very well-written note from a citizen of Lexington, about lands located between the Mississipi [*sic*] and the Tumbigbee [*sic*] Rivers and between latitudes 32° and 35°."[63]

Dr. Samuel Brown, a brother of the Senator Brown who had helped Dufour look for land around Lexington in 1798, is mentioned as the person who either wrote the letter or inspired its author.[64] He knew the area, as by his marriage in 1808 he had become a planter in Natchez on the east bank of the Mississippi, and after his wife's death, near Huntsville, in northern Alabama, east of the headwaters of the Tombigbee.

During the meeting of January 2, 1817, chaired by Lee, several society members volunteered similar information that they had in turn obtained concerning this area which was at that time the American Southwest. After a thorough examination of the choice of land, the members present agreed that it would definitely present advantages if located in latitudes 33° or 34° north. There they would find, in a temperate climate, a healthy country with a navigable river that could easily be reached by land or by sea. This country, covered with lush natural vegetation, benefited from the proximity of the settlements in Tennessee and New Orleans. Improving the region was already "in the provident and benevolent plans of the various branches of the government of the Union."

Geographic coordinates situated this land in the eastern part of the Mississippi Territory, which the Americans had created in 1798 after the Treaty of San Lorenzo (1795) set the boundaries between the Spanish colonies in America and the United States. Lying between the thirty-first and thirty-fifth parallels, bounded on the west by the Mississippi, on the east by Georgia, and on the southeast by Spanish Florida, this territory was subdivided into twin parts, the future states of Mississippi and Alabama. The navigable river common to both was the Tombigbee, with its source in the one and its course in the other until it joined the Alabama River; from there it flowed toward Mobile Bay on the Gulf of Mexico via the Mobile River. The location and natural conditions of the region were thus favorable elements, additionally falling into a particularly auspicious context. The Indians who had been living there for centuries had just ceded the area to the whites.

In March 1814, the defeat at Horseshoe Bend on the Tallapoosa River in Alabama was the final blow for the Creeks who had made war in 1813, but also, not long thereafter, for the Choctaws, Chickasaws, Cherokees, and moderate Creeks, despite their having joined General Jackson's Tennessee volunteers. On August 9, 1814, Jackson, at the Treaty of Fort Jackson, obtained for the United States twenty million acres, that is, two-thirds of the lands belonging to the Creeks, which cleared about three-fifths of Alabama of its native inhabitants. The last pockets of resistance were wiped out by a few new treaties, among them the one known as the Treaty of Fort St. Stephens,[65] signed on October 24, 1816, with the Choctaws: they ceded a territory of about three million acres east of the Tombigbee for which the United States agreed to pay $10,000 in goods to be delivered immediately, and $6,000 per year for twenty years, a total of $130,000.[66] The Choctaws crossed the river and settled to the west, leaving behind land for the taking.

The society decided that its vice-president and board would petition Congress on its behalf for a grant of lands to be selected between latitudes 32° and 35° on either bank of the Tombigbee River or any other part not yet bid for. The choice covered an extensive area, and it is not certain that at this date the society members had already thought of the spot that was selected when the first party of settlers arrived six months later at the confluence of the Tombigbee and the Black Warrior Rivers, latitude 32.5° north and longitude 87.8° west.

DRAWING UP PRINCIPLES OF GOVERNMENT

Four months after the publication of the founding charter, the project of an agricultural settlement of émigré families had taken shape. The Colonial Society now

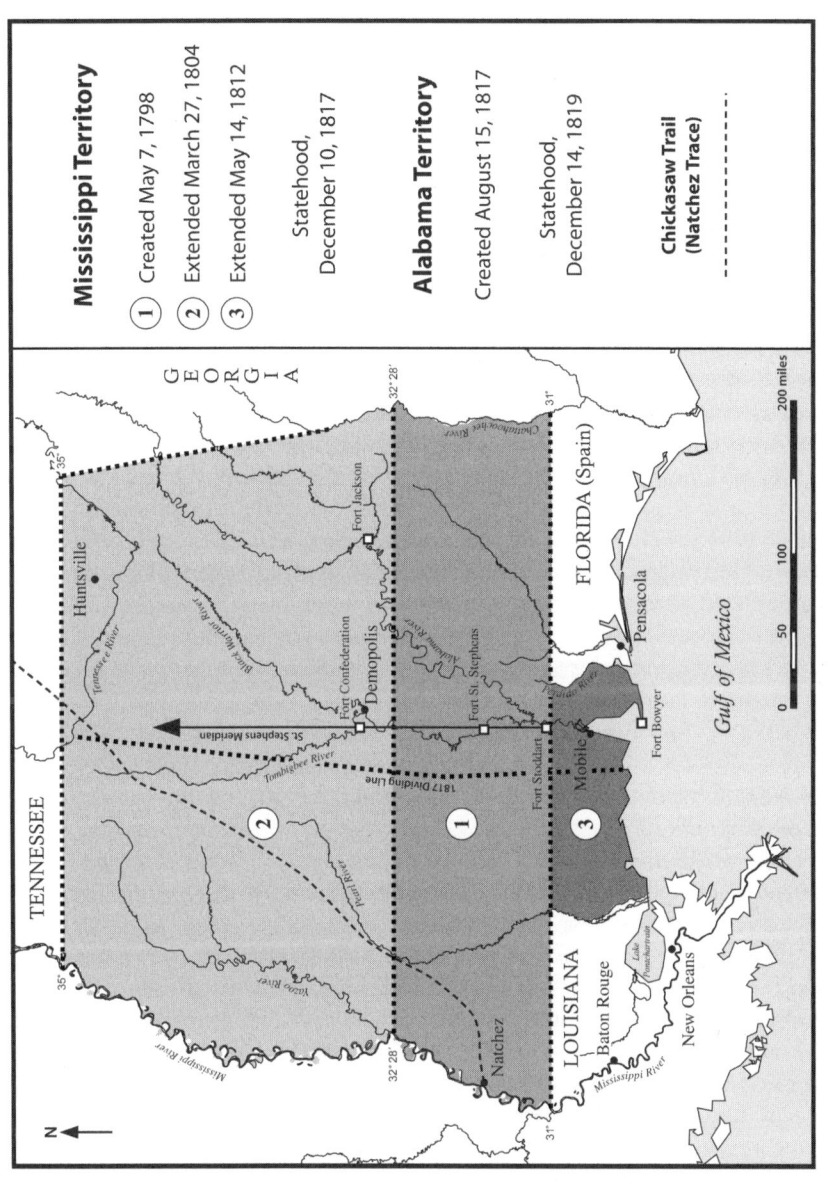

Map 3. The Alabama and Mississippi territories (1798–1817).

officially existed, with a name, an elected president and board, internal statutes, regular meetings, a goal, and a destination. There remained the well-being of the future colonists. The society was not expecting "permanent happiness" in its coming community experience, but it could at least give itself the means to happiness with "wise regulations."

THE SOCIAL CONTRACT

Rousseau's influence, which could be sensed in the August declaration of intent, remained strong when the time came to engrave permanent sentences onto new tables of the law. Among Rousseau's works, *Émile* and *Julie* were not the only bedside books of the society's thinkers; there was also *The Social Contract*, the founding text of modern political law. Published in 1762 in Amsterdam, it did not become widely known until the French Revolution borrowed some of its vocabulary and principles: law as the expression of the general will and sovereignty as residing in the nation, associated with three essential objectives, equality, liberty, and truth. In 1794 Joseph Lakanal, a member of the Convention's committee for public instruction, expressed what the era owed to Rousseau, whose ashes were being transferred to the Panthéon: "It is the Revolution that has explained *The Social Contract* to us."[67]

The society members could take from this book passages perfectly suited to their situation, particularly in the sixth chapter of book 1, entitled "The Social Pact":

> Now, as men cannot create any new forces, but only combine and control those that do exist, they have no other means of self-preservation than to form by aggregation a sum of forces which may prevail over the resistance, to put them in action by a single motive power, and to make them work in concert.
>
> This sum of forces can be produced only by the combination of many; but the strength and freedom of each man being the chief instruments of his survival, how can he pledge them without doing harm to himself, and without neglecting the concern he owes to himself? This difficulty, applied to my subject, may be expressed in these terms:—
>
> "To find a form of association that may defend and protect with the whole force of the community the person and property of every associate, and by means of which each, joining together with all, may nevertheless obey only himself, and remain as free as before." Such is the fundamental problem of which the social contract provides the solution.[68]

For Rousseau, the only possible foundation of legitimate political societies was the contract of association and not the pact of submission defended by Hobbes. Political association alone should orient the forces and constitute the common power, without submission to a third party, while satisfying two conditions: to ensure the preservation of each individual by protecting his person and property, and to preserve his liberty. On its own scale, the Colonial Society of French Emigrants was already a body existing and living in democracy, with members, meetings, resolutions, an elected board, commissioners, and a president. As with all political bodies, the society now had to "endow it[self] with movement and will by legislation,"[69] to give itself laws for its own preservation, in order to better submit to them.

But what "extraordinary man" should be entrusted with the task of drawing up the conditions for the society? In the case of the best rules "most suitable to nations, a superior intelligence would be necessary . . . whose happiness would not depend on us, and who would nevertheless be quite willing to interest himself in ours," wrote Rousseau.[70] The legislator must be independent of the men to whom he gives laws: "When Lycurgus gave laws to his country, he began by abdicating his royalty. It was the practice of the majority of the Greek towns to entrust to foreigners the framing of their laws. The modern republics of Italy often imitated this usage; that of Geneva did the same and found it advantageous."[71]

This restrictive clause ruled out the society's talented legislators: Lakanal, whose avowed admiration for Rousseau went back to the Revolution and had continued during the Empire, when he prepared an edition of his works; Pénières, who had named his oldest son, born in Paris a few days before Rousseau entered the Panthéon, Émile,[72] and his daughter Virginie in homage to Bernardin de Saint-Pierre.[73]

There was in the United States a man who met Rousseau's criteria, a polymath, independent of the society members but a Francophile interested in their lot: Thomas Jefferson, the author of the Declaration of Independence, whose advice the French revolutionaries had sought when writing their own Declaration. A former president of the United States, he was, in 1817, in the words of Rousseau describing the ideal legislator, "an authority that is a mere nothing."[74] Without political power, the sage of Monticello was not without influence, thanks to his continuing relationship with Madison and Monroe, his successors in the presidency. Nor was he without opinions: his aversion to the Bourbons was well known, as was his friendship for the exiles, with many of whom he corresponded. As a final advantage, he liked wine and had favored the introduction of vine cultivation. His advanced age and the weariness of a long political career were, on the other hand, reasons he had already given to justify his retreat from political life.

THE COLONIAL PACT

During their meeting on January 2, 1817, the society members decided to write to Jefferson, asking him to draw up "the basis of a social pact for the local regulations" of their association, so that they could present to future centuries "a monument to the bliss that could be attained by a group of true friends of liberty."[75] In a general sense, the society wanted to pool all its individual faculties in the service "of the protection and the promotion of each of its members in particular, while living in the most perfect independence as to the exercise of their political rights, the disposal of their property, their industry, and the most unbounded expression of their opinions." The society's vocabulary and principles were indisputably borrowed from Rousseau. Martin Ducolombier, who had become its second vice-president, and Parmantier, its secretary, drew up the January 5, 1817, resolution. At their request, Jefferson addressed his answer of January 16 to William Lee, the first vice-president, in Washington.[76]

Jefferson was touched by the confidence he inspired, and his desire to merit it was equaled only by his friendship for the exiles, whose difficulties adapting in a country so far and so different from theirs he understood. He gave his support to their enterprise but refused the task of proposing regulations, which they alone would be capable of giving themselves as they became used to living together, supposing this even made sense. Within what context would these regulations fall? Creating a new state was out of the question, because the area granted to the colonists would of necessity be too small. If they were to be functioning as a state's county or a simple voluntary association, it was no longer a matter of actual laws, but of submission to the legislation of a state or to general law. No matter what the scenario, the colonists could not help but be subject to the laws of the United States, whose lands they were going to occupy. Jefferson concluded that these considerations should be fundamental to the society.

The resolution's authors had used Rousseau's ideas to persuade Jefferson to offer them principles of government. But Jefferson, who had read Rousseau, refuted the necessity that the legislator be either foreign to or independent of the inhabitants for whom he was to legislate; on the contrary, "No member of a foreign country can have a sufficient sympathy with these. The institutions of Lycurgus, for example, would not have suited Athens, nor those of Solon Lacedaemon. The organizations of Locke were impracticable for Carolina, and those of Rousseau and Mably for Poland."[77]

The society members' reasoning is difficult to follow, but their goal could not have been to receive from Jefferson the foundations of a separate constitution—only, as said above, of a colonial pact for the local regulations of their association; the adjective "local" in itself shows the modest nature of the society's aims. There were in its ranks many jurists, former legislators, magistrates, members of the Council of State, lawyers, sub-prefects, and senior civil servants who had been active during the Revolution and Empire, plus Frenchmen who had emigrated years earlier, were often naturalized, and if not, very well acquainted with the American political system or probably at least conscious of what they would be allowed or not allowed to do. What they were sure of was that by consulting Jefferson they associated him de facto with their cause and expected positive reactions to this. The French-language *Abeille* and the English-language newspapers *Niles' Weekly Register* and the Washington *National Intelligencer* did indeed pick up the news and gave the project the best possible publicity.[78]

9

A Gift from Congress

Among the *splendid fooleries* which have at times amused a portion of the American people, as well as their representatives in congress, was that of granting on most favorable terms to certain emigrants from France, a large tract of land in the Alabama territory, *to encourage the cultivation of the vine and olive,* passed the 3rd of March, 1817.

This tract contains 92,000 acres, and was *sold* at $2 per acre, payable without interest, in 14 years—in truth, much better than a mere *gratuity* of so much land considering the license of *selection,* and which could not, at this time, probably be purchased of the *proprietors* for less than two millions of dollars.

—*Niles' Weekly Register,* Baltimore, August 8, 1818

We do not have the text of the Colonial Society's petition for lands, but this is scarcely a hindrance, since we know its driving principles and the names of its signatories. The important thing is to understand that, in a generally favorable climate, there was every chance the petition would succeed. The "era of good feelings,"[1] introduced by the outgoing Madison administration, was not confined to politics, but spread its optimism to other sectors of intellectual, socioeconomic, and religious life: the process of transforming America was under way.[2] As for the outlawed men, people continued to pity their lot and link their distress to that of Napoleon confined to St. Helena by what was considered a most unjust decision by the English government. The Emperor—whom America could have welcomed— and his partisans enjoyed the friendship and compassion of many Americans, from ordinary citizens to their leaders. Monroe's triumphant election to the presidency on December 4, 1816, over Federalist candidate Rufus King, opened future prospects to the French émigrés with the chance that Franco-American relations would improve. A federal commitment to the exiles was less likely to offend the sensibilities of the French diplomatic corps and its superiors in Paris.

The latter were quite happy to see the exiles leave the East Coast. What with pruning vines and picking olives, their contribution to the colonization of the South would keep them occupied and turn them from the reprehensible goals that the French ambassador in Washington, Hyde de Neuville, attributed to them: getting involved in the struggle for independence of Spain's South American colonies, reviving the Bonaparte dynasty somewhere in the Americas, with the Emperor himself, whom they would free, or his brothers Joseph or Lucien, or else his son, Napoleon II. Hyde had recurrent nightmares over these ultrasensitive problems that his country had created for itself by expelling people right and left or pushing them into exile with its reactionary policies. In the light of the situation and wiser for his own experience as a banished man during the Empire, he kept repeating what a bad idea exile was (forgetting in passing that he had voted for the so-called "amnesty" law of January 12, 1816, which was really a proscription law).

The exiles now had an opportunity to mend their ways and put an end to rumors; it was important not to miss it and thus prove that those who trusted them were right.

AT A CROSSROADS

It will be recalled that on March 12, 1815, after failing to rouse the garrisons in the department of Aisne to revolt, the officers involved in this aborted rebellion had to choose between two diametrically opposed directions: fall back into rank, or rally to Napoleon in his reconquest of power. In choosing to rally to the Emperor, Generals Lefebvre-Desnoëttes and the Lallemand brothers suffered defeat, disgrace, and exile. In the thirty or so months that followed, they found themselves, along with hundreds of other Bonapartists who had crossed the Atlantic, faced with a comparable dilemma: keep out of the limelight in America and hope for a royal pardon, or resume politico-military intrigues centered on freeing Napoleon. The straight and narrow path or the dangerous road?

Until Napoleon's death in 1821 brought them to an end, rumors about plans of his escape from St. Helena fed diplomatic correspondence, the press, and public opinion on both sides of the Atlantic.[3] The mere mention of St. Helena sent the political world into turmoil, so dread had become the fear of seeing its prisoner escape and reappear center stage on the international scene. Traumatized by the precedent of the island of Elba and its dramatic consequences, the governments and chancelleries in Paris, London, Vienna, and Madrid worried or played down the risks to put their minds at ease, but let nothing pass. In England, concerned with a recession and joblessness exacerbated by demobilization, the threat of Napoleon's escape received the least coverage, so great was confidence in the defensive measures put into place, on and around the island, to wipe out any risk. In March 1816, the French ambassador in London, the Marquis d'Osmond, emphasized the precautions taken to ruin any vague attempts on the part of the restless "seditionaries" in America. They were not a cause for worry; a vessel sailing near the island had not the slightest chance: "This fact can calm the fervor of people like Renaud de Saint-Jean d'Angély and elsewhere," he concluded.[4] A year later, in April 1817, Sir Charles Bagot, the British ambassador in Washington, was even more categorical in stating that rescuing Napoleon from St. Helena was the craziest idea he had heard since arriving in the United States. It was nothing but a matter of reveries with which these hotheaded French officers kept boredom at bay.[5] At about the same time, Baltimore's *Niles' Weekly Register* was telling its readers the same thing in an article listing the impressive means being used on St. Helena to dissuade anyone from attempting the impossible.[6]

Hyde de Neuville, who had been asked during the first Restoration to draw up a report on the risk of an imperial escape from Elba, did not feel at all confident, especially since that report had not been taken seriously. On January 26, 1816, the instructions from Richelieu, prime minister and minister of foreign affairs, to Hyde backed him up in the vigilance he was again to observe: Did the Americans have relations with St. Helena, and was a venture or intrigue being planned in the United States with the goal of helping Napoleon escape? Joseph Bonaparte's conduct was thus to be closely watched in case he was an intermediary between his brother and the French exiles in America.[7] If the ambassador learned important facts, he was authorized to send a ship to France as quickly as possible to report them.[8]

Even before his ambassador left for America, Richelieu's fears were supported by those of the French ambassador in Madrid, who in turn got his information from the Chevalier Luis de Onís, the Spanish ambassador in Washington.[9] He thought that Joseph Bonaparte, Generals Grouchy and Clauzel, and several others were calling together the banished men and the most unruly members of their party. Joseph, said to be enormously rich, had supposedly been contacted by an American privateer named Carpenter, who, for 100,000 piasters, was offering to kidnap Napoleon from St. Helena.[10] France did not doubt that the American government would take the necessary steps to block such arrangements, but it used the occasion to summon its diplomatic corps in the United States to keep a close eye on such plans coming from French exiles, whether motivated by interest or conviction.[11]

Hyde de Neuville did not disappoint Richelieu. In his very first dispatch of June 22, 1816, reporting his arrival in New York, he wrote that four schooners with their crews, more than three hundred men and artillery, had left Baltimore in the greatest secrecy.[12] Thomas Taylor, a privateer captain from Buenos Aires, "extremely daring and a very faithful admirer, like all pirates of his sort, of the prisoner of St. Helena," was involved, as well as some French officers: César Fournier, formerly of the imperial navy, and Colonel Jacques Roul. It was claimed that the sole purpose of the expedition was to join the rebel Bolívar, but this was only to divert attention from their true destination: St. Helena.[13] Hyde had decided to assume that any mysterious expedition had this goal. Thus, on the following July 12, he suspected three slave ships from Baltimore of two aims: first, to go "prowl around St. Helena to attempt to abduct Bonaparte by surprise," and then, if they failed, "to go to the coast [of Africa] to purchase Negroes" and sell them back in Havana.[14] No sooner at his posting, the ambassador was obviously zealous in carrying out his mission. It is useless to dwell here on a subject treated elsewhere (see note 2), by listing supposed, fanciful, or serious escape plans, none of which in any case came close to succeeding. Let us rather relate it to the matter of the emancipation of the South American colonies from Spanish authority.

Rear Admiral Cockburn, back in England in the summer of 1816 after taking Napoleon to St. Helena and then supervising him on the island until Hudson Lowe arrived, expressed these two contradictory assertions: while "common vigilance" would prevent any escape, attempts to free him would nonetheless be fomented in America.[15] The desire of many Americans to see Napoleon freed and the state of emancipating effervescence of Spanish America made the continent an ideal parade ground for the Bonapartists and other potential saviors of the great man. He, who had missed the boat on the île d'Aix (off the coast of Rochefort), where he had told his entourage that he wanted to retire on the banks of the Mississippi or the Ohio,[16] a year later assured General Gourgaud, a companion in exile, that "sooner or later" they would go to the United States (or England).[17]

Thanks to clandestine communications between St. Helena and the two shores of the Atlantic, the Emperor's hopes were known, and he himself knew that his friends were busy trying to fulfill them—as were his enemies to thwart them. In January 1817 he received a message from Joseph informing him that a group of rebels had offered him, Joseph, the crown of Mexico, which, according to Montholon, another companion in exile, they offered the Emperor shortly thereafter:[18] a Bonapartist empire in Spain's American colonies was a possibility. It was all the more so because there was no lack of Frenchmen to support Joseph in this plan, or of generals to lead them, Clauzel or the elder Lallemand, of whom the Emperor said, on St. Helena, that he had no equal in conducting a hazardous enterprise.

The hope that was stirring on St. Helena weighed on those who thought it could be carried out, since anything could be expected of the Emperor—not on the British cabinet, confident in the security measures deployed around their prisoner, but on the French and Spanish governments, being fed alarmist dispatches by their diplomatic representatives in the United States. Hyde de Neuville was among those most worried: an alliance between the rebels against King Ferdinand VII of Spain and the French officers gathered around Joseph could lead to the deliverance of Napoleon and his return to center stage. There, in South America's Spanish colonies where the revolutionary spirit was spreading, the Emperor was capable of single-handedly carrying out "a great revolution."[19] France ought therefore to examine carefully anything that might divert the French officers from these subversive machinations back onto the right path.

On January 10, 1817, in a long report to Richelieu on the general situation of the United States, Hyde told of the unusual plan of a French colonial settlement being considered for the region of the Ohio and Mississippi Rivers.[20] It was to be organized by Lee, the former American consul in Bordeaux, along with a committee composed of Garnier de Saintes and Dirat, both excepted from the amnesty law, and men named Martin and Parmantier, of whom he said he knew nothing. As for the plan, Hyde did not know what would come of it, for it was already said to be compromised by dissention among an excessive number of leaders. In any case, he hoped it would be carried out and thought France should contribute to it financially, for two reasons: first, the settlement would promote French economic and cultural expansion in developing regions; next, it would be a way to restore the dignity of men whom France had only half condemned in banishing them.

The king had the opportunity to show his generosity in saving from despair and ruin men who, in their exile, were getting each other worked up, harming themselves, and spreading regrettable reports and bad ideas in their new country. Hyde thought that in the long run banishment had far more serious drawbacks than house arrest under the supervision of the mayor or a good police officer: in their own towns, if they did not return to better feelings and the right path, the guilty and lost would at least be forced to be quiet and remain calm; abroad, on the contrary, they did damage to their native country.

Richelieu answered on April 18 with tempered enthusiasm. While he somewhat cynically admitted the relevance of the economic argument, he was doubtful of the appropriateness of official aid. This would associate the French state with actions that the Colonial Society might decide to undertake but that were not necessarily in line with its objectives, particularly toward the Spanish colonies. Furthermore, as the Restoration was stingy with regard to the veterans of the Republic and the Empire who had remained in France,[21] it was inconceivable to send a budget across the Atlantic to military men struck from the ranks of the army. The minister, however, saw no major obstacle to the establishment of a French colony in the United States.[22]

While Richelieu remained reserved, he did not intend to hinder the society's plan, which could be a real chance for exiles wanting to redeem themselves. Moreover, this tacit agreement reinforced the position of the United States, freed from the threat of a diplomatic quarrel.

THE LAW OF MARCH 3, 1817

Even though things seemed to be getting off to a good start, the Colonial Society officially sent its secretary, Parmantier, to Washington, to second William Lee in his lobbying of the members of the Fourteenth Congress, in session since December 2, 1816.[23] Many other society members came to support them and show the congressmen that their vote and the fate of the petition they were bringing was important to them: Lajonie said that he remained in Washington for two weeks, where he had "had the honor of shaking Madisson's [sic] hand," probably during the last two weeks of his presidency, which ended on March 3, 1817.[24]

There is no detailed record of the three-week discussion in Congress between those for and those against the bill.[25]

On February 10, 1817, the Senate Committee on Public Lands[26] brought to the floor a bill "to set apart and dispose of certain public lands, for the encouragement of the cultivation of the vine, and other exotic plants thereon." The text was not immediately approved and had to be amended before passing by a large majority on February 21 under a new heading that added the cultivation of the olive to that of the vine. It was sent to the House of Representatives that very day, where it received two readings.

The Speaker of the House had for the past six years been Kentucky's Henry Clay, who had supported Madison during the War of 1812 and been a member of the commission that negotiated the peace treaty of Ghent. When he returned, he declined the offer of a diplomatic or cabinet post, wanting to consolidate his strong position in Congress before becoming, he hoped, Monroe's secretary of state. While in Europe, he had visited Paris and met Lafayette. The latter had, on December 26, 1815, written him a letter on the state of France that closed with praise of Lakanal, who was asked to send it to him.[27] Lakanal did this when he reached the United States, but it was only in July 1816 that Clay told Lakanal, then living in Kentucky, that he had received it. In February 1817 Lakanal took the opportunity of another letter to pay him a glowing tribute, with the idea of winning his vote and that of his friends.

On the third of the same month, Lakanal had sent Parmantier, who was working in Washington, a copy of Clay's letter of July. The society's secretary used the letter as an introduction to Clay, who received him kindly on his first visit, and even more enthusiastically after receiving another letter from Lakanal that Parmantier, accomplished lobbyist that he was, said he would be happy to translate for him. Reporting this visit to Lakanal, he assured him that Clay, never having received a more flattering and better-written letter, said to him with a smile: "You deserve that I use my influence for you. You did not then go to bed? You are worthy of being the friend of Mr. Lakanal. I will do all for his friend."[28]

On February 25, Parmantier thought the bill would be successful, awaiting the expected agreement of the House of Representatives and the signature of President Madison, of whom the petitioners were "very sure," as they were of his predecessor Jefferson and his successor Monroe.[29] Thus was confirmed the positive impression that Chaudron had expressed to Lakanal seven weeks earlier on the nice way things were going for the French settlement.[30]

The bill went back and forth between the Senate and the House. On February 27, Lewis Condict, a congressman from New Jersey,[31] brought it to the floor as amended by a House committee and it was sent back to the Senate the next day. On March 4 the Philadelphia daily *Aurora* carried details on the debate.[32]

Among its opponents, the newspaper listed Cyrus King (an Anglophile Federalist and half brother of Rufus King, who had been badly defeated in the last presidential election), and Bolling Hall, who had been in favor of the war against England.[33] Hall objected to the first section of the bill, which withdrew from sale to individuals four townships of vacant public lands located in a given part of the Mississippi Territory. Elected by the state of Georgia but living as a planter in east-central Alabama since 1808, Hall might have been irritated at the proximity of French colonists whom he saw as the fortunate beneficiaries of unfair favoritism. He had long shared his discontent with adversaries of this sort of plan, which exposed public lands to speculation and diverted them from better use. Relinquishing a part of the national territory to people who could profit handsomely from it was considered a crime against the republic. The country's public lands history, filled with wild schemes for decades, inspired a legitimate distrust, but this was forgetting that the French first had been its victims years earlier in the mess of the Scioto affair. In any case, the notion remained that the French had benefited from special treatment. Several months later the *London Traveller* pointed out, concerning their settlement along the Mobile River: "This is one of the most extraordinary speculations ever known even in America, fertile as it is in spirited adventures. It cannot fail that such a colony, planted in such a situation, must produce singular advantages to the American states, as well as to the settlers themselves."[34]

In favor of the project, according to the *Aurora,* was Speaker of the House Henry Clay, a Francophile who came through for the people who had solicited his vote and, backing him, men who were to hold high positions in American political life in the second quarter of the nineteenth century: secretary of state, vice-president, and president of the United States. These included John C. Calhoun, whose father had fifty years earlier settled Pastor Gibert's Huguenots near his plantation; John Forsyth; William Henry Harrison, the former governor of the Northwest Territory; Thomas B. Robertson; his brother John, as faithful a Jeffersonian as Samuel Smith, a congressman from Maryland and an influential debater in Congress, whom we saw at the side of the Baltimore postmaster on the occasion of a troublesome toast to the glory of the proscribed men and the shame of the Bourbons.[35]

The plan's defenders countered the risk of speculation with a humanitarian argument—the duty of hospitality—and an economic one: the opportunity of seeing specialized branches of agriculture and industries derived from them prosper in their country. This would certainly increase the value of these frontier lands.

Hall's motion to delete the first section of the text was soundly defeated. After several amendments were passed following a stormy debate, the bill was read a third time the next day in the House.[36] It was then sent back to the Senate, which accepted the proposed amendments on March 1, 1817, and passed the bill by a very large majority.[37] On March 3 the secretary of the president of the United States came to Congress to read Madison's message approving the grant of public lands for the encouragement of the cultivation of the vine and the olive. It was one of the last acts he signed into law. On March 4, 1817, Monroe was inaugurated as president. The act was now a law of Congress. Six days later, writing from Baltimore, Jacques Lajonie presented it in his own way:

> We are situated between the thirty-third and the thirty-fourth degrees latitude. The temperature is like that of Montpellier. Our river is the Tombecby or the Mobile, which flows into the Gulf of Mexico 80 leagues from New Orleans. It is parallel to the Mississippi and to the east of it. Our town will be one hundred leagues from the mouth of the river. It will bear the name Demopolis and our republic that of New France. The squares and the streets will bear the names of the famous battles of the French against the famous

allies. We shall govern ourselves according to a municipal code dictated by the famous Jefferson, ex-President, and the best members of the society. We shall have a representative in Congress. Our land is suitable to tobacco, cotton, vines and even sugar, and consequently every kind of grain.[38]

The law was composed of three sections. In the first, against which Hall had argued, the secretary of the treasury was to designate and reserve four townships, each six miles square, of vacant public lands of the part of the Mississippi Territory that had become a territorial district by the law entitled: "An act for the ascertaining and surveying of the boundary lines fixed by the treaty with the Creek Indians, and for other purposes." The second section set their price at two dollars per acre, the total payable in fourteen years, without interest, after conclusion of the sales contract between the Treasury Department and the agent or agents of the recently emigrated French, associated to form a settlement in the United States, of the age of majority, and in number at least equal to the number of half-sections of 320 acres contained in the townships—no one being permitted to possess more than one whole section. Finally, the third authorized the Treasury Department to begin allotting the lands, and to stipulate by contract the obligation to cultivate the vine and other crops that it would deem suitable. The issuance of title deeds was dependent upon respect of the conditions; no title could be granted before all the land in the said townships was completely paid for.

The conditions obtained by the French long aroused irritated reactions. In October 1817 the *Huntsville Republican* published a letter from an Alabama resident denouncing the "special and peculiar privileges" granted by Congress to these foreign capitalists, to the detriment of Americans of modest means.[39] In April 1818 the *Alabama Republican,* in turn, published a letter protesting the fact that these foreigners were given the opportunity to select the best lands and the best site for their settlement on the Tombigbee.[40] It was obvious that the French would never respect the conditions set them, would plant no vineyards, but would sell their allotments at a high price. *Niles' Weekly Register* dealt the finishing blow the following August, declaring that the grant was "in truth, much better than a mere gratuity," when it could already be valued at two million dollars.[41] The notion that the Pactolus River of Asia Minor, famous for the gold washed from its sands, watered the grant rather than the Tombigbee immediately aroused excitement, among the French as well as the Americans. In July 1817, subscribing for a quarter share, Montulé congratulated himself on an investment offering every advantage: "Obviously this is a gift from the government, in view of the length of time allowed for payment [fourteen years], which will have long been more than realized by the time it is due."[42] The following November, Skinner, the Baltimore postmaster, who had no more seen the lands than had Montulé, raised their value from two to ten dollars an acre.[43]

Historians have repeated time and again that the sale of the 92,000 acres at two dollars an acre payable in fourteen years without interest "was almost a gift."[44] The price of the land (much lower than the reality of a growth market), the duration of credit (three and a half times that in use since the Harrison Act of 1800), plus the fact that it was without interest are objective elements of a generosity that might easily be attributed to reasons that were primarily altruistic and economic: the country's natural duty of hospitality and its solicitude for exiles touched with Napoleonic radiance, the colonization of a region cleared of its indigenous population, and the development of new crops profitable to the national economy. Each party, the French and the Americans, committed to the other in an equitable exchange: work in exchange for land and the promise of riches to be shared.

In their respective works cited several times in the present study, *The Emperor's Last Campaign* (English translation, 2009) and *Bonapartists in the Borderlands* (2005), Argentinean historian Emilio Ocampo and his American counterpart Rafe Blaufarb each prefer to defend a different point of view, both centered on the idea that there was not a just reciprocity of interests. According to Ocampo, the French were concerned with their interests to the detriment of those of their hosts; according to Blaufarb, the Americans put their interests ahead of those of the émigrés. Let us examine their arguments before debating them.

CONTROVERSY ABOUT THE REASONS FOR THE VOTE

For Ocampo, the goals announced by the Colonial Society were false leads to divert attention from its true intentions, which were completely the opposite. The issues raised at the beginning of this chapter mark out the wrong road that, as some worried, the French exiles could take toward the Spanish colonies and St. Helena. Joseph Bonaparte and the generals who arrived in 1816—Grouchy, Clauzel, Lefebvre-Desnoëttes, and the younger Lallemand—strengthened by Grand Army officers who were daily joining them, wanted to settle somewhere on the Mexican frontier, or along the Ohio and Mississippi Rivers where the Emperor had said he wanted to live after his second abdication. It was therefore in order to obtain land for the society that, aided by William Lee, they were thought to have solicited Henry Clay, the Speaker of the House.

A Bonapartist Empire in America

Already in November 1816, Onís, the Spanish ambassador, claimed to have seen through the "perverse designs" of the Bonaparte family, which, after ravaging Europe, was planning to scatter over America the "seeds of its iniquities," with the tacit consent of certain powers that wished to weaken Spain and encourage the independence of its colonies.[45] He accused Joseph of playing the lead role in a drama that included a troupe of actors as broadly composed as it was varied in its personal, collective, political, and lucrative interests. French, Polish, and Italian veterans marching under the imperial banner, like officers Jacques Roul, failing in Buenos Aires,[46] or Raymond de Latapie and Philippe Gustave de Pontécoulant, in Brazil.[47] Leaders of independence movements like Chilean José Miguel Carrera or Francisco Javier Mina, who was fighting for a free Mexico. People of varied backgrounds, moving in shady circles: shipowners and merchants from the north-eastern United States, ready to furnish money, ships, and supplies; pirates, filibus-ters, and cosmopolitan adventurers: Frenchmen Louis-Michel Aury and brothers Jean and Pierre Lafitte, based in Texas, the Scot Gregor McGregor, who had been active in Venezuela, and the Spaniard José Álvarez de Toledo, wanting to liberate Mexico. These examples, adduced by Ocampo in a gallery of countless portraits, testify to a confused situation in which, further muddled by rumors, it was not al-ways possible to know who was on whose side or who wanted what in order to go where. Joseph could turn this confusion to his advantage by accepting the crown of Mexico that the rebels, or the generals led by Clauzel—"in the delirium of their exalted arrogance"[48]—were supposed to have offered him. Lucien was not forgot-ten; he was to rule Peru.[49] And of course the plotters' ideal candidate for the throne of their American empire was the exiled Napoleon himself.

Whichever Bonaparte was crowned, first the countries over which he could reign had to be conquered, starting with the viceroyalty of Mexico, offering its nearby frontiers to the Bonapartists and their allies working to liberate them-selves from the Spanish government. It was in this conquering perspective that the French colony to be established in the southern United States had its place. Grapes and olives did not in the least interest those who were supposed to produce them; what interested them was the location of the lands and their market value. The Rousseauism of a return to nature, the exoticism of plants, and the rustic re-training of soldiers were nothing more than mollifying themes hiding ambitious geopolitical motives.

Attacking Mexico from the United States seemed the obvious plan, except that this was against the new neutrality bill submitted to Congress by the Madison ad-ministration[50] and risked starting a war with Spain, something the United States did not want. The solution was to go through West and East Florida, as Colonel Aaron Burr had explained in his 1810 memorandum to Napoleon, asserting that without them there could be no hope of success either in Mexico or in Texas.[51] To get around the obstacle of American neutrality, the Floridas had to be conquered to make them independent of Spain, and then, after all of Spanish America had been liberated, sold to the United States. In this strategic view, it was essential to occupy Pensacola, since expeditions could ideally be launched against Mexico and its cousin colonies from its admirable natural harbor. Consequently, the proximity of Alabama made the port of Mobile and the sites being colonized along the Tom-bigbee very important, to serve as starting points for assaults on Pensacola.

The Bonapartists looked for an ally among the patriots, to seize the Flori-das. In mid-February 1817, Generals Lefebvre-Desnoëttes and Henri Lallemand were said to have gone to New Orleans for this purpose to meet with Francisco Javier Mina, based in Galveston, but he refused to discuss a raid on Pensacola when he learned that the Floridas would afterwards become American.[52] Later, when Gregor McGregor arrived in New York, his help was solicited: his recent past in the service of Bolívar and Miranda in Venezuela made him a prime can-didate. Regnaud de Saint-Jean d'Angély, sent by Joseph, supposedly talked to him about the plans being fabricated by the French generals exiled in America: put to-gether a squadron to go free Napoleon, take him to the Rio de la Plata and then to Chile, where he would establish an empire.[53] McGregor considered this impos-sible; he agreed to the less fanciful idea of seizing Spain's Amelia Island, northeast of East Florida, declaring it independent, and then asking Clauzel, who was wait-ing in Mobile (where he would pretend to set up the French colony that Congress had permitted by its law of March 3, 1817), to take Pensacola. Caught in a pin-cer movement, the Spanish would quickly surrender. After this, the plan would re-main the same: the Floridas would be the launching pad of expeditions to free the Spanish possessions in South America. This is what McGregor, a longtime Brit-ish spy and informer, reported to Bagot, his ambassador in Washington. He ex-plained that the Bonapartists' chief goal was Mexico, that "all the French officers in the United States were ready to join in the undertaking and that the settlement recently made by French Refugees upon the Tombigbee River in the Mississippi Territory was connected with this project."[54]

The fate of the enterprise was linked to two laws passed by Congress on March 3, 1817. The unfavorable one, the neutrality law, forbade any armed at-tempt against Florida and Mexico from the United States, while the favorable one granted the French émigrés 92,000 acres of land somewhere on the Tombig-bee, between the thirty-third and thirty-fourth degrees latitude. The interest of the grant was both geographic and economic: to assemble, quite legally, in one place relatively close to Mobile and Pensacola, a large number of Frenchmen; and, spending hardly a penny, to have at their disposal land likely to be worth a great deal in a developing region, whose sale or exploitation would enable them to pur-chase supplies, arms, and munitions.

The Spanish ambassador, Luis de Onís, was not taken in, thanks to informa-tion from Álvarez de Toledo. In 1816, Toledo, who had supposedly convinced Joseph that the Americas would welcome him if he accepted a liberal constitution, was a penitent rebel back in Madrid seeking the pardon of King Ferdinand VII. He claimed that the true objective of Clauzel, the younger Lallemand, and Des-noëttes, whom he called the leaders of the undertaking, was to set up a military port rather than a colony of farmers near the Spanish border. If the American gov-ernment was not in cahoots with them, it was at least guilty of a cover-up.[55] On March 3, 1817, Onís understood that to save Mexico the Floridas would sooner or later have to be sacrificed.[56] (It was he who, in the name of Spain, ceded them to the United States in the Adams-Onís Treaty, signed in Washington on Febru-ary 22, 1819. This did not prevent Mexico from declaring independence two years later.)

Meanwhile, the neutrality law threatened the French in pursuit of their in-trigues, while the law that had given them lands did not counterbalance it. If, thanks to his money and desire to be crowned king of Spain and the Indies, Joseph was truly, above the officers, at the head of plans against Spain, he was taking a huge risk in case the United States managed to establish this fact. Proof of his im-plication could make him lose the advantages of his position in his host country or even have him deported. Then, in late April 1817, General Charles Lallemand disembarked in Boston. Impetuous as ever, he gave second wind to the veterans' plans, working hard to recruit them locally: a year later many would find themselves on the road to ruin in the Texan Champ d'Asile. The elder Lallemand's arrival marked a turning point in the history of the French presence and action in the United States after 1815.

It was a turning point, for, as will be seen in chapter 12, Lallemand would, in the six months following his arrival, attempt to lead the members of the Colonial Society astray with a mirage of an exotic military oasis far headier than the rural, settled way of life awaiting them in the South. This fact is indisputable. Ocampo's analysis, making Lallemand a starting rather than a turning point in the development of the Colonial Society, is less so. The difference is substantial, for the society's entire philosophy is at stake. I have favored the sincerity of the initial step taken by the society's founders, which Ocampo, conversely, refutes. He sees the dice as loaded from the very start; the social and economic, even philanthropic motives of the society hid rather disreputable geopolitical ulterior motives: obtain lands and use them for more ambitious ends than to plant vines and olive trees, such as the conquest of the Floridas and Mexico or the deliverance of Napoleon. Ocampo relies particularly on the appointment of the elder Lallemand as president of the Colonial Society in November 1816,[57] and thus on the strong influence of this "redoubtable" general:[58] he who, at Rochefort, had urgently advised Napoleon to go to America, would not rest until he had him come there from St. Helena. As to vines . . . However, he was not the president of the society at its beginnings, but one year later in circumstances that we shall see were quite turbulent. In the autumn of 1816, Garnier de Saintes was president, while Lallemand was still in the Ottoman Empire after making a fruitless round of the pashas and sultans of the distant Middle East. The society supposedly chartered a ship to send to the Levant to bring him and Savary to the United States.[59] Certainly the society, of which his brother Henri was an important member, was interested in him, but it is doubtful that Lallemand played an actual role in its birth and in the determination of its goals.

What of the link Ocampo sees between Henry Clay's strong support of the French request for a grant of land in Alabama and his commitment to the emancipation of the Spanish colonies? What correlation is there between the two, other than the sympathy the Speaker of the House showed toward victims of injustice and fighters for liberty? In March 1818, during the extensive congressional debate on the opportunity to quickly recognize the Spanish-American republics, Clay led the partisans of recognition because, beyond any other possible consideration, the new republics were going to become, he said, the sanctuary of freedom and refuge against all persecution, as the United States already was. Clay was badly beaten when the issue came to a vote and his amendment was rejected, to the satisfaction of Secretary of State John Quincy Adams and John Forsyth, his principal adversary, the realism of whose arguments blurred Clay's more idealistic vision of a North-South American future.[60] Finally, while Clay and Forsyth were on opposite sides on this question, one year earlier both had voted for the bill in favor of the French.

Returning to the society's intentions, it would be unfortunate simply to accept their innocence to the detriment of Ocampo's argument. The Argentinean historian has discovered enough evidence in unpublished or obscure archives to call for a close examination of the goals of the leading French exiles (this does not mean French officers enlisted on-site, like General Brayer or the hussar Georges Beauchef[61] in Chile). Before the elder Lallemand joined them, Joseph, Grouchy, Clauzel, the younger Lallemand, and Desnoëttes had contact, at one time or another, with partisans of the independence of the Spanish colonies, and exchanges on the means and manner to help Napoleon escape from St. Helena. Among many other cases, the undeniable exchanges between Grouchy and Carrera in the summer of 1816 prove this: the former financed the latter in his purchase of two ships in Baltimore and asked him to include his two sons in the expedition.[62] Better yet, Grouchy proposed to the Buenos Aires government that he take command of the rebel army against Spain: South America would be, he wrote, "the new theater of glory" for Napoleon's officers exiled in the United States. He nonetheless promised to leave after liberating it from Ferdinand's tyranny. His only condition was that 120,000 duros be deposited in escrow to compensate him if his French properties were confiscated.[63] A single condition, but it changed everything.

These important Frenchmen exiled across the Atlantic seem often to have been irresolute men, incapable of truly espousing the patriotic cause because this could ruin their position in the United States or their chance to return to France. In short, they had more to lose than to gain. From this viewpoint, Grouchy's case is so exemplary that Hyde de Neuville remained doubtful when his informers warned against him: why would Grouchy have risked displeasing Louis XVIII? Received into American society, owning lands in New York, a large landowner in France for which he was homesick, concerned with his sons' careers, aspiring to a royal pardon—what could he hope from active participation in the independence of the Spanish colonies? Similar arguments apply to Clauzel, Desnoëttes, and the younger Lallemand. And what of Joseph Bonaparte? He was not in despair because he could not return to France as long as the Bourbons were in power: did not the wealthy, bourgeois life that awaited him on his estate of Point Breeze in Bordentown suit him better than that of king? American historian Patricia Tyson Stroud gives us the portrait of a Joseph fulfilled in his American private life, among his works of art, influential friends, and mistresses, as he had never been in his public life during the Empire.[64] Napoleon on St. Helena was convinced of this. In Spain, he confided to Bertrand, Joseph had been the worst head of state possible. He had always preferred his freedom, pleasures, women, books, and the amenities of social life to work. Despite his intelligence and talents, the torments of a possible second reign were not for him, nor the role of leading the Spanish colonies to freedom.[65] Could the French officers in exile, many of whom had fought in Spain and knew how incompetent he was in military matters, think differently? His wealth, on the other hand, was a powerful magnet. Among the "maniacs"[66]

who had come to the United States in the summer of 1817, Jacques Louis Galabert,[67] an officer and adventurer who must have known Joseph, as he had carried
out several missions in Naples and then Madrid, is supposed to have told Luis de
Onís that none of the émigrés liked him, only his money: a credible remark, even
though it was made by this man with a troubled past to a foreign ambassador.[68]

While it can be assumed that during meetings with former imperial dignitaries, Regnaud de Saint-Jean d'Angély, and the generals, Joseph did not merely sigh
when old memories were recalled—that he had contact with patriots and that he
was concerned about ways of freeing his brother—it must be noted that he was
never officially reproached with a breach of his duty of reserve, nor with known
participation in any Bonapartist plot. He was thus exonerated when the secret of
a Napoleonic Confederation was discovered in 1817. This vast conspiracy, modeled on Aaron Burr's, had designs on Mexico and the territory west of the Mississippi, whose crown would go to Joseph, alias the Count of Survilliers. Although
the compromising documents were addressed to him and Clauzel, Joseph was in
no way responsible. The plot was conceived by Lakanal alone, led astray in a moment of frenzy.[69]

Authenticated documents of this kind would have strengthened Ocampo's argument, often consisting of rumors or information coming from individuals whose
word or writings are difficult to verify—Toledo (a traitor), McGregor (a spy), Roul
(a fantasizer), Galabert (a possible double agent)—passed on by ambassadors un-
decided between the seriousness of their revelations and their charlatanism, or
the neo-Orleanist press, for instance, reporting that Stephen Girard had given
$15,000 "to finance a filibustering expedition against Mexico, supplemented with
very considerable sums from Joseph Bonaparte and 'other distinguished Frenchmen.'"[70] Unto those that have, shall more be given, which makes it easier to attribute to the former king Joseph or the banker Girard the strongest support possible, though both were so stingy, or to the Dupont manufacturers the intention of
subsidizing illicit operations.

A greater number of convincing facts might also have given more weight to
Rafe Blaufarb's argument justifying the political risk Congress was supposedly
taking by proving the bill's opponents and skeptics right, and incurring the displeasure of others by granting its favors to the French.[71] Blaufarb understands the
affective and economic reasons for Congress's positive vote, but subordinates them
to ulterior motives of a military nature, aligned more directly with American interests than those advanced by Ocampo.

Obvious Strategic Interest

General Humbert's presence at Jackson's side at the battle of New Orleans had
called attention to the weakness of the southern coastal defenses; General Bernard's and Captain Poussin's appointments to the army engineers had demonstrated the American government's wish to remedy the situation by constructing
a network of coastal fortifications and roads that would enable it to repel another

invasion of its territory, especially one originating in Spanish Florida.[72] At stake was the stability of the arc of frontier settlements linking Georgia to Louisiana as well as access to New Orleans, whose control was indispensable to the economy of the Ohio and Mississippi Valleys exporting through the Gulf of Mexico. Also at stake was the port of Mobile and its backcountry, which the Americans had taken from the Spanish in 1813 and which was the natural outlet for products coming down the Tombigbee and Alabama Rivers to the Mobile.

For the time being, neither vanquished England nor Spain, deprived of the income from its empire, was in a position to be intrinsically dangerous, but their ability to react in an international context where everything could change dramatically was not to be underestimated. Finally, there was the threat of Indians and runaway slaves, of whom the British had made fearsome use to sow terror from Florida to Georgia and Alabama in 1814. The peace treaty of Ghent had driven out the British expeditionary corps, but it had not disarmed the blacks or the Indians, incited by their former allies to continue their raids on American territory from their base on the Apalachicola River in the Spanish zone. Jackson ordered General Gaines to send his troops across the frontier to wipe out this base, known as Fort Negro, which was done on July 27, 1816. But in 1817 these cross-border guerilla operations were continuing to harass the pioneer settlements of the Southeast and encourage slaves to desert. Measures had to be taken to secure the interior of the country and reinforce its defense against the Floridas and the ever possible risk of another invasion from the Gulf of Mexico. From this viewpoint, an influx of population and the militarization of the colonists in the Mississippi Territory could resolve the problem.

In 1810—that is, twelve years after the creation of the Mississippi Territory—there were only 6,420 whites in Alabama, chiefly settled on the best lands along the waterways;[73] in 1813, when it was annexed, the port of Mobile had only three hundred inhabitants. The extremely sparse settlement made the region vulnerable to repeated hostile incursions. Opening millions of acres of land ceded by the Creeks and Choctaws to the public changed the situation by attracting speculators and cotton planters from neighboring states. Attributing a share to French émigrés was probably part of this wish to accelerate Alabama's development in order to consolidate the territorial expansion of the Union and enrich its economy.

But before the arrival of tens of thousands of American emigrants who increased Alabama's population thirtyfold and thus settled the Indian problem,[74] recourse to colonists "competent to its defense" had been considered, as General Jackson suggested to James Monroe on November 12, 1816,[75] and then, two years later, recourse to the militia, which the state's first governor thought superior to the regular army for its protection.[76] Men who could handle both arms and the ax or plow would be these required colonists, if they also had experience in military command and strategy. These soldier-farmers that the country needed to preserve its territorial integrity were coming to the United States after returning from European battlefields under the Napoleonic eagle.

Their settlement on the Tombigbee one or two hundred miles above Mobile could serve American interests in blocking raids from Spanish Florida and controlling navigation on a waterway vital to the development of the region down to its maritime outlet, whose forced Americanization was awaiting international recognition. Spain, less ready than any other country to accept the annexation of Mobile at its expense, saw, in the creation "of a colony of Frenchmen not far from the Mississipi [*sic*] and Western Florida in the region called Tombegbée" a sign that the American government wanted, "in order to achieve its views of expansion," to remain in possession of this part of Florida, as the colony had access to the sea only through Mobile Bay.[77] Therefore, one year after Congress passed the law of March 3, the Spanish ambassador in Paris tried to persuade the prime minister to protest along with him against an intolerable situation: the presence of people "known by their aversion to the august house of Bourbon" near a region extorted from Spain and adjoining the Spanish possessions on the coast of the Gulf of Mexico.

Spain deplored the fact that the United States had taken advantage of Spain's state of exhaustion after six years of devastating war in Europe and had spurned its indisputable rights in America. It begged France, in the name of their related Bourbon monarchies, not to leave Spain isolated, but plead its cause to a United States that was ignoring Spain. Florida would inevitably soon be taken over. While the methods of doing this had yet to be specified, landmarks could be set out, like the French camp on the Tombigbee, that would eventually make Spain give in. Blaufarb concludes that the congressional decision to respond favorably to the society's petition came from its concern to settle the problem of Spanish Florida by both demographic and military pressure.

Humanitarian Duty

While this reason certainly must be included in the considerations that motivated the American government and Congress, it nonetheless seems that its importance should be qualified. To the best of my knowledge, no document, whatever its origin—French or American, public or private—explicitly shows a cause-and-effect relationship between the necessity for an organized and armed colonial body in the Alabama Territory and the granting of four townships to French émigrés thought to be predominantly former officers of Napoleon, and this neither before the vote nor after. In November and December 1816, when they were corresponding regarding this subject, neither General Jackson's recommendation to have recourse to competent men for the defense of the territory nor Monroe's to extend pioneer settlements to it makes mention of the French. It can simply be supposed that the subsequent discussion of the society's petition took place within the general context of these sensible recommendations, given the sensitivity of the frontier area concerned. It should be stressed that the choice had not initially been guided by the national interest, but by a citizen of Kentucky interested in the lot of the French and the kind of natural conditions necessary to their enterprise. Furthermore, while the townships were granted near the Tombigbee River, their exact location had not yet been determined, and the choice of the site was left to the assessment of a member of the society, and so not controlled by American officials. On March 20, 1817, after confirming to Lakanal the success of the French exiles' request to Congress, Henry Clay added: "In the choice of the individual that shall be commissioned to select the land which the colony is authorized to purchase, great care should be had that it be a person of sound judgment who knows the region well or is in a position to know it well before determining the site."[78] The French choice of Alabama seems simply to have coincided with American territorial stakes in the region, given the rather modest contribution of military men to the functioning of the society.

Granting lands to the French on the belief that they would furnish a locally effective fighting force implies that veterans of the imperial army were legion in the United States. However, before returning to this topic in the next chapter, it is essential here to put into perspective both their number and their role in the society at its beginnings, that is, from its founding to Congress's vote of the law, for it is the situation preceding the vote that must be considered in justifying its success, not factors after the event.

At no time did officers compose more than about 20 percent of the society, and a great number of them were not involved in its preparatory work, since they arrived in America many months after this was completed. For that matter, with the exception of Desnoëttes and the younger Lallemand, the exploring generals, no officer is named along with the former Convention members and Domingan refugees, the primary movers of a colonial plan whose possible military function was never mentioned, whereas its commercial, agricultural, and manufacturing functions were repeatedly emphasized. The recruitment of members, carried out actively by William Lee among others, aimed at enlisting skilled workers, and it was partly on them that the society, backed by the *Abeille,* which echoed its goals, counted to win congressional support. Nor were there officers among the lobbyists with Lee and Parmantier. After all, what influence could they have in a country that did not want to incorporate them into its army, or if so, only exceptionally, as was the case with General Bernard? Grouchy was not in Washington with the society members to celebrate the congressional vote and shake the president's hand, but Lajonie, whose presence had nothing to do with his already distant past as a lowly second lieutenant in the cavalry, but with the sincerity of his commitment to serving the society, was.

In none of his letters, either before or after the time of the vote, does Lajonie speak of the colony from this point of view, except on August 19, 1818, when he was commissioned as a company lieutenant, First Battalion, Ninth Regiment of Governor Bibb's Alabama Territory militia, Desnoëttes being its captain. Lajonie was not a war hawk: "I do not give a damn for their service and their militia. I want only enough of it to be at peace, as one is generally everywhere in the United States. No more war except for defense."[79] But the defense in question concerned neither the Choctaws whom the French considered their friends, nor members of other tribes who never bothered them, but the Americans themselves, who were occupying French lands and refusing to leave. The Spanish objections to the founding of a colony near West Florida, mentioned above, were baseless. The fact that they very quickly disappeared from diplomatic exchanges between Spain and France makes their use to confirm a military background for the colony even less convincing.

Given the petition's economic cast and the large number of civilians involved, it is difficult to believe that the American elected officials based their decision first and foremost on the Colonial Society's military and defensive potential once it was ready to get to work.

The lands thus granted were in a barely pacified region with almost no white population, but with lush vegetation, with no better means of communication than rivers and therefore subject to the caprices of weather, far from any large urban center, being more than a thousand miles from the federal capital by land and more than two hundred miles by water from Mobile, the closest seaport. Moreover, if the principle of a gift is that it is given without compensation, Congress was not really giving one to the French émigrés, since it expected in return the planting and propagation of new crops that were to contribute to national prosperity. But the medium-term success of these delicate, specialized crops, demanding much preparation and care, was difficult to predict. Congress knew this. On two occasions, in 1806 and 1813, it had to consider the request of an extension of due date for payment from the Dufours and their Swiss associates in Indiana, who had explained their difficulties by the special nature of winegrowing. Finally, Congress was taking precautions against the risk of a quick resale of the lands with two very strict preventive clauses: first, none of the beneficiaries of the law could receive an individual property title for his allotment as long as all the lands had not been paid for, which assumed solidarity among hundreds of proprietors and the actual development of all the lands; and second, none of the beneficiaries could obtain more than one whole section, that is, 640 acres, which thwarted the plans of potential speculators, or in any case reduced their possibility of speculating on public lands. This minimal purchase to which they were held could not satisfy émigrés with means and landowning ambitions, like Joseph Bonaparte and Réal in the Northeast. General Clauzel, another member of this group, had already in April 1817 sent a set of questions to William Lee. Not knowing the answers, Lee reformulated them in English for Josiah Meigs, commissioner of the General Land Office in Washington:

> Can I purchase of the Government of the U States a thousand or two acres of land in the neighbourhood of this French settlement?
>
> Must I purchase them at public sale on the spot or can I treat with the Govt for these lands at private sale?
>
> What are the conditions of sale? If sold at auction.
>
> What are the conditions if sold to me by treating for them with the Government? When can I purchase them? Must I wait till the whole Territory bought by Jackson is surveyed before I can purchase or can I purchase immediately of the U States?[80]

Meigs answered Lee that very day that General Clauzel could purchase public lands adjacent to the townships allotted to the French émigrés as soon as their
site was chosen. But this could be done only at the Land Office of the Mississippi
Territory, at auction, the lands, sold by quarter section of 160 acres for two dollars
an acre, going to the highest bidder, one-fourth payable in cash, the balance outstanding in three payments, at the end of the second, third, and fourth years. He
hoped that a considerable part of the lands ceded to Andrew Jackson by the Indians would be surveyed and ready for sale within six months. In other words, as
long as they followed the rules applying to everyone, nothing prohibited Clauzel or
other French émigrés from purchasing public lands outside the townships granted
by Congress.

When all is said and done, if Congress had favored the émigrés with advantageous financial conditions, it had not done so lightly, conscious of dealing with
lands belonging to the people of the United States and of arousing the jealousy
of American colonists, but taking precautions against the risk of land speculation
ever prevalent in the country: the Scioto Company affair and Gallipolis, Ohio, of
which the French had been the victims, were fresh in everyone's memory. Considering the unusual nature of these new colonists—or at least some of them, the
Napoleonic officers—the American authorities remained on the alert. They were
right to do so, for in the fall of 1817 some general officers, first among them the
elder Lallemand, diverted the society from its original manufacturing and agricultural vocation, despite the fact that they were outnumbered by civilian members.

10

A Family Affair

A number of Frenchmen, among them some of the most enlightened men of the century, have formed, in Philadelphia, a company under the appellation Colonial Society, for the purpose of creating an establishment on the banks of the Ohio and the Mississipi [*sic*]. About 100 persons are already members of this company, which is composed of naturalists, farmers, manufacturers, artists and artisans.

—William Lee, New York, November 14, 1816

Jefferson had turned down the Colonial Society's request to draw up its rules, but he did not doubt that its members would succeed in their settlement and bring happiness to their descendants. The society seemed to be established upon a firm foundation with encouraging signs of its development guaranteeing its outcome: its builders were experienced men and their plans aroused increasing interest. Six months after its creation, there were some four hundred members who had already subscribed or promised to do so. Coming from a broad range of geographic, religious, social, and professional backgrounds, most were, however, not there by chance. The recruitment of members owed a great deal to a sort of family co-optation, as many were united not only by their status as émigrés but by kinship ties. In any case, the American citizen Lee, wishing to aid the progress of his country, was pleased that it could benefit from the contribution of these men in the sectors of agriculture, crafts, and industry. There was no mention of military men.

REGISTERING MEMBERS

The applications of the society's original members stretched out over a period of about a year, although most had joined by the spring of 1817. The Scioto Company had, during the Revolution, been more efficient because its headquarters were in Paris and its canvassing limited to France. The Colonial Society covered a broader field: it met in Philadelphia, corresponded with its vice-president in New York and later in Washington, and recruited in Europe and the United States. The time it took for mail and people to get from one place to another influenced the rate of enrollments, whose number and chronology are known to us: they followed one another as ships arrived in America and information was spread by newspapers, letters, and word of mouth.

In September 1816, at the latest, when he received a listing of the future citizens of Demopolis, Lakanal answered that he would be pleased "to be on this honorable list," probably in the company of some of the best-known émigrés.[1] In his October 1816 call for subscriptions, Dirat stated he had the names of the twenty future inhabitants of Proscripolis; in his November call, Lee raised this number to one hundred.[2] Membership had increased fivefold in five weeks. Had the business sense of the former consul, who invited interested Frenchmen to come to his New York home for information, led him to inflate the numbers in order to create a current that would shortly sweep up Lajonie?

In January 1817, just after arriving in the United States, Jacques Lajonie heard of the plan for a colony that included important French individuals. He still needed to get precise information, but he already thought he would subscribe as a colonist.[3] No newly arrived émigré could miss running into his compatriots in the consulates, boardinghouses, hotels, taverns, stagecoaches, steamboats, private homes, or offices of French immigrants, all meeting places where news was exchanged. As mentioned earlier, while in Burlington Lajonie met a French "ex-paymaster" whose commercial plans for Saint-Domingue did not withstand the promises of a brother who told him to join the colony of which Grouchy, Regnaud, and Pénières were members: "At the same time the letter told him that if he wanted to subscribe, there was no time to lose. This last bit of advice was important for me, since I had been planning on letting those who were the most eager take the first steps. But particular and certain interests for the first subscribers oblige me after this to be approximately the _hundredth_."[4]

Note how the information circulated and how Lajonie received it, first by letter between two brothers, then by oral communication during a chance encounter. The colonial plan thus spread among the émigrés who became its enthusiastic promoters, sending a great many letters to their relatives, friends, former colleagues, or fellow soldiers who had remained behind. During the crossing, those who were leaving with their membership ticket already reserved or with the idea of applying in America had time to make converts among the other passengers.

The press played its part in the general mobilization of French émigrés. American papers, closest to the information, wrote of the colony being formed and regularly ran news about the famous exiles, those whom Lajonie called "important people." Their membership made the project both serious and impressive. _L'Abeille,_ the official publication of the Colonial Society in Philadelphia, had the greatest readership for the threefold reason that it circulated in many other cities (Montreal, Boston, New York, Baltimore, Norfolk, Richmond, New Orleans, St. Louis), was published in French, and defended the cause of the outlawed men against Bourbon tyranny. A week before Lajonie calculated, in his letter of January 31, that he was in hundredth place, the newspaper reported on people's eagerness to join:

Several members of the Colonial Society of French Emigrants have shared with us letters from France and Belgium which promise complete success to these courageous founders. As soon as they have obtained the favor of a grant, men of all mechanical trades, landowners, farmers, and manufacturers intend to bring their financial means and their industry to a country where all these precious advantages are guaranteed by law. The figures we have been given list no fewer than three hundred individuals disposed to join the founders, and more than one hundred applications have already been submitted to the Society's officers.[5]

Three hundred new enrollments whose confirmation was expected were thus to join the hundred already registered. They came from France, Belgium, and Italy, where among the applicants for American exile, potential subscribers were increasing, solicited by letters from the United States or notified by the press. Because they were censored, French newspapers did not dwell on the situation of the exiles across the Atlantic, publishing only a few innocuous letters.[6] The opposition papers published by French exiles in Belgium, on the other hand, often featured a subject that was dear to them and related news concerning their distant compatriots. The *Mercure-Surveillant* of December 16, 1816, reprinted recent news from Great Britain from December 10: "American newspapers through November 14 have been received. Every day French émigrés are arriving. A society has just been created in their midst which is going to found a new colony on the banks of the Mississippi."

But for a long time, letters remained the principal medium of transmission, even among exiles in Europe itself. Thus, on July 2, 1817, former Convention member Félix Desportes,[7] writing from the spa town of Wiesbaden, Germany, informed Countess Thibaudeau, in exile in Prague, of the creation of a French colony in the United States and outlined for her how to go about joining:

> In New York, Regnault de Saint-Jean d'Angély is registering the future buyers and is undertaking to represent them in their transactions with the United States: all one needs to do is send him the authorization. One can become the proprietor of 640, 320 or 160 acres of land, besides a lot in town 100 feet wide and 200 feet deep. Regnault, who knows what he is doing, affirms it is a deal that will increase in value thousands of times within ten years. The craze for distant possessions has gone to all the Wiesbado-Gallic heads, whether cut off or about to be, but you will not catch me reaping my harvest on the other side of the world; we have too fine a crop to bring in right here near us.[8]

At this time the information could still be put to use, but it was a bit late. From the summer of 1816 to January 1817, the society claimed to have enrolled four hundred members, 25 percent of them definite and 75 percent to be confirmed when they actually arrived in the United States. The number that could be admitted was not limitless, but there was really no numerical restriction, as long as the 92,160 acres of land, divisible into allotments of 640, 480, 320, 160, 120, and 40 acres, had not been distributed.

The arrivals of latecomers spread out over the entire year of 1817 and continued into 1818, with the last three generals arriving during this period: the elder Lallemand in Boston in April, Vandamme in New York in July, and Rigau in Philadelphia in November 1817. (In 1818, General Gourgaud, back in Europe after keeping Napoleon company on St. Helena, might have joined them: according to Pierre Fougnet, Lefebvre-Desnoëttes had received a letter from him in which he was "asking for land in the colony.")[9] When all is said and done, it took over a year to bring in all the members of the Colonial Society, particularly those in the public eye. But the last arrivals, counterbalanced by those deleted from the list, did not modify the initial estimate of a base of four hundred members. On October 25, 1817, the reporter of the last general meeting held in Philadelphia noted the official number of 315 members who participated in the drawing of allotments, to whom must be added those who would have forty-acre reserve allotments.[10]

Taking into consideration all the changes in membership that occurred until 1829, a total of 460 individuals who were members at one time or another can be identified (this list, with nineteen items of information possible for each person, is given in the appendix). The year 1829 has been chosen because of a symbolic exchange: the departure from the French colony in Alabama of Jacques Lajonie, ex-second lieutenant in the dragoons, who was one of its leading citizens for twelve years; the arrival of Joseph Étienne Gardien, ex-second lieutenant in the royal guard and great-great-grandfather of Texan Kent Edmond Gardien Jr. (1924–2006). Separated by a century and a half, Lajonie, a privileged eyewitness, and Kent Gardien, an enlightened descendant, each played an essential role in perpetuating the history of the Vine and Olive Colony.

DEFINING THE MEMBERSHIP

The first printed list of the members of the Tombigbee Company, one of several successive names of the society, was published in Washington in 1818.[11] It was followed after 1820 by new lists of names, revised and augmented, but incorporating without change the original transcription errors—uncertain spellings, a frequent absence of given names, and total absence of punctuation[12]—that were long reproduced by historians in their publications.[13] Gardien was the first to revise the lists, correct the truncated or garbled family names, and identify a great many of them belonging above all to the groups of military exiles and Domingan refugees, easier to locate than the others.[14] He then drew up a biographical repertory, a model I have followed and considerably expanded, based on detailed research into French and foreign, public and private, associative and individual sources.[15] In my own repertory, every subscriber has the most detailed possible individual entry, including official public records, occupation, military career, date of and reasons for emigration, settlement in America, naturalization, stay in Alabama, Texas, or other states, return or non-return to France, and American and French descendants.

The congressional law had imposed on future society members the requirement to be French, of age, and of recent emigration, that is, assumed to have left their country after 1814. None of these criteria was respected. The society welcomed foreigners, Europeans, Americans of Anglo-Saxon or French origin, minors, and immigrants, often naturalized, particularly Domingan refugees, long settled in the United States.

The reality of the society of the French who had emigrated from their homeland to the United States was complex. In the first place, the members were not all French or were no longer so due to naturalization; furthermore, they did not all physically emigrate: some were born of immigrants, Creoles, or children of former colonists. The society was certainly French in essence, but it was also multinational.

A total of 126 members were said to be of nationalities other than French: 38 Europeans and 88 Americans. Most of the Europeans had served the Empire, with the exception of the Swiss Dufour and Siebenthal families growing grapes in Kentucky and Indiana. Despite their small number (10), Americans of Anglo-Saxon origin could not be on the list; though struck from it and replaced, they remain in our repertory because of their initial participation in the project.[16] Much more numerous were their compatriots of French descent, born in Louisiana like planter Valcour Aimé, or naturalized citizens of their adopted country like the former colonists of Saint-Domingue and Cuba, who had chiefly settled on the East Coast, from Pennsylvania to the Carolinas, and in Louisiana. Between eighty and one hundred members were from the Domingan refugee community, many of whom were, in 1815, living in the United States as American citizens. Though the door was not absolutely closed to foreigners, naturalized French, and those who had emigrated years before, it was open in priority to the émigrés of the Restoration for whom it had been created and who composed two-thirds of the membership.

France in all its geographical diversity met in the society. Within the territorial framework of the 22 regions and 92 departments of present-day France, it has been possible to identify precisely 227 places of birth (out of a possible 322) in 21 regions and 57 departments.[17]

With 122 listings to 105, southern France comes out just ahead of the north, predominantly due to two areas: the southwest, most particularly the department of Gironde, from which many people emigrated to the Caribbean and Louisiana in the eighteenth century via La Rochelle and Bordeaux,[18] and the southeast, open to the Western Hemisphere through Marseille, along with Corsica, birthplace of Bonapartists attracted to the United States by Joseph—Colonna d'Ornano,

Table 1. Distribution of 444 Colonial Society members by continent and country of birth

Europe	361	Metropolitan France	322
		Overseas France (Guadeloupe)	1
		Italy (contemporary)	19
		Kingdom of Piedmont-Sardinia	9
		Duchy of Parma	1
		Duchy of Tuscany	1
		Duchy of Savoy	1
		Earldom of Nice	1
		Papal States	1
		unknown	5
		Switzerland	8
		Canton of Vaud	5
		unknown	3
		Germany (contemporary)	4
		Prussia	1
		Duchy of Berg	1
		Duchy of Württemberg	1
		City State of Bremen	1
		Poland	3
		Warsaw	2
		Palatinate of Krakow	1
		Spain	3
		Monovar (Alicante)	1
		Malaga	1
		Toledo	1
		Ireland	1
Americas	83	United States of America	26
		Pennsylvania	11
		others	15
		Former French colonies	56
		Saint-Domingue	47
		French Louisiana	8
		Quebec	1
		Dutch Guyana	1
Total	444		

Table 2. Distribution of 227 known birthplaces of Colonial Society members in metropolitan France divided by a line running from the northwest (Nantes) to the southeast (Annecy)

SOUTH

Region		Department	
Aquitaine	42	Gironde	30
		Lot-et-Garonne	7
		Dordogne	2
		Landes	2
		Pyrénées-Atlantiques	1
Rhône-Alpes	19	Rhône	9
		Isère	6
		Ain	1
		Ardèche	1
		Loire	1
		Drôme	1
Provence-Alpes Côte d'Azur	18	Bouches du Rhône	14
		Var	4
Charentes Poitou	15	Charente-Maritime	9
		Vienne	4
		Charente	2
Midi-Pyrénées	11	Hautes-Pyrénées	5
		Haute-Garonne	3
		Ariège	2
		Aveyron	1
Corse	8		

NORTH

Region		Department	
Ile-de-France	31	Seine(Paris)	22
		Val d'Oise	3
		Yvelines	4
		Essonne	1
		Seine-et-Marne	1
Bourgogne	10	Bourgogne	1
		Yonne	4
		Saône-et-Loire	3
		Côte d'Or	2
Champagne-Ardennes	9	Aube	6
		Ardennes	2
		Haute-Marne	1
Haute-Normandie	9	Normandie	1
		Seine-Maritime	7
		Eure	1
Bretagne	9	Bretagne	1
		Ille-et-Vilaine	5
		Morbihan	3
Pays-de-Loire	9	Loire-Atlantique	8
		Vendée	1
Franche Comté	7	Doubs	5
		Jura	2

Region		Department	
Limousin	4	Corrèze	2
		Haute-Vienne	2
Auvergne	3	Allier	2
		Cantal	1
Languedoc-Roussillon	2	Hérault	1
		Aude	1
	122		

Region		Department	
Centre	6	Indre	3
		Loiret	3
Lorraine	6	Vosges	3
		Moselle	2
		Meuse	1
Alsace	3	Bas-Rhin	2
		Haut-Rhin	1
Nord-Pas-de-Calais	3	Nord	3
Basse-Normandie	3	Calvados	2
		Orne	1
	105		

Peraldi, Ilari, Luciani, Robaglia, and Sari. Île-de-France, seat of the nation's capital, and its neighbors Burgundy and Champagne furnished nearly half the northern contingent, far outstripping the west, despite its ports of Le Havre, Cherbourg, Saint-Malo, Lorient, and Nantes. The society did not really interest Normans or Bretons, represented by its secretary Parmantier from Lorient, but the people of the south. This is partly due to the fact that military men and Domingan refugees formed the core of the émigré group: the west had furnished fewer officers to Napoleon's armies than any other region, and the southwest had, more than any other, supplied the colonizing influx across the Atlantic.

In addition to these regional disparities, urban France is more strongly represented than rural France: 86 society members were born in 24 department prefectures (with Paris, Bordeaux, Lyon, Nantes, and Marseille in the lead), and 26 in 18 sub-prefectures, making a total of 112, which comes to half of the total number of listings on the preceding table. Two-thirds of the 115 remaining were born in the main towns of administrative districts (arrondissements and cantons), and the final third in simple rural towns and unincorporated villages. This urban origin was not the best asset for a Colonial Society devoted to agriculture.

Even if it appears at this stage that more members came from southern than from northern France, and a great many more from cities than from the country, the multiplicity of geographic origins prevents us from drawing up a typical profile. Vast differences separate Parisians from provincials, city dwellers from villagers, inhabitants of ports from the country people of the interior.

Religious affiliation does not help us get a clearer picture. Parish registers show that almost all of the members listed were Catholic, but Protestant registers are often missing, and baptisms or church marriages are not irrefutable proofs. A great many members, particularly from southwestern France, were Protestants, but their ancestors may have abjured or stopped practicing their religion openly when it was forbidden, as did the Lajonies in the Protestant stronghold of Gensac in Gironde. It is thus difficult to determine who was Catholic and who Protestant, even though there was a decided majority of the former. In some cases, attendance at religious services in the United States gives a better picture of the émigré's religious origin than do French archives, though it must be remembered that quite a few émigrés adopted the Protestant religion of their American wives. In any case, unlike the eighteenth-century migrations for religious reasons, such as that of Pastor Gibert in South Carolina, here religion had nothing to do with the course of events. The Swiss winegrowers from the canton of Vaud were the children of a pastor, but their emigration was motivated by the desire to go and propagate vine cultivation, not their Calvinist faith. Two of the society's members were churchmen, but rather unconventional ones. Mathieu Bernard Anduze from Aveyron was ordained as a priest in the United States and long served in Louisiana. Juan Rico Vidal, a Spanish Capuchin and former liberal member of the Cortes, the Spanish parliament, abandoned his monk's cowl for a general's uniform—the better to charm, marry, and abandon the widow of an Alabama colonist.

Young Men, Fathers, Sons, Brothers

We encounter the same problem with regard to the émigrés' average age, the arithmetical result of 302 available birth dates, but without significance, due to the varying reasons, conditions, and times of their exile. What connection could there be between a Domingan colonist who took refuge in America in 1792 and a second lieutenant born after this date who emigrated in 1817? Since the law of March 3, 1817, stipulated that society members had to be of legal age (twenty-one) to have the right to an allotment in the colony—which was not always the case—the émigrés' ages have been calculated as of this date.

From bottom to top of the population pyramid, three generations separate the youngest member of the Colonial Society, Louis Constantin, born in Lorient in 1802, the natural son of a quartermaster sergeant, from the eldest, Jean Thomas Carré, born in Normandy around 1744, a colonial planter recycled into educating young men in Philadelphia. Between those under twenty and the septuagenarians, both numerically in the minority, we find four groups, each smaller than the one preceding it. The largest group, at 37 percent, consists of those between twenty and twenty-nine: sons of Domingan colonists born on the island before its implosion or shortly thereafter on the East Coast of the United States; military men of the second half of the Empire, children of the Revolution, draftees, guards of honor, and lieutenants, many of whom came under the enemy's fire during the last campaigns in France and Belgium; tradesmen and artisans, sometimes demobilized soldiers, who had their fill of the Bourbons and were motivated by the hope for a better life. Next, at 27 percent, come those from thirty to thirty-nine: Domingan colonists and officers who experienced the entire Napoleonic epic. The 20 percent between forty and forty-nine are less important because of their numbers than because of the exceptional presence of six generals. The same is true of the 13 percent in their fifties and sixties, which include the three former Convention members and other Revolutionary figures, in addition to a few Domingan colonists like brothers François Barthélémy and Charles Louis Coquillon, born in the mid-eighteenth century, plus the fathers and uncles of the first group's refugees.

Thus the society was composed of individuals who were very diverse, with the exception of gender. It was strictly male, with only ten departures from the rule. There were two unmarried women: Léontine Desportes, chambermaid to the Countess Desnoëttes, and Eugénie Le Grand de Boislandry, Victoire George's niece; they later married members François Violle and Isaac Butaud, respectively. Two were married: Rose-Victoire Brun to Sylvain Godon, a Philadelphia mineralogy professor committed to the Pennsylvania Hospital for insanity; and Victoire Le Grand de Boislandry to Édouard-Côme George, supercargo, who was mentioned in chapter 6.

Six female members were widows. Two of these had been married to officers: Pierrette Julienne Basire, from Dijon, whose husband, General Pastol de Keramelin,[19] was killed in Silesia in 1813; and Louise Adèle Gertrude de Sevré, a Domingan refugee, married to Adjudant-Général Henry David in Philadelphia in 1804[20]—and later to Frédéric Ravesies. The other four were Françoise (Guessy) Dagneaux from Quebec, widow of a planter killed in Saint-Domingue; Joséphine (Verrier) Delaunay, born in Saint-Domingue, widow of a Bordeaux tradesman who died in Philadelphia in 1813, and remarried to Jean Claude Benoît de Boutière; Constance (Ogé) Démérest, also born in Saint-Domingue, widow of Charles Démérest, who died in the United States circa 1816; finally, Angélique (Thibaudeau) Audibert, whom we met as Garnier de Saintes' neighbor on the banks of the Ohio.

Besides their masculinity, multiple family ties connected the society's members to one another. Members often joined along with one or several close relatives. Family ties, either direct or by marriage, linked at least half of them according to six principal scenarios. First, fraternal ties were the most numerous: twenty-nine groups of siblings connect sixty-two brothers, from the four Follin brothers from Saint-Domingue—the largest family—to the two Generals Lallemand—the most famous, along with Marshal Grouchy's sons. Second came father-son relationships: fourteen fathers and seventeen sons. Grouchy, Joseph Martin, and Claude Payen each having two; one mother, Joséphine Delaunay, had a son, Pierre. Third, seven uncles had six nephews, Joseph Jeannet being the nephew of brothers George Nicolas and Louis René Jeannet; one aunt, Victoire George, had a niece, Eugénie. Fourth, there were ties between fifteen brothers-in-law, some having two: Louis Descoins-Belair with brothers-in-law Guillaume Promis and Simon Pothier, or Frédéric Ravesies with George Richard and François Plaideau, all from Saint-Domingue. Fifth, fathers-in-law and sons-in-law form nine doublets, two of which included Jean François Rolland (Prefect Réal's nephew), as his daughters married Camille Arnaud and Alexandre Ambroise Germond in New York in 1818. Finally, bonds of marriage united six couples, only one, Victoire and Édouard George, predating the creation of the society.

These kinship ties between two individuals, father-son, brothers, uncle-nephew, and so forth were the most common, but there were also relationships between three persons—General Clauzel was Constance (Ogé) Démérest's cousin and Louis Edmond Bourlon's uncle—or even entire families, often from Saint-Domingue: Chaudron-Stollenwerck, Ducoing-Fontanges, Bayol-Noël, Brugière-Teisseire, and Chapron-Teterel.

THE SOCIO-PROFESSIONAL CONTEXT

The diversity of social origins and occupations is the outstanding characteristic emerging from a study of 344 members and the fathers of 151 of them, extended, when possible, to other relatives or relations, particularly grandparents and godparents, who are often valuable social indicators. The size of the group furnishes useful statistics, favoring, however, distribution by sectors of activity rather than socio-professional classification.[21] While there are occupations whose name alone suggests the importance of the person exercising them—architect, judge, physician, officer—there are a great many for which this is not enough to indicate their status. The knowledge that a given person is a farmer does not tell us whether he owns his own farm, is a sharecropper, or is a day laborer; that another is a cabinet-maker does not tell us whether he is an employee or the boss, whether he is an artisan, offering only his technical skill, or an artisan shopkeeper, selling in his shop the production of his workshop. It is difficult to mark a boundary between crafts and trade, as well as between retail trade and commerce. How many tradesmen, designated as such, could claim the superior title of merchant?[22] Neither the milliner nor the dealer in porcelain, but the clothier, the wine merchant, the forwarding agent, and the shipowner. Specifying the place of society members in the social hierarchy would thus necessitate research going beyond the scope of this study, from the analysis of occupations and economic activity to the perusal of marriage contracts, estate inventories, and inheritance declarations—useful for those who were relatively well off. Sources are so numerous and so scattered that the task is impossible. The discussion is therefore confined to case studies, with particular attention to public records, in an attempt to define as closely as possible the social and occupational origins of the society members. The complexity of French society and the entanglement of its various stations—nobility, bourgeoisie, and working class—is mirrored in the Colonial Society of French Emigrants.

Social Classes

Some 50 French members and 5 foreigners—that is, 12 percent of the 460 listings—claim connections to nobility, either by title, name, or ancestry. The abolition of privileges makes identifying nobles more difficult, but a closer look reveals that in most cases these connections link only to doubtful or recently created nobility. After 1808, eleven of them, sons of grocers, clothiers, and gamekeepers, were made knights, barons, or counts of the Empire; also a baroness, Julie Basire, the widow of General Baron de Pastol de Keramelin and Adèle de Sevré, widow of Chevalier David. This new nobility did not owe its legitimacy to bloodlines, but to merit; it had coats of arms and had been confirmed by the Bourbons. The nobility of the others, most of whom had neither quarters, nor an authentic title, nor the least mention on any peerage list, but the nobiliary particle "de" or a hyphenated name as distinctive sign, was of no better quality. Society members bore the enhanced names they had inherited: Pierre Pascal Saint-Guirons des Traverses took the name of his father, Gabriel, a lawyer and crown prosecutor in Roquefort (Landes) in southwestern France; Charles Brossier Devanjeu's name comes from the farm of Petit-Vanjeu that his physician grandfather in Berry had used to inflate his family name; Joseph Martin du Colombier was the son of Antoine Martin, a planter ennobled into du Colombier thanks to the coffee produced by his slaves in Saint-Domingue; René François Delacroix Louvrais, an officer in the merchant marine, was the son of René Delacroix *sieur* (lord) de la Louvrais thanks to an obscure country estate in Brittany, as Jacques Lajonie was *sieur* de Lapeyre, the name of the vineyard-surrounded house where he was born, and before him, his father and grandfather who called themselves bourgeois. When they did not inherit it from a parent, often a landowner enriched through trade, they created a name to distinguish themselves from their parents: Lecoq Dumarselay, grandson on his father's side of *Noble Homme* Mathurin Lecoq, a wealthy merchant in Nantes, and on his mother's side of *Noble Homme* André Gaudin du Breüil and Renée Freneau de la Couronnerie; Garesché La Poterie and his brother Garesché Maisonneuve, sons of Jean Garesché du Rocher, a Domingan planter and merchant, and Elisabeth de Brossay. Examples abound. But these flattering genealogies should fool no one. They reflect only an apparent, at best minor nobility, far removed from the high aristocracy that had emigrated to the United States a quarter century earlier.

Counterexamples are rare. Emmanuel de Grouchy, first of all, a count of the Empire but a genuine marquis whose grandfather and father, both equerries, were lords of various holdings in Normandy, the one having been page to the Princess de Conty and the other a member of the Grande écurie du roi—the king's Great Stables (a vast establishment housing not only the king's hundreds of horses, but also the equerries, stablemen, musicians, and pageboys). In 1785 the future marshal married Cécile Félicité Céleste Le Doulcet de Pontécoulant, the daughter of a marquis and lieutenant general; Louis XVI signed the marriage contract and gave them a gift of four thousand livres.[23] Another was René de David de Perdreauville, born in Versailles and a queen's page, son of the governor of Marie Antoinette's pages, who had once before emigrated to Philadelphia with his sister Marie Louise Antoinette, where in 1801 she had married the Count de Dreux-Nancré, who had fought beside the Revolutionary troops in the Battle of Yorktown. Their support of the Empire might well have been seen by the Ultras as an inconceivable denial of their class. As these two cases are exceptions to the rule, it remains difficult to dissociate the noble portion of the Colonial Society from the bourgeois portion in which it originates.

But within the bourgeoisie itself it is even more difficult to distinguish the various levels that composed it, and to suggest trustworthy percentages, given the methodological problems laid out above. An examination of the occupations of 151 fathers of members shows that two-thirds were artisans and tradesmen in France and the colonies, but this usually does not enable us to determine their status in their profession. It is certain that a good number of these people were wealthy bourgeois, rising little by little toward nobility by fortune, public office, and marriage, as had the ancestors of General Lefebvre-Desnoëttes. His grandfather Charles Lefebvre, *sieur* du Pond, a bourgeois cloth merchant in Laigle, Normandy, married Marie Marguerite de Malleville de Monpoignant. His father, Jean Charles Lefebvre-Desnoëttes, was a king's councilor, pensions controller for the city of Paris, a member of the Directoire de l'habillement des troupes (Council of Troop Outfitting), a bourgeois cloth merchant with a shop, called Prince-de-Conty, in the rue Saint-Honoré near the Pont Neuf.[24] It is easier to place into this upper category of the bourgeoisie the professions serving the king in the civilian (prosecutor, judge, etc.) and military (officer) sectors, and in those belonging to the liberal arts (physician, architect, lawyer, etc.).

Fewer than 20 percent of the members' fathers considered came from the comfortable bourgeoisie composed of merchants, well-to-do shopkeepers, master artisans, physicians, lawyers, officers, and landholders. Bourgeois and notables in their towns or villages, some became pillars of the imperial regime in their local political and administrative life and sent their sons to lycées, military schools, or the regiments of honor guards in the closing years of the Empire.

The case of George Philippe Batré, assessor and clerk of the royal court in Stettin, Prussia, where he was born, illustrates this perfectly. Established as a merchant in Bordeaux around 1780, he soon married the daughter of a local bourgeois with a dowry of thirty thousand livres.[25] Ruined by the Revolution, he reestablished himself as president of an arrondissement of Bordeaux and then as a prefecture councilor during the Consulate. As head of the National Guard delegation and member of the electoral body of the department of Gironde, he attended Napoleon's coronation. The Empire's success assured his own. In 1812 he was among the one hundred most highly taxed residents of his department. He swore loyalty to the Emperor when he returned from Elba and was rewarded by a position as provisional director of the poorhouse. The royal edict of July 12, 1815, dismissed him as a civil servant of the usurper, which eventually led his sons to emigrate—not the eldest, who had just served four years in a regiment of dragoons, but the younger ones. Charles, twenty-two, led the way to America, followed by Adolphe and Frédéric. The latter, an ex-honor guard, died in New Orleans in 1818, but his brothers became notables in the French merchant and Freemasonic community of Mobile, Alabama, where they settled.

The bourgeoisie constituted the educated portion of the Colonial Society at a time when the quality of studies often depended upon social level and wealth. This reality persisted despite the efforts of the Convention's Committee of Public Instruction to establish a national education system, the creation of the *école centrale*[26] during the Directory, and that of the lycées during the Consulate.[27] The latter gave access to the prestigious higher education institutes, but chiefly to sons of well-to-do parents, as room and board were costly. Achille Chapotin, the son of a lawyer from Auxerre, went from the lycée Napoléon in Paris to the imperial École polytechnique and then to the Applied School of Artillery and Engineering in Metz. Jacques Lajonie, born into a rural bourgeois family that had rallied to the new political ideas, went from the lycée of Bordeaux to the Special Imperial Military School of Fontainebleau. His letters from America demonstrate the writing abilities that could be attained by émigrés coming from the same mold or an equivalent environment, as has been seen in the case of Édouard de Montulé or Guillaume Tell Poussin. The hundreds of other letters collected in connection with this history separate, along a social fault line, educated émigrés, politicians, civil servants, officers, engineers, merchants, artists, and so forth from their uneducated companions. The irreproachable handwriting and prose of landowner Lecoq Dumarselay stand out against the clumsiness of Bellemère, a stocking manufacturer from the department of Aube, where his grandfather had been a wheelwright. Lajonie's inventiveness contrasts sharply with the naïve style and almost phonetic spelling of François Constantin or Antoine Fougnet, who emigrated at the age of fifteen with almost no schooling. Good writing abounds in the breeding ground of correspondences because their authors wrote—which does not mean that these authors were necessarily the most numerous in the Colonial Society.

In the society's melting pot, common people are intermingled with the minor nobility and the bourgeoisie, but they remain anonymous. Whereas the archives of the upper classes reveal a great deal because their members produced paper in their private and public lives, those of the working class remain mute, often reduced to simple public records with no systematic indication of occupation and sometimes only an X as signature. Some wrote French with difficulty, and others spoke it no better, expressing themselves only in dialect. These were people from towns and villages, sometimes on the fringe of the lower middle class, shopkeepers, small manufacturers, and retail artisans. There remains the case of farmers, of whom research in archives has turned up only sixteen.

The members of the Colonial Society thus came from the various levels of French society, in metropolitan France and the colonies. Some remained in the class of their origins, while others took advantage of the upheavals of the Revolution and Empire to leave it and rise to a new social level. This would be evident in their marriages if they were studied; it is evident in their occupations. Most of the members did have one thing in common, however: economic fragility—property lost (Saint-Domingue) or seized, absence of income (no more pay), scant savings, business misfortunes (personal management or the economy). Very few émigrés had left behind an estate that in itself would make their return to France worthwhile.

THE RANGE OF OCCUPATIONS

Research into the professions, jobs, and occupations, gives us a list of about eighty, exercised by 343 members, grouped into ten principal categories. When there is a choice due to personal situations that had changed with time, the member's last known activity at the time of his emigration is used. Lakanal, for example, began his career in religion and education, pursued it in politics, then settled into a bursarship at a lycée, before ending as inspector of weights and measures. Jean François Roland de Bussy, the son of an infantry officer, began his in the army and finished in the police, where he was a commissioner and then secretary-general of the Paris Police Department, whose prefect, Count Réal, was his uncle.

During the Restoration, purges cost thousands of civil servants their positions, and demobilization sent home even more military men. The task of classifying members suffers from this transitory passage between an imperial past and an uncertain future. On exiles' passports this leads to statements that are often incomplete or inexact, hiding the bearer's real situation, especially when he had played an important civilian or military role in the service of imperial power. Roland de Bussy, mentioned above, is listed as a landowner.[28] A great many officers or noncommissioned officers are registered as merchants going to the United States on business: Charles François Génin, said to have been a naval gunner, is listed as a dealer in lace, sailing to New York from Le Havre.[29] Despite these reservations, table 3 makes it possible to identify several outstanding features of the composition of the Colonial Society.

Three facts stand out. The first is that there are five times as many civilians as there are military men. The second is the negligible number of members with a background in agriculture. The third is the preponderance of the sectors of commerce and industry. It is worth noting that the society's advertising never boasted of the presence of former officers in its ranks. The American customs service labeled them unproductive; they were not, from a strictly occupational point of view, attractive elements or guarantees of economic success. Lee, the society's American vice-president, preferred to enhance its image by listing among the first one hundred members only useful or productive working individuals, manufacturers, artists, artisans, naturalists, and farmers.[30] Among those expected from Europe in January 1817, the *Abeille* mentioned "men of all mechanical trades, landowners, farmers, and manufacturers" who intended to bring their financial means, their industry, and their skill to the United States, where these precious advantages were "guaranteed by law."[31] In October 1817, the transformation of the Colonial Society into the Tombigbee Agricultural and Manufacturing Company attests, if need be, to the industrious calling of the French colony.[32]

A Clutch of Colonels

Methodological concerns and documentation gaps complicate the attempt to arrive at an accurate census of the military men. A consultation of three essential sources concerning officers has resulted in identifying almost all of them, for how can an officer have neither a personnel file at the defense ministry, nor a file as a member of the Legion of Honor in the national archives, nor his name registered at one time or another on the list of those injured in the imperial wars? The probability of escaping one of these three possibilities is negligible. On the other hand, it is impossible to enumerate the noncommissioned officers and private soldiers. This would require a perusal of army recruitment files, regimental muster rolls, and pension fund records. With this reservation, the military contingent constituted only from 20 to 25 percent of the Colonial Society, a fact that, as Blaufarb points out, contradicts a historiography obsessed with the presence of high-ranking figures and a few others less so but who, far from their homeland, claimed for themselves a rank that was no longer or often had never been theirs.[33]

The king had used his legitimacy to invalidate the usurper's promotions and set the officers back to their rank prior to the Hundred Days. At best, the demoted men accepted this grudgingly; in exile, like Grouchy clinging to his marshal's baton, they paid no attention to it. The list of military men who emigrated to the United States is diminished by the Bourbon Restoration: of fifty-six officers recorded, more than half are junior officers (second lieutenants, lieutenants,

Table 3. Occupational classification of 343 Colonial Society members (1814–1817)

	No.	%		No.	%
Agriculture	16	4.70	Wine merchant	1	
Farmer, husbandman, plowman	9		Tobacco Merchant	1	
Planter	2				
Winegrower, winegrower & nurseryman	5		**(Liberal) professions**	36	10.50
			Architect	1	
			Art		
Domestic staff	2	0.60	Naturalist artist	1	
Maid/lady's companion	1		Painter	1	
Valet/estate manager	1		Miniaturist painter	1	
			Musician	1	
			Education		
Crafts, trade, & industry	59	17.20	Teacher	1	
Armorer/gunsmith	1		French teacher	1	
Textile artisan	1		Spanish teacher	1	
Baker, bakery shop owner	2		Language teacher	1	
Hatter	1		Music teacher	1	
Pork butcher, pork butcher shop owner	2		Master of studies	1	
			Dancing master	1	
Carpenter	2		Riding master	1	
Chocolate maker & tobacconist	1		Law		
Apprentice chocolate maker	1		Lawyer	3	
Hairdresser, hairdresser perfumer	2		Man of law	1	
			Notary	1	
Confectioner	3		Engineer	3	
Shoemaker	1		Topographic engineer/cartographer	2	
Cutler	1		Health		
Distiller	3		Physician	4	
Cabinetmaker	1		Health officer	3	
Bottler	1		Surgeon	3	
Building contractor	1		Practitioner	1	
Stocking maker	2		Pharmacist & physician	1	
Music box maker	1		Veterinarian	1	
Tinsmith	1				
Horse hair curler	1		**Public service (civilian)**	11	3.20
Engraver	1		Member of council of state (ex-minister)	1	
Clockmaker, jeweler, gold/silversmith	14		Inspector of weights and measures	1	
Printer	1		Examining magistrate	1	
Mechanic	1		Presiding judge of the customs court	1	
Tawer	1				
Bandbox maker	1		Prefect of police (mbr. council of state)	1	
Locksmith	1				
Tailor	3		Secretary of the prefect of police	1	
Tanner, tanned hide seller	4		Secretary of the minister of police	1	
Dyer	1		Head of combined tax administration & manager of the Bank of France	1	
Cooper	1				
Glassworker (manufacturer) & deputy	1		Sub-prefect	3	

Table 3. *Continued*

	No.	%		No.	%
Merchant marine	7	2.00	**Public service (Military)**	65	19.00
Ship's captain	4		(*see table 4*)		
Supercargo	2				
Officer of the merchant marine	1		**Miscellaneous**	7	2.00
			Clergy	2	
			Employee	2	
Commerce & business	129	37.60	Student	3	
Tradesman, shopkeeper, merchant	121		**No occupation**	11	3.20
Broker	1		Landowner	3	
Grocer	1		No indication	8	
Bookseller	1				
Dealer in lace	1				
Milliner	1				
Dealer in porcelain	1				
Wine merchant	1				
Tobacco merchant	1				

and captains); only fourteen are field officers and generals (major to lieutenant general). Furthermore, it will be recalled that many foreigners and even Frenchmen were enrolled in allied armies, chiefly those of Poland, Spain, Italy, and Naples, at a rank that was overvalued compared to what they would have had in the French army, and which the Restoration contested.

Grouchy, the much debated marshal, was the society's Napoleonic eagle. Although he fiercely defended his honor, damaged at Waterloo, his reputation had suffered, not only with the French diplomatic corps in America, which ignored the marshal's brevet he was issued during the Hundred Days, but also within the community of exiles, which was divided concerning him. More intent upon obtaining a royal pardon than upon settling permanently in the United States, he remained on the sidelines of his compatriots' work. This was not the case with his peers at the top of the military hierarchy in exile: Lieutenant Generals Clauzel, Lefebvre-Desnoëttes, and Vandamme and Brigadier Generals Rigau and the Lallemand brothers. Younger than Rigau, more thoughtful than Vandamme, less impetuous than the elder Lallemand, Clauzel and Desnoëttes were the men the society could truly count on.

Behind these seven came two colonels and two lieutenant colonels. In a surge of filial solidarity, Colonel de Grouchy had come to join his father in exile for a year. In his flight to escape imprisonment, Colonel Douarche had left behind a destitute wife and two children. "A man of honor before the enemy,"[34] Douarche well deserved his colonel's rank after twenty years of service, but not the money taken from various regiments to settle his gambling debts. Convicted of embezzling funds from the Fifth Foreign (Belgian) Regiment during the retreat from Waterloo, he was compelled to use two-thirds of his half pay to reimburse them, which

Table 4. Classification of sixty-five of the Colonial Society's military men according to their rank (February 1814)

Army			Navy	
Officers			Officers	
General officers	7		Lieutenants	2
Field officers	7		Employee in naval	
Junior officers	32		administration	
Officers of unknown rank	10	56	Ship's steward	1
				3
Guards of honor	4	4	To whom can be added:	
Employees in army	2	2	*Presumed officers and*	28
administration			*noncommissioned officers*	
		62		

prevented him from satisfying the demands made by officers of another regiment he had robbed in Spain two years earlier. He attempted a diversion via Bourbon Island (the present-day Reunion in the Indian Ocean), claiming that very important business matters awaited him there, but the minister refused his permission. Douarche escaped his obligations by crossing the Atlantic.

The two lieutenant colonels were the Italian Forni, the last artillery colonel in the Guard of Murat, king of Naples, and Arnaud Texier de la Pommeraye, who had served as infantry colonel in King Joseph's army in Spain, before obtaining the retirement pay of a lieutenant colonel by a royal decision of 1814. Having spent thirty of his fifty years in the army, Texier resumed active duty during the Hundred Days; the prefect of police issued him his passport for America in August 1816.

Not one of the other twelve French officers calling themselves colonel in the United States actually was one. Consul General Pétry was not taken in by this usurpation of rank. "Every day," he pointed out to his minister, "we learn that those who call themselves colonels were captains at most, and that the captains were only lieutenants, soldiers, or servants."[35] They apparently took advantage of the local custom of granting themselves a higher rank "in a country where, as you can see for yourself, everyone keeps for the rest of his life the military rank that he held for a few months, if only in the militia," wrote Victor Dupont de Nemours to Hyde de Neuville.[36]

Colonel Voerster had enlisted in 1806, the year his native duchy of Berg was ceded to Napoleon. He transferred his allegiance to Louis XVIII as infantry major on the staff of the royal guard, but in the spring of 1817 he boarded the *Amazon* in Amsterdam, bound for Philadelphia. A shipmate was Colonel Jordan, a member of one of the first families of the Palatinate of Krakow and a lieutenant of the Polish lancers in the cavalry of the Old Guard when he was injured in the battle of Arcis-sur-Aube on March 20, 1814. The society had another Polish colonel in the person of Jean Schultz, a cavalry major during the French campaign and then light horse commander of the Guard on Elba. After his return to Paris, the Emperor promoted him to captain of the Guard's lancers with the rank of infantry major.[37] Schultz followed the Emperor to Waterloo, where he was injured, and then to England, from where, denied the opportunity to accompany him to Saint Helena, he sailed to Malta with Lallemand and Savary. After many and varied incidents, he returned to France in 1817, but obtained a passport and soon left for America.[38]

The island of Elba was the breeding ground of the Colonial Society's "colonels." In addition to Schultz, three other officers served the Emperor there. The infantryman Michel Combe, a lieutenant in the foot grenadiers of the Old Guard, had already given him a toe, lost to frostbite in Russia.[39] Napoleon promoted him to captain on Elba, then in Paris to major in the Guard with the rank of lieutenant colonel. The Restoration had rescinded his promotion after Waterloo, where he had distinguished himself, and heaped on him the humiliations that may have been responsible for his later irascible character and certainly prompted him to leave the country. He was welcomed in America as a brave man.

The son of a clockmaker, Combe had risen from the ranks. Nicholas Raoul was a graduate of the prestigious École polytechnique and a general's son.[40] A lieutenant in the Guard's artillery in 1812, he fought in Germany, Russia, and France before following the Emperor to Elba, where he headed the military engineers. A promotion to major in the Guard with the rank of colonel in the regular army rewarded his loyalty, but injured at Waterloo, taken prisoner, freed on parole, he was stripped of all pay and of his recent promotions because as a French officer he had served in the independent state of Elba. Upon this, he resigned and went to Rome, where he had been offered a position as tutor to Napoleon Louis, Grand Duke of Berg and second son of Louis Bonaparte and Hortense de Beauharnais, who had ruled Holland during the Empire. The French ambassador to the Vatican soon accused him of being an emissary of the Bonapartes,[41] but above all as the man who had taken the Marquis de Sinibaldi's money and wife, Teresa, née Alvora. After his return to France, gossip in Paris regarding his Roman adventures thwarted his repeated requests to be reinstated to his army position. In late 1820 he went into exile in America with his Italian mistress, Teresa, and two of her children.

Least colonel of all, naval lieutenant Louis Taillade left without his wife and daughter, because of whom he had settled in Elba.[42] A Breton seaman sailing in the Mediterranean, he had entered Napoleon's service in May 1814 when he was given command of the schooner *Bacchante,* part of the Elba flotilla. To thank him for transporting the reconquering army to the French coast, the Emperor promoted him to commander on March 20, 1815. Outraged that Taillade's promotion was the same date as Napoleon's return from Elba to Paris, the Bourbons struck Taillade from the naval lists the following July 29 and put out a warrant for his arrest. He took passage on the brig *Saunders* in Leghorn and, in July 1817, arrived in Philadelphia, where he received aid from the elder General Lallemand.[43]

Like Taillade and Colonel Parmantier, Colonel Cluis had come to America without his wife but, like Raoul, with his mistress. Before rising to his ephemeral rank of captain in the national police (gendarmerie) during the Hundred Days, he had been a cavalry lieutenant. Colonel Charrassin, a lieutenant in the Old Guard's lancers, had become a last-minute captain on July 5, 1815, by order of the commission of the provisional government. Colonel Gruchet, another cavalry captain, claimed to be a major, perhaps promoted during the Hundred Days, and an aide-de-camp on the staff of General Belliard after Waterloo. He had fled France in 1816 to avoid persecution "for having served under Murat," he said.[44] Marshal Soult had added Colonel Galabert to his staff in Spain, had promoted him to major in 1814, and later to lieutenant colonel.[45] Louis Jacques Galabert—adventurer, secret agent, poet, sailor, and survivor of Saint-Domingue and the Revolution—had safely returned from a sea voyage around the world to enter upon a military career later in life. He missed Waterloo because Soult's order to join him is said to have reached him after the battle. Relegated to non-active duty and sent to live in Pau as a half-pay, he resigned and sailed for America via Liverpool.

This officers' portrait gallery does not show the military side of the society in the best light, with its true-false marshal, one general who was a brute and another a fanatic, a genuine colonel in flight for embezzlement, and a clutch of questionable ones. Concerned with their personal futures or with self-promotion, these important figures turned their backs on the society and were not the driving force that humbler members—like Lajonie, whose letters show his naïve delight at having his name associated with theirs—would have had the right to expect. The society relied on ordinary people, enlisted men, merchants, artisans, and former Domingan colonists.

Roots in the Soil

The relatively modest military participation is the first lesson learned from our analysis of the society's composition, and the slim representation of the rural sector is the second. A mere sixteen members, fewer than 5 percent of those identified, were engaged in working the land as day laborers, farmers, and winegrowers, like the Swiss Dufour family and Siebenthal brothers: five genuine viticulturists who never came to Alabama.[46] It should not be deduced from this that the society was not at all connected with the rural world, which at the time was that of the vast majority of the population. Some of the wealthiest members were landowners who carefully saw to the management and expansion of their estates, where they rested from their military campaigns or their political activities. They were conversant with agricultural matters and often took an interest in them. Sent to Dordogne on a mission in 1793, Lakanal had set up a Center of Rural Economics in Bergerac, a sort of model farm where he taught the art of agriculture. In 1806 Garnier was vice-president of the Agricultural Society of Saintes; in 1813 Pénières received a prize of five hundred francs from the Society for the Encouragement of National Industry for growing and grafting walnut trees in his native Corrèze, and in 1816 William Lee described him to President Madison as an experimental agriculturalist and naturalist whose participation was a guarantee for the society.[47] Lee thought the French colony would soon surpass the settlements of Vevay and New Harmony in Indiana.

But, more importantly, most of the members had their roots in the soil, particularly that of the winegrowing regions abounding in France. There were the appropriately named Plantevignes, two brothers, merchants who had emigrated from Bordeaux to New Orleans in 1816 and whose ancestors had distilled brandy near Cognac. Bugey, Roudet, and Vial from Dauphiné; Lajonie, Mangon, and Fougnet from the Bordeaux country; Manoury from Champagne; and Jame from Burgundy, to cite a few examples—all had been born in the heart of the vineyards and grown up to the rhythm of grape harvests. Jean Baptiste Jame's story is worth telling.[48]

Born in Chagny (Saône-et-Loire) in 1767 to an old family of wine merchants, he studied at the Collège d'Autun, which Joseph Bonaparte also attended. Jame became his friend and then Napoleon's (charmed by one of his sisters) before becoming their right-hand man. He participated in the expedition to Egypt, provisioning it with Chambertin wine,[49] then went into finance, becoming the head of the tax administration, manager of the Bank of France, and Joseph's general manager in Paris. At the height of his success, Jame mysteriously disappeared on Christmas Eve 1813, abandoning his family, a fortune of one million francs, a château on the Saône River that he had decorated in "return from Egypt" style and whose estate he had considerably extended, with farms, meadows, a stud farm, and a merino sheep-breeding establishment.[50] He also left behind in Côte d'Or exceptionally fine vines in various vineyards in Vosne, Flagey, Nuits, and Boncourt, including in particular a plot producing today's great Romanée wine. He went to America via Switzerland and Belgium under the name Chevalier James de Bellièvre.[51] In 1817 he was in Baton Rouge, Louisiana.

Wealthy merchant, banker, landowner, winegrower deluxe, and gentleman farmer Jame was the exception. However, their horizons broadened by their military service, many members who were no longer farmers had remained close to their rural origins and were still capable of attending to field work.

The same could be said of the refugees from Saint-Domingue, planters dispossessed of their acres of indigo, coffee, or sugarcane, more than of their children born before the slave revolt or shortly thereafter in the United States. A few had started anew on the North American continent, in the Carolinas or Louisiana, but all had retained a nostalgia for their lost paradise and their taste for land, and were looking to regain both. In the United States, most of the colonists had gone into the wholesale or retail business, dealing in dry goods, wines, and liqueurs imported from France, or heavily into related enterprises such as equipping ships, consignment, and brokerage. In the large American ports, the refugees formed partnerships among themselves and with other émigrés, with frequently changing corporate names: in Philadelphia, Lapeyre, Farrouilh & Co.; Garesché & Ravesies; Curcier, Ravesies & Co.; Chapron, Frenaye & Co.; Ducoing & Lacombe. Often founded during the 1790s, these Franco-American commercial firms had not broken off contact with France, where their close relatives were often their special correspondents. After 1815 they were all the more eager to welcome the new arrivals, as these had made the journey on ships whose cargo was coming to them from Bordeaux, Nantes, or Le Havre. Older and new émigrés made common cause.

These refugees did not live by trade alone, but also by their expertise in the domains in which they excelled, such as clockmaking, jewelry, and fine metalwork. Simon Chaudron headed the list, which included Bréchemin, Castan, Gallard, Gouiran, Meslier, Promis, Robin, the Stollenwerck brothers, and others.

Dozens of others came from France to reinforce these merchants and artisans needed by the society: first, a commercial army of apprentices, shop assistants, traveling salesmen, whose competence and experience are difficult to assess; then a cohort of craftsmen making, processing, and selling products and consumer goods in foods, clothing, furniture, construction, woodworking, leatherworking, and metalworking. The sum total of jobs in crafts, shipping, and commerce puts them in first place, with 57 percent of the occupations recorded, far ahead of the 19 percent who were military men and very far ahead of the 10 percent in the liberal professions (the arts, education, the law, engineering, and health). Among the artists were painter Joseph Halma, miniaturist Binsse de Saint-Victor, the talented naturalist artist Charles Alexandre Lesueur, musician Mansuis L'Huiller, and architect Prosper Baltard, the brother of the future designer of Les Halles, Paris's central food market.[52] Chapotin, Debrosse, and Pichot were lawyers. The health professions—physicians, surgeons, and health officers—included Bulliard, Canobio, Combes, Formento, Lefeuvre, Mocquard, and Violle, as well as pharmacist Ducommun and veterinarian Moynier.

In its heterogeneous composition and varying motivations, the Colonial Society closely reflects this quite remarkable French immigration to the United States between 1815 and 1818. In it we find, along with refugees from Saint-Domingue who were often naturalized, Napoleonic officers and civil servants exiled by the Bourbons, tradesmen and artisans come to seek their fortunes in a country in need of skilled labor, but very few farmers. In any other context, nothing would ever have brought together people with such widely different backgrounds and careers: a gulf separates the general from the stocking maker, the sixty-year-old Convention member from the young apprentice chocolate maker, the imperial general's widow from the wife of a mineralogy professor committed to an insane asylum. On the other hand, the family ties that linked a great many members gave the association a certain necessary cohesion. The creation of the Colonial Society of French Emigrants was thus both a matter of circumstance and a family affair.

IV

French Lands in Alabama

11

Routes to the South

Since May 6, we have been reconnoitering the country to the east and west of the Mississippi; this reconnaissance was long and difficult, given the rigors of the season and the nature of the country. The maps of the interior of this country are so faulty that they are not as useful to us as we would have wished.

—General Bernard, New Orleans, August 13, 1817

In the first half of April 1817, in Philadelphia, the Colonial Society deliberated over how to get to its future place of settlement and retained three options accessible to each other at various connection points. None of the proposed routes—land, maritime, and river—was without its risks, but on the scale of pitfalls and discomfort, the first easily outdistanced the other two. It took six weeks for Israel Pickens, a former congressman from North Carolina, to travel from Georgia to his post as registrar of the Land Office in St. Stephens on the Tombigbee River. He and his family arrived safe and sound, meeting with neither misfortune nor hostile Indians; on the other hand, they did suffer from the state of the road, which was at times indescribably bad and swampy.[1] One could not count on making it unscathed through a long, solitary walk along a "nearly impracticable road"[2] surrounded by dangers of which the Indians were actually not the worst.

THE LAND ROUTE

With the purchase of Louisiana, the need had grown for a land link between America's Northeast and its periphery bordering the Gulf of Mexico. A two-section route already existed: the Great Valley Road linking Philadelphia to Nashville, and the Natchez Trace continuing to New Orleans. In 1803 the postal service could travel it in two weeks, changing horses and men at regular intervals. But as it was difficult and not overly safe, it was imperative that a more direct alternative be found, passing between the Appalachians and the Atlantic, from Athens to western Georgia to New Orleans, through the territory of the Creeks, who granted right of passage. In 1806 Congress authorized the opening of a road that considerably reduced the distance: from the Oconee River in Georgia, it passed through Coweta (a large Indian town near present-day Columbus) to Fort Stoddert, located between Mobile and the confluence of the Alabama and Tombigbee Rivers, and then on to the Mississippi. Despite faster mail delivery, this new road never managed to replace the one via Knoxville, Nashville, and Natchez.

When the probability of a war with England and an attack from the south became a certainty, moving soldiers took priority over moving letters. In 1811 the American army opened the Federal Road that followed the postal route from Fort Stoddert to Milledgeville on the Oconee, but while the postal road went north from there, the new road continued toward the east, where troops could be recruited and resupplied—where also, in Georgia and the Carolinas, land-hungry colonists were growing impatient. But this new road, compared by a historian to the Appian Way[3] and created in anticipation of a war, caused another. The Indians, who had refused the soldiers passage, never accepted their invading presence; they made war and were defeated. The military road became a passage for waves of emigrants, and in 1815 and 1816 Congress voted funds to repair, maintain, and improve it.

This route from Washington to New Orleans via Mobile had, for the Colonial Society members, the advantage of distance. It is not known how many took it, but they were fewer than the risks incurred. In spite of progress in travel in these regions, crossing river valleys (the Tallapoosa, Coosa, Cahaba, and Black Warrior) was no picnic; they were long considered lands barely touched by civilization. To natural obstacles—forests, marshes, and impassible streams—were added human risks, for a journey through the Creek nation, signatory of a fragile peace treaty, remained a hazardous undertaking.

Captain Charassin, a former lancer of the Old Guard whom Montulé met in New Orleans,[4] had ventured onto this road with three companions, among them probably Captains Humbert, an infantryman, and Pfeuty, an artilleryman in the Guard. In 1816 they had traveled together on the *Jeune-Henriette* from Antwerp to Charleston,[5] then likely first to Philadelphia, where Charassin and Humbert bought shares in the Colonial Society before attempting to enter the service of the partisans of independence for the Spanish colonies in Mexico City. This is thought to be the only reason that they went from Charleston, where they had returned, to New Orleans, reaching it after an eventful odyssey from which these veterans of imperial campaigns emerged without mishap.

The fact remains that what was a perilous expedition for these seasoned men was not something that just anyone could undertake, particularly if burdened with a family and baggage. As the French colonists were aiming to be pioneers, not explorers, most chose one of the two less adventurous options. The more logical and rapid one was to go to the ports of Mobile or New Orleans by ship. The less risky one allowed the traveler to avoid the uncertainties of a sea voyage and to discover the interior of the country via the Ohio and Mississippi Rivers. Depending upon their choice, the travelers formed two sets, subdivided into independent groups from just a few to several dozen individuals, and their departures were spread out over a period of months and years. So it was not at a given time and place that the colonists all met together on the same ship. The first departures by land and river took place after April 20 and those by sea in early May 1817.

THE RIVER ROUTE

In September and October 1817, American newspapers published the information that an Indian agent, now in Washington, had several times while coming up the Ohio encountered more than one hundred French people heading toward the new settlements on the Tombigbee.[6] Favoring the river route also meant underrating the dangers of navigating on rivers of a size unknown in France.

ON THE OHIO AND THE MISSISSIPPI

In 1804 the French botanist Michaux had published an account of his journey down the Ohio, saying that its banks, from Pittsburgh to Louisville, would within twenty years become the most densely populated region of the United States.[7] In July he went from Wheeling to Marietta and Gallipolis by canoe, and was impressed by the dangerous acceleration of the current when approaching the islands dotting the Ohio.[8] Michaux was young and accustomed to physical activity, unlike Garnier de Saintes, who was twice his age and had spent most of his life in the inland department of Charente.

We left Garnier in his cabin on the banks of the Ohio, in New Albany, Indiana, where he was living a hand-to-mouth existence selling alcohol and tobacco. He was waiting for his son to continue his journey with him. The consul in New York was told in May 1817 of the arrival in Baltimore of Athanase Garnier, accompanied by Alphonse Delaroderie, said to have ridden with Murat's Neapolitan army. It was said that these two ex-officers intended "to settle near the Mississippi, on the Tombigbee."[9] While waiting to go sail down the Ohio with his companion to join his father, Athanase Garnier stayed with John Latour in Baltimore for a few days, the very man who had hosted Jacques Lajonie—and like the latter, young Garnier became enamored of Elisa, the family's daughter.[10]

Finally reunited, the Garniers embarked for New Orleans on a "skiff with wheels,"[11] that is, a steamboat, which was shipwrecked in late 1817, probably after a boiler explosion, between Louisville and New Madrid, where the Ohio joins the Mississippi.[12] Garnier de Saintes, it will be recalled, had done everything he could to avoid going to the United States. Had he had a premonition of his American fate? "It is thought that Monsieur Garnier de Saintes and his son have drowned in the Mississippi, journeying from Louisville to New Orleans; as yet the only proof is the boat on which they left, which has been found,"[13] the French minister of police told the family, after being informed by the consul in Philadelphia who himself had learned of it in American newspapers.[14]

Thus no route to Alabama was without its dangers, but Lajonie and Poussin, who encountered the same rapids, were able to get through them. Just before leaving Baltimore, Lajonie summarized his itinerary: "I am making my journey from here to New Orleans by land and fresh water. The stagecoach will take me in six days to Pitzburgh, which is 90 leagues from Baltimore, to the west and a bit to the north. There I shall embark on the Hoyo or Oyo, on stam-boots [*sic*]. Twenty-five days will suffice to cover five hundred leagues by water, first on the Oyo and then on the Mississippi."[15]

He waited to get to New Orleans before relating a journey whose timetable he had scrupulously respected from the day of his departure, April 20, 1817. Since the French émigrés who chose the same route traveled either alone or in small groups, many journeys took place at the same time or successively. Poussin, who had sailed to America one month before Lajonie did, had met him in Washington in March 1817, along with Parmantier and also Bernard, whom he was to join in New Orleans to carry out the topographical mission entrusted to them by the government. He assessed the hardships of the journey, considered at the time to be "a long, difficult and dangerous enterprise," requiring lengthy preparation.[16] Traveling across the Allegheny Mountains by carriage, on foot, and on horseback, Poussin made his way to Pittsburgh. Lajonie left Baltimore a month later in a stagecoach, the ordinary way in which "gentlemen" traveled. As far as Chambersburg he passed through rich grain country, but after that was disenchanted with the dreary countryside and the unspeakable roads whose jolts bruised him and his trunk too. He saw large, well-situated villages—Bedford, noted for its mineral waters, Somerset, and Greensburg—but between these oases, an almost empty land with widely scattered houses and few cultivated fields; a sloping country with fir and cedar forests followed by oaks and sugar maples, poor soil, less game than one might have thought, but astonishing fauna. Lajonie was neither a Michaux nor an Audubon, but he was delighted with the natural wonders he saw and noted his observations with simplicity. Then came Pittsburgh, with eight thousand inhabitants.

Despite its extraordinary trade and industry, Lajonie predicted for Pittsburgh a military rather than an economic future because of the heights that hemmed it in and destined it to the role of a fortress. In view of the recent English attacks on the United States, this notion was not as strange as it may at first seem. Poussin, on the other hand, was struck by Pittsburgh's cosmopolitanism, "the big general store of the western lands," where emigrants newly arrived from Europe with "dialects as varied as their customs" mingled with slave traders leading their "herd of Africans" south, "red men" come to trade their furs for alcohol, blankets, lead, and powder, New Englanders led by their spirit of adventure in search "of *good lands* and a *home* suiting their views of work and speculation," and finally, true Yankees, an Anglo-American character type, peddling the iron implements of American industry.[17] A fortress for one, a tower of Babel for the other, Pittsburgh was for both the necessary stopover to take on supplies before going down the great rivers of the West. The most usual way was to buy or charter a boat, then let the force of the current carry it to its destination, which Poussin said could be reached in sixty days. Lajonie did it in half the time, rowing and using a sail.

The boat he bought with three Americans took them to Natchez in twenty-five days, giving Lajonie time to confirm that the Ohio deserved the name "beautiful river" given it by the French. He marveled at enchanting settings, fertile land, and colossal dimensions: the Dordogne valley was a fragment of the Ohio valley, the Dordogne itself a brook compared to the American giant, and the towns along its banks showed promise of one day rivaling the cities of Asia: Pittsburgh, Cincinnati, Steubenville, Louisville, and Gallipolis were the largest, followed by Marietta, Vienna, Point Pleasant, Portsmouth, Alexandria, Linsetown, Augusta, Wilkinsonville . . .

The Mississippi did not diminish his love for the Ohio, for its waters were cloudier, obstructed with trees that were like floating obstacles, and its plain was often flooded. The first urban site below their junction was New Madrid, destroyed by an earthquake in 1812. Natchez, perched on the east bank, was finally a town, but its port would never permit it to grow. It took only two days on a steamboat to reach New Orleans, one hundred leagues downstream, passing through Baton Rouge, a little town just beginning to grow, which would prosper due to the fine position that it occupied on the river and its good lands that stretched very far to the east.[18]

"I have just arrived from my great journey," Lajonie wrote from New Orleans on June 25, 1817. He had accomplished his peregrination in exactly thirty-five days, just like Poussin, who had arrived a month earlier and had since been crisscrossing the Mississippi's bayous in his flat-bottom keelboat.

Stopover in New Orleans

New Orleans impressed travelers who arrived there after a tedious river voyage. A "forest of ships of all sizes"[19] arose from the Mississippi along an artificial earthen levee that prevented the river, which was higher than the level of the city, from overflowing.[20] Lajonie was struck by this characteristic, considering its drawbacks from the point of view of the future colonist he intended to become: "All of Louisiana is a flat region that does not suit us. . . . When it rains, the water leaves the river and spreads into the woods or cypress swamps (low terrain, unfit for cultivation). Without the levees lining the Mississippi, this river would overflow constantly. . . . When the Mississippi is full, the streets of this city are two or three feet lower than the level of the river water. At first glance it looks as if the ships are on the levee."[21]

His journey over, the amused curiosity that had been his until then gave way to the critical observation of a man whose primary focus had now become the conditions of his settlement and his economic future. Montulé, on the other hand, made the most of his status as a simple visitor to describe the city's architecture and its urban layout. But they both lingered over the French, English, and American inhabitants, alongside of whom lived Chickasaw and Creek Indians: "The savages . . . are generally well built. They go naked, even on the street. However they have a little apron one foot wide and one and a half long. Sometimes they have a Roman-style blanket that serves as a bed when necessary. They do not wear hats. All of them are brown. Their skin is the color of an old sack. The women carry their burdens, but they also handle the money. They trade in furs. They change dwelling places like the French change styles."[22]

The Indians' space was shrinking in a Louisiana in full demographic expansion; the 1820 census listed more than 150,000 persons, 52 percent of them black.[23] This numerical equality mirrors that of its largest city, where exactly half the 27,000 inhabitants were of each color.[24]

Although New Orleans's commercial prosperity regularly continued to bring new inhabitants—black slaves, Europeans, and emigrants from the eastern part of the country—there was one factor strong enough to dissuade some people: yellow fever, the summer killer that emptied a portion of the city, sending its inhabitants to inland plantations.

Practicing autosuggestion, Lajonie never missed the opportunity of a letter to France to convince himself and those he was urging to join him that the climate was not dangerous. In June 1817: "You must not be worried to know that I am in a city that is reputed in France to be the scourge of the human race. It is hot and that is all." In August: "There is no illness, no fashionable fever. They said that in Havana, on the island of Cuba, there was yellow fever. With good food, one lives here as elsewhere." In September: "Do not be frightened if you hear about illnesses. They are no more dangerous here than in several cities of France. With a good diet and little to do during the month of August, one lives as well and as long in New Orleans as elsewhere." He assured them that good food and the fresh air brought by the wind that had been blowing since the end of August were the best remedies against yellow fever. The real summer scourges were the frequent strong hurricanes that threw the ships onto the city's levee.[25]

Lajonie could not have been ignorant of the menace the illness was for newly arrived people, or of the epidemic that claimed more than eight hundred victims that August and September. After working relentlessly in the city for two months, Poussin and General Bernard, whom he met there as arranged, were conscious of having "both had the good fortune" to escape it.[26] Yellow fever, or American typhus, was indeed the murderous scourge[27] that caused "terrible losses" almost every summer,[28] most particularly in this summer of 1817, when the epidemic, the sixth of its kind since the first significant one in 1796, claimed many more lives in proportion to an ever increasing population. Doctors noticed that yellow fever first struck recent arrivals, but they concluded that it attacked people who had not had the time to adapt to the excessive heat or who had opened themselves to "imprudent fatigue, or intemperance or melancholy."[29] This was the opinion of the Société médicale that held its first meeting in New Orleans in August and waited until the end of the epidemic to publish a report on its probable causes: abundant rain, stagnant water, summer heat, and the concentration of non-acclimated foreigners.

Even if no one linked water with mosquitoes and yellow fever, the matter of water recurred frequently among the scientists and politicians concerned with disease prevention. In 1811 New Orleans had granted engineer Henry Latrobe the privilege of supplying the city with water by means of a station modeled on the one the senior Latrobe had already built in Philadelphia. Henry's death of yellow fever in September 1817 prompted his father to resign from his position restoring the federal Capitol and come to New Orleans to complete the work and thus help stamp out the scourge that had caused his loss.

Latrobe thought the scourge could be, if not eradicated, at least moderated by appropriate measures,[30] except that he did not understand the role of mosquitoes, impressed though he was by their noisy swarms in New Orleans. On September 3, 1820, three years to the day after his son, Latrobe died of yellow fever. Until the beginning of the twentieth century, New Orleans remained the Necropolis of the South.[31]

First Contacts. Spared by the fever, Lajonie endeavored to get himself recognized as a serious-minded individual—adventurers were overabundant in this part of the country—and to contact the people easily recommended to him since Louisiana was populated by Frenchmen, many from southwestern France, who had emigrated directly from the mother country or from Saint-Domingue. After Curcier in Philadelphia and Latour in Baltimore, he was interested in anyone he met in respectable circumstances who might simplify life for him. Before leaving the East Coast, Lajonie had obtained names of men in New Orleans to whom his mail could be forwarded: the architect Lafon, from Languedoc;[32] the merchant Du Bourg Sainte-Colombe, a native of Bordeaux and a Domingan refugee;[33] and Derbigny, a lawyer from Picardy who was a judge on Louisiana's supreme court and a very influential figure.[34]

Barthélémy Lafon had emigrated as a very young man. An architect trained in the neoclassical style, he had succeeded in gaining a following for his houses by combining French principles with the local constraints of terrain and climate.[35] His talents had also made him a cartographer,[36] and then head engineer in charge of fortifications during the War of 1812.[37] At the height of his career, the tide turned. His character, taste for political racketeering and intrigue, and questionable connivance with the Barataria pirates and Lafitte's enterprise in Galveston are said to have eventually discredited him.[38] Lajonie knew nothing of this, interested in him chiefly as a man well acquainted with a region he had surveyed meticulously. In June 1817, Lajonie complained that he found no mail waiting in his office at 60 St. Louis Street,[39] but did he even see Lafon, on his downward path of politico-piratical tribulations and land deals?[40] It is known, however, that he received a letter from Latour at Pierre Derbigny's.[41]

Lajonie went to the Hotel Trémoulet, the last large French hotel in a city that was quickly Americanizing. The establishment occupied the upper part of a building adjoining the former French Government House on the southwest corner of Jackson Square (the place d'Armes) where Decatur Street (Old Levee) and St. Peter intersect.[42] The building was owned by the French consul's wife, and rented by the Trémoulet couple, hotelkeepers who had immigrated years earlier from Gascony in southwestern France.[43] Latrobe took board and lodging at the hotel, the first travelers came upon after landing, since Old Levee ran along the river. French was spoken there, which Latrobe welcomed since his own had gotten rusty with long disuse. He left mixed impressions of his six-week stay.[44]

Trémoulet, he says, had begun as a cook, then acquired success and respectability as proprietor of the city's two largest establishments. Ruined three times due to his naïve generosity, he had failed in an attempt to start over in Havana. Since then, he was poor and embittered against the American government, which he said he did not like. As for Madame Trémoulet, she was a most energetic woman, but also famously cruel toward her slaves—a shrew who whipped them herself or had them whipped until they bled when work was badly done. The place was not elegant. The technician in Latrobe admired its French construction, its appearance, and its location (his room looked out over the masts of docked ships), but the guest deplored an interior that was the filthiest he had ever seen. Besides, the regulars were not of the best society, at most they were decent people, among whom two or three might have good manners. All spoke French and sang in French, fast and very loudly, when they were not shouting, so that the hotelkeeper readily admitted that he and his compatriots were "a bit noisy." Latrobe did not like the French.

Another account gives a less murky version of the place, frequented by the elite of the bourgeoisie come to dine à la carte in quiet rooms and by travelers served table d'hôte at scheduled times.[45] In any case, one had to show one's credentials as a worthy member of New Orleans's French society. The hotel's restaurant was called Le Veau qui Tête, and Trémoulet was its inspired chef. His portrait and that of his wife give an idea of the models that conforms to this second version of the hotel, but it must be admitted they were painted in their glory years.[46]

In January 1819, when Latrobe took a room there, the Bonapartists who frequented the hotel, along with merchants and planters, were the disenchanted remnants of the unhappy Champ d'Asile experiment. Two years earlier, the hotel was still the meeting place of officers filled with faith in the future.

On the day he arrived at the hotel in May 1817, Montulé was told that Generals Lefebvre-Desnoëttes and Lallemand junior were there, but the first had left the day before and the second was to return in a few days. Montulé says little about Lallemand, except that in exile he was not well off, which had greatly surprised the Americans. But General Humbert, who was also staying there, was even less so, despite a pension from the French state. Montulé noted the lively eyes, strong voice, and brusque gestures of this rather elderly looking fifty-year-old. "Of the thirty persons eating there, fifteen or twenty had left France because of their opinions," he remarked, adding that in a country "where one can talk about politics and the government as freely as one does in France about styles and horses," these men could pour out their feelings and shout "Long live the Emperor," with the city echoing their sentiments.[47] In 1816, Montlezun, an ultra-royalist passing through New Orleans, deplored the fact that, out of twenty thousand inhabitants, there were not twelve partisans of royalty: "There is not a country where the banished revolutionaries might find an opinion more in keeping with theirs."[48]

The Cult of the Emperor. Lajonie had not noticed that people around him were dying of yellow fever, but he did see that in New Orleans everyone was caught up in the Bonapartist contagion: "The inhabitants of this city love Napoleon. I shall make a log of their demonstrations of love for this prisoner. . . . There will be twenty dinners (800 people) on Sunday in Napoleon's honor."[49] Without going over the past three years, when the fleur-de-lis escutcheon on the French consulate's door had been vandalized, the return from Elba had been applauded at the Orleans theater, "pantagruelic orgies" had taken place,[50] and Napoleon's bust had been paraded in great pomp down the city streets, Lajonie could simply open the newspapers to find displays of devotion to the Emperor and exasperation with the king—this in a city founded under the Bourbons and sold by the First Consul. In early May of that year, the city's principal theater had staged *La journée des trois empereurs ou La veille de la bataille d'Austerlitz* (The Day of the Three Emperors or The Eve of the Battle of Austerlitz), in an atmosphere of indescribable delirium that scandalized the consul of a France at peace and allied with the United States. The Emperor's flag, cockade, and name were acclaimed by a huge crowd, while the king and the royal family were insulted.[51] What a ridiculous, indecent, and outrageous scene.

The celebration of the Emperor's birthday on August 15 was the high point of the Napoleonic festivities. The day before, a Frenchman invited his friends in particular and Napoleon's friends in general to his home to celebrate the great man's day: "Two demijohns of good wine will be drunk to the health of the gallant men who remained faithful to him and to the memory of those who died on the field of honor, fighting for France against its cruel enemies the Allies."[52] Two days later, on a Sunday, a magnificent display of fireworks was to feature a life-sized Napoleon, lost in thought on his island of Saint Helena."[53] At about the same time, letters from Bordeaux were published, denouncing the tragedy that had overtaken the country: trade wiped out, poverty of the working class, hunger uprisings, the country subjected to the English who were policing France.[54] To supply their anti-Bourbon columns, the New Orleans editors drew on the *Abeille Américaine,* where Chaudron and Dirat's corrosive humor scored against the king, Bordeaux, and England.[55] The inhabitants' sympathy for the Bonapartist exiles was therefore largely due to their own uprooted past and to veneration for Napoleon, the most illustrious of France's outlawed men.[56]

In addition to New Orleans's consoling atmosphere, there was also the natural solidarity of those most favored toward those whom fortune or talent had not served as well. Thus Lajonie was looking forward to General Bernard's return from Mobile to the Trémoulet Hotel in late July. In Washington, where the men had met earlier, the general had offered to help him in any way he wished as soon as he got there. Bernard could sympathize all the more with Lajonie's monetary difficulties, because he admitted that he himself was not paid enough by the government to meet the "extraordinary expenses of this country."[57] Unlike Lajonie, who had other concerns, Montulé took advantage of his meeting with such a man to question him at length about the Emperor, with whom he had daily rubbed shoulders until he left Malmaison for Rochefort. Because he did not approve of his system, Bernard, he says, did not like him as much as one would imagine, but he was attached to him by a sort of unreasoning charm "that has fascinated in his favor the eyes of so great a number of French people."[58] Besides his closeness to the Emperor, Montulé was impressed by Bernard's incorporation into the American army, calling it "a kind of political phenomenon,"[59] for he knew of the opposition it had aroused, despite the need to recruit an experienced army engineer to fortify the unprotected coasts of the United States.

As the situation was particularly crucial along the Gulf Coast, which had just been shown vulnerable, Louisiana and Alabama were General Bernard's first sphere of activity. After traveling through the interior of the country via Nashville, where he met General Jackson, he had arrived in New Orleans in May 1817. With Poussin's help, his mission was to undertake a systematic hydrographic study of the region to the east and west of the Mississippi, including Mobile Bay and the rivers that flowed into it, the Tombigbee and Alabama.[60] So Lajonie would naturally listen to Bernard, who spoke highly of the land in this part of the South just opening to colonization and also of the depth of Mobile Bay.

An Associate and Indentured Workers. Lajonie could hope for no more than minimal help from Bernard, who would leave as soon as he had completed his mission. However, his information on the geography of the French grant encouraged him in his ideas of settlement and taking on indentured men. With this in mind, he had to find trustworthy correspondents in New Orleans and Mobile, the obligatory ports for the commercial ventures he planned to develop in the Tombigbee colony. The ideal once again would be to associate his interests with those of someone who came from his own region in France, a compatriot more likely than anyone else to look out for them. One month after his arrival, an "honest and worthy" man offered his services.[61] He was the ideal person.

Jean Quessart Sr. came from a family of the rural lower middle class in the area around Libourne, west of Gensac, where Lajonie was born. He had emigrated to Saint-Domingue and then taken refuge in Cuba, from where he was expelled in 1809. He came to New Orleans at the very latest in 1815, the date of his marriage to Elisabeth Berquin, a twice-widowed Domingan Creole. The couple lived on Burgundy Street with eight slaves,[62] and Quessart had a tobacco and chocolate establishment at 7 St. Ann Street.[63] He was a good contact who had kept his commercial and Freemasonry ties to the Bordeaux trade. He was interested in the Tombigbee grant, acquiring forty acres, and agreed to become Lajonie's correspondent. They had the time to settle the details of their collaboration, since Lajonie had to wait five more weeks for the indentured men he had sent for from Gironde.

The American ship *Rebecca* had left Bordeaux on June 16, 1817, and reached New Orleans on August 26.[64] The passengers included eleven indentured persons, a young woman and ten men, whom Lajonie welcomed with the joy of an exile seeing his homeland again: "I embrace a part of myself, I embrace affection and friendship. All of our compatriots have arrived in good health and are already enjoying cherished liberty."[65]

These indentured workers were heirs of the "thirty-six months," named, in the seventeenth century, after the length of their indenture in the French pioneering enterprises of America. Coming from the region around La Rochelle, their port of embarkation, most were poor young men from rural backgrounds. In the eighteenth century, Bordeaux's move to preeminence as the French port for departures directly to the Caribbean islands and the progressive replacement, on the plantations, of unskilled whites by slaves changed the picture. As they were now leaving from Bordeaux, the indentured people came from the neighboring ports and large towns inland located along the Garonne, Dordogne, and Charente Rivers. Rather than agricultural workers, they were artisans in textiles and clothing (shoemakers, weavers, and tailors), wood (carpenters, cabinetmakers, and coopers), iron (wheelwrights, ironsmiths), and so forth. As to the reasons prompting them to leave, they resulted from "the irreducible whims of fate and individual and family psychological factors: a large family, a father or mother's death, the promise of acquiring land or better wages, the example of a brother or a relative who had settled in the islands, the appeal, above all, of resounding success for small-town people, or quite simply a taste for adventure," writes Jacques de Cauna.[66]

By the nineteenth century the situation had again changed. The system of indenture itself no longer existed since its suppression by royal decree (1771), and the island of Saint-Domingue, which had become Haiti in 1804, had lost its function of receiving emigrants from the Aquitaine region. We can nonetheless find in the origins of Lajonie's hired help the characteristics of their predecessors.

They were young men. Montfrand, the oldest, was thirty, and Gigon, the youngest, thirteen; the average age of the eight on whom we have information was twenty-one. Solidarity of origin went hand in hand with religious fraternity: all were born in eastern Gironde, living near one another and never far from the Dordogne River, in Gensac, neighboring Coubeyrac, Sainte-Foy-la-Grande and the adjoining town of Pineuilh, and Bazas. All were Protestant. They came from modest backgrounds with parents who worked in textiles or agriculture, which is what they themselves did, one being a tanner, another a dyer, a third a mason, and four farmers. There were family ties. The Montfrands and Labrousses of Gensac were related; the Fougnets, a brother and sister, were Lajonie's cousins and neighbors, and the unifying element of the group formed to answer his call for workers. Lajonie was conscious of his responsibilities and guaranteed them his help: "my boat, our table and my friendship will be at their service"; "I shall be a Croesus of happiness."[67] Not all, however, were ready to follow him, and several attempted to find their own employment when they arrived. Montfrand, the tanner, found work outside his area of specialization. Lauveau was hired by Macarthy, a merchant. Labrousse did not find a dyer's workshop in the city.

On September 11, 1817, after two weeks of rest, the indentured hands and Lajonie boarded the schooner *Victoire*, bound for Mobile. They stayed there long

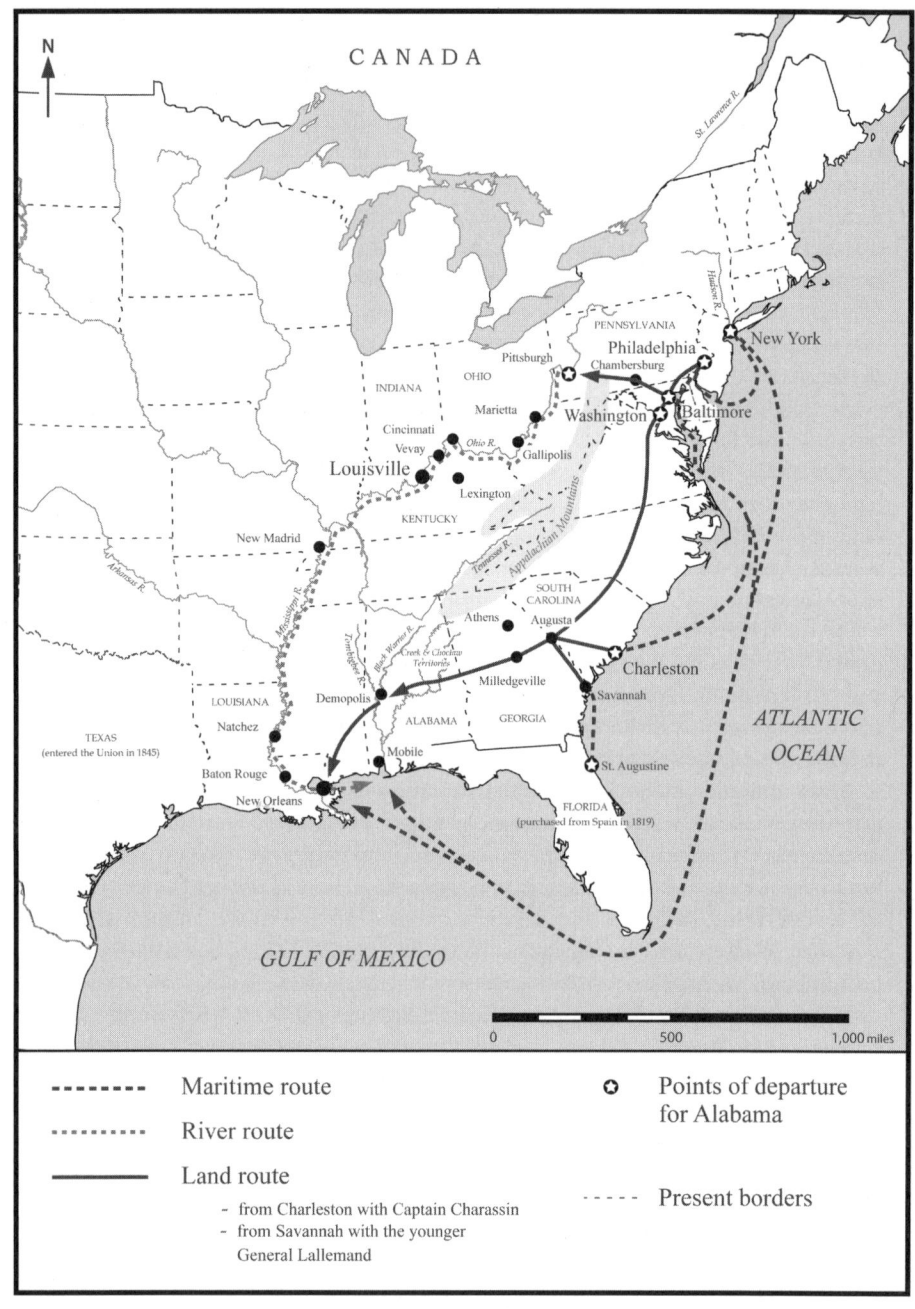

Map 4. The routes to the South taken by the French emigrants (1817–1821).

enough to load food, merchandise, double-barreled shotguns, and various other things they had brought onto another ship. Lajonie calculated it would take two weeks to get to their destination, fifty leagues from the sea, since brigs, he had been told, could sail upriver into the interior. Leaving in the second half of September, he arrived around October 12. In all, it had taken him five months and two weeks to accomplish his journey begun by stagecoach in Baltimore, then continued on a boat equipped with sails and oars, a steamboat, a schooner, then a brig, and, at the end, probably on foot or horseback. This had not enabled him to be the first to arrive in the colony, but the society members were not in a race. The expedition that had taken the sea route from Philadelphia to Mobile had arrived three months earlier.

THE MARITIME ROUTE

The vanguard of the Tombigbee Society left Philadelphia around May 4, 1817. The *Mayflower* of the French colony was a small one-deck schooner, manned by just a few sailors, the *McDonough*,[68] whose captain, John MacCloud, was an experienced commander. The twenty or so passengers[69] were led by two men in the name of the society: Nicholas Simon Parmantier, the society's secretary; and Benoît Marguerite Poculot, a nephew by marriage of Charles Villar, the Society's general agent. Poculot, descended from an old family of Lyon textile merchants, may have emigrated with his wife, Louise Bichat, and their eighteen-year-old daughter, Marguerite.[70] Parmantier was forty-five and Poculot almost forty, but several of their companions were only half their age:[71] Jacques Le François from Rouen, Parisian architect Prosper Baltard, Nantes surgeon Antoine Marie Mocquard, and George Noble Stewart, the youngest, chaperoned by his mother, the widowed Helena Counsell Stewart, and his stepfather, Michel Le Bouteillier, a ship's captain.[72] The latter, eldest of the group at fifty-one, had spent half his life in Philadelphia, where he had obtained American citizenship in 1798 and eight years later the hand of Helena Counsell, the recent widow of an Irish Catholic émigré sailing in the U.S. Navy.

Other passengers may have included Michel Mestayer, a merchant from Gironde and a Freemason; Maurice Laurent, an officer from Saint-Domingue; and finally William Tabele, an American broker from Philadelphia, his wife, Martha, and their three children.[73] This group's composition reflects that of the society in its variety of ages, origins, and professions. Even so, the large proportion of American citizens must be noted, and, as a corollary, the small proportion of recently emigrated French passengers, and most particularly, of military men.

On May 25, 1817, after an uneventful three-week voyage, the ship reached Mobile Point at nightfall as the wind was rising, and the captain was said to have had an obsolete map, according to passengers who held MacCloud responsible for the ensuing grounding.[74] In a letter written the next day, Parmantier, on the other hand, opposes the captain's calm competence to the irresponsible conduct of some of the passengers at the moment when the ship was in danger of sinking, which might explain their unfavorable recollections of the captain.[75] Whatever the case may be as to these contradictory versions, they were about to be shipwrecked off the shore of the promised land that no one on board had ever seen.

Neither the cannon fired from Fort Bowyer nor the coastal lights set out to guide it prevented the ship from running aground. A boat came to the rescue, risking the lives of its occupants, and brought the women and children safely to shore, while the men stayed aboard the *McDonough*, which finally floated free and reached the port of Mobile. The survivors' gratitude toward their rescuers was equaled only by their subsequent gratitude for the inhabitants' comforting hospitality.

Parmantier drew contrasting conclusions from this eventful arrival. On the negative side, it was important to remember the maritime route's risks up to the actual landing in Mobile, and this rift in the colonists' cohesion when faced with their first trial. On the positive side, it was important to recall first the people of Mobile's fine welcome, then the size of their bay, promising the economic development of the port and its region, to which the French émigrés would continue to contribute.

Putting In at Mobile

The French had founded Mobile in 1702, moved it to its present site in 1711, and given the newborn city clothes too large for it: the title of capital of a Louisiana stretching to the snows of Canada. When New Orleans took this title from it in 1722, Mobile went into a period of decline that, a century later, prompted the *Gazette de la Louisiane* to say that during this period the town had lived in obscurity and attracted few people because it was inaccessible to large ships.[76] Alternately French, English, and Spanish, Mobile had become American in 1813: a second birth certificate, judging by the consequences this had on its growth.

Parmantier noted that lacking a real quay, the port of Mobile was in the process of constructing one that, once finished, would be quite long, extending into the deep water of the bay, even at low tide. The city itself was expanding rapidly. Built along wide streets, the wood houses, already numbering about one hundred, were going up quickly to lodge a population of one thousand to fifteen hundred souls,[77] and receive the visitors who had settled upstream.[78] Arriving shortly after his leader, Lajonie described the town in identical terms and had great expectations for its commercial relations with France:

> The port of Mobile has long been considered nearly impracticable by the French, who do not realize that in the last ten years Mobile has become a pretty little city and that the end of the river whose name it bears is already tolerably well settled. Because of this ignorance, which will not last long, the inhabitants of this vast region who draw directly on France will have far more advantages than those who buy in New Orleans. Large brigs can not only go up to Mobile easily at all times, but even to Fort Stephens in high

water. This fact, which is now well known only by the Americans who do all the trading in this city, quite encourages me.[79]

Many French people did indeed settle in Mobile in the following years, among them colonists headed for the Tombigbee, most of whom passed through Mobile, either sailing directly from Philadelphia, or from New Orleans, the transit port for travelers coming down the Mississippi or arriving by sea from France.

In late August and early September 1817, Mobile and New Orleans newspapers reported that several French emigrants had just come through Mobile "on their way to the Blackwarrior," where they were to choose a site for a settlement: "This country is said to be extremely fertile and good for vine growing; and as our government always encourages enterprises of this nature, we can soon hope to harvest right here wines of as good a quality as those of foreign countries."[80] On October 9, the same papers reprinted an article from the Washington *National Intelligencer* announcing that Generals Clauzel and Desnoëttes had sailed from Philadelphia on a ship filled with passengers going to Mobile.[81] The port of Mobile was the obligatory port of entry for the Tombigbee and Alabama's interior regions, which repaid it in a few years: "The Creek River and the peopling of Alabama with cotton planters, have been the good genii of Mobile, and, like the hand of a magician, at once erected into a sea-port of the second order."[82]

THE FINAL APPROACH

The passengers of the *McDonough* spent five days in Mobile, during which the people of the city showed interest in their project and gave them assistance before the final leg of their journey. All understood that this was a matter of general prosperity, that the colonization upstream along the rivers flowing into their bay would benefit their demographic and economic growth. Three men—Gibson, a government agent; John Toulmin, a brother of Judge Toulmin; and Addin Lewis, the port's customs collector—treated them with the greatest possible consideration and introduced them to the inhabitants who could best inform them concerning the country through which they were preparing to travel.[83]

Up the Tombigbee

On May 30, 1817, Lewis put the revenue cutter at the disposal of the colonists, whose upriver foray into the country would be made, as far as St. Stephens, under the United States flag, noted Parmantier. The journey up the Mobile began among lands that were constantly flooded, then dry areas with sandy and clayey soil, and finally, just before the confluence of the Alabama and the Tombigbee, a marshy region that did not really correspond to the French word *marais,* marsh, erroneously used to designate it, for the abundance of oaks, chestnuts, and nut trees gave it an appearance of perpetual greenery.

The next day they stopped at Fort Stoddert, constructed in 1799 at Ward's

Bluff, the present-day Mount Vernon, as a military post on the new border with Spanish Florida as well as a colonization hub.[84] Judge Toulmin, who lived nearby, received a delegation of the colonists and expressed his wishes for their success.[85] A Unitarian pastor, he had himself been obliged to flee England for America, where he held important positions in Kentucky before being appointed judge of the Mississippi Territory and then the first federal judge of the Tombigbee district in St. Stephens. He had a reputation for integrity, generosity, and hospitality.[86]

The judge took Parmantier to Fort Montgomery, on the Federal Road a short distance up the Alabama River. There Parmantier met Toulmin's son-in-law, General Edmund P. Gaines, recently appointed as a commissioner to negotiate with the Creek Indians.[87] A field worker, he had spent the major part of his career leading reconnaissance and topographical surveys in the Mississippi Territory and beyond.[88] Gaines offered to show Parmantier the maps in his possession and asked his officers to give him information on navigable waterways, the soils along their banks, roads, and the kind of people who had already settled in this immense territory. They then came to discuss the advantages and drawbacks of the sites the colonists might consider settling.

The Tennessee Valley would indisputably have come first because of its fertility, if the sale price of its products were not compromised by the difficulty of getting them to market, particularly to distant New Orleans. The choice should be one of the many rivers flowing directly to the sea, with preference given to two of them: the Alabama had broad, fertile banks, but they were subject to flooding, and its water level in the dry season was too low for river transport; as for the Tombigbee, its banks rose higher and higher as one went upstream, but it was navigable all year long as far as the mouth of the Black Warrior for barges with a fifteen-inch draft. Above St. Stephens, several sandbars reduced the depth of the river in periods of drought, but these could be removed.

The lands lying between the Tombigbee and the Black Warrior, at their confluence, were rich prairies, and the banks of the latter were of excellent quality after Nawna Folia, thirty miles above the confluence. The quality of the land along the Cahaba was praised, but this river had the disadvantage of not being in immediate communication with the eastern part of the Tennessee, as was the Black Warrior, its natural canal.

Parmantier came out of this exchange of views with a much clearer idea of the place where he was going to settle, as well as with letters of recommendation: several from Judge Toulmin for friends in St. Stephens, and one from General Gaines for his brother, Captain George Strother Gaines, the government agent to the Choctaw at Fort Confederation on the Tombigbee.

On June 7 Parmantier continued up the Tombigbee and stopped at Carney Bluff (Fort Carney). Major Josiah Carney was an old friend who had emigrated from North Carolina in 1810 with his family and slaves[89] and settled along a convex meander of the river, where he built an impregnable enclosure.[90] There, by chance, Parmantier met Captain Gaines's father-in-law, who had just been recommended to him and who had the same surname as his son-in-law.[91] This man,

Young Gaines, graciously offered to lodge the travelers at his plantation below St. Stephens so that their guide could comfortably pursue his exploratory mission. This was done two days later, after a stop at Jackson, a town that already had one hundred buildings.[92]

The group of colonists must have gone ashore two miles south of St. Stephens, on the opposite bank where Gaines's land stretched between two parallel streams perpendicular to the Tombigbee.[93] When he returned to get them a month later, Parmantier wrote an encouraging description of it for the colony's budding viticulturists: "This farm, although on the highlands, shows what may be done by a skillful cultivator. It consists of 156 acres of land, and the manure from 20 cows maintains it in a state of fertility which delights the beholder. In an orchard of 10 acres there are more than 200 peach, plum and apple trees, and a vast number of wild vines, loaded with grapes! This grape, which is a blue Muscadel, is extremely abundant throughout the whole country. If circumstances permit, I will see this fall what can be extracted from it."

Until now, Parmantier had seen mainly marshy areas, sand, clay, and mud. But just beyond Carney's, he had noticed that the countryside was changing, and what he saw delighted him: "Between the region of swamps, which extends to within about a mile from the river, and the ferruginous hills, there is a middle region rising by a gentle ascent whose soil is a blackish earth, thickly spread with small flint stones, or round quartz. It is the same sort of soil of which I have seen the vine dressers so desirous in the southeast part of France, where the grapes are very heavy, as for instance, on the banks of the Izere."

Parmantier had stopped at St. Stephens. Located on a height away from the river, it had originally been a defensive site occupied a century earlier by the French, where the Spaniards had in 1789 built Fort San Esteban de Tombeckbé, on Hoe Buckintoopa Bluff. Ten years later they had left[94] and abandoned the fort to the Americans, who created a town at the site in 1807 and briefly made it the administrative seat of the Alabama Territory.[95] Two centuries later, at the archaeological site of Old St. Stephens, with not a vestige visible among its trees and shrubs, it is difficult to believe the contemporary records emphasizing the tremendous growth of this town whose elevated position up the Tombigbee made it grow "with a rapidity beyond that of any place, perhaps in the western country."[96] The forty-five-by-ninety-foot lots sold for two hundred dollars, but some went for ten times as much. Houses were going up every day, and there was a shortage of laborers to build them. A school with professors of merit had eighty students: an example to follow for the future French settlement, remarked Parmantier. The terminus of navigation on the river, St. Stephens, which had become a communications hub with construction of the Federal Road, held, at the time, a monopoly on trade, but the French, if they settled at the next confluence, might well contest this.

Armed with Judge Toulmin's introductory letters, Parmantier made the rounds of St. Stephens's influential citizens. He began with Magoffin, the registrar of the Land Office, who gave him the maps and reports of his surveyors, which corroborated the data he had already gathered.[97] He nonetheless pursued his inquiries with "Colonels" Wharton, Fisher, Dinsmoor, and Dale and a civilian, Malone, all well established locally.[98] Silas Dinsmoor, the government agent to the Choctaws from 1802 to 1814, was well acquainted with the geography of the area in which they lived. "Big Sam" Dale was equally authoritative. Like Davy Crockett and Daniel Boone, this woodsman, who knew every inch and every Indian of the forest, was a frontier hero, renowned for his exploits during the war against the English and the Creeks.[99] After the war, he was a tax collector, before serving as a delegate to the convention that divided the Mississippi Territory. These experienced men confirmed what Edmund P. Gaines and his officers had already told Parmantier, namely, that more than twelve hundred colonists had already settled between the thirty-third and thirty-fourth degrees of latitude, from the banks of the Black Warrior twenty miles to the east. Parmantier was even introduced to an agent of a Carolina company whose members were preparing to settle as squatters on the lands being considered by the French.

The White Bluff

On June 9 at the earliest, Parmantier, Poculot, and several others left St. Stephens in search of the ideal site. Their exploration lasted about four weeks. It reached the goal toward which all the information had led them, and that was in turn recommended by George Strother Gaines, whom they met at his trading post on the upper Tombigbee, at Jones Bluff. The place they chose, known as White Bluff, was located slightly below the confluence of the Tombigbee and the Black Warrior. It was one of the most beautiful locations that he had ever seen, wrote Parmantier on July 14, 1817:

> We have explored the country situated on the eastern side of the Tombigby, above the line called the Old Choctaw boundary, and we have resolved to fix ourselves on the spot known by the name of the White Bluff, about three-fourths of a mile below the junction of the Black Warrior and the Tombigby rivers, as part of our grant. It remains to say in what shape the four townships are to be laid out, and this we will do as soon as the meridian line shall be marked as far as the said Bluff. The season is already advanced, and no resource would be left to a number of individuals during the ensuing winter, if the benevolent intentions of the government towards us are not administered with some celerity. White Bluff is one of the finest situations I ever saw in my life, and the lands lying around it are of the very first quality. Nature here offers us everything. If we know how to profit by these advantages we must be happy.

Hudson, the Photographer. In the early years of the twentieth century, Robert Lee Hudson left his job as electrical engineer to work with his father in the New York Racket Store, founded by him in 1895, on Washington Street in Demopolis.[100] When his father died, Hudson did not return to his original profession, but kept the store—and that for several decades. It was a "country store" like those found in all small American towns, and in Demopolis itself, born a century earlier, when Francis Bierne, a veteran of Napoleon's Russian campaign, sold razors, combs, toothbrushes, shoes, shirts, cotton thread, shawls, curtains, watches, candles, plates, glasses, coffee, jam, sugar, oil, Jamaican rum, Madeira, knives, guns, powder, saddles . . . With a few exceptions, Hudson still sold just about everything, at low prices, groceries along with household and farm utensils, which guaranteed him faithful customers, most of them black, during the Great Depression.

Bierne the soldier and Hudson the storekeeper died one and a half centuries apart in the town, one its cofounder and the other its photographic memory during the 1940s and 1950s.[101] From his Kodak to his laboratory, Hudson was a virtuoso of snapshots and the darkroom, reproducing with an admirable sense of framing, contrast, and perspective the Demopolis and surrounding countryside of the postwar years: inspired portraits, urban architecture, economic activities like wood and cotton, blacks in the fields, the natural environment, a rare shot of the town under snow, prairies, trees, flowers, chaotic vegetation, a remainder of wild early landscapes, the white cliffs from all angles, and the two rivers in the winter at flood stage, submerging banks so high they had been thought beyond reach.

Hudson's photographs show us what the storekeeper who preceded him, Bierne, saw when he arrived at the place that the American Congress had granted to the French and that Parmantier had selected for them—before the construction of two dams, the first in 1904, the second in 1954, changed its face, raising the water level first twenty and then approximately forty feet.

Two hundred fifteen miles above Mobile and two miles below its confluence with the Black Warrior, the Tombigbee begins a wide, perfectly regular curve to the east. The concave part of this semicircle consists of a smooth, high, chalky white wall, rising at a seventy-five-degree angle, whose summit, sixty to seventy feet or more above the river, is entirely covered with dense green growth. For the French navigators who were already used to the Tombigbee, this type of cliff was not new; some that were even higher had already impressed them—but, aside from its majestic beauty, this was theirs, marking the end of the journey begun on June 18 at Waterloo.

With the erection of the dams and the mutilation of the cliffs, some of the magic is gone, but there remain some thirty chalky feet that, on fine October days, their base bathed by the waters of the Tombigbee, still glow in the afternoon sun.

12

The Promised Land

[In the village] the most charming harmony of social and natural life prevailed. In a corner of a cypress grove, in what had once been the wilderness, new cultivation was coming to life. . . . Everywhere the forests were delivered to the flames and sending dense clouds of smoke up in the air, while the plow went its slow way among the remains of their roots. Surveyors with long chains went about measuring the land. Arbitrators were establishing the first properties. The bird surrendered its nest, and the lair of the wild beast was changing to a cabin. Forges were heard rumbling.

—Chateaubriand, *Atala,* 1803

After July 1817, one arrival followed another, so that by the beginning of the following year, according to Pénières, more than 150 colonists to whom Congress had granted land were clustered on the White Bluffs.[1] Depending upon the conditions of their emigration and their reasons for joining the Colonial Society, their presence in Alabama was either the fulfillment of a dream, the default choice of a new existence, or a reflection of the simple desire to get going. The composition and motivation of the Demopolis colonists differed from those of their predecessors in New Bordeaux, Gallipolis, and Vevay in that there was no true unity. Considering the difficulties those settlements had encountered, the task of these ill-assorted French people in a distant, barely stabilized frontier region would be even more arduous. But it was up to them to reflect on the circumstances of the failure of the Ohio group and the success of the Indiana settlers in order to try to carry their own project in Alabama to a successful conclusion. This they did not do.

TAKING POSSESSION OF THE LAND

The French colonists' first concern was to found a town. In this they already distinguished themselves from the Swiss colonists, who had not put the cart before the horse, but had made clearing the land their priority. While Vevay, after a long gestation, came into being at the exact location planned for it, Demopolis was long a name without a concrete existence. Furthermore, the name Demopolis was not immediately agreed upon. For a time it competed with Proscripolis and, when this possibility was abandoned, with the name White Bluff, designating its natural site, and finally that of Aigleville.

FROM DEMOPOLIS TO AIGLEVILLE

White Bluff, majestic and favorably located a mile below the confluence of the Tombigbee and the Black Warrior, became the emblem of the French colony. On August 16, 1817, Nicolas Simon Parmantier and Benoît Poculot signed the first official letter from "Demopolis on the White Bluff."[2] Not a single colonist contested its choice: "There cannot exist in any part of the world a nicer, more pleasant site and a richer soil. We shall place our town near an immense prairie on a little peninsula formed by the Takalouze and Tombegbée, two beautiful rivers that with the Alabama form the Mobile," wrote Jean Augustin Pénières in June 1818.[3] "The site is so beautiful, so well placed for a town, that one cannot make a bad bargain, above all when one intends to be a merchant there," added Lajonie in 1819.[4] Lin Troy, a former French civil servant owning two 120-acre allotments, exclaimed in 1820 in a New Orleans newspaper: "This region is the most beautiful and healthy, not only of the United States, but of the entire World."[5]

Demopolis and White Bluff soon had a rival. Just as there had been a drift from Proscripolis to Demopolis reminiscent of Gallipolis, there was one from Demopolis to Aigleville reminiscent of Aiglelys, a name suggested by aristocrats for their own settlement next to the Gallipolis colony. The *lys* ("lily," the emblem of French kings) paid homage to royalty; Aigleville ("Eagleville") to the Empire and the United States. Demopolis, a place dedicated to the Emperor's memory, with a street system celebrating his victories, from Marengo Square to Austerlitz, Wagram, and Friedland Streets, could instead be called Aigleville, referring to both the Napoleonic eagle and the American eagle, the nation's emblem.

The name Aigleville appears in the December 18, 1817, issue of *L'Abeille Américaine*.[6] A month later, in a letter from Demopolis, Lajonie confirms it: "Pleasures are still too rare in Aigleville (the final name given to our town)."[7] The name was subsequently Americanized to Eagleville or Eagle Ville, used as such in 1818 by John Melish, the first to map the Alabama Territory independently of the Mississippi Territory and of the southern United States (see map 5). It also occurs on the oldest known map of the town: a single straight line, Main Street, perpendicular to the curve of the Tombigbee, approximately at its center, as shown in map 6.

With the birth in 1818 of another Aigleville near the Black Warrior, it is quite difficult to distinguish it from Demopolis/Aigleville on the Tombigbee. This toponymic confusion reflected a fluctuating situation resulting from the fact that the first colonists had settled on lands that had not yet been surveyed by government agents, nor submitted for the approval of the secretary of the treasury and the president of the United States, nor shared out in the drawing of allotments in Philadelphia.

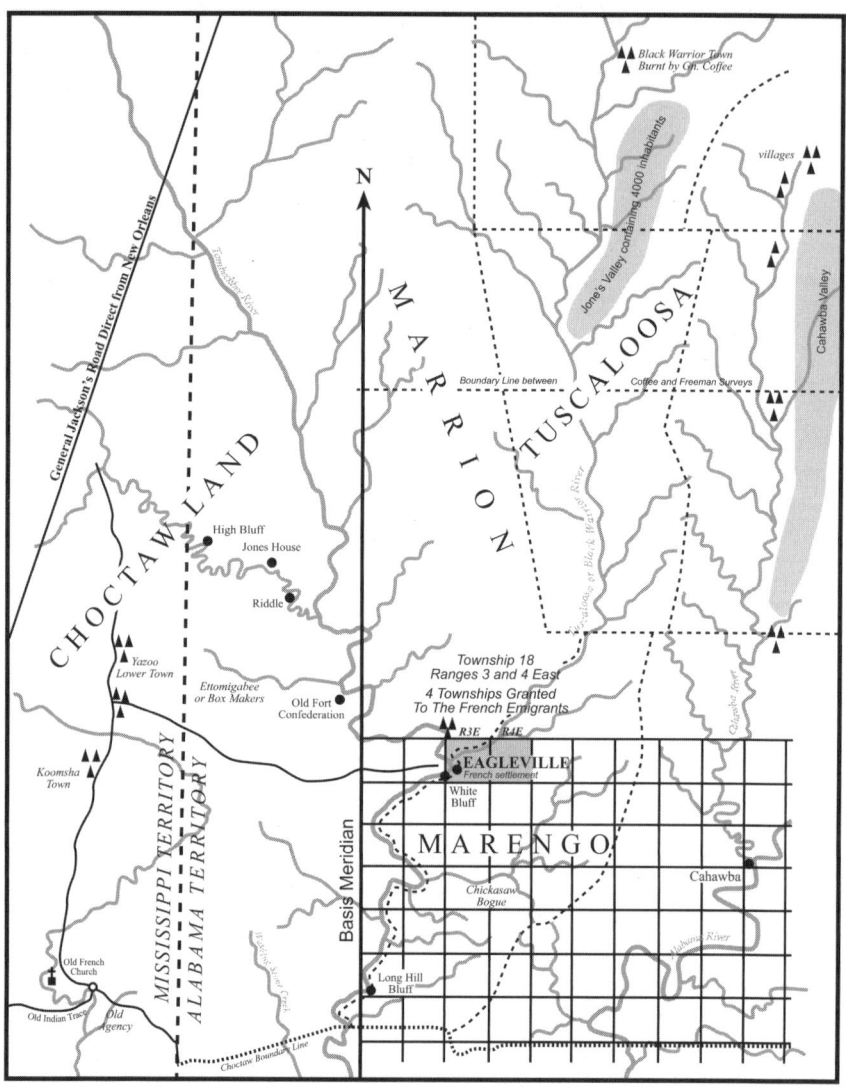

Map 5. A detail based on the first map of Alabama (1818) by John Melish.

A Topographical Error

As no other choice seemed better or even possible, they had settled out of love for the location and the necessity to attend to the most urgent things first. So it had been decided to grant those arriving at the very beginning sixteen acres upon which to build themselves shelters and begin planting food crops whose harvest

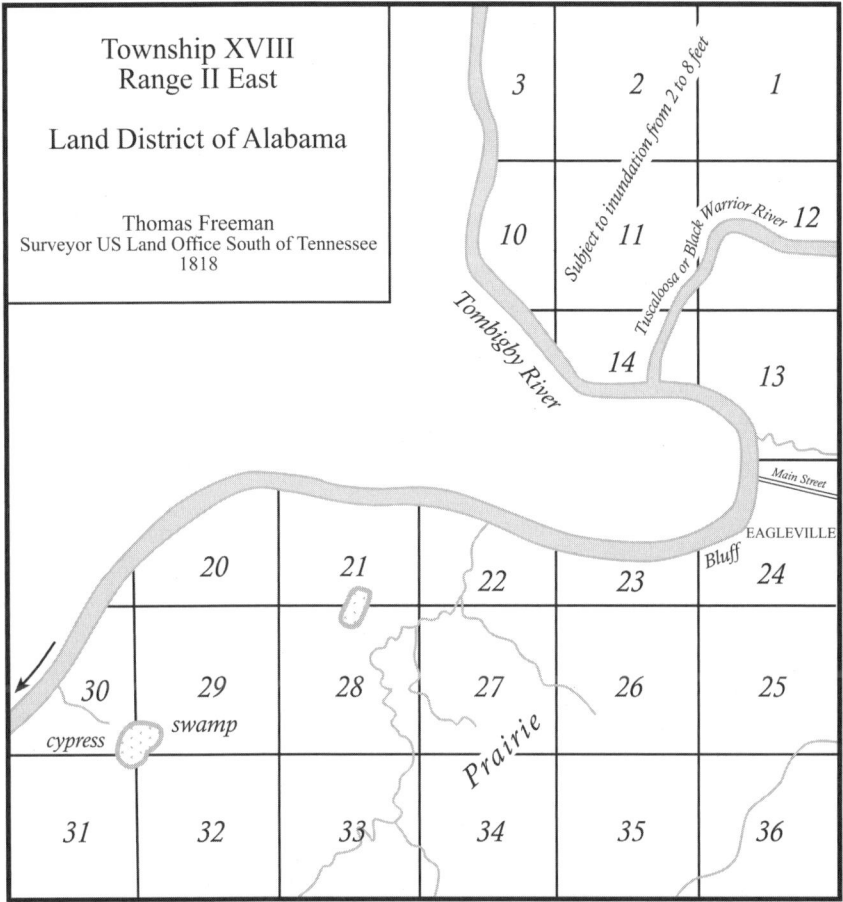

Map 6. Eagleville, Township 18, Range 2 East, from a map by Thomas Freeman (1818). The Basis Meridian was also known as the St. Stephens Meridian.

would be indispensable for the survival of the group that had made a point of staying together. The colonists therefore occupied land located along or near the Tombigbee and in the narrow space stretching toward the east to the Black Warrior, forming the peninsula mentioned by Pénières and that others called a "bend," because of the river's curve. It was planned that this area would be the colonists' landing and meeting place, but it was only one element in the geographic expression of the four townships to be selected. On August 16, 1817, Parmantier answered the letter of June 16 from Major Thomas Freeman, chief surveyor of public lands south of the state of Tennessee, based for the past six years in Washington, Mississippi Territory:

How happy I feel to have anticipated all what [*sic*] you have the kindness to advise me in the choosing of the tract of land granted to us by act of Congress of the 3d of March last in having taken possession in the name of the society of the spot known under the name of the White Bluff, just at the junction of the Black Warrior into the Tumbigbee very near the line of the 33d degree of latitude, with the intention to extend our ground 12 miles above this place on the last river in as much as the township line will allow it and the same or thereabout in ascending the Black Warrior. I am also extremely thankful for your kind recommendation to the deputy surveyor. It will be very useful to us. All our friends will be apprised of your liberal disposition towards the association and will participate in our feeling of gratitude. I trust our request has been handed to you and dare to hope that a choice has been made by you of a deputy Surveyor to operate on this spot. It is an operation very much wanted and for which we beg leave to solicit the most earnestly execution.[8]

According to this letter, it would seem that there was agreement between the government's and the society's agents regarding the choice of White Bluff, and only regarding this precise site, the location and boundaries of the townships being as yet nothing but an idea in the mind of Parmantier. He wanted to take as the base of the townships a line below the junction of the two rivers, incorporating White Bluff, and then extend the colony twelve miles to the north, which would allow for two townships, each with six-mile sides, to be side by side with two others. Parmantier's intention was to integrate White Bluff into the French colony and probably to have the Tombigbee as the western boundary of the two western townships. This configuration would allow for optimal use of the Tombigbee, along which would lie quite a few plantations, and the Black Warrior, which would flow through many others. But this prospect was dependent upon official topographic confirmation, which explains Parmantier's specific request that Freeman send a Land Office surveyor.

With these expectations, Parmantier sent him a map delimiting the area. The map and drawing of the course of the Tombigbee, which was its major thoroughfare, were not the result of a recent land survey, but came from a survey by Captain George Strother Gaines that Freeman said was thought to be "tolerably correct." On October 9 he sent the map from St. Stephens to his superior, Josiah Meigs, the commissioner of the General Land Office in Washington, D.C., with a very instructive letter:

I take the liberty of enclosing you herewith a rough sketch of the position occupied by the French emigrants under an act of Congress, granting them four townships, etc. The principal agents of these people express anxiety to have the limits of their lands or grant, designated by the surveyors. The act in their favor says "any four contiguous townships, each six miles square," to be designated by the Secretary of Treasury, under the directions of the President of the United States.

The position they have chosen, being on the bank of the Tombigby river, and the opposite side being Indian territory, the townships adjoining, and covering the position selected, will be all fractional. The agents of the emigrants, have, no doubt, made known to the Secretary of Treasury, the location they have made, and wishes respecting it. Under these circumstances, I have to request the favor of you, sir, to let me know whether the location made by the agents of the emigrants, will be confirmed by the President or not; and if it does, in what manner the legal quantity of lands shall be laid off for them? I made this inquiry in consequence of a pressing application made to me by the principal agent of the emigrants, to have the limits designated for them.[9]

Freeman immediately understood that there was a discrepancy between the terms of the act of Congress and their interpretation by the society's agents on-site. While they had some freedom in selecting the location of their colony, they had to comply with the conditions they had accepted. The townships were to be contiguous and measure six miles on each side, and thus, notwithstanding the fact that the general layout could be determined by the colonists, they were to form a block of 144 square miles divided into as many sections. Now, the French proposal did not take into account this requirement of uniformity, and resulted in six fractional townships, which must have been apparent on the map sent to the American government. They were broken up, for one thing, because the Tombigbee had been chosen as their western border, instead of the straight line required by township surveying, and for another, because the opposite bank of the river was not yet open to colonization because it was in Choctaw territory (until 1830). The French plan showed a total misunderstanding, or perhaps total ignorance, which amounts to the same thing, of how American land was surveyed and integrated into national geography as more territory was acquired. The rule they were supposed to apply was neither more nor less than what was applicable to everyone on the frontier.[10]

The error could have been corrected quickly if Freeman, who should have been aware of the lack of experience of the French, could have gotten them back on track by return mail by asking them the pertinent question he asked Meigs: "How shall the four townships, or 144 sections, be laid off and designated, for the emigrants at this position?" Freeman was not a novice but a recognized scientist who had won fame with the first topographical survey of Washington, the new federal capital, and then through exploratory missions, notably along the Red River, before devoting himself to the survey of the lands below the line separating Tennessee from Alabama, which he himself had drawn. An upright civil servant, he fought all his life against land speculators. In this individual case, given the treasury secretary's instructions of June 6, 1817, concerning the surveying of the northern half of Alabama (with its Tombigbee and Black Warrior watersheds), which did not mention the French townships,[11] did he imagine that the French were getting preferential treatment? Did he overestimate the level of the relations between their agents and the treasury secretary, or underestimate their lack of precision? He first referred the problem to Meigs, who submitted it to Treasury Secretary William Crawford, who, on November 10, 1817, clarified for them his instructions on the surveying of the French townships:

> The townships must be contiguous and they must form a component part of the general plan of the survey of the Said territory, so as not to form fractions of townships. To this end it is necessary that the standard lines established in those parts of the territory already Surveyed, be extended, as soon as the selection is made by the agents of the association, so as to embrace the townships selected. The selection and the form, whether square or oblong, of the four townships, shall be at the discretion of the agents of the association, subject nevertheless to the principle of contiguity, and to the general connection with and conformity to the subdivision of the whole territory into townships, so as to avoid the creation of fractions of townships.[12]

Crawford was reestablishing order, simply complying with the congressional act and the general dividing up of the territory that were supposed to be in agreement. The surveying system for Alabama, including the French colony, had been drawn up starting with two basic lines: a meridian passing through St. Stephens at 88° west longitude, and a parallel passing north of Mobile at 31° north latitude and separating the United States from Spanish Florida. The townships went north starting from the thirty-first parallel and were divided into east and west ranges on either side of the St. Stephens meridian. Thus, Township 18, Range 3 East began 102 miles north of the thirty-first parallel (which was the southern boundary of Township 1) and was located to the east of the St. Stephens meridian. It was all a matter of geography and degrees. The French agents were neither geographers nor careful in their calculations.[13]

On November 12, two days after receiving the treasury secretary's instructions, Meigs sent him the French proposal, pointing out that their selection would not correspond with his instructions.[14] On the fifteenth, Meigs sent Freeman a copy of Crawford's instructions and directed him to follow them.[15]

If most of the blame for not conforming to the law or misinterpreting it falls on Parmantier, can some of the responsibility be placed on the American government? It does seem surprising that eight months went by after March 3 before such precise instructions were sent to the head of the Land Office in Washington, Mississippi Territory. But the government could not really intervene before the first colonists arrived at the site, which took five months (an understandable amount of time, given their journey), or before Crawford was given the complete list of the colonists, identifying their allotments by number and size, which took three more months. The drawing of allotments in Philadelphia did not take place until October 27, because precious time was lost due to a dispute among society members and ensuing mediation by the treasury secretary.

BACK IN PHILADELPHIA

Parceling out the grant and operations relating to the land drawing stretched out over a two-month period filled with misunderstanding, absurdity, conciliation, and newfound unity—or so it seemed. The *Abeille Américaine* and the French diplomatic corps reported on the various episodes, each in its own way, the one optimistic, the other pessimistic. Without seeking to deny the difficulties that were holding up the application of the law in the Colonial Society, the French émigrés' newspaper attempted to play them down and preached unity. Though deploring them, the French consulate in Philadelphia also magnified them to paint a dark picture of the exiles' conduct on American soil, most particularly that of the military men. These two views, less divergent than it might appear, allow us to understand what was happening in the society, both in Philadelphia and Washington, from late August to late October 1817.

The Military Takeover Attempt

On August 30, 1817, the *Aurora* published a notice addressed to the émigrés by Louis Marie Dirat, the association's secretary. This was a serious directive, without the sarcasm that often characterized his writing:

> The shareholders who have subscribed for half or quarter shares of the lands granted by the United States Congress to the French émigrés, are asked to notify the executive committee's office, 96 Chesnut [*sic*] Street, of the name or names of the subscribers of their acquaintance, holders like they of fractions of shares, with whom they wish to associate to form an entire share, which must be entered under a single and same number, at the drawing of allotments in a general meeting. They are advised that if they have not indicated their wishes in this respect before next September 3, the company's executive committee will automatically do this.

After this clarification, the *Abeille* and the *Aurora* reminded interested parties of the official drawing in their respective issues of September 4 and 5. The shareholders of the Tombigbee Agricultural and Manufacturing Company were informed that the general meeting for the drawing of allotments would be held on September 9, at 5 p.m., in the hall housing Mr. Auriol's Dance Academy, behind the Pennsylvania Bank. This hall was chosen because it was large enough for the expected crowd. Those unable to come could be represented by a society member. Absences were numerous. Many were already on the Tombigbee, and others, like those of the Lefebvre-Desnoëttes group, had left before the drawing, trusting their representatives and the organizers, who seemed to be in control of the event. Entry into the hall was limited to those with a card bearing their name, issued to shareholders by the office of the executive committee at 96 Chestnut Street.

An informer nevertheless managed to get in. Transmitted by the consul, the contents of his report reached the minister of foreign affairs in Paris: the meeting had been a fiasco, and the drawing was postponed because of military men who had repeatedly interrupted the proceedings and caused trouble.[16] The officers had stood apart from the merchants and other civilians, who were "consequently separated from them." This division, added the consul, signaled the fact that they intended to take charge of the deliberations. Already a year earlier, Pénières had expressed reservations concerning the plan for a community system, fearing that the strong would want to impose their law on the weak, a law dictated by the power of money or that of military authority. His fears were now justified.

The session opened with a declaration of war. Contrary to the expectations of the majority, they were asked to "decide whether the French who had taken refuge in this country before the Restoration had the right to be members of this association or this company." If the law of March 3 was strictly applied, it indeed concerned only the "recently emigrated" French; that is, it might be understood as applying to those who had arrived in the United States after July 1815, to the exclusion of "former French people" who had come to the United States earlier, even much earlier, like the Domingan refugees. This was the point of view of General Vandamme, a colonel who was Vandamme's aide-de-camp, the elder General Lallemand, and Colonels Galabert and Grouchy, whom the consul mentioned in this order, not by rank, but according to the degree of virulence they had exhibited. Arriving in the United States later, at least after the law had been passed, these men had had no part in the creation of the society, nor in its meetings or deliberations. As they were not the only ones in this situation, they represented the other officers who had also arrived after the fact and for whom there were no allotments left in the grant, since the subscription had taken place before their arrival. In asking this thundering question right at the outset, the military men were attempting to take over the meeting. The consul reported that faced with their bad faith and ingratitude, the "former French" were justified in retorting: "You asked us to sign the petition to Congress requesting this concession. It was granted in answer to this petition and perhaps to our names. You asked us to subscribe and we agreed. What right do Frenchmen, who were not here at the time and were consequently unknown, now have to strike us from a list to which we were the first subscribers and signatories?"

The Bonapartists had been neither the instigators nor the managers of the colonial project, unlike the former Convention members and the Domingan refugees. It took "hotheads" like Vandamme, Lallemand, and Galabert to dare challenge and deny the validity of the participation of the former refugees, in their presence, in their own city, under their own roof, in a country that had become their own. Not content with wanting to oust them, they insulted them. Vandamme's bad character had crossed the seas, and already at La Fère, Major Pion des Loches had said that Lallemand lost his temper when he did not get what he wanted. These generals did not appreciate resistance. They insulted the civilians and lost their tempers: "The retorts got these military men all worked up; from insults and coarse, demeaning name-calling, they went on to threats. General Vandamme was heard to shout angrily that on the Tombigbee they would be the masters, and threaten to take the saber to civilians or drown them."

The threatened members did not let themselves be intimidated. Right was on their side, and many of them had fought the English during the War of 1812. They were not afraid of Napoleonic soldiers who had been defeated at Waterloo. They answered "firmly and courageously that here they had joined the heads of battalions, that they were their equals, and they left their names and addresses for duels the next day."

Villar, a moderate man, had succeeded Garnier as president of the society. He was unable to restore order, impose silence, or read a letter from the secretary of the treasury, who was to settle the matter. The confusion, agitation, and turmoil reached such a point that some people decided to put out the lights, which put a de facto end to the meeting and, at least for that day, to the quarrels.

None of the military men challenged to a duel showed up to "take up the gauntlet" the next day, September 10. So a dozen Frenchmen each decided to make a formal complaint to the authorities against Vandamme for threatening them, but they were dissuaded by compatriots "who pointed out that they should respect the French character." They refrained from teaching a lesson to a general who was usually "so brave at the head of an army," but the damage was done. As soon as this scandal with all its details became public, it diminished the military men's reputation among the country's inhabitants and among those society members who had not been able to attend the meeting; worse, it broke the tie between the general and field officers and the French merchants: "The latter had pitied them, taken them in and given them good dinners. Today they say that their fate is well-deserved and that the King was right to exile them. Some inhabitants of the country have the same opinion," noted the consul, irritated to see that some of the French did not agree.

In addition to the dissension between the military men and the society members they had wanted to exclude, there was discord among the soldiers themselves.[17] During a subsequent meeting of the society in mid-October 1817, the younger Lallemand hurled "the coarsest insults" at Taillade, who arranged to meet him in New Jersey for a duel. Informed of this, the local authorities had the two men arrested and set a very high bail to ensure they would not fight. But worse than the amount they had to pay, they were mortified to be taken to the town hall by policemen through a crowd of onlookers. The charges were taken no further on condition that it not happen again. "This circumstance brought about a reconciliation, they say, that will last until the subject of their old quarrels is reawakened by some other small, less important incident concerning the honor of both." The consul added a chagrined note, not yet realizing how right he was: "Such is the example the French are setting here; such is their conduct in a hospitable land. Your Excellency will judge for himself what opinion the country's inhabitants must have of their political and private conduct and how they must judge the reasons that brought them to the United States." But nothing was yet lost. The exiles invoked their solidarity and attempted to calm things down.

Mediation by the Secretary. The society's executive committee was caught in the crossfire between the original subscribers, who clung to their legitimate benefits, and the latest arrivals, who contested them. Unable to deal with the situation, the committee suggested submitting the problem to the American government and asked Villar, its president, to consult with Treasury Secretary Crawford, whom they considered the "supreme Judge" in the matter. Several Frenchmen went to the capital, commissioners accompanying Villar, and others, representing their own interests and those of their partisans. It is unclear who was really with whom, as the military men themselves did not form a united group.

Concerned with the turn taken by these Franco-French quarrels, President Monroe asked William Lee to investigate, for the elder Lallemand, whose attitude and plans were puzzling, had become suspect. Lee uncovered schemes that were incompatible with the spirit of the law. Drawn up on September 27, his report was submitted to Secretary of State John Quincy Adams the next day,[18] and to the president the day after that:

[The French officers] represent that though they have ample funds in Mexico for all their purposes, they are in want here of the means of putting their plans in execution. For the purpose of obtaining the means, they have been endeavoring to force upon the company formed for making a settlement on the Tombeeby [*sic*], about an hundred officers as subscribers, for whole, half and quarter shares of the four townships granted by Congress to the French emigrants. These shares when obtained, to be placed in the hands of certain merchants in Philadelphia, who are to advance them 50 or 60,000 dollars thereon, which, they calculate, will be sufficient to begin their expedition with, but in this they will be disappointed, for it appears that Mr. Villar, the President of the Tombeeby Society, having obtained some hints of their plans, communicated the same to Generals Clausel [*sic*], Desnouettes, Vandamme, Grouchy, and Count Real, concerned in the association, who have taken measures to prevent the mass of these officers from becoming subscribers to their company, as well as to shut out the possibility of those who have heretofore subscribed, of obtaining titles to their shares in these townships, without which no transfers can be made and of course no facilities obtained.

All the French officers of distinction except the Lallemands disapprove of this project. Genl. Vandamme censured yesterday Genl. Lallemand and Colo. Galabert in so pointed a manner, before Mr. Villar and Colo. Taillorde [*sic*] (who were sent here by the Tombeeby Company to confer with Mr. Crawford) that a serious quarrel like to have ensued.

It appears certain that Joseph B. has pointedly refused all aid and assistance to this and the like schemes; that he has been solicited in every way and all means used, to induce him to patronize these adventurers without success, on which account they are liberal in their epithets against him.[19]

This project concerned the independence movements in Latin America in which Lallemand planned to participate by raising troops in the American West, under the command of French officers. Lee had obtained his information from good sources, but the steps taken by the generals cited seem doubtful. Clauzel and Desnoëttes had left Philadelphia for the Tombigbee in late August, and Vandamme's and Colonel Grouchy's behavior at the meeting of September 9 did not suggest opposition to Lallemand. Everything was thus still hazy, but becoming more worrisome by the day. On September 29 Crawford informed Adams that Henry Clay had just expressed doubts about this enlisting of recruits by the French emigrants. By this time, Crawford had already met with the French, since, according to Adams, he was displeased that the commissioners had not yet left.[20]

So Crawford had already seen Villar, who was not yet aware of these revelations that cast an unfavorable light on Lallemand's intrigues. Crawford, wrote Villar, had received him like a father, listened with kindness and interest, deplored the division among the French, and finally had as a friend offered to intervene if this could not be resolved.[21]

Villar accepted his offer and, in the name of their common interests, conferred with Henri Lallemand, who was in Washington. On the day following the interview, both received a letter from Crawford in which he proposed that the executive committee be reorganized, and called for reconciliation and sharing. The first subscribers, that is, the signers of the petition on the basis of which the grant had been made, were not to be excluded, but should have beeen willing to make sacrifices to satisfy the latecomers who would be admitted: everyone's interests were to be taken into account, as members of a family, brothers. The consul thought that the journeys to Washington by the committees of both parties, differing in opinions as well as in feelings, had restored calm and peace.[22]

From Compromise to Concord. The new executive committee shrewdly introduced itself to its members under the "respectable aegis" of the secretary of the treasury, considered the founder of the colony: the committee was the result of Crawford's wishes, not of a base internal maneuver, which gave weight and legality to the changes that it had to get everyone to accept. For the first time, the Colonial Society was headed by a military man, Charles Lallemand, who became its third president. This appointment showed that the government had not intervened in the society's affairs; otherwise it is not certain that this choice would have been approved—unless Lallemand had made amends to the authorities and promised to abandon his plan to militarize the townships in order to raise money. Villar, his predecessor, who did not seem to see it as a repudiation, fell back to the rank of vice-president. Finally, a former tax administrator, Jean Baptiste Durand, whom the French diplomatic corps distrusted, was also a powerful new man in the society.

The reconciled assembly met on October 26, 1817, with General Lallemand presiding. Several days earlier the *Abeille* had praised the "spirit of order, justice and impartiality" he had displayed during the executive committee's preparatory meetings when "with a judicious and fraternal remark" he had wiped out all the distinctions that might divide the members. He no longer thought of imposing his contingent of officers upon the society.[23]

The meeting, attended by some two hundred people, offered "the image of the most perfect harmony," and only applause was said to have interrupted the executive committee's officers during their three-part presentation.[24] Lallemand spoke first, organizing his speech around three chief points. He began by expressing gratitude to Congress and its law, "the kindly act of legislation directed by principles of generosity," and homage to the country offering rest, freedom, and "all the aid of fraternal hospitality, all the gifts of national munificence" to men oppressed by power. He then stated his regret for the unavoidable debates and objections raised during the disastrous meeting, for the errors committed by preceding administrations, which he praised in passing for their founding role, and for his own mistakes (but who could doubt the purity of people's intentions?). Finally, he asked the assembly to rally around the leader, Charles Lallemand: "Unite with me."

Lallemand manifested his wish for appeasement by reducing to mere incidents the affronts that had nearly resulted in bloodshed and by paying tribute to the men to whom the society owed its birth: he had to get them to forgive the scornful and brutal behavior of his men. After playing his part of unifier, he handed the meeting over to his colleagues. Villar, showing their agreement, spoke in his turn of the September 9 meeting to sympathize with the extremely difficult position of those whose rights had been challenged and made reference, in Lallemand's mollifying way, to annoyances and "somewhat bristling nuances" that had not been able to tarnish a "serene, cloudless core." He then told of his mission in Washington, before turning to Durand for the technical explanations preliminary to the drawing.

THE DRAWING

Driving home the message of union, Durand imitated the preceding orators. He too understood everyone's worries and claims, but he now wanted to leave behind the few disagreements that might have divided the society members. This time, things were starting on a better footing. For one thing, all decisions would be submitted to the authorities; for another, the members had taken encouraging steps toward each other: "We have seen in the first shareholders of land a consideration that added value to their relinquishing a portion, and in the claimants, a fairness and moderation that alone could respond to the sacrifices they expected and asked for." In other words, they were meeting halfway to satisfy one side without permanently displeasing the other. Today it was a matter of defending "property common to illustrious unfortunate men."

Technical Particulars. The reporter opened his explanations with the equation born of the congressional law granting a quantity of 92,160 acres of land divided into 144 sections of 640 acres (or one square mile) each. This equation could not

Table 5. Distribution of 82,280 acres in 315 shares

	Number of shares	Number of acres	Total in acres
	85	480	40,800
	20	320	6,400
	54	240	12,960
	85	160	13,600
	71	120	8,520
Total	315		82,280

be changed. What remained was to integrate into it the number of shareholders, which had increased during the preceding eight months, from March to October 1817. The strife of September 9 concerned this problem, since the military men had thought they could resolve it by excluding those who were not strictly speaking included in the law, that is, those who emigrated before 1815.

The lists of subscribers had been "religiously verified." A few of the first subscribers were excluded because the law eliminated them, but Durand reiterated to the others the assurances that had already been given them. In the name of acquired rights, taking away from the first shareholders the total of their shares was out of the question; they would be left as found on the original lists. However, he explained, to reconcile the equally just demands of the French who had arrived later with the land that remained, a division had been drawn up that closed the gap between the latecomers and those who had come earlier. Each person would have his share, but unfortunately it would not be the same for all, said Durand: "All, undoubtedly, will not have an equal amount of land, because it would not have been proper to dispossess long-standing legal proprietors who may seem to us to have a right due to having come first; but this arrangement of an unequal parceling out does not prevent us from regretting that we cannot include all subscribers in an equal classification."

Having determined the number and names of the members, it was now necessary to justify the method used to determine exactly what area of the available lands could be granted to each person, according to the inegalitarian criteria previously referred to. The 144 sections of 640 acres each could not be used as such. Subtraction and addition had to precede division and distribution. The first subtraction: that of an entire section reserved for the location of the town to be founded. The second subtraction: that of 144 allotments of 16 acres each surrounding the town to be given *immediately* when people arrived so they would have means of subsistence and could regroup before their later dispersion. These 144 allotments multiplied by 16 acres make a quota of 2,304 acres, the equivalent of 3 and 384/640ths sections. The first addition: the total of the two subtractions—that is, 4 and 384/640ths sections. The third subtraction: this sum subtracted from the 144 sections or 91,160 acres, giving 139 and 384/640ths sections or 89,216 acres. This is the base on which the proportionate calculation given in table 5 was made.

The total of 82,280 rather than 89,216 acres is due to the choice of land parcels of various sizes, from 120 to 480 acres. To allow for a greater number of shareholders, there were no shares totaling the maximum possible of 640 acres (as had perhaps originally been the case), and fewer acres had been given to more shareholders, but the inequality in ranking was flagrant, since the 85 largest shares equaled almost as much as all the others put together. Furthermore, each shareholder was granted, in proportion to his allotment, a number of acres of the 144 sections surrounding the town, plus one lot in the town itself. Thus each shareholder ended up owning either 12 acres near the town and a lot 100 feet wide by 200 deep if he had a 480-acre section; or 6 acres near the town and a lot 100 feet wide by 100 deep if he had a 360- or 240-acre section; or 3 acres near the town and a lot 50 feet wide by 100 deep if he had a 160- or 120-acre section.

The Duty of Mutual Aid. Once the distribution was made to the 315 shareholders on the list of the drawing that was to take place on October 25, there remained a reserve of 6,936 acres resulting from the difference between the 89,216 acres available and the 82,280 acres actually used. This reserve, according to the executive committee, would allow them to grant "allotments to the French people who might still come, having the same rights as we to belong to the same family as well as to enjoy the same benefit." Once the drawing was made, it would have been difficult, without this reserve, to satisfy the "just demands" of new arrivals who might apply, even though the deadline for subscriptions had been extended to October 24—that is, the day before the drawing. As people did apply during the following two weeks, well before mid-November there remained only 1,707 acres, which could then be granted only on-site. Besides being allotted to the last "arrivals," this remainder would have another use: to compensate the shareholders who drew badly situated or marshy and hence unfarmable lands, or who would lose some area to roads: "None of us owes personal sacrifice to assure the common advantage."

This philanthropic character of the society was confirmed by the creation of a relief fund with the positive balance of $1,584 that was in the general treasury in November 1817. The society managed a "common treasury" consisting of "settling rights" paid by the future colonists in order to obtain their permanent certificate of ownership. The money in the relief fund that was not used to meet the expenses constantly facing the society would go to those who, when they arrived in the colony, did not have sufficient means to start their farm. As proof that the society used the dollars it received from its members wisely, the relief money was not simply given to the needy, but lent, to be reimbursed after the second year's harvest. Before dealing with the procedures of the drawing itself, Durand concluded his report with a lyrical flight on family, union, happiness, mutual affection, and a new existence that nothing could now destroy.

This was the impression within the society, but also outside it. Consul Pétry, who could not be suspected of wanting to paint an idyllic picture, added his final touch to the *Abeille*'s account of the meeting that he sent to Paris: "This Association seems very wisely and well organized. One can only wish it success, which will be the result of the harmony that will reign among them. May they one day believe that some of them earned the Government's severity, that others feared it based on nothing but their own opinion, and finally that the King in his mercy opens his arms to receive the penitent and that the return of his children to their homeland is the desire most ardent and dear to his heart! Most of them already regret being so far from their homeland."[25]

After these speeches of reconciliation, the drawing took place calmly and peacefully, according to what Pétry wrote to the minister. In accordance with the plan of land division that had been adopted on the American model, each subscriber was to draw a number corresponding to a section number in one of the four townships. These numbers ran from 1 to 137 (or 140 in other documents). Each number applied to several shareholders, from a minimum of two (e.g., section 1: twice 320 acres, or section 22: 480 and 160 acres) to a maximum of five (section 2: five times 120 acres and 40 acres in reserve) according to how the section had been divided, apparently arbitrarily, as a trial, before the drawing, in a prelude to an actual distribution on-site. As for the reserved lands, they would later be designated by the letters of the alphabet from A to Z and then AA to KK, giving 37 supplementary 40-acre shares.

At the time of the drawing the numbering of the allotments was abstract. The numbers did not designate any location, since the government's surveyors had not yet drawn up a map of the townships. The executive committee regretted that it could not satisfy both a curiosity it found natural and each person's desire to know the location of his allotment. After the drawing, the shareholder could say only with whom he would share his section and who his neighbors would be. The committee had decided to start at a known point of a given grid and give it the first number of the sections, continuing the series uninterruptedly. This work then had to be brought into line with that of the Treasury Department in Washington so that no one would be wronged by the result of the drawing.

The drawing seems to have gone perfectly smoothly. On the other hand, after examining the composition of the lists of the drawing, one might well have doubts concerning the impartiality of choices when the submissions were finally recorded. They had not been entirely brought into conformity with the law, as had been announced. One has only to look at the case of the three Dufours. As Swiss citizens, they had nothing in common with the refugees from Saint-Domingue, nor with the Bonapartist exiles. But their role as precursors in the career of American winegrowers had given them rights among the colonists, along with 480 acres that the law did not give them.

The inequality deplored by reporter Durand was not due to chance, but the result of a concerted policy. The shareholders with the largest farms all belonged to the group of most important exiles, former Convention members, Domingan refugees, and officers. Among the last group there is a hierarchy in land ownership: 480 acres to generals, 320 acres to colonels. The number of acres situates the receiver economically and especially socially, for the conditions of land acquisition, at two dollars an acre payable within fourteen years, was in principle not impossible to meet. Lajonie, a simple second lieutenant discharged for medical reasons, received 480 acres, but his situation was different: he had worked in the society before the law was passed. Furthermore, thanks to family ties one could better one's position—thus running counter to the idea of sharing vaunted by the society when requests poured in at the time of the drawing. Grouchy and his two sons found themselves at the head of 960 acres; Garnier and his son or Pénières and his, of 720 acres. Pénières had sent for his nephew and intended to send for his two brothers.

The drawing of October 25 put an end to seven weeks of conflict among the shareholders, but the consul, while wishing them well in his letter of November 24, could not help but worry about their union's chances of survival: "Can [unity] be expected of individuals whose political opinions are as varied as the impressions they have received, either through education, rank, or their situation in life?" The consul's question identifies an essential factor. What did military men coming out of a quarter century of battle, civilians, famous, educated, and wealthy men, and nobodies embarked on the same venture have in common? What aspirations could be shared by men who had played a major political or military role in France during the Revolution or the Empire and men seeking a peaceful farm for themselves and their families? Could the future be the same for long-standing refugees and brand-new Bonapartists who had no intention of growing old far from their homeland or feared having to exchange their sabers for spades?

On October 27, two days after the drawing and shortly after escaping a duel and prison, the younger Lallemand married Harriet Girard, Stephen Girard's niece. He could either go to the colony on the Tombigbee with her or rent out his land; in any case, he could take advantage of the conditions of an American life that seemed most auspicious. He preferred to follow his brother, who had more glorious plans for himself and the officers than to go plant young vines in the wilderness or live in bourgeois boredom. The elder Lallemand, one of the Emperor's last companions before St. Helena, was a hothead whose life could not come to a halt on the banks of a river whose name may have been exotic, but that was so distant and peaceful: "I have more ambition than can be satisfied in this colony on the Tombigbee," he wrote to his younger brother.[26] His membership in the society and his speech at its last meeting were not sincere. Lallemand's attempt to use the society for his own ends had failed in part, but he would not give up. By luring officers away from the Tombigbee colony where they could have gone and settled, he was one of the people responsible for the French emigrants' failure, indirectly in Alabama, and directly in Texas, where he led hundreds of men, seventy of whom had subscribed for a share in the society. On December 17, 1817, under the leadership of Generals Lallemand and Rigau, the schooner *Huntress* set sail in Philadelphia for the Gulf of Mexico, with a cargo of men, provisions, arms, and munitions. On January 16 the ship was in the Galveston harbor. Ashore, the men met the pirate Lafitte, went up the Trinity River, and, in the spring of 1818, organized themselves into four cohorts in the fortified camp of the Champ d'Asile, living from fishing and hunting to save on supplies and salt meat. Soon on the road to ruin, abandoned by the elder Lallemand, and evicted by the Spanish, who worried about this armed force near Mexico, the colony fizzled out before the end of the year. Many colonists died; others got back to New Orleans as best they could.[27] The tragedy experienced by these refugees aroused a surge of passion in France along with a wave of subscriptions to send them financial aid.[28]

The End of the Land Survey

In his letter of November 10, Crawford informed Meigs that the president and vice-president of the association, General Lallemand and Charles Villar, had submitted to his department an authenticated copy of their proceedings and a list, seen and approved, of the names of the French emigrants receiving an allotment of the land under the act of March 3. Crawford, who had until then had a ringside seat at the French squabble and had undoubtedly developed a wait-and-see attitude, could now expedite things in the field: "You will therefore without delay instruct the Surveyor of the district in which the land intended to be set apart shall be located, to cause the same to be surveyed in the same manner that the other public lands are surveyed, and that the sections and quarter sections be numbered in all respects in the manner invariably practiced in the other public lands in the Alabama territory," he wrote to Meigs. The French had been given preferential treatment by the vote of Congress; they could not be given it again by tailor-made rules which, furthermore, their pathetic conduct in Philadelphia during the past two months did not justify.

If the secretary of the treasury is held blameless, how could one not also exonerate Freeman, who was merely his subordinate? Prudently, Freeman had notified the secretary of the French agents' rash arrangements, assuming that there might be a special agreement with the émigrés, since their very colony was an exception to the normal distribution and sale of lands. Freeman was doing his job, conscientiously surveying the areas placed under his control, but in this Alabama district things were not progressing as fast as he would have liked, due to difficulties he listed for Meigs on October 10, 1817: the high cost of supplies from neighboring states, the outrageous wages and loose living of the laborers working for the Land Office, and the rarity of good surveyors who often resigned, weary of the work, physical woes, and wages that did not cover expenses. Yet Freeman did not despair. He considered what he had accomplished and hoped to have surveyed most of the best lands by May or June 1818, if the Treasury Department provided him with the funds he so badly needed for his huge task: "Twenty Thousand dollars to enable me to keep the surveyors as far as practicable constantly at work during the Winter and Spring."[29]

Freeman eventually went into the field himself, setting out in early November 1817 on "a painful and disagreeable journey thro' the Alabama District . . . for the purpose of visiting arranging and hurrying the Surveyors employed in the district, and also for the purpose of viewing and selecting the most proper & eligible Sites for towns therein, in compliance with instructions from the Secretary of the Treasury of the 6th of June last enclosed from your office to me."[30] He set out on a journey to the French townships with old instructions just when instructions meant specifically for them were leaving Washington, D.C.

In November 1817, Lajonie reported that the land had been surveyed, but he did not say how or with what result, believing that news of Desnoëttes and Villar's departure for Philadelphia was more important. He was wrong: "The distribution has not yet been made because the government's surveyors have only recently finished their work in the place where we were to choose. Messrs. Lefebvre-Desnouettes and Villar left for Philadelphia in order to put an end to the haggling, speculation and intrigues that are creeping into the affairs of the Society whose principal Committee is still in Philadelphia. The first month, we hope to see these gentlemen with full power to conclude with us who hold and _claim we should hold_."[31] Lajonie was wrong because this survey did not conform to Crawford's instructions of November 10, which Freeman did not receive until he returned from his three-month circuit in mid-February 1818, along with other letters from Meigs dated October 28 and November 13 and 15, 1817. On February 15 he answered Meigs: "On my arrival here I found your letters . . . with enclosures from the Secretary of Treasury on the subject of the survey of the lands to be set apart for the French Emigrants—And requesting that I should forward without delay returns of the Surveys already made. All of which shall be immediately and punctually attended to—and should have been answered a month sooner had I not been detained nearly the whole of that time in the woods confined to my tent with a number of sores & several Boils."[32]

What had been done had to be redone, and could already have been redone if Freeman's illness had not delayed operations by a month. While the government instructions were traveling and Freeman was suffering in his tent, the French colonists, left to their own devices on the Tombigbee for months, were in a state of complete uncertainty. Thus Lajonie, building his house on his sixteen-acre plot, was not certain "that it is situated on lands sold by the government to the Society. This is a consequence of the ignorance of our commissioners concerning the laws of the government."[33] With the building finished and occupied since early January 1818, he expresses his disillusionment:

> Our society, in spite of the fact that it includes important people, is swarming with schemers. This is not anything to worry about, but unfortunately fate designated particularly these people as exploring commissioners of the colony, and, what is more, a permanent Committee was created in Philadelphia, precisely 400 leagues from the promised land. In a reversal of things, an immediate consequence of intrigue, the Committee has changed, but our commissioners remain intact, as well as their plan of colonization, which is as contrary to certain laws of the land as they are profligate with the Society's money.
>
> It is to be hoped that the presence of several members of the new committee will hasten the excessively slow and uncertain progress of our dear explorers. . . . We are on our allotment of 16 acres while waiting for the al-

lotment in town and the large allotment. The first allotment will fall to us at the end of the week by drawing; as to the second, we are as uncertain of the time that will give us possession as our commissioners are of obtaining the site of the town which is a fraction larger than the four quarters sold to us by the government. This uncertainty is troubling the spirits of the colonists.[34]

Lajonie had identified the nature of the problem as well as the troublemakers whom they could judge on-site: the colony's exploring commissioners, "schemers" guilty of skimming through the congressional act, ignoring the rules of the grid system used in surveying American territory, misjudging Anglo-Saxon insistence on following rules, and acting according to their own personal criteria. The French sufferings were caused, not by the American government, as has often been written, but by timing problems, an unfortunate game of mail tag between surveyors and official instructions, the slow separation of an independent Alabama Territory from the state of Mississippi, and their own leaders' inexperience and irresponsibility: Parmantier, dazzled by the beauty of the site, but forgetting elementary principles he should have followed because the future of hundreds of individuals depended upon them; the senior Lallemand, busy trying to hijack the society in aid of his Texan dreams, in which he did no better than at La Fère. Naïveté on one hand, pipe dreams on the other, on the banks of the Tombigbee or in Philadelphia salons, all converged toward the failure of the enterprise, for the colony never recovered from this initial muddle and the time lost. White Bluff, the colony's natural jewel, the ideal place for the town and the center of its future commercial and administrative activity, was located outside of any possible grant. What had been done, collective clearing or individual construction, had been a waste. With disaster in the offing, Washington's pressure to proceed with the surveying of the French colony's lands bore fruit. Several days after returning to St. Stephens, on February 18, 1818—the congressional vote was approaching its first anniversary—Freeman assured Meigs that the following April he would go to supervise and move operations along in the French townships:

> I have directed two of my best Surveyors to proceed immediately to survey the Townships embracing the location made by the French Emigrants on the Tombigby River—They are directed to extend the Township lines from the first Standard line (meridian between Ranges 4. & 5.) West, to the Tombigby River and to complete the Survey of these Townships in all respects as the other public lands have been Surveyed for Sale, Agreeably to the Secretary's instructions on that subject.
>
> To cause these lands to be surveyed in the same manner that the other public lands are Surveyed, and that the Sections be numbered in all respects in the manner invariably practiced on the other public lands in Alabama District.
>
> The Townships thus directed to be Surveyed are Townships N° 17, 18, 19 & 20. in Ranges N° 4, 3, 2, 1. East—The Tombigby and Black Warrior Riv-

ers, being Navigable streams, and passing thro [*sic*] these Townships, will of course produce fractions of Townships and Sections on the margins of both Rivers—These Townships shall be prepared for your Office in the usual way, and a separate Sheet or map of them shall be prepared for use of the Secretary of the Treasury.[35]

All of this information could have satisfied the Senate, which, by a resolution of December 31, 1817, had asked the president about the progress of the colony whose creation he had advocated nine months earlier. None of the action-packed episodes occurring since the previous summer could have escaped his notice. But in his message of March 19, 1818, Monroe, unable to furnish these details that had not yet reached him, simply gave them the secretary of the treasury's report.[36] This mentioned that "the French emigrants have deposited in this department a list containing three hundred and fifty names, and embracing an allotment of the lands contemplated by the said act, by which no individual is to receive more than four hundred and eighty acres, nor less than one hundred and twenty," and that in consequence of the president's approval, instructions had been sent to the commissioner of the General Land Office to begin the surveying of lands, the result of which was unknown to him as of February 19, 1818.

The Grant's Boundaries. On August 3, 1818—that is, more than a year after the Parmantier group arrived at White Bluff—Freeman informed Meigs in Washington of the completion of the surveying operations in the French townships and in neighboring townships that would be put up for sale.[37] As indicated on map 7, the colony was located in Townships 18, Range 3 East, and Townships 18, 19, and 20, Range 4 East. This was nothing like the plan Parmantier had submitted to Freeman in October 1817. The colony had lost all physical contact with the Tombigbee, and the Black Warrior flowed only through Township 18, Range 3 East. The worst thing was that about two-thirds of the land between the two rivers, where White Bluff and Demopolis were located, was conclusively outside the grant, belonging to Township 18, Range 2 East, which could be only a fraction of a township.[38]

In addition to the official map of the colony, of which many copies were printed, there exist several on which their owners wrote comments: one belonging to Édouard Paguenaud, a civil engineer from Bordeaux, now in the collection of the Alabama State Archives in Montgomery;[39] one belonging to Jacques Lajonie, another exceptional document preserved in family archives in Gensac, traced from the original onto a sheet of brown flimsy to save money: "I am going to copy the map, which costs too much to send it to you (the booklet costs 40 francs)."[40] The two maps are similar in presentation, information, and place-names, but they differ on one point. While Paguenaud, on his copy of the official map, indicates the townships as they are legally, Lajonie adds an excrescence of four square sections into Township 18, Range 2 East, corresponding to the location of Demopolis, forbidden to the French, but where he had acquired, outside the boundaries of the colony, particularly advantageous sites.

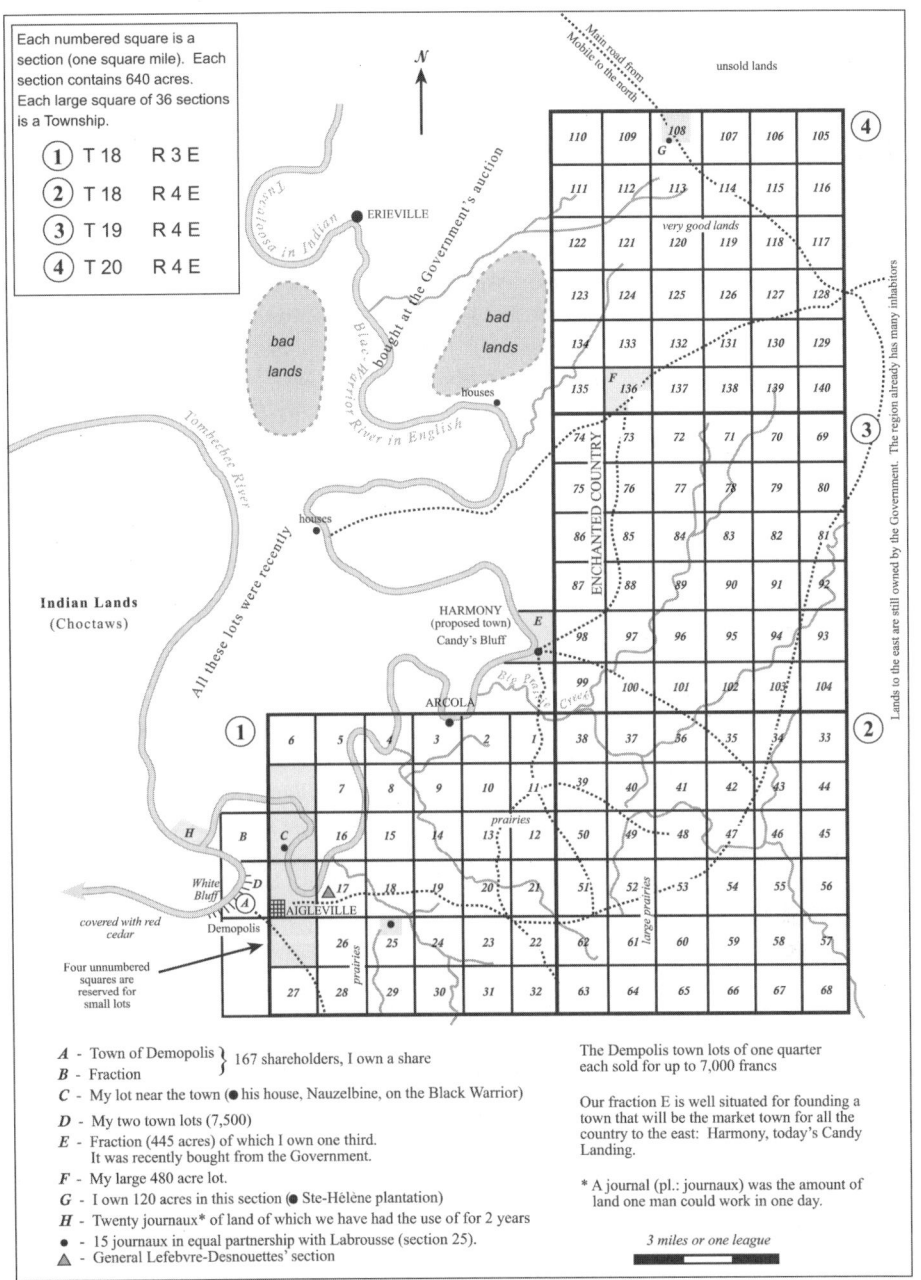

Each numbered square is a section (one square mile). Each section contains 640 acres. Each large square of 36 sections is a Township.

(1) T 18 R 3 E
(2) T 18 R 4 E
(3) T 19 R 4 E
(4) T 20 R 4 E

N

Main road from Mobile to the north

unsold lands

Tuscaloosa in Indian

ERIEVILLE

bought at the Government's auction

Black Warrior River in English

bad lands

bad lands

houses

Tombecbee River

Indian Lands
(Choctaws)

All these lots were recently

houses

ENCHANTED COUNTRY

HARMONY
(proposed town)
Candy's Bluff

ARCOLA

Big Prairie Creek

very good lands

Lands to the east are still owned by the Government. The region already has many inhabitants

110 | 109 | 108 G | 107 | 106 | 105
111 | 112 | 113 | 114 | 115 | 116
122 | 121 | 120 | 119 | 118 | 117
123 | 124 | 125 | 126 | 127 | 128
134 | 133 | 132 | 131 | 130 | 129
135 | 136 F | 137 | 138 | 139 | 140
74 | 73 | 72 | 71 | 70 | 69
75 | 76 | 77 | 78 | 79 | 80
86 | 85 | 84 | 83 | 82 | 81
87 | 88 | 89 | 90 | 91 | 92
98 | 97 | 96 | 95 | 94 | 93
99 | 100 | 101 | 102 | 103 | 104

(4)
(3)
(1)
(2)

6 | 5 | 3 | 2 | 1 | 38 | 37 | 36 | 35 | 34 | 33
7 | 8 | 9 | 10 | 11 | 39 | 40 | 41 | 42 | 43 | 44
prairies
16 | 15 | 14 | 13 | 12 | 50 | 49 | 48 | 47 | 46 | 45
H | B | C | 17 | 18 | 19 | 20 | 21 | 51 | 52 | 53 | 54 | 55 | 56
White Bluff | D | AIGLEVILLE | | | | | | large prairies
A | Demopolis
26 | 25 | 24 | 23 | 22 | 62 | 61 | 60 | 59 | 58 | 57
prairies
27 | 28 | 29 | 30 | 31 | 32 | 63 | 64 | 65 | 66 | 67 | 68

covered with red cedar

Four unnumbered squares are reserved for small lots

A - Town of Demopolis } 167 shareholders, I own a share
B - Fraction
C - My lot near the town (● his house, Nauzelbine, on the Black Warrior)
D - My two town lots (7,500)
E - Fraction (445 acres) of which I own one third.
    It was recently bought from the Government.
F - My large 480 acre lot.
G - I own 120 acres in this section (● Ste-Hélène plantation)
H - Twenty journaux* of land of which we have had the use of for 2 years
● - 15 journaux in equal partnership with Labrousse (section 25).
▲ - General Lefebvre-Desnouettes' section

The Dempolis town lots of one quarter each sold for up to 7,000 francs

Our fraction E is well situated for founding a town that will be the market town for all the country to the east: Harmony, today's Candy Landing.

* A journal (pl.: journaux) was the amount of land one man could work in one day.

3 miles or one league

Map 7. The four French townships in Alabama. Map traced from the original, completed and captioned by Jacques Lajonie in 1821.

The four French townships have the shape of a reversed capital L whose base, located three miles south of Demopolis and stretching twelve miles east beyond Faunsdale, would today parallel U.S. Route 80 to Selma. From the southeast corner the boundary went north for eighteen miles, then west for six. Finally, two-thirds of the western side, from the northwest corner (where Sawyerville is now located) to the angle formed in the south where it meets the two townships at its base (a mile and a half from the river's curve called Candy's Landing), faced the Black Warrior without ever touching it.[41] It was within these boundaries, on an area of 92,160 acres, that the French colonists were to plant cotton, grapevines, and olive trees in what was to be, according to Lajonie, New France.

NEW FRANCE

This was only a tiny part of a vast region that was to become famous under various geographic and economic designations: Blackland Prairie, Cotton Belt, and Black Belt.[42] The French townships represented 144 square miles, that is, 2.4 percent of the 6,000 square miles of a diagonal strip of land stretching from western Mississippi Territory slightly southeast, and then straight through the center of its eastern portion. For the most part in the Tombigbee and Alabama River watersheds, bordered by the Fall Line Hills in the north and east, the Interior Flatwoods in the southwest and the Chunnenugge Hills in the south, this diagonal strip is a northern subdivision of Alabama's largest physiographic region, the Coastal Plain, which reaches the Gulf of Mexico in Mobile after gradually falling from an altitude of two to three hundred feet. Thus the Black Prairie's chief surface feature is its low elevation, to be found also in the French colony where the land is often undulating, but always between one hundred and two hundred feet above sea level. It is therefore not its relief that makes the region unique, but its geological bedrock, responsible both for the properties of its soil and, with the help of climate, for the composition of its vegetation.

In the Heart of the "Black Prairie"

The substratum of the Black Prairie is composed of a layer of sedimentary rock dating from the Cretaceous period when the sea covered the region and the dinosaurs were dying out. This sediment, white with shiny blue-gray flecks, the remains of fossilized one-celled living organisms such as phytoplankton, is rich in minerals. The fact that it is almost half calcium[43] justifies the use of the terms "Selma Chalk," and, more simply, "chalk" or "lime rock," to designate this acid, soluble, fine, and friable sedimentary rock that has been made into cement in Demopolis since 1901.[44]

While the limestone's white cliffs attracted the colonists and the stone itself stimulated the region's industrial activity in the twentieth century, it had also created the subsoil conditions for agricultural development, after the clayey portion bound to the sedimentary layer had produced a layer of black or very dark arable soil enriched by surface vegetal decomposition. This black soil, heavy or hard and cracked depending upon the weather, sometimes hard to work, acid in forested zones and alkaline in prairie zones, is always productive.[45] But the name Black Prairie is misleading, for the arable surface does not consist entirely of black earth, and the prairie, in the strict sense of the word, shares the terrain with broad-leaved and coniferous forests. What is true two centuries after the French arrived was even more true then, since geographers estimate that at the time prairie covered only from 6 to 10 percent of the region, and in the 1830s, after some land had been cleared, from 23 to 33 percent.[46] This, then, was an intermittent prairie and not a vast lawn upon which the pioneers could simply have set themselves down. But this intermittence also created the richness and beauty of the region. Traveling from Columbus, Mississippi, to his Lowndes County mission in the heart of the Black Prairie in 1822, missionary-linguist William Goodell and his companions succumbed to nature's charm:

> As you approach . . . from the east, there opens unexpectedly to view an extensive prairie, which contains several thousand acres, and which appears to be without a single stone, or tree, or fence, except now and then a small cluster of trees at great distances, like the little isles of the sea. Casting your eye over the prairie, you discover here and there, herds of cattle, and horses and wild deer, all grazing and happy. The grass, which will soon be eight feet high, is now about eight inches, and has all the freshness of spring. The oak . . . with the sycamore and mulberry, borders the prairie on all sides. Flowers of red, purple, yellow, and indeed of every hue, are scattered, by a bountiful God, in rich profusion, and in all the beauty and innocence of Eden . . . and their fragrance is, as if the very incense of heaven were there offered. "This," said one of the missionaries, "is the Lord's plantation."[47]

Though more circumspect than the missionaries as to the reality of divine intervention, Lajonie nonetheless shared their idyllic vision, jotting "enchanted country" onto Township 19 on the map he had drawn for his family, and noting on the other townships the presence of "very good lands," "prairies," and "large prairies," as he regularly did in his letters. Concerning the colony's lands in general, he wrote that the larger part of them, "grassland interrupted by lines of oak trees," looked delightful if one thought only of the pleasure of having many herds and flocks. This picturesque countryside was well worth a journey: "Come to enjoy our delightful prairies dotted with the most beautiful cedar groves." Concerning the soil of his 480-acre allotment, he was happy that "half is in prairie with excellent soil, the other, covered with black oak," or once more, "2/3 of which are in fine prairie dotted with groves of red cedar that are a refuge for our fine deer."[48]

This alternation of woods and prairies characterized the Black Prairie's countryside, even if wooded regions covered the major part of its area with a great variety of species, such as willow, ash, elm, cypress, several kinds of oak, hickory, and red cedar.[49] The latter proliferated in places where the layer of earth, thinned by erosion, barely covered the limestone bedrock. In Demopolis, for instance, chalky white patches and a multitude of tiny granules are scattered throughout the green ground cover, and the shallowest trench soon hits stone. On his map, Lajonie mentions a large space "covered with red cedar" along the Tombigbee just below its curve. The association of prairie and cedars embellished the French townships with names like Prairieville and Cedarville, separated by two streams flowing into the Black Warrior: Little Prairie Creek and Big Prairie Creek.

Lajonie and Goodell state that the countryside was not smothered by forests: in their midst, like so many islets of respiration, prairies abounded and were sometimes quite large. Their accounts are corroborated by contemporary official Land Office notes. For example, concerning a certain line (line P–Q) of Township 18, Range 3 East, a parity of prairie and woods is mentioned: "South ½ prairie, North ½ woodland," and concerning an adjoining area (line Q–R), more diversity: "Level land, oak, hickory, gum and some prairies, light soil."[50]

So the colonists did not come into an impenetrable jungle, but into an environment that nature and humans had already partly cleared. While scientists have shown that geological influence on vegetation via the soil's composition played the major role in forming the prairies,[51] they do not exclude the influence of thousands of years of human occupation. By burning the forests to flush game or plant crops, the native Chickasaw and Choctaw populations dotted the forested area with clearings that eventually became permanent prairies. But according to Sam Dale, the woodsman whom the Parmantier group met in St. Stephens, the Choctaws had a less rational but more poetic explanation, attributing the prairies' origin to the presence of a race of cannibal giants said to have lived there at the dawn of time. This is the version he gave John Claiborne, Mississippi's first historian: "[These giant cannibals] utilized the mammoth as their burden bearers. They kept them closely herded and as they devoured everything and broke down forests, this was the origin of the prairies."[52]

After occupying the land, there remained the matter of climate. Arriving in mid-July, the first colonists immediately found out about it. As previously mentioned, one of the criteria in selecting the region had been the supposed similarity of climatic conditions in the southeastern United States to those of southwestern France. The statistics of the Southeast Regional Climate Center give us weather conditions for part of the twentieth century as reported by the two stations most representative of the area settled by the French: Demopolis in the southwest and Greensboro, twenty-five miles to the northeast.[53]

Summers are very hot, with an average temperature of 80.5°F, and the mercury exceptionally rising to 107°—which is not as high as in twenty-seven other American states. It is a damp heat, with a humidity rate of 75 percent from April through October. Alabama's location in the southeastern part of North America on the Gulf of Mexico is responsible for this humid subtropical climate: weather and climate are controlled by the combination of hot, humid, maritime air masses from the Gulf of Mexico in the summer and masses of cold, continental air coming down from Alaska and Canada in the winter, but also in early spring. In Demopolis, a blue sky is part of daily life, with a 63.5 percent rate of sunshine from April to October. Spring and autumn temperature averages are close at 64° and 65°. Nothing suggests a harsh winter, but it can get cold: January is the coldest month of the year, with an average temperature of 45.5°F. Thus there is an average temperature range of 35 degrees.

Snow is rare in Demopolis and Greensboro, but it does rain and freeze. Rainfall is heavy: fifty-four inches yearly, on the average, with the most rain in January and March; while summers can be very dry. The first white frosts of autumn can arrive in early November, and late frosts can be dangerous until the end of March and even April. There are from 220 to 250 frost-free days.

This presentation of the natural environment offered the colonists shows that the choice of this region of the United States looked promising. Despite exposure to sudden climatic changes, tornadoes, frost, and scorching heat waves, it was still a temperate area with relatively moderate winters and summers. Its geological history had endowed it with a unique subsoil and a thick, rich soil to which its lush vegetation testified, and the existence of numerous open prairies was an invitation to colonization to which the native peoples had long ago responded. The Black Prairie's limestone bedrock was not only covered with black soils and red cedars, but with an age-old cultural sediment that the French came to discover.

After taking a tenant's look around and making a beginning of taming the natural environment, the colonists still had to sign the government's contract that would, after fourteen years of good and loyal agricultural service, give them access to the permanent status of proprietors.

SIGNING THE CONTRACT

Like the Senate one year earlier, the House of Representatives passed a resolution on December 10, 1818, asking the secretary of the treasury for an inventory of the French colony created by its act. What about the reservation of the four townships? Had a contract for them been signed with the agents of the French emigrants, and if so, on what terms? Were these agents living with the other colonists, and what progress had been shown in the growing of grapevines and olive trees?

Crawford reported that agents, duly appointed by the French emigrants, had come to his department of their own accord with many documents attesting to the positive development of the colony, among them several maps and a list of the names of the allottees. But it became apparent during this meeting that the "chart of reservation" submitted to him had been certified only by the head clerk of the inspector-surveyor of public lands (Freeman) under whose direction the reservation had been made, but that no report had been sent by the inspector in person to the commissioner of the General Land Office, who was his superior in Washington, D.C. No contract could be signed between the secretary of the treasury and the agents of the French emigrants as long as the required certification had not arrived.

The responsibility of the French agents, among whom was Villar, is thus once again involved in this administrative hiatus, although it is unclear whether Freeman shares the blame. Did they want to hurry things along by short-circuiting Freeman, busy in the field? Or were they dealing with a head clerk who was as scatterbrained as his predecessor, M. Pease Jr., appointed in the fall of 1817 to replace him in his absence and who, according to Freeman, turned out to be hopeless? Freeman did not send the documents to Meigs until January 12, 1819: "Sir. Herewith are transmitted to your department, Maps and descriptions of 11. Townships of the Alabama district embracing the Location and settlement of the French Emigrants at the confluence of the Black Warrior and Tombigby Rivers, with a connected map of that part of the District."[54] Did this supplement a preceding mailing certifying the reservation of the townships, or had an arrangement been concluded? Three days earlier, on January 9, the contract had been signed by Villar for the colonists and by Crawford for the federal government.

While Crawford might have been criticized for his belated instructions regarding the surveying of the French lands, it must be granted that this time he made things very clear in eight carefully worded articles.[55] After the reminder of the $184,320 (92,160 acres at two dollars an acre) due the Treasury Department before the end of fourteen years (by January 8, 1833), the first four articles stated that (1) by three years from the date of signature, each allotment be occupied, either by the title holder, or by others appointed by him; (2) within fourteen years, at least ten acres of land be cleared and under cultivation in every quarter section (the 160-acre quarter sections were "taken together," as the contract was collective and not individual); (3) within seven years, at least one acre of vines be planted on every quarter section, also "taken together"; and (4) within seven years, at least five hundred olive saplings be planted in the four townships, unless it were demonstrated to the president that it was impossible to grow them.

The last four articles, administrative rather than technical, stated that (5) the association's agent submit to the Treasury Department an annual report indicating the real number of homesteads, the variety of plants grown, and the progress and prospects of grapevines and olive trees; (6) the list of associates submitted to the Treasury Department was recognized, with the exception of several persons listed by name, whom others, designated or to be designated, were to replace;[56] (7) emigrants validly registered on the list and settled before August 1, 1818, on lands they had improved before knowing which ones (whether small or large allotments) would be officially assigned them, would have the right to keep them in proportion to the area of lands finally received, unless within six months the new, legal owner made them an offer of the value of the improvements made, in which case the first occupant was to leave the property and go to the allotment granted him in the drawing (if no offer was made, the legal owner would have the right to the allotment granted to the colonist occupying his land, area for area); and (8) the lands not yet allocated could be allotted to new arrivals from France who had no land, subject to the approval of the secretary of the treasury and to their actually settling on the land.

It is important to recognize the moderation beneath the dry legal language of the contract's terms. Neither the guarantees demanded nor the chronology is excessive: occupying each allotment by 1822; farming ten acres of land per quarter section by 1833—but according to the collective method of reckoning chosen, the colonists in the aggregate had to have a minimum of 5,600 acres under cultivation (560 quarter sections multiplied by ten acres) over the colony's entire territory, not necessarily ten acres in each quarter section—that is, 6 percent of the total area; planting grapevines on one acre of land per quarter section by 1827, that is 560 acres, again according to the collective principle, or 0.6 percent of the area; finally, planting, by the same date, five hundred olive trees, that is, one tree for every 184 acres.

Expressed in these proportions, each of the first four articles was the least one could expect of an agricultural colony based on the growing of grapevines and olive trees and registered as such by Congress. While not giving the colonists a blank check, the secretary of the treasury was not holding a knife to their throats either. But the historiography of the colony has forgotten how moderate were these conditions put to the French, focusing instead on the obligation of solidarity with which the colonists had to comply according to the act of March 3 and which the contract reinforced. The colony had been a collective grant; in order to avoid speculation, no member could reap individual benefits. This obviously put those who, like Lajonie, worked the hardest at a disadvantage, but it must be recalled that if this vision was not in accord with the intentions of someone like Garnier de Saintes, advocating the respect of the freedom of the individual within the group, in spirit it was, fundamentally, not all that far removed. It was basically a matter of passing from theory to practice.

DISTRIBUTING THE ALLOTMENTS

Along with the general map of the townships that served as a reference and the list of the allottees at the time the contract was signed, there were four diagrams, one for each township (see maps 8–11). Each 36-square-mile diagram is subdivided into 36 squares, called sections, each measuring 1 square mile, or 640 acres. This makes for a total of 144, only 140 of which are numbered from 1 to 140 in Arabic numerals, four squares being taken out of the distribution and reserved for the site of Aigleville and the small town allotments located around it. This subdivision of each township into square sections of equal size and the numbering order are the normal way in which townships were divided. In the case of the French colony, the first square (no. 1) is located in the northeast corner of Township 18, Range 3 East, the second (no. 33) in the northeast corner of the adjacent township, the third (no. 69) in that of Township 19, Range 4 East, and the fourth (no. 105) in that of the last township (see map 7). Each square section is in turn subdivided into allotments numbered from 1 to 347, in the same order as the numbering of township sections, plus 37 allotments, 26 of them labeled from A to Z and the other

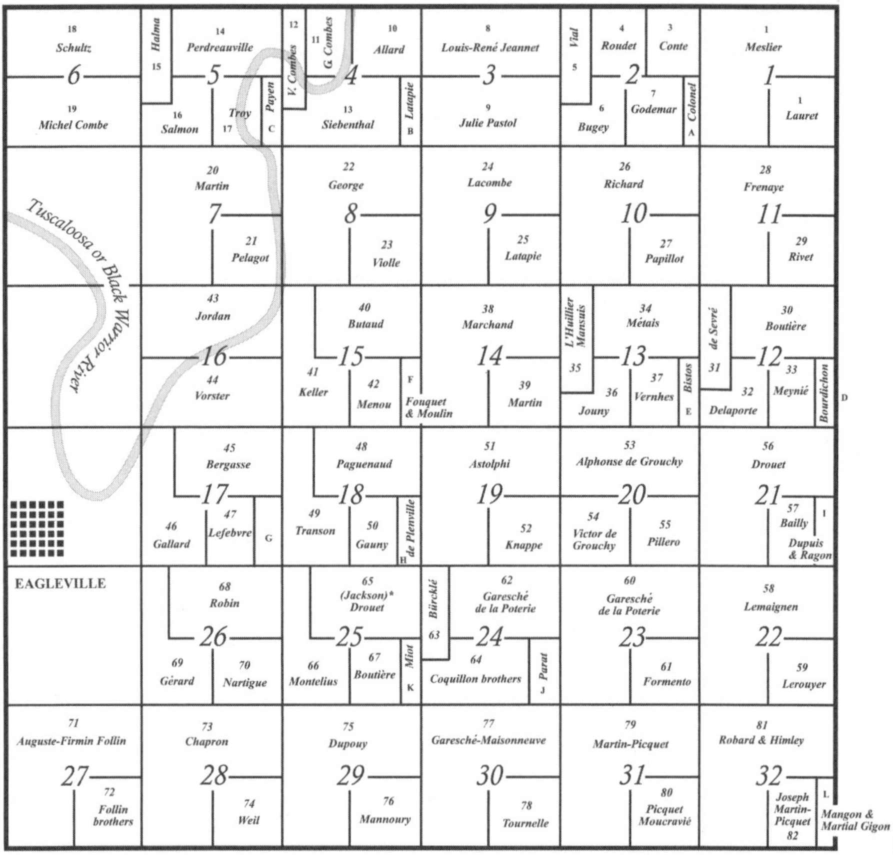

Tuscaloosa or Black Warrior River

18 Schultz	Halma	14 Perdreauville	12	11	10 Allard		8 Louis-René Jeannet	Vial	4 Roudet	3 Conte	1 Meslier
—6—	15 —5—		V. Combes	G. Combes —4—		Latapie	—3—	5 —2—		7 Godemar	Colonel	—1—
19 Michel Combe	16 Salmon	17	Payen C	Troy	13 Siebenthal	B	9 Julie Pastol	6 Bugey		A	1 Lauret

20 Martin	22 George	24 Lacombe	26 Richard	28 Frenaye
—7—	—8—	—9—	—10—	—11—
21 Pelagot	23 Violle	25 Latapie	27 Papillot	29 Rivet

43 Jordan	40 Butaud	38 Marchand	L'Huillier Mansuis	34 Métais	de Serré	30 Boutière
—16—	—15—	—14—	35 —13—	37	31 —12—	33
44 Vorster	41 Keller	42 Menou	F Fouquet & Moulin	39 Martin	36 Jouny	Vernhes	Bistos	32 Delaporte	Meynié	D Bourdichon

45 Bergasse	48 Paguenaud	51 Astolphi	53 Alphonse de Grouchy	56 Drouet
—17—	—18—	—19—	—20—	—21—
46 Gallard	47 Lefebvre	G	49 Transon	50 Gauny	H de Plenville	52 Knappe	54 Victor de Grouchy	55 Pillero	57 Bailly	I Dupuis & Ragon

EAGLEVILLE

68 Robin	65 (Jackson)* Drouet	Biirckié	62 Garesché de la Poterie	60 Garesché de la Poterie	58 Lemaignen
—26—	—25—	63 —24—	—23—	—22—
69 Gérard	70 Nartigue	66 Montelius	67 Boutière	K Mint	64 Coquillon brothers	J Parat	61 Formento	59 Lerouyer

71 Auguste-Firmin Follin	73 Chapron	75 Dupouy	77 Garesché-Maisonneuve	79 Martin-Picquet	81 Robard & Himley
27—	28—	29—	30—	31—	32—
72 Follin brothers	74 Weil	76 Mannoury	78 Tournelle	80 Picquet Moucravié	Joseph Martin-Picquet L 82	Mangon & Martial Gigon

* 65. Jackson's allotment, transferred to Drouet.

Map 8. Township 18, Range 3 East, with the names of subscribers.

11 from AA to KK, making a total of 384 allotments, each allocated to a colonist, alone or in association. It will be recalled that the numbered allotments vary in size, from 480 acres, that is, three-quarters of a section, for the largest to 120 acres for the smallest, that is, 40 acres less than a quarter section. These 40 acres are allocated to what are called reserve allotments.

Each section can thus be cut up in different ways, from a simple division placing two colonists into one section (480 + 160 or 320 + 320) to a multiple option (120 + 120 + 120 + 120 + 120 + 40). Maps 8–11 are derived from the collection of Hamner Cobbs, editor of the local weekly *Greensboro Watchman* after World War II. They should be consulted in conjunction with the list of colony members in the appendix, as each name corresponds to a section number and to that of the individual allotment.

95 Garnier Jr. —38—	93 Durand 37—	91 Promis 36—	89 Jamet 35—	86 Lecampion —34—	83 Auzé brothers —33—
96 Pénières Jr. / 97 Widow Audibert / Mignon M	94 Robaglia	92 Desmare	90 Rigau	87 Brechemin / 88 Humbert	84 Brugière / 85 Barraud
98 Nidelet 39—	100 Galabert —40—	102 Anduze 41—	104 Gubert 42—	106 Douarche —43—	108 Villar 44—
99 Cousin / Roudel N	101 Petitval	103 Frédéric	105 Moynier	107 Gruchet	109 Pagnerre
121 Besson 50—	118 Poculot —49—	116 Vandamme 48—	114 Frenaye 47—	112 Pagaud 46—	110 Dirat 45—
122 Lemeusnier	119 Baltard / 120 Mocquard	117 Angeli / Fouquet P	115 Laurent / Delaunay O	113 Fallot	111 Mondin
123 Lemeusnier 51—	125 Gen. Rigau 52—	127 Texier de la Pommeraye —53—	130 Martin du Colombier 54—	132 Ravesies 55—	134 Debrosse 56—
124 Germain	126 Mariano	128 Haraneder / 129 Métayé / Pénard & Rougier Q	131 Campardon	133 Bordas	135 Merle
149 Le Bouteillier 62—	146 Kimbal & Antoine / 145 Tulasne —61—	143 Joseph Lakanal 60—	141 Duval 59—	138 Davis —58—	136 Ladurelle 57—
150 Plantevigne	147 Billington & Antoine / 148 Boiteau / Putek S	144 Léontine Desportes / Desaffres R	142 Baclé	139 Firmin / 140 Montallegri	137 Canobio
151 Moncravié —63—	154 Cluis 64—	156 Garnier Sr. 65— / Macré 159	158 Wells & Leclerc —66—	162 Fontanges 67—	164 Descoins Belair 68—
152 (Brown) Bringier / 153 Monnot	155 Ruffier	157 Simon	160 Dumas / 161 / Dulmazieau / David T	163 Victoire Godon / Blancon & Taverly U	165 Sagnier

Map 9. Township 18, Range 4 East, with the names of subscribers.

When, shortly after the contract was signed, the maps with numbers and the list resulting from the drawing in Philadelphia—reviewed and corrected by the government—were published in a booklet sold, according to Lajonie, for the equivalent of forty francs, the French colony's territorial and human foundations were conclusively in place. But, while some seven months had gone by between the émigrés' first meetings in Philadelphia and the congressional vote, twenty more had elapsed between the publication of the act and the signing of the contract, for the reasons described above. Efficiently conducted at the beginning, arousing enthusiasm and attracting members, the matter had subsequently bogged down, dragging along from delays to errors, squabbles to separations, accords to discords, arrivals to departures. Uncertainty weighed too long on the colonists' shoulders, and many lost patience. When the contract was signed, the colony had already lost some of its colonists, who were consequently never very numerous, even at the beginning.

Map grid (Township 19, Range 4 East):

178 Roland **74** 179 Pichon	176 Charassin **73** 177 Vasquez	174 Carré & Ducommun **72** 175 Génin	171 Gatti / 170 Sari **71** 172 Ilari / 173 Millon	168 Clauzel **70** 169 Blaquerolle	166 Ch. Lallemand **69** 167 Valcour Aimé
180 Charreton-Raspiller **75** 181 Grillet	182 Texier **76** 183 Martinet / 184 Vitalba	186 *Cavaroc* / 185 Jogan **77** 187 Brugière / 188 Chapon / Mahé V	189 Dubarry **78** 190 / *Saldaignac* / *W. Labrousse* (192 *Onfroy*)	191 Descourt **79** 193 Pochard / 194 Fux	195 Stewart & Raoul **80** 196 Gilbert
211 Pothier **86** 212 Schubart / 213 Neel & Pfister	209 Lefrançois **85** 210 Groning	207 Widow Delaunay **84** 208 Castan (204 / *Tasca*)	203 Bonnot **83** 205 Blandin / 206 Azan Brossier Devanjeu / X	200 Nardel / 199 Richard **82** 201 Chauveau / 202 Plaidaut	197 Sévelinge **81** 198 Mane
214 Beyle **87** 215 Malczewsky	216 Teterel **88** 217 Pagnerre	220 *Lesueur* / 219 George / 218 Dubocq **89** 221 Dor / 222 Maillet	223 Stollenwerck brothers **90** 224 Vallot	225 Mathieu **91** 226 Allain / *Mayer* Y	227 Jeandreau **92** 228 Caillebaux / 229 *Nelson & Butaud* Z / Constantin & Deschoulles
243 Melizet **98** 244 Corso	241 Martin du Colombier **97** 240 Desplans	238 Chaudron **96** 239 Gilbal	236 Barbe / 235 Baumier **95** 237 Estribaud / 238 Desormes	233 Grouchy **94** 234 Deschamps	230 Taillade **93** 231 Olivieri / 232 Luciani
245 Hamel **99** 246 Havard	247 Pénières-Delzors **100** 248 Fanchon	249 Lecoq Dumarselay **101** 250 Godat / *Morin* AA	251 Forni **102** 252 Guillot	253 Badaraque **103** 254 Conté	256 Pascal / 255 Desfouch **104** 257 Fouasche / 258 Bernard

Map 10. Township 19, Range 4 East, with the names of subscribers.

The Number Present. For lack of accurate data, it is difficult to determine the number of colonists on the banks of the Tombigbee and the Black Warrior during the first two years. From July to December 1817, the founding half year, groups arrived one after another: the twenty or so of Parmantier's vanguard in July, Lajonie's nine or ten in October, and about forty in the Clauzel-Desnoëttes group in November are our least imprecise bits of information. Of the hundred reported by newspapers to be going down the Ohio in September and October, how many reached Demopolis and how many preferred the big city of New Orleans? Of others, arriving, like Pénières, alone or in small groups, we know neither their number nor the chronology of their journey, except that people came steadily, according to Lajonie: "Our colony is growing every day" (Demopolis, October 28, 1817). "Joseph [Bonaparte] is sending many officers and workmen here" (Aigleville, February 12, 1818). "The town of Aigleville is growing daily" (October 1, 1819). His

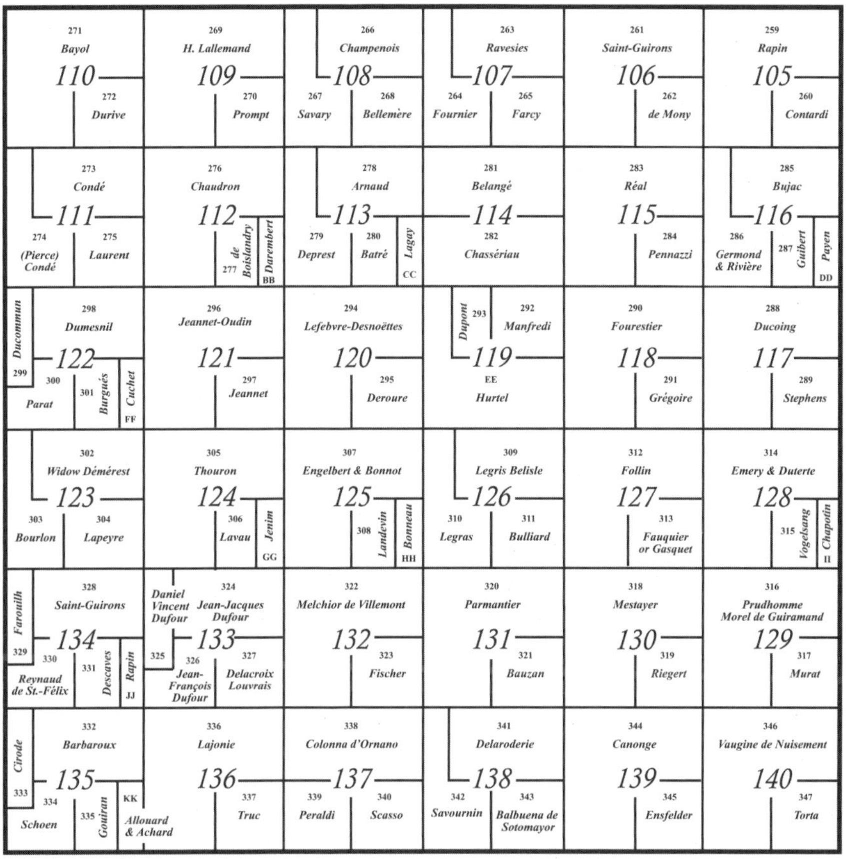

271 Bayol 110 — 269 H. Lallemand 109 — 266 Champenois 108 — 263 Ravesies 107 — 261 Saint-Guirons 106 — 259 Rapin 105

272 Durive — 270 Prompt — 267 Savary | 268 Bellemère — 264 Fournier | 265 Farcy — 262 de Mony — 260 Contardi

273 Condé 111 — 276 Chaudron 112 — 278 Arnaud 113 — 281 Belangé 114 — 283 Réal 115 — 285 Bujac 116

274 (Pierce) Condé | 275 Laurent — 277 de Boislandry | Darembert BB — 279 Deprest | 280 Batré | Lagay CC — 282 Chassériau — 284 Pennazzi — 286 Germond & Rivière | 287 Guibert | Payen DD

Ducommun | 298 Dumesnil 122 — 296 Jeannet-Oudin 121 — 294 Lefebvre-Desnoëttes 120 — Dupont 293 | 292 Manfredi 119 — 290 Fourestier 118 — 288 Ducoing 117

299 | 300 Parat | 301 | Barguès | Cuchet FF — 297 Jeannet — 295 Deroure — EE Hurtel — 291 Grégoire — 289 Stephens

302 Widow Démérest 123 — 305 Thouron 124 — 307 Engelbert & Bonnot 125 — 309 Legris Belisle 126 — 312 Follin 127 — 314 Emery & Duterte 128

303 Bourlon | 304 Lapeyre — 306 Lavau | Jenim GG — 308 | Landevin | Bonneau HH — 310 Legras | 311 Bulliard — 313 Fauquier or Gasquet — 315 Vogelsang | Chapoïn II

Farouilh | 328 Saint-Guirons 134 — 324 Daniel Vincent Dufour | Jean-Jacques Dufour 133 — 322 Melchior de Villemont 132 — 320 Parmantier 131 — 318 Mestayer 130 — 316 Prudhomme Morel de Guiramand 129

329 | 330 Reynaud de St.-Félix | 331 | Descaves | Rapin JJ — 325 | 326 Jean-François Dufour | 327 Delacroix Louvrais — 323 Fischer — 321 Bauzan — 319 Riegert — 317 Murat

Cirode | 332 Barbaroux 135 — 336 Lajonie 136 — 338 Colonna d'Ornano 137 — 341 Delaroderie 138 — 344 Canonge 139 — 346 Vaugine de Nuisement 140

333 | 334 Schoen | 335 | Gouïran KK — 337 Allouard & Achard | Truc — 339 Peraldi | 340 Scasso — 342 Savournin | 343 Balbuena de Sotomayor — 345 Ensfelder — 347 Torta

Map 11. Township 20, Range 4 East, with the names of subscribers.

New Orleans correspondent Quessart wrote in the same vein: "I also see with plea-sure that every day new families are going to settle on the Tombigbee" (February 7, 1820). But it is difficult to distinguish between reality and the desire to create an image of demographic prosperity for the colony. At least at first, when Lajonie hoped his family would join him, it was not in his interest to describe to them a region to which no one was coming or that everyone abandoned after a brief visit.

The first trustworthy information is given in a petition addressed to the gov-ernor of the Alabama Territory, undated but surely from 1818, by twenty-three signatories. In this petition, which will come up again later, we find the sentence: "On the lands granted by Congress to the French association, we already Count upwards of one hundred and fifty individuals living at Aigleville. (White Bluff)." The association of these two names would indicate that the petition was submitted before the colony left Demopolis, which was for a time called Aigleville. However that may be, it is doubtful that there was a rush of French people to Alabama, since, long after Parmantier's arrival, 150 persons represented perhaps only a quarter of the colonists potentially hoped for, women and children included. An advertisement dated December 13, 1817, and reappearing at intervals in the *Philadelphia Aurora* into the spring of 1818 is particularly revealing in this respect.[57]

Entitled "Fine Farm Gratis," it states that a number of French emigrants (with no indication of how many) knowing nothing of agriculture wanted to share their Tombigbee allotments with one hundred honest families who did. It would not cost the interested parties one cent, on the sole condition that they settle on the allotment and farm it. Such a gift, in a region whose rich soil and healthy climate had no equals in the entire world, was explained by the need to attract to these lands a vigorous and active population. The landowners, conscious that the success of their colony depended upon attracting settlers, would prefer large families. Requests were to be addressed to Vital Marie Garesché, 229 Market Street, Philadelphia, or Charles Villar, head of the French association in Aigleville, the chief town of the new settlement, where he would be in early March 1818. The fact that this advertisement was reprinted over a period of several months would tend to prove that candidates did not rush to Alabama, despite the attractive offer, and suggests that the grantees had not done so either.

However many people were there, they all had to leave their cleared and parceled-out lands and the town allotments acquired after a local drawing in mid-January 1818. Lajonie, for instance, was ideally situated on White Bluff, near the place where the cliff, opening to let the Tombigbee flow to the Black Warrior, permitted the landing of merchandise and passengers. The colonists had to abandon their lands outside the grant, leave infant Demopolis, and go found a new town within their allocated boundaries. It is understandable that some had already become discouraged and left. For the more tenacious it was, in the worst cases, only one more exile.

The colonists were leaving a remarkable site, but the town of Demopolis that they had conceived had only reached the embryonic stage. In their few months of presence and staggered arrivals, they had only prepared its terrain, sketched out its plan, marked out roads, and put up a few cabins—but this was already a lot, since it was useless. Maps of the colony indicate Demopolis as a simple dot, an x, a letter, or at best, the line of its main street. Part of what had been built could be salvaged and taken to the new site on the southwest side of a curve of the Black Warrior, a mile and a half east of Demopolis, on one of the unnumbered sections set aside for this purpose. Nothing remains from the founding period of Demopolis.

Lajonie's map, dated the second half of 1819, shows a square, a half mile on each side, roughly subdivided into sixteen units, while Paguenaud's printed one indicates forty-nine. According to Lajonie, who experienced its development, only the northern part of Aigleville was built, consisting of a horizontal line of houses (where later the Southern Railway and the St. Louis–San Francisco Railroad were to intersect). On a map with a scale of 1:24,000, the site of Aigleville can be seen next to present-day Demopolis, bordered on the west by Front Street, on the south by Arcola Street, and on the north by the railroad. Like the original Demopolis that disappeared without a trace, Aigleville is said to have left archaeologists only three bricks thought to be French-made because of their characteristic size. On the other hand, unlike Demopolis, its name has given rise to a very interesting collection of illustrations, somewhat similar to Currier and Ives in style, part of which is in France at the Musée national franco-américain du château de Bléran-court (the National French-American Museum at the Blérancourt château) in the department of Aisne, and the other in the Alabama Department of Archives and History in Montgomery.

TERRITORIAL AND ADMINISTRATIVE ORGANIZATION

The belated organization of administrative structures appropriate to the Alabama Territory was another destabilizing factor for the French colony. The Alabama Territory had been detached from the Mississippi Territory, of which it had been a part since its creation, by a congressional law of March 3, 1817, as the western part of the Mississippi Territory was to become a state by a law passed on March 1. But the umbilical cord linking the two parts of the large territory thus split was cut long after passage of the law, a section of which stipulated that the Alabama Territory could come into being only after the announcement of the writing of a constitution and the creation of a government in Mississippi. On August 16, 1817, the governor of Mississippi issued the written announcement, addressed to President Monroe, who waited to receive it before appointing William Wyatt Bibb as first governor of the Alabama Territory on September 25.[58]

A Virginian by birth, a Georgia physician by profession, and a Jeffersonian Democrat by conviction, Bibb had just spent four successive terms in the House of Representatives before moving to the Senate, where he had replaced William H. Crawford, who was appointed ambassador to France in 1813. Having resigned his seat in November 1816, Bibb was free. The date of his appointment is considered that of the birth of the Alabama Territory. Bibb did not arrive in Alabama until late 1817, settling in St. Stephens. Although there was no real urgency for the French, as local affairs continued to be supervised by the former secretary of state of the Mississippi Territory, this seven-month period of administrative vagueness, when Alabama was no longer a part of the Mississippi Territory but had not yet become one in its own right, was probably harmful to their interests.

The French had set a date too early, before anything at all could have been organized. On February 16, 1818, in one of his first mailings as official governor, Bibb suggested to Secretary of State John Quincy Adams that he have printed a map of the territory with its principal rivers.[59] That had not yet been done. If the governor himself felt a bit lost in his new territory, it is understandable that the French did so before him, even if this does not exonerate them completely. There were maps of the southern United States, but none of just the Alabama Territory. Cartographer John Melish made up for this lack in 1818, publishing the first.[60] Along with the map itself, a few statistical and geological remarks described the territory, whose population, estimated at 29,683 inhabitants according to the last census, was said to be increasing rapidly. There were rich mineral resources of iron, coal, limestone, and chalk.[61] The soil was generally good, producing cotton, tobacco, and wheat. The Alabama River was navigable to Fort Jackson, the "Tombeckbee" to "Eagle Ville" and then Cotton Gin Port, the Black Warrior to the river's northern rapids. As the name "Eagle Ville" shows, even though translated, the French settlement was not forgotten: on the map, at the confluence of the two rivers, the town is associated with White Bluff, the "French Settlement," and the "4 Townships Granted To the French Emigrants." According to the map, the French colony was divided between two newly created counties, bordered on the west by the Tombigbee and Choctaw territory: Marion County stretching far to the north, and Marengo County stretching south to the borders of Clarke and Washington Counties. St. Stephens was the county seat of the latter and the capital of the Alabama Territory.

Marengo County. Marengo County was created soon after Governor Bibb's arrival. It resulted from an "Act of the Alabama Territorial Legislature" of February 7, 1818. Abner Smith Lipscomb, a young jurist from South Carolina who studied law with John C. Calhoun,[62] is said to have chosen the name Marengo for the county. In 1811, with a passport from the governor of Georgia to travel through the Creek nation, Lipscomb had stopped at St. Stephens, where he practiced law and, in 1813, married Elizabeth Gaines.[63] She was a daughter of Young Gaines, who four years later took in the Parmantier group, and sister of Anne Gaines, the wife of Captain George Strother Gaines, then stationed at Fort Tombeckbee. Through the Gaines family, Lipscomb had contacts with the French, to whom he paid homage.

Bonaparte's victory over the Austrians at Marengo on June 14, 1800, had been hard-won but resounding. Exiled officers, such as Captain Lefebvre-Desnoëttes, had participated in it. The Tombigbee was not the Po, nor Alabama the Piedmont, but Marengo, a village on an Italian plain covered with grapevines, was honored by American geography. The Bavarian village of Hohenlinden succeeded it in honors. Since the victory of Marengo was not decisive enough to obtain peace, General Moreau struck the final blow against Archduke Jean of Austria at Hohenlinden on December 3, 1800.

When the time came to select Marengo's county seat, Screamersville, a locality south of the French colony, was selected.[64] The court was officially established on December 13, 1819, and the community was named "Town of Marengo." But a commission entrusted with selecting a new location chose the neighboring place called "Old Town," and Allen Glover, a commission member, advanced the funds to purchase the site in December 1820. Three years later, on December 19, 1823, at the same time that it became the county seat, the town was named Hohenlinden, a bit later shortened to Linden. The next year, the town was surveyed, and lots were laid out and put up for sale by George Noble Stewart. In the 1840s, Linden boasted the court staff, a man who rented out horses, a barber, a schoolteacher, two shopkeepers, and more than 160 other residents. Its central location in the county had favored it, while Aigleville, far off in the northwest, was already struggling to survive.

While Linden kept its status, the shape and size of Marengo County were constantly changing from its creation until the end of the nineteenth century, leaving little resemblance between the original territory and its modern version.[65] These variations determined in which county the French colony was located. The 1818 reduction of Marion County in favor of Marengo and Tuscaloosa Counties briefly placed the entire colony into Marengo. But the following year it was for the most part in Greene County, created from the northern part of Marengo.[66] After 1819, the colonists were therefore very unevenly shared out between the extreme north of Marengo and the southern half of Greene. The next important modification came in 1867 when parts of Greene and Marengo were taken to create a new county, Hale, with the Black Warrior as its western border. Its seat was Greensboro, a town that had been in Greene County when the county seat there was Erie and later Eutaw.

Greene County. Erie owed its name to pioneers from Erie in Ireland and dominated the Black Warrior from the height of its cliffs, as, more modestly, Demopolis did the Tombigbee. About four miles southwest of Sawyerville, Erie was thirteen miles west of Greensboro, half of this distance passing through French Township 20. The site of Erie, located on the farm of a certain Parkel, was chosen in 1819 as the future county seat because it was an ideal place to load cotton onto boats. Thirty years later, the proprietor of the large warehouse there stated that it was the only landing on the Black Warrior not subject to flooding.[67] After the courthouse was built in 1820, Erie developed around cotton farming that used it as its river port, but the town suffered from a shortage of drinking water and its slightly off-center location, which made its access even more difficult because in the wet season the roads were impassable. Finally, in 1838, the seat of Greene County was transferred to Eutaw, farther west. The county officials left Erie, the town went into a slow decline, the post office closed, its name appeared on Alabama maps for the last time at midcentury, and it died.[68]

Greensboro was luckier. Long after it was founded, the town benefited from the creation of Hale County to become its permanent county seat. Three brothers from Georgia are said to have been the first settlers in the region of the future site of Greensboro;[69] in 1817 they were joined by a great many American settlers who built a village they called Troy, a Trojan answer to Athenian Demopolis. Its inhabitants were scarcely more fortunate: once Alabama was a state, they learned they were living on the section that the law set aside for public schools. As the French had moved from Demopolis to Aigleville, these families and shops, among them that of Jason Candy, Troy and Greene County's first storekeeper, moved to the present site of Greensboro, which they also called Troy or New Troy until 1823. This time they took precautions. The area was surveyed in 1820 by an expert, John C. McAlpine, and duly registered.[70] On its 160-acre quarter section, the town grew rapidly. By 1821 there were hotels, saloons, and stores where one could purchase all kinds of merchandise; the following year, symbolic of this American-style growth, two lawyers set up shop. The inhabitants asked that their town be incorporated in Greene County as Greensborough (Greensboro). The state of Alabama made its existence official on December 24, 1823.

The administrative and judicial structures that were to favor the development of the region in which the French colony was located were thus in place within just a few years. However, it is noteworthy that not a single important town, however small it still may have been at the time, was founded within the strict boundaries of the four townships: Demopolis and Linden (Marengo) and Erie and Greensboro (Greene) were located outside their borders. Among the towns that survived the colony, neither Prairieville nor Cedarville, respectively located in Townships 18 and 19, Range 4 East, ever really developed.

13

The French and the Others

I do not know whether Monsieur Lajonie has described to you the region in which he is living. . . . Your friends live in a region which, four years ago, was inhabited only by savages and the wild animals that they hunted.

—Jean Quessart, New Orleans, June 16, 1819

I shall [negotiate the bill of exchange] only after receiving news from Monsieur Lajonie, seeing that he will perhaps give me a counter order for that one, unless he plans to buy black help.

—Jean Quessart, New Orleans, October 23, 1819

The colonists could have enjoyed their new environment right away, if one complication had not followed another. The trials of the journey and then of marking the boundaries of their lands made many realize how far they were from cities and the conditions of a comfortable existence. However, while not an Eden, for the thermometers soared in the summer, it did offer advantages: abundant rain, fertile soil, a long growing season, vast prairies, forests filled with game, and a handy river connection to a seaport. They were not the first to recognize these assets; the native peoples had enjoyed them long before the arrival of the French, and Anglo-American farmers, pushing in with their slaves, also wanted the land. Within a few years, power relationships developed between the native peoples on whose former territory they found themselves, the squatters against whom they had to defend it, and the blacks whom they needed to work it. The French colonists lived with these three groups on terms peculiar to each of them: trade with the Indians, harmony or strife with the Americans (who did not accept their presence), and exploitation of the blacks.[1]

NATIVE PEOPLES

After a heavy rain, the sandy banks of the Tombigbee and the Black Warrior, newly collapsed and furrowed, reveal vestiges of very ancient human occupation. A pair of boots and a bit of attention are all that is needed for one's foot to disturb or one's eye to catch on stones whose form and marks are the result of handiwork rather than natural erosion: large or small fragments with sharp edges, scraps from the chipping of arrowheads, or arrowheads themselves, the reward of amateur collectors. Away from the banks, little knolls, burial mounds, also reveal human presence. The French were coming into possession of a territory of which they were neither the first occupants nor the first colonizers.

CHOCTAW COUNTRY

This land belonged to the Choctaw Indians, whose ancestors, adopting a sedentary lifestyle, had, during the prehistoric Mississippian Period a thousand years earlier, reached a degree of development unequaled in North American aboriginal culture. Its principal feature was the erection, around a central plaza, of truncated earthen pyramids on which the temples and houses of chiefs were built. One of its largest sites resembles on a very rustic scale those of the Mayan civilization: called Moundville, it covers an immense plateau overlooking the Black Warrior.[2] The town, which must have had a population of three thousand at its apogee, extended its influence along the valley of the Black Warrior to its juncture with the Tombigbee—the site of Demopolis. The disappearance of the Mississippian culture, followed, in the sixteenth century, by the sudden arrival of conquerors Ponce de Léon and later Hernando de Soto, marked for the Indians, chronologically, the beginning of their history, and ethnologically, the beginning of their destruction.

The American Indians living in the Mississippi Territory probably descended from this culture. A custom that they inherited and that is said to have distinguished the Choctaw tribe, one of the largest in the region,[3] gives evidence of this. Some of the various etymologies suggested for the tribe's name claim that it comes from the Indian word *chahta*[4] or the Spanish *chato*, meaning flat, a reference to the Choctaw practice of flattening babies' heads, as the inhabitants of Moundville and the culture associated with it had done before them.[5] The Choctaw idea of beauty surprised the Europeans, as did their funeral rites, which consisted in letting corpses decompose before placing the bones into a coffin box.[6] The practice of head flattening may have given its name to the tribe; that of gathering the bones of the dead gave its name to the Tombigbee River: *Itombaigabee*, composed of *itombi* (box, chest, coffin) and *ikbi* (maker), coffin maker.[7]

The Choctaw nation was one of the powerful Indian tribes of this part of North America, stretching from the south-central part of present-day Mississippi to southwest Alabama, related in language to the neighboring Chickasaws to the north and Creeks to the east. Together with these two nations and the Cherokees and Seminoles, the Choctaws formed what were in the nineteenth century called the Five Civilized Tribes. They lived on hunting, gathering, and the products of their agriculture in a vast region that they did not even begin to populate: these tribes and their satellites probably did not exceed 100,000 in number, a quarter of them Choctaws,[8] friendly, peaceable people who preferred growing maize, sweet potatoes, squash, beans, melons, and pumpkins to the hunting and fighting favored by their Chickasaw cousins.[9]

Relations between the tribes deteriorated when the Europeans involved them in their struggle to control American lands. In the Great Lakes area, the English had allied themselves with the Iroquois against the French and the Hurons; in the south, they allied themselves with the Chickasaws against the French and the Choctaws. In taking on the quarrels of the foreign competitors who were asking for their support in order, in the end, to despoil them, the Indians were duped: their human dignity vanished in the alcohol that was sold them; guns and barter with pelt and fur traders accelerated the disappearance of game and aggravated intertribal conflict. Their friendship or hostility toward this or that foreign nation also led to expulsion from their ancestral lands, accomplished within two generations after American independence. This context allows us to understand the particular case of a fort in Choctaw territory that flew four different flags on the banks of the Tombigbee River.

Fort Tombecbé

Successively French, English, Spanish, and American, with a new name at every change in nationality, this outpost in Choctaw country was designed to strengthen the military, diplomatic, and commercial ties between the tribe and each of the countries wooing it. But after eighty years of existence, it gave its name to the treaty that rang the death knell of Choctaw hopes of keeping their lands—on part of which, with Indian hearths still warm, the French emigrants were settling.

When, in 1732, Louisiana was going through difficult times at the hands of an Indies Company incapable of governing it, the king of France appointed Bienville as governor.[10] It was urgent to settle the Indian problem and check British expansionism. Alliances with the tribes the English were attempting to bribe had to be renewed, and the Chickasaws, England's strongest allies for the past thirty years, had to be fought. To prevent the creation of a triple entente between the Choctaws, Chickasaws, and English, which would spell disaster for French territorial continuity in America, Bienville decided to establish a fort in Choctaw country, to serve as a trading post and military garrison, thus controlling Choctaw loyalty and Chickasaw aggressiveness. It was completed in 1737 at the place called White Rock Bluff, part of today's Jones Bluff on the west bank of the Tombigbee. The location, as bucolic as it was strategic, was reminiscent of White Bluff, several days' navigation downstream. The fort occupied the summit of an eighty-foot limestone cliff overlooking the river that flowed at its feet and gave it its name, Fort Tombecbé. For twenty-five years it successfully acted as a "window for the French"[11] into the Choctaw nation with which there was ongoing trade, guarded against the Chickasaw danger, and blocked English colonial expansion. In 1763 the Treaty of Paris transferred the fort, renamed Fort York, to the English, but after a series of problems that convinced them they were wasting money on tribes that would never be satisfied or pacified, they left the site. In 1772 French voyageur Bernard Romans found it already fallen into ruin.[12] Twenty years later the Spaniards rebuilt it.

While French Louisiana had had to counter the menace of English expansionism, the Spanish possessions feared the far more dangerous risk of invasion by land-hungry people like American freebooters. In 1791, Carondelet, the new Spanish governor and intendant general of Louisiana and Florida, devised a plan whose two essential points were an alliance with the Indians and the construction of new forts, creating a buffer zone between his provinces and the American advance. The Indian nations, threatened by the same peril, went along with the plan: in 1793 the Choctaws signed the Treaty of Boucfouca with Spain, ceding the land of the former French fort on the Tombigbee for the construction of a Spanish fort and trading post.[13] With the Chickasaws, Choctaws, Cherokees, and Creeks on one hand and the Spaniards on the other, having approved the creation of an Indian confederation under the king of Spain's protection, Fort Confederación was born.

As Fort Tombecbé had been for the French, Confederación was for the Spanish a key element in their advance defense system and a privileged place to meet with the Choctaws. But just as Fort Tombecbé had fallen victim to an international treaty signed thousands of miles away, Fort Confederación fell victim to the Treaty of San Lorenzo, which in 1795 pushed the Spanish south of the thirty-first parallel. The colonial authorities, who had succeeded in gaining the Indians' confidence and had kept the Americans at a distance, applied the treaty grudgingly. Their days in this part of America were now numbered. The fort was abandoned in 1797, and the Choctaws and other Indian nations were handed over to the Americans. The arrival in Natchez the following year of the first governor of the Mississippi Territory, created to the north of the thirty-first parallel, started the countdown for the Choctaws, who, in treaty after treaty until 1830, were to lose all their lands east of the Mississippi. The forces that had insidiously been eroding the foundations of aboriginal culture for three centuries took thirty years to undermine them so that they collapsed completely.[14] In 1816, Fort Confederation was one of the stations on the Choctaws' way of the cross.

Vanishing Lands

For a long time, relations were good between the Choctaws and the Americans, whom they had supported in their War of Independence. The Hopewell Treaty, signed in South Carolina in 1786, had made the friendship official: in exchange for sixty-nine thousand acres of land—a negligible number in a territory of millions—the Choctaws accepted the protection of the United States, as did the Chickasaws, who transferred their allegiance from the British to the Americans. Washington's Federalist administration thus inaugurated a policy of pacification and paternalism, developed further in the Jeffersonian era—except that, between the two, during Adams's presidency, the creation of the Mississippi Territory radically altered the situation.

Until then, the Indians had recognized American sovereignty, but, while not an abstract concept, it was not truly noticeable. From 1798 on, the Indians found themselves under the direct supervision of the whites, who took charge of their future in a muddled chain of command going from the president, according to the

Constitution the ultimate person responsible for Indian affairs, to the men in the field. Instructions held that it was imperative to conciliate them and relay to them the message that the strictest justice would always be observed in their regard. Like his predecessors, Jefferson advocated the need to civilize the Indians through education, modernization, and intermarriage, which would accelerate the process of acculturation: civilization, agriculture, husbandry, and manufacture became the leitmotifs of official thought.[15] To this economic-philanthropic concern was added that of detaching the Choctaws from the Spanish magnet and thus avoiding at all costs an alliance that would be disastrous to American interests. National security and humanitarianism guided Jefferson's choices, but Guice calls his policy republican with a small "r," for it reflected one man's philosophy and not the wishes of a land-hungry white population indifferent to the fate of the natural occupants of these lands, providing they went away. Between these opposing perceptions, between presidential sincerity and public opinion, the latter carried the day and the Indians were duped. This in spite of doing their bit, for the Choctaws truly turned toward the United States, commercially as well as militarily.

Choctaw Allegiance. The Indians were the victims of traffickers who illegally sold them alcohol and bought their furs at a low price. To remedy this and also to remove the Indians from Spanish influence, Congress, in 1795, voted funds to establish trading posts administered by the War Department, which were to be centers of fair exchange. In an extension of the system, new posts were created in 1802, among them one for the Choctaws in St. Stephens on the site of the former Spanish fort San Esteban.[16] The manager who arrived in 1803 was during the following year assisted and in 1806 replaced by George Strother Gaines. The merchandise (clothing, cloth, blankets, iron, lead, guns, powder, hatchets, tools, household utensils, etc.) was quite similar to what white slave traders exchanged for captives on the African coast. Since the object was to strengthen ties with the natives, agents were advised to see to the quality of the articles and lower their price if necessary. Barter was also on an overvalued footing, particularly when it came to the item most frequently traded, deerskins, on which the government incurred heavy losses due to lack of demand. But this was the price of peace in the region. Friction between the Choctaws and the whites, reduced to cattle theft or alcoholic violence condemned by both sides, never weakened the Indians' loyalty to the Americans, who did not lose sight of what was most important.

As their liaison officer, Gaines won the Choctaws' trust and persuaded them to join the Americans against the Creeks: Pushmataha, their chief, came to St. Stephens himself to enlist his warriors. Unlike the Choctaws and the Chickasaws, the Creeks had never fallen into line. Their siding with the English during the War of Independence had led the victorious government to open their lands to the colonists' rapacity. After that, the tribe split between the more moderate White Sticks and those who refused cohabitation with the whites and domestication via civilizing measures, the Red Sticks.

The latter's defeat at Horseshoe Bend in 1814 was decisive in the process of expelling the Indians from their lands, "the turning point in their ultimate destruction," in the words of Robert Remini.[17] The subsequent Treaty of Fort Jackson, one of a long series between the American government and the Indians (friend and foe alike), was the most consequential.[18] The Choctaws alone had already signed five, from 1786 to 1805, which had taken from them more than 7.5 million acres of land.[19] Concentrated in the southern part of the Mississippi Territory, along the thirty-first parallel, these lands formed the buffer zone desired by Jefferson for the strategic reasons mentioned above. For Jefferson, considerations of national security were more important than the need to satisfy land-hungry colonists, but for the Indians, dispossessed of their property without any significant material or financial compensation, this did not make any difference.

In this sense the Creek War was a decisive event. To the political motives—punish the Creeks and move them a bit farther from Spanish temptation—Andrew Jackson added greed for land: "Actually he was less the government's agent than the agent of westerners," writes Remini.[20] After this, it was impossible to slow the advance of white settlers who couched their requests for land in patriotic terms: the intensive and sensible farming of lands that had until then been so little or so badly cultivated by the Indians, they said, would be an essential contribution to the nation's economic development and prosperity. This time it was no longer problems of frontier security that were advanced—the Treaty of Ghent with the English and Spanish loss of power having reduced the threat—but the need for domestic tranquillity.[21] With the Choctaws, the government returned to ideas of paternalism, education, evangelization,[22] and commercial loyalty in an attempt to pacify and assimilate them a bit more, without really wanting them to blend into the white landscape.

So it was decided in 1815 to move the St. Stephens trading post upstream on the Tombigbee "for the purpose of placing it further from the White settlements, and nearer to the Choctaw Nation."[23] Moved in the winter of 1815–16, the post was set up some hundred yards from the site of Fort Confederation, on a little stream, Factory Creek, joining Itombaigabee Creek (now called Jones Creek) where it flows into the Tombigbee.[24] It opened successfully in the spring and that autumn even became a post office, with Gaines as postmaster. The isolation of the place made it necessary to establish a connection with St. Stephens, so as soon as he arrived, Gaines saw to the construction of two barges to be used to bring in manufactured items and send out the Indians' skins and furs. Depending on the load and the boat used, a round trip took from three to seven weeks. This trading post, which was to be a place of meeting and exchange with the Choctaws, went beyond its conciliatory mission, as a treaty was signed there that was very advantageous to the United States: the Treaty of Choctaw Trading House.

It was signed on October 24, 1816, not at Fort St. Stephens (a name frequently given as that of the treaty) but at the trading post moved near Fort Confederation, in the heart of the Choctaw nation.[25] Its terms will be recalled. The United States bought from the Choctaws three million acres for $10,000 in kind plus $120,000 dollars payable within twenty years. Shortly thereafter, on March 3, 1817, it granted the French 92,000 acres of land—that is, $1/32$ of those three million acres—for $184,000 payable within fourteen years. These figures put into perspective the "gift" the American government had given the French.

FACE TO FACE

The Choctaws were forced to cross the Tombigbee and occupy its western bank, the small share left them in the state of Alabama. Fourteen years later, on September 15, 1830, the Treaty of Dancing Rabbit Creek expelled them once and for all, as well as from the state of Mississippi, where they had considerable land.[26] Meanwhile, for thirteen years they lived alongside the French colony without creating any problems, judging by Lajonie's, or later Lakanal's, kindly allusions. Both document a friendly relationship.

Well before then, French voyageurs had described the Chaquetas (or Chactas) in pejorative terms: the men were lazy, the women were ugly, and their sexual, alimentary, and funerary practices were disgusting.[27] Impervious to the mysteries of the Christian religion, they were depraved, coarse beings from whom nothing could be hoped for, even with regard to the French, whom they thought for a time of leaving in favor of the English. In short, there is a wide gap between the impressions of these travelers and the idealized version of professional literary hacks.

Several hundred of these "Chacta savages" were living peacefully, said Lakanal, some two hundred paces from his dwelling in Mobile and furnishing him with excellent game.[28] Lajonie also likens the Indians to savages, since that is what he most frequently calls them, along with natives, Indians, and Choctaws, but without ever denigrating them. On July 28, 1817, he considers that if things went badly for lack of funds, he could go "over to the savages" and survive there, even with difficulty, rather than take charity. In November he confirms that there is no reason to fear them, that peace from the natives is assured, which permits trading with them: game, fish, and poultry for rum or money; "the Indians sell us chickens and turkeys in the winter: 35 to 50 sols apiece, a leg of venison, 25 to 35 sols";[29] for a watch and 25 francs, an Indian horse, "good only for riding" (July 4, 1818). Great lovers of horse meat, according to Milfort in his *Coup d'œil rapide*,[30] the Choctaws had, through contact with the French, broadened their tastes to chicken, pigs, and also cattle. In June 1818 Pierre Fougnet planned to go to the Indians to get cows, which General Desnoëttes and Poculot had recently done: "They made a few small acquisitions, but only for themselves, except two cows for Le François, that cost him twenty gourdes more with their calves" (June 20, 1818). Everyday relationships with the Choctaws enabled some colonists to learn, if not their language, in any case the rudiments "of savage" that were essential for dealing with them. As early as 1817, Lakanal the scientist took an interest in the different Indian nations in America,[31] and in 1834 he stated that he had written a paper on the Alabama Choctaws and a dictionary of their tribe.[32]

But what Lajonie liked even more about the Choctaws—and this made him a true colonist—was their land: not the land he occupied and which had been theirs so recently, but the land across the river. This interest was probably due to the lo-

cation of White Bluff and therefore of Demopolis-Aigleville, which the French
had not been able to integrate into their colony for the surveying reasons resolved
in the preceding chapter. On October 1, 1818, he wrote:

> The government is about to negotiate to purchase the lands of the Choc-
> taw. This nation has never been at war with foreigners; they like the French
> very much because formerly they were treated well by them. The Tombecby
> River separates us from them and they are here every day to sell us game or
> to buy rum. Aigleville will prosper when we have these lands, that is when
> the American government has managed to purchase them. There may be
> some difficulties. The Choctaw are very attached to their lands, 100 square
> leagues, 30,000 inhabitants, 10,000 warriors. They detest the Americans.
> We are their brothers.[33]

Lajonie's words convey the ambiguity of the French colonists' position with respect to the Americans and the Indians. On one hand, his "when we have these
lands" is without qualms associated with the policy of territorial expansion at the
expense of the Indians; just before arriving, on July 20, 1817, he had been ready
to take possession of their lands: "In another few days I shall be able to say legitimately (the savages have sold to the United States): 'My lands of Nauze.'" On
the other hand, he understands the hatred of the Indians for the land-stealing
Americans and their traditional feeling of fraternity toward the French. Americanophobia was a feeling shared by the Choctaws and French, targets of the jealousy and hostility of many American colonists. This is shown in an excerpt of a
letter written by Pierre Fougnet on July 14, 1818: "Before we left we were talking
about this Indian nation and we were hoping that we should be far away from it.
But chance wanted the contrary, that we be very close. These people are not at all
fearsome, on the contrary, they are our best friends. There is not as much to fear
from them as from people around us."

In 1802, addressing Choctaws camped near Natchez, William C. C. Claiborne,
the second governor of the Mississippi Territory, had expressed to them his hopes
for the future: "You and the white man face to face."[34] Because of the dichotomy
between Jefferson's wish and the public interest, this direct, face-to-face communication never took place, but all other things being equal, the face-to-face encounter between the American and French colonists was also a failure.

PIONEERS

In the first half of the sixteenth century, the conquistadores were the first to explore the southeastern part of North America, in search of gold that was nowhere
to be found, for the Indians had none and had no idea what it was. Three centuries
later, their descendants were subjected to the pressure of new conquerors, but this
time on their most sacred possession, the land, of which chief Tecumseh said that
he could no more sell it than the air or the sea.[35]

The Land Rush

The sixty-four hundred white people living in Alabama in 1810 were evidence that the creation of the Mississippi Territory twelve years earlier had not been enough to attract crowds. These pioneers, coming down from Tennessee or up from Mobile, had settled on the best lands along the major rivers, often, as we have seen, atop bluffs for protection against Indians and floods. To these colonists must be added whiskey traffickers and other renegades fleeing justice in the woods, as well as squatters, intruders, and border jumpers[36] occupying Indian lands not yet surveyed or put up for public sale. These individuals compromised the double (and ambiguous) policy of pacification and expansion without stimulating the local economy, concerned only with living from one day to the next or, at best, with bridging the gap between two harvests of corn planted on poorly cleared land. But by 1813 their numbers had forced the Choctaw agent to complain of the whites crossing Indian lands in defiance of the law.[37] For a long time, the army and the Indians had been trying to straighten things up by clearing out the intruders and burning their cabins, but to no effect: the colonists' pressure was stronger than they. This pressure grew after 1815, and Alabama's demography took off. Its white population increased thirteenfold from 1810 to 1820, and thirtyfold from 1810 to 1830.[38]

Several factors explain this massive arrival of whites in Alabama: the end of the wars against the Creeks and the English in 1814 and 1815; the cession by the Choctaws, Chickasaws, and Creeks, in treaty after treaty, of almost all their lands in the state by 1820; and the progressive sale of these lands by the government to speculators and especially cotton planters rushing in from neighboring states.

The cultivation of long-fibered cotton had long prospered along the coasts of Georgia and South Carolina, while short-fibered cotton, more difficult to process manually, grew at higher elevations. The 1793 invention of the cotton gin, a machine that did the work of fifty slaves, benefited the regions growing short-fibered cotton, making them the principal cotton-producing area in the United States. The subsequent cotton boom was hard hit by the effects of maritime insecurity on export curves, but as soon as normal trade with Europe resumed, the price of cotton took off with renewed vigor, giving a boost to the calling of planter. In Georgia and South Carolina, however, the supply could not keep up with the demand. Either there was not enough land, as most of it was in the hands of old, established families, or the soil was depleted by overexploitation. In late 1815, the sale of millions of acres in Alabama, the fertility of the lands, and the quality of the network of navigable rivers leading to a seaport were all features that attracted settlers from the old states of the South to Alabama.

Those leaving came chiefly from the piedmont rather than the coastal regions where the planters' situation did not force them to the sacrifice of a physically difficult emigration.[39] After selling their land, they set out in wagons with their families and baggage, their slaves herding along their animals. Lajonie gives us a striking description:

Thousands of Americans go two, three, four and sometimes five hundred leagues from their usual dwelling to lay the foundations of a new establishment. Sometimes the head of the family has seen the region to which he is going, but often he knows it only very imperfectly. These families almost always leave separately, each one for itself. When roads go to their goal, they take them and leave them if they take them too far out of their way. Reeds, woods, thick [illegible word], streams, rivers, nothing stops our caravans, almost always composed of the head of a household with a troop of Negroes. The children and their wives follow the wagon or large cart resembling an ambulance. The men are on horseback, leading the cows and the pigs. If during the first year the land does not meet their expectations, they break camp and go farther. (June 7, 1818)

With the first public land offers in 1816–17, immigrants poured in and, even before the large sales of 1818, upset the local economy with a corn shortage that caused a near-famine among the Indians and improper occupation of lands from which the squatters often had to be evicted in favor of the legal buyers. The pioneers, who came into Alabama by the many land routes—among which the Federal Road was one of the most important—or the port of Mobile, often selected their terminus according to their point of departure.

Colonists from the Carolinas and Georgia were the most numerous, followed by their neighbors from Tennessee, Virginia, and Kentucky. Some counties were occupied mostly by immigrants from a given state, like those in the Tennessee Valley in the north by Tennesseans; most of the others offered an intermingling from all the southern and border states. It would seem that in Greene[40] and Marengo Counties, where the French colony was settling at the time, the majority were from the Carolinas.[41]

The number of colonists is as interesting as their origin. Colonists did not rush to these counties as they did to others. The 1820 census lists 1,891 whites in Marengo County, the second from the last of the twenty-four counties that turned in their forms, and 2,878 in Greene County, putting it in fifteenth place. During the next decade the number of whites increased by 250 percent in both counties taken together, but Marengo remained far behind the counties along the Tennessee River.[42] The *Historical Atlas of Alabama*, giving the population, date, and mapped location of settlements in the nineteenth century, permits an assessment of the changes in population density with the identification and chronology of the principal historical sites still in existence or that have disappeared.[43] Its maps and list of sites show that settlements were spaced out along both banks of the Black Warrior (located outside the French colony), that there were, moreover, relatively few before 1820 or even 1830, and that, consequently, the French colony was not located in an overpopulated region. Thus its existence should not have been considered bothersome by the American colonists. But Lajonie, while continuing to admire, lucidly and critically, their efficiency and pragmatism, sometimes gives evidence of the contrary.

"The Devil with the Americans"

Lajonie's initial enthusiasm survived his disillusionment with the Americans, whose materialistic nature he denounced: "The devil with the Americans," he grumbled (December 1, 1821). But this did not prevent him from praising their merits, turning their faults into virtues, or comparing them favorably to his compatriots: courage, work, money, and patriotism were the essential values that were already making the United States a great nation. He thought that France would do well to take a leaf out of the US's book. The courage he admired was first of all that of colonists who did not hesitate to leave everything to begin anew elsewhere in order to better their lot, in spite of great distances to travel and obstacles to overcome. Lajonie, who saw entire families traveling from neighboring Georgia along the roads of Alabama, deplored the comfortable sedentary existence of his compatriots: "Ah! poor Frenchmen! If under a leader, however admirable he may be, you had to vanquish so many difficulties, you would a thousand times over curse the day on which you were born. It is with such people, my good friend, that a country prospers and becomes rich" (June 7, 1818). While they may have inherited this courage from their migratory parents, the colonists not only kept it but reinforced it with their natural predisposition to work: "Our Frenchwomen pride themselves on working! The poor women! I would like to see you replace an American countrywoman in her family and take care of eight to twelve children, dress them, make cotton cloth, put shoes on their feet, wash their clothes, feed them. Ah! What a lot of work!" (November 1817).

The lack of skilled labor meant that all had to do their share and that manual laborers were paid very high wages: "People with a trade all make a fortune in America," observed the newly arrived Lajonie (January 20, 1817). Since courage and work had the same goal, money, they would be nothing if they were not permanently underpinned by good business sense, acquired in childhood and a factor of future success: "The Americans who know how to conduct their affairs give each of their children, however little they may be, either little pickaninnies, or sows, or cows, or chickens, etc., so that their children learn domestic economy at an early age. With their little flocks and herds they buy, advised by their mothers, what they need for their upkeep" (September 18, 1821).

There was nothing comparable in French upbringing, according to Lajonie, who wished for such precepts for his own children. This individual state of mind was also that of the community and the nation:

> Come, discontented Frenchmen, come and learn to do without superfluous things, colonial goods, come and learn to encourage the arts, do come, and you will see that the American, who is dominated by patriotism, finds good only what his compatriots harvest or manufacture. You have disdained using grape sugar, beet sugar, the finest that can be made; the American takes pride in consuming only the national sugar that everyone knows as brown sugar, raw sugar that cannot be refined. . . . He can do without your wine; he drinks water and sometimes whiskey, the spirituous liquor of the country.

He can bring down your manufactures by putting high duty on your products and by paying his laborers well; no national product pays taxes or entry duty, and the government does more: it pays to encourage the exportation of salt meat and refined sugar from the country. With such principles, a nation is great and important and its citizens are admired by their neighbors. (August 3, 1817)

While admiring these principles because they most effectively develop the economy and power of the country that advocates and practices them, Lajonie nonetheless denounced their excesses: "Without wanting to put you off America, it is indispensable that I warn you that the Americans' favorite phrase is make money. Do not expect to find here the pleasures, even less the comforts of France, in a new establishment like this one. So if we want to make money, we must work hard, at least the first five or six years, at the risk of catching a felon, a fever, and scabies" (January 10, 1818).

And on another occasion: "One must see America to know the mentality of its inhabitants. Make money is the rallying cry of all Americans. We are going to imitate them" (February 12, 1818). American-style success meant amassing money and accumulating land; the price of this success was relentless work to the detriment of the pleasures of existence. But this will to succeed through work and money, although respectable in itself, was often tainted with less noble sentiments or behavior: credit, cupidity, speculation, and bankruptcy, of which Americans made a game (January 2, 1821), and constant, ruinous recourse to the law to settle disputes in a country where "everything is cheap _except lawyers,_" he wrote on April 12, 1821—after stating earlier that it was better to be a lawyer than a farmer if one wanted to become rich in the United States. Added to all this was envy, tinged with a bit of xenophobia, when foreigners' business ventures might outdo those of American citizens. Such feelings were directed against winegrowing, the reason for which the French had come: "I should like you to know that we had grapes, among other bunches two very fine and very good, which I removed from the voracity of the young people. We picked them on August 15th. I took them to the election office, which refused my vote as not being American. Setting down these two beautiful bunches of grapes, I told them: 'Here are my qualifications' and I withdrew. They produced a marvelous effect, given that several Americans hope that we shall not succeed in getting the vines to grow! Many people are jealous of us" (August 15, 1820).

Failure of the vines meant for the Americans that the contract connecting the French to their government would be canceled and that, consequently, they thought, tens of thousands of acres that had wrongfully escaped them would be freed. While the colonists had first wanted the Indians' lands, whose extensive farming they called wasteful, they now cast an eye onto those of the French, whose advantage was perceived as all the more unjust because they were incapable of improving them. Many American colonists simply seized the French allotments either before the arrival of their legal owners, after which lawsuits were necessary to get them off, or after noting the prolonged absence of those who never journeyed to Alabama.

Between his genuine admiration for the Americans and his rejection of a mentality he found so disconcerting, it is not always easy to determine the relations that Lajonie and his peers had with their planter neighbors. Undoubtedly because they were foreigners, not naturalized and not citizens, many of them felt marginalized by American society and perhaps closer to the Indians, whom they called their brothers. On December 23, 1822, after six years in the United States, Lajonie expressed his bitterness: "You know that I detest American morality; they have no honor; they are all merchants." However, thanks to their regular correspondence between France and Alabama beginning in 1829, we know that Lajonie had struck up a friendship with one of his neighbors, Colonel Alexander McAlpine, a wealthy planter who had arrived in Greene County in 1819.[44] McAlpine was very well disposed toward the French, regretted the Lajonie family's departure, and hoped they would return.

How can we allow for both Lajonie's sincerity and his bitterness at not succeeding like the Americans? From the outset, his stated objective was to follow their example, to speculate on land, to earn money through agriculture and trade: "Landowners now have immense revenues, but in order to become one, hard work and manpower are needed," he wrote on June 27, 1817. Hard work and manpower: his own, those of the indentured men, and those of slaves. While Lajonie reproached the Americans for their harshness and lack of morality in business, he never objected to their practice of slavery, which he adopted without turning a hair, as did all the other Frenchmen who had the means to purchase black labor.

SLAVES

On the occasion of the 150th anniversary of the founding of Demopolis, a Birmingham daily headlined an article "Helpless without Slaves."[45] The author subscribes to the historiographic tradition that attributes the French colonists' failure to the fact that they had no slaves to help them because they were too poor to buy any and, above all, because slavery was against their moral code.

SHADOWS AND ENLIGHTENMENT

On February 4, 1794, the French Republic had abolished slavery by a vote of the National Convention. How could one think that the abolitionist former Convention members now in America, or generals who had risen from the ranks during the Revolution, like Clauzel and Desnoëttes, could have compromised themselves in such a vile activity? The touching traditional depiction of the soldier-farmers of the Vine and the Olive and the Champ d'Asile featured their richly colored uniforms in the midst of fields with palm trees for exotic background, but forgot the blacks. Black slavery and men who had unleashed the winds of liberty were simply incompatible.

The spectacular character of the revolutionary abolition had undoubtedly made people forget what followed. The altruism of the Convention members had in part drawn its strength from the spirit of the Enlightenment; the realism of the French merchants and the refugees from Saint-Domingue drew its strength from economic necessity and convinced the First Consul that the "deplorable times of demagoguery" were over. On May 20, 1802, the obedient Legislative Body reestablished slavery, to the delight of its partisans and the indifference of others.[46] Passage of the law really made sense only if Saint-Domingue was reconquered, but the military expedition, which included several future American exiles, failed. With the Grande Isle lost, the peace of Amiens broken, and the Royal Navy mistress of the seas, the slave trade and slavery disappeared from the field of imperial preoccupations. They reappeared with the Restoration.

The French went back and forth on this subject in 1814 and 1815, with England as arbiter. That nation had abolished the slave trade in 1807, and the future king of France in exile had agreed to do the same once on the throne; but he did nothing, and it was Napoleon, no more motivated but more opportunistic, who abolished it during the Hundred Days. When he returned, Louis XVIII was told to do away with the trade, which he did half-heartedly.[47] In short, the France of those years was a proslavery nation, resistant to the idea of abolishing the slave trade, first of all because the idea originated with the English and was thus hateful, and second because, though faint, there was still hope of recovering Saint-Domingue (France did not recognize the independence of the island, now Haiti, until 1826, in exchange for an indemnity of 150 million gold francs). In such a context and given their personal position, the exiles would not have been sensitive to an issue that was not mobilizing public opinion. Nor did their arrival in America awaken them to the problem.

In fact, the first compatriots the exiles met in Philadelphia were the Domingan refugees, for whom the loss of their plantations had been a calamity and the freeing of their slaves, a heresy. These people could not be expected to persuade the newcomers that they should do without slavery when growing vines and olive trees in Alabama. In returning to the land, the former colonists' only goal was to re-create a system like the one they had been obliged to leave twenty-five years earlier. This idea was not preposterous in a territory lacking a real tradition of slavery but with a promising future in a very favorable regional geography.

In the years following 1815, the French discovered a country deeply rooted in slavery, particularly in the southern states, where the cultivation of tobacco, indigo, rice, cotton, and sugarcane required an abundant supply of slave labor.[48] After two centuries of slave trade, there were 1.7 million blacks in the United States, 87 percent of whom were slaves—that is, about 18 percent of the total population.[49] The number of slaves had doubled between 1780 and 1810 and then increased very rapidly until 1820, despite the fact that the trade had been banned for more than a decade. Almost all lived in the states of the South.

Because of an anomaly that has always raised the same questions, the Americans, who had since their Revolution—the major event of their history—become "the most liberal, the most democratic, and the most modern people in the world,"[50] had missed the opportunity of becoming the first nation to abolish slavery on its entire territory. How can these superlatives be conciliated with the negation of the rights of millions of individuals forgotten by political, social, and economic progress? How can one speak of democracy when men of color are excluded from citizenship? American democracy was a democracy of whites, and many of its great founding names came from the South. Before the Civil War the Americans elected seven southern slaveholders to the presidency and reelected five of them:[51] Washington freed his three hundred slaves on his deathbed; as for Jefferson, he owned some two hundred.

And yet, even before the abolition of slavery by a French Revolution freed of monarchy's yoke and acting in accordance with its principles, the American Revolution, freed of the colonial yoke, had made the United States the first abolitionist nation—but only in the North, where, well before the War of Independence, the Quakers' humanist convictions[52] and the ideas of equality of the religious movement known as the Great Awakening had been the driving forces behind the antislavery movement.[53] The North American rebels continued the trend: George Washington's Continental army enrolled black slaves, promising them freedom, Vermont (1777) and Massachusetts (1783) immediately freed their slaves, and Pennsylvania did it gradually (1780). The United States could then have given new impetus to this first great wave of emancipation—unique in the Western world—and extended it to the entire country, but this did not happen. Why not?

The Enlightenment had certainly enlightened the Founding Fathers, but with its political and constitutional proposals rather than the antislavery argument. This explains why the French philosophes' ambiguous position[54] found its way into the Declaration of Independence and the Constitution. Jefferson, who drafted the former, evokes the equality of men at birth and the inalienability of the individual rights of life, liberty, and the pursuit of happiness, but could blacks, thought of as secondhand creatures of the divinity and slaves by nature, lay claim to these? The color of their skin, visible and indelible mark of a biological and intellectual inferiority, forever excluded them from equality to whites. Slavery was an evil, emancipation a hope. But how could one go about it with irresponsible beings? "For men probably of any color, but of this color we know, brought from their infancy without necessity for thought or forecast, are by their habits rendered as incapable as are children of taking care of themselves," Jefferson wrote in 1814, repeating the old refrain of the blacks' incapacity for freedom.[55] In the same letter, Jefferson stated that he had very early wished for the emancipation of slaves, but the tasks he had been assigned since the American Revolution had distanced him from these questions so that until his return to Monticello in 1809, he had "little opportunities of knowing the progress of public sentiment here on this subject."[56] The arrival of time for reflection could have led him to free his own slaves, but he did not. As early as 1776 Jefferson could have inscribed his name in the abolitionist pantheon, since he had included in the Declaration of Independence's list of American grievances with the king of England a paragraph opposing the slave trade, but North Carolina and Georgia had succeeded in eliminating it. This augured ill for the nationwide application of antislavery intentions. Thus the Constitution, coming eleven years after the Declaration of 1776, did no more to condemn the practice of slavery than had the earlier document, and this for reasons of state: the young nation's union of the North and a South weighing in with its mass of slaves. Disorganizing a century-old system was out of the question; the Union took precedence over any other consideration, even the most praiseworthy. In 1787, Congress made do with an ordinance forbidding slavery in the Northwest Territory—and created a de facto boundary between the slave states and the others along the Ohio River. That same year, the authors of the Constitution made a gesture by shifting the discussion from the abolition of slavery to the end of the slave trade planned for 1807, a final extension of twenty years for importing shiploads of slaves. It was thought that the end of the slave trade would bring about that of slavery, but this was a wait-and-see attitude. In 1804, although the northern and central states had all passed laws of gradual emancipation, the American movement was on the wane: the abolitionist cause was not a national priority, and the antislavery initiative passed to the British. In spite of new organizations denouncing the slave trade more than slavery itself, the fight lost strength. Struggling to survive, the United States did away with an internal bone of contention by passing the law of March 3, 1807. This law outlawed the slave trade but changed nothing as to slavery itself, whose control remained a power reserved for the states. The southern states, where the number of slaves was increasing at the speed of the spread of cotton and sugarcane along the Mississippi River, had no intention of following the abolitionist path of their northern counterparts, and thus experienced another half century of slavery-based prosperity. It was not until the 1850s, after the Second Great Awakening and the activity of William Lloyd Garrison had renewed the abolitionist cause, that the emancipation of blacks became a major political issue and led to the Civil War.

But when the French arrived, the country was still very far from this turmoil, and everything was being done to avoid it. In 1816 the American Colonization Society sought less to abolish slavery than to free the slaves in order to send them back to live in Africa, and to prevent the illegal introduction of blacks by capturing the slave traders at work, even non-Americans. In 1820 the United States declared the slave trade an act of piracy and sent out naval cruisers to suppress it. The Americans had not yet signed any convention with the European countries engaged in the international hunt for recalcitrant slave traders, but they took it upon themselves to seize foreign offenders. In May 1821 the capture off the coast of Africa of four French slave ships by the USN *Alligator*[57] angered French authorities and their ambassador in Washington, who began talking war. It was easy for Hyde de Neuville on this occasion to denounce the contradiction of a country that was internally proslavery and externally philanthropic: "The Americans constantly accuse us these days of favoring the slave trade: according to them it is our white flag that covers this odious trafficking. These ardent friends of humanity who live surrounded by slaves . . . use the most odious sophistry to prove that they have the right to enslave their fellow human beings."[58]

Continuing to condemn the United States, Hyde cited the September 11, 1821, *Moniteur Louisianais*, stating that "all along the coasts of the state of Louisiana, slaves are brought by dozens while our warships are cruising around Africa to prevent the slave trade."[59] In such a context, between the French who made an outcry when their slave ships were meddled with, and the Americans who were laying down the law far from their shores but tolerating slavery within their borders, the immigrants had few sure points of reference except what their consciences might have indicated. But conscience abdicated before an institution that was so intimately linked to the culture of the South.

"NEGROES, NOTHING BUT NEGROES"

Just about all the French colonists who were in a position to have slaves had them, by mere virtue of their presence in Alabama, from Lefebvre-Desnoëttes, a lieutenant general and count of the Empire, through less prestigious names, to Joseph Lakanal, founder of the Institut de France. Only lack of money prevented slave ownership, not morality or religion, no matter what the colonists said. "People generally and rightly speak out against the enslavement of blacks," wrote Lakanal in 1817.[60] Years later, however, he acquired "ten good strong slaves" for his Mobile plantation, justifying himself by the fact that he was treating them "like unfortunate friends," and that, after all, he was merely following American laws.[61]

Colonists ideologically resistant to owning slaves could base their position on the Scriptures, preaching love of one's neighbor as oneself, or on the declaration of the rights of man and of the citizen, decreeing that men were free and had equal rights. But the ideals of the Revolution had since been flouted, and a literal, rather than humanist, reading of the Bible made it possible "to accept slavery in this world and wait for freedom in the world above."[62] Proslavery people found arguments in the Old Testament, and the position of the principal Christian denominations, which did not see slavery as a practice incompatible with evangelical values, supported them in their dogma. The Protestants were in the forefront of the abolitionist struggle, but being a Protestant rather than a Catholic was in no way a valid reason to denounce slavery. The French were members of various denominations, and some were agnostics. This made no difference. Economic reasons won out over humanitarian duty.

A comparison of the attitudes of two colonists, one Protestant, one Catholic, toward slavery bears this out: Jacques Lajonie, with deep Huguenot roots, and Victoire George, née Le Grand de Boislandry, a member of the Norman Catholic nobility who was in Stephen Girard's circle in Philadelphia and who settled on the banks of the Black Warrior in January 1820.[63] Their letters cannot say enough about the need to have slaves.

At no time did Lajonie, for whom it was a new experience, or Victoire George, who had lived with the system in Saint-Domingue, have second thoughts about the slavery they planned to practice. Only the neophyte Lajonie let slip two sentences, written a year apart, in which he ventured a critical judgment. In the first, touching on the fact that the slave state of Maryland neighbored Pennsylvania, where blacks were free, he wrote: "This proximity often gives the Negroes an opportunity to slip out of the barbaric yoke of the whites" (April 27, 1818). In the second he pointed out the American "contrast" between the concept of a republic and the purchase of blacks (March 2, 1819). But these reservations did not in any way defer his adherence to the "peculiar institution"; at best, they raised the hope of humane treatment of his own slaves.

This silence regarding the iniquity of the world of slavery speaks all the more loudly, since their mentions of blacks are frequent during the first years, when it was opportune to acquire some. The question arose at the very outset, even before they were in a position to do so. In January 1817, barely ashore in America and knowing nothing of Alabama, Lajonie sketched his three-part "plan for country life": "I should like to buy land in a region where blacks are slaves, where one could grow vines, which are very rare in the United States, and where the French language were on a par with English." He was most probably thinking of Louisiana, where slavery was doing well, thanks particularly to the French—Creoles, émigrés, and Domingan refugees—who were ardent proponents of the system, with jurist Pierre Bourguignon Derbigny in the lead.[64] Appointed secretary of the Louisiana Legislative Council at the time of annexation, he went to Washington, where on July 4, 1804, he made a speech in Congress asking for the immediate reopening of the slave trade. Victoire George was not to be outdone: in her very first letter from Mobile, where she had just arrived, she said she had found a black woman for sale.[65]

Nothing dampened their resolve to have slaves, because on virgin lands under the burning Alabama sun, necessity knew no law. They needed a workforce able to clear and cultivate the land without complaining about the climate, working conditions, or remuneration. It was difficult for whites, whether indentured French or American day laborers, to live up to such demanding terms and conditions. The first whom Lajonie brought from New Orleans to Demopolis quickly made him regret having brought them, as they were either overwhelmed with all sorts of physical ills (felons, boils, scabies, fevers) or proved disloyal and ungrateful, ready to follow the colony's idlers. He very soon had to consider sorting them out: "I am going to reward those who conduct themselves well [Labrousse and Mangon] and get rid of two who conduct themselves badly. I shall not have many left, so much the better, less profit, fewer troubles" (November 1817). He was not objecting to the principle of indenturing but to the quality of the indentured men and therefore proposed a more rigorous selection process, eliminating people they did not know, adventurers, and military men in favor of young people known to them, neighbors or relatives, single or couples, country folk who were "a bit dumb" and even deaf. The important thing was that they were well furnished with clothing and tools and had a useful trade: carpenters, cabinetmakers, masons, or blacksmiths.

On February 16, 1818, the *Marie-Thérèse* brought two new helpers from Bordeaux to New Orleans: Antoine Fougnet, called Titi, fifteen years old, the brother of Pierre and Élisabeth, called Élise or Lise, and Pierre Ragon, called Pierrillot, twenty-two, from Belvès-de-Castillon.[66] The first was satisfactory despite his youth, but the swift desertion of the second, forcing Lajonie to devote time to bringing him back to the fold, confirmed his initial feeling: "The indentured men make a sensitive heart too sad. In spite of the fact that we have lost nothing at it, this kind of speculation is not to my taste. One always meets only with ingratitude" (November 1817). To make up for their failings, Lajonie hired an American to do some plowing, at eighty-five francs per month, plus room and board: "You can see the expenses in this country and what the revenues must be in good years" (March 2, 1819). But recruiting local laborers was not a cure-all, because, in short supply in this vast region, they preferred working their own land rather than hiring themselves out, and were therefore so expensive that few colonists were able to afford them.

Victoire George, who did not have a network of indentured men and was thus even more dependent than Lajonie on free European-American laborers, complained: "We absolutely need hands, and the whites are ruining us. With the best food and eighteen gourdes a month, a day laborer costs at least two hundred fifty gourdes a year, and brings in perhaps only two hundred," she wrote from Demopolis on February 9, 1820, adding a month later: "It is becoming more and more obvious to me that all the white men who come here expect wages that are impossible for us to pay. They all want to become farmers in their own right, and they do this if they can, so that we are reduced to using the hands of our two Gentlemen, or to giving wages too exorbitant for the land they cultivate." In short, if one left them a clear field, the domestics and day laborers would completely devour them, complains Victoire George, insisting: "We must absolutely have Negroes"; to which Lajonie echoes: "Negroes, nothing but Negroes" (June 13, 1822).

How could one do otherwise? By working oneself. But only to a certain point and for a certain length of time. Victoire George admired her son, who did not balk at working: "We have fifteen acres cleared and my son is going to clear twenty-five more. His ambition is to show his father a beautiful farm. He is actually working more than a day laborer. People have no idea how difficult it is to clear land: they have to see it to believe it. I hope that our labors will be rewarded" (September 7, 1820). Three months later, however, she had lost her illusions: "My children and I see that without Negroes we will only wear ourselves down and vegetate. Even to grow vines one needs hands used to moving the earth; in a climate like this, one cannot grow cotton without them and that is the chief revenue of this region" (December 14, 1820). Lajonie agreed. The life of planters without blacks is too hard, dreary and difficult (November 22, 1819), whereas it is easy with blacks (January 2, 1821) particularly for women, who have almost nothing to do when they have black women (September 18, 1821). He begs his sister Nauzille, ready to join him, to bring enough money to buy herself "a Negress, *at least, at least one*" (April 1, 1820). This is what he was then thinking of doing himself: "We need at least a little Negress. I do not find it at all pleasant when Mademoiselle Elise is indisposed to be at one and the same time cook, housekeeper, hewer of wood, bearer of water, miller, baker, carpenter, carter and manager. I do not hate work, yet I could find tasks more noble than that of milking cows, which is what I do every morning" (August 16,

1819). In sum, buying a few slaves was necessary, "in order not to have to do everything ourselves" (April 1, 1819).

Unreliable indentured help, ruinously expensive day laborers, exhausting work, a blazing sun, endangered health, a life without pleasure, an impoverished future—such was the world of whites without blacks. And blacks, thought to be tough and better acclimated, were viewed as having been born for just this purpose, to relieve whites of their difficulties. Better than beasts of burden and draft horses whose illnesses were very costly, complained Lajonie and Victoire George, likening blacks to animals in plantation inventories. In 1826, after selling part of his American property, Lajonie figured that he still had "two Negroes, two horses, a few cows and hogs that I could not do without. Their value, counting the tools and household utensils, is about 1,200 gourdes cash" (December 17, 1826). This balance sheet was a far cry from Lajonie's hopes as he counted his chickens nine years earlier—long before they hatched:

> Let us buy 20,000 francs worth of land, use 30,000 francs to become well established, buy 20,000 worth of Negroes whom we will marry off and put 30,000 more into the Bank of the United States. In twenty years, our land will be worth 40,000 francs and our Negroes 30,000 and we will have profited from the 30,000 earning 8% per year depending upon the political situation. Thus our children will have 30,000 francs, and if by necessity or by speculation we sell our own produce and if we raise a large number of animals, we can even add to the capital. Those are more or less the advantages that planters have. (January 1817)

A year later, commercial setbacks and the hitches related to the distribution of allotments prompted Lajonie to abandon business and land transactions and return to the point of view about which he continually wrote his family: "We should seriously consider putting capital into Negroes. They are the best speculation for a planter" (January 19, 1818). To this he soon added speculation on farm animals. Blacks were not only slaves whom Lajonie needed to relieve him in his work, but also objects of speculation. In May 1818, writing from New Orleans, he again gives his thoughts on the subject:

> If peace prevails for six years, which one can assume for America, capital used now for Negroes will be doubled at the end of that time, and the work of the Negroes will have produced, in addition to the upkeep of the house, a revenue equal to that first laid out. I think therefore that with a layout of 50,000 francs, thrift and hard work we shall in six years have a capital of 150,000, provided, that is, that business does not change too much and that there is no mortality in the black race. Just count the production of twenty sows and twenty cows at the end of six years and you will agree. Cotton is at 30 sols; an acre can give 300 pounds. A good black can, with a horse, cultivate 20 acres. 6,000 francs that this Negro will bring you clear because he

will grow besides enough maize and potatoes for himself. Six Negroes, 100 acres and six horses give 30,000 francs per year net.

If the work of slaves was in itself a source of revenue, trading in them was another, either buying them cheap and reselling them at profit, or breeding them. Lajonie contemplated both.

BUY OR BREED?

Alabama was not the first place where Lajonie meant to buy slaves; its recent opening to colonization had not yet permitted a sufficient influx of them from neighboring states. With almost none present when the Mississippi Territory was created, the number had risen to twenty-six hundred by 1810, and after 1815 had grown steadily to reach nearly forty-two thousand by 1820.[67] But this increase did not mean one could draw from them at will; almost all the blacks, who had come with their masters, migrating with families and animals from Tennessee, Georgia, and the Carolinas, were therefore not necessarily available for sale except at prohibitive prices. But there was no lack of supply zones on the American market.

Lajonie first thought of New Orleans, which he knew from having stayed there, particularly with Jean Quessart, a merchant who knew a great deal about slaveholding. As mentioned earlier, Quessart was part of the wave of Domingan refugees driven out of Cuba in 1809 who came to Louisiana with their slaves. Along the river—that is, from Baton Rouge to Natchez—French planters had rapidly expanded sugarcane growing and, along with it, the number of slaves. In 1803, when Louisiana became American, half of its inhabitants were black and one-third were slaves.[68] By 1820 its population had increased fivefold, and the fifteen thousand slaves made up 36 percent of the population. In New Orleans itself there was one black for every white and more than one slave per four inhabitants. It was easy to draw from this mass of slaves at auctions whose dates, locations, and nature were announced in the *Ami des Lois* and the *Courrier de la Louisiane.*

On October 20, 1817, at Maspero's Exchange,[69] three young, strong, and robust black men, raised by the seller and never having had another master, were offered against payment within six months.[70] On the twenty-eighth, at the same place, six slaves, all French and of various ages and both sexes, guaranteed hard workers, healthy and without blemish, were up for sale. From cane cutters to mulatto laundresses, supply met demand. Slaves were not cheap. Bidding sometimes went very high: "One Negro costs between 4,000 and 10,000 francs, depending upon his merits. Carpenters are the most expensive. They have been known to sell for 15,000 francs," exclaimed Lajonie (September 1, 1817). But some could be had more cheaply: "Two good Negroes sell for up to 20,000 francs in this city. Sometimes, however, one can manage to get two tolerable ones for 10,000; two young Negroes of fifteen knowing no trade cost 4,000 francs each" (July 15, 1817).

Their price was beyond his means, and Lajonie left New Orleans without slaves. It was never ethical reasons that kept him from having them, but lack of money.

Between November 1817 and July 1818 he regularly mentions a sum of twenty-six thousand francs that a certain Barthe owes his family and that he asks them to discount as quickly as possible in order to have the liquid assets necessary to buy slaves. And in her first letter from Demopolis, Victoire George appeals to Stephen Girard for help: "To improve our lands, we would need two thousand gourdes to buy Negroes. Please tell me if you could advance me them, to be repaid from year to year from the income you manage for me; this would be a very great favor you would do me" (February 2, 1820). Not hearing from their respective sponsors, Victoire George and Lajonie tried to convince them by suggesting more attractive options for the purchase of slaves south of Pennsylvania, where slavery had long been abolished, in Maryland and Virginia.

His stay there with Latour must have given Lajonie some ideas on the subject. Baltimore, he writes, is the place to buy slaves, and what is more, "fairly cheaply" (January 10, 1818). If his family had decided to join him there, Lajonie could have combined business with reunion: "My journey to go meet you would have been doubly advantageous; Negroes bought in Baltimore for 4,000 francs sometimes sell for 6,000 francs in New Orleans. We need some if we want to have a worry-free life. This revenue is often very lucrative" (April 7, 1818). This solution, thought of very early, remained a constant: "Aside from a few small drafts on New Orleans, if you have one to draw for a rather large sum in order to purchase our Negroes, do it on Baltimore where I shall go to buy them myself. One can get the least expensive Negroes in Baltimore because this city is in a state where the slavery of blacks reigns supreme and also adjoins the state of Pennsylvania where blacks are free" (April 27, 1818). Repeating his advice, he asked them to send thirty thousand francs in drafts on a Baltimore bank, where he would be able to buy the "young Negroes" he needed (May 1818). This option was to have priority until 1820, when he abandoned it because his family was no longer coming. Generals Clauzel and Desnoëttes had the same idea.

Actually, their initial intention had been to recruit "redemptioners" emigrating from Germany and Switzerland according to a system similar to indenture: the ship captain who gave them passage expected payment within a certain amount of time—generally fairly soon—after arriving in the United States. These people were not slaves, but in practice the very harsh terms of the contract often resulted in their sale and enslavement.[71] According to the French consul, Clauzel had, in August 1817, chartered a broken-down schooner in Philadelphia "to transport to Tombigby German families that he and Desnoëttes" had recruited to work their lands in Alabama.[72] Without further detail, Lajonie announced that the generals had arrived in the colony with "indentured help for next to nothing" (February 12, 1818). Were these "redemptioners"? The Marengo and Greene County archives attest to the presence of Germans and Swiss in 1818—Schneider, Breton Büetschi, and Breitling—whose descendants still lived in Demopolis in 2008.

Other Europeans, also driven to America by poverty, joined the colony. In 1819 in Boston, Lin Troy, who taught the art of penmanship at the Carré & Sanderson boardinghouse in Philadelphia,[73] hired four Irish domestics for eighteen months to plow his two 120-acre allotments.[74] Putting them to work on January 1, 1820, Troy hoped that eighty acres of land would be under cultivation that year and would produce at least forty-eight hundred bushels of maize, from which at fifty sols per bushel, at least, he expected a revenue of twenty-two hundred piasters.[75]

Like all the other French colonists, not only Lajonie and Victoire George, the two generals confirmed the fact that farming in Alabama without slaves was almost impossible. In the summer of 1818 they boarded a ship in Mobile for the East Coast and, after having to put in at Havana, arrived in Baltimore on October 8: "They are coming here to buy Negroes and agricultural implements. They are to leave again shortly for Mobile," wrote the consul.[76] (The generals, who were ill when they disembarked, took advantage of this to inquire whether their return to France was conceivable, given their good conduct, but it was still too early.) Three years later, in 1821, Lefebvre-Desnoëttes was still in Alabama, building slave-supported castles in the air with Lajonie: "We were calculating with General Lefebvre the other day that 50,000 francs would be enough to buy 30 Negroes, one third of each age, and he told me that with this number of slaves, a gentleman I know made 12,000 to 15,000 francs in revenue from his harvest alone in an ordinary year" (September 18, 1821).

In that year, 1821, while Charles Villar, the former president of the French association in Philadelphia, had decided that Tennessee was the place to purchase "two vigorous Negroes for 1100 piasters,"[77] the younger General Lallemand, at first distracted by his marriage to Harriet Girard, then by the Champ d'Asile, and finally by the publication of a work on artillery, remembered his grant in the colony. Victoire George learned that he planned to go to Virginia to buy slaves 50 percent cheaper than in Alabama, where he would bring them. She had Girard ask him to get her some for two thousand dollars (February 8 and 18, 1820). She idealized Virginia as the best market for slaves. But they were still expensive, according to Lajonie, who stated that it was nonetheless possible to get some there "for 3,000 and 4,000 francs, and Negresses for 2,000 and 3,000" (March 2, 1819), because there were many, half a million. The end of the legal slave trade in 1808 had not caught the state of Virginia short; on the contrary, at the Philadelphia Convention of 1787, its representatives had recommended the trade be abolished—to increase the value of their own slaves, not to save those to come.[78] Along with Maryland and North Carolina, Virginia encouraged childbearing to replace the trade. In 1974, economists Robert Fogel and Stanley Engerman aroused controversy by refuting the interest southern planters would have had in resorting to the sale of slaves to make up for the loss of agricultural capital in areas where the soil was becoming depleted, and thus the organization of systematic slave breeding for the domestic market.[79] It is true that until 1860 there was an interregional redistribution of slaves from the Old South to the virgin lands that were increasingly being offered to colonists even farther south and west, but the vast majority of the slaves concerned undoubtedly migrated with their owners.[80]

What is astonishing in the case of Lajonie, who had no past experience of slavery, are his inclinations. In the first letter in which he considers using them, he already mentions the idea of having slaves in order to "breed" them. That is why, in later letters, he often states his preference for young slaves fit for reproduction: "The purchase of eight or ten Negroes of twelve to fourteen years of age, will work the best because when the indentured hands' time expires we shall have Negroes who will be able to work well and also have little offspring, so that after ten years of a rather active life, but also rather pleasant, we should have tripled our capital in Negroes" (April 27, 1818). And again: "Do try, my dear friends, no matter what the cost, to make enough money to buy eight or ten young Negroes, male or female who at the age of 16 or 17 years, will replace the indentured help whose time will have expired. At that age they will be fully able to cultivate our lands and to have children who will be a great product for us. Everyone, particularly Monsieur Quessart, approves of this plan" (May 6, 1818). It would be anachronistic to think of this as a scientific breeding project. Lajonie simply wanted to grow his capital in slaves by natural economic means, like Covey, a Maryland farmer too poor to build a herd of human livestock in any other way, who had, writes Frederick Douglass in his *Narrative,* bought his first slave in 1833 to make of her a "*breeder*": "The children were regarded as being quite an addition to his wealth."[81]

As for Lallemand, in the end he canceled his journey to Virginia in favor of Florida, where he was offered land "at a very low price and long terms," but as the validity of the deeds seemed suspect to him, he preferred to come to the French colony in September 1821. He stayed at Desnoëttes' plantation and toured his allotment. Lallemand was favorably impressed, finding the place good for settlement on the condition of spending some money. After gathering information, he was convinced that "a hard working planter with twelve Negroes, animals and implements for plowing, would shortly be very comfortable, and eventually rich or at least quite wealthy." But he also discovered "that someone with many fewer Negroes will only vegetate, as the revenue would then not be large enough for the expenses. Negroes here cost in the range of six or seven hundred dollars, and at that one rarely finds one or two to buy, so I must go to Philadelphia."[82] Lallemand announced that he and Desnoëttes would arrive in the North in late October to buy slaves there. The time it took to travel was conducive to reflection, for Lallemand gave up the idea of Tombigbee, convinced that he could not prosper enough there to provide for the material well-being of his family.[83]

Victoire George had not influenced his decision. On March 2, 1820, she had already thought of writing him "how much he could pay for Negroes nearer to here than Virginia, consequently easier to bring and much less costly." For it was in Alabama that she and Lajonie bought their slaves. In December 1820, George Gaines, who had left his trading post at Fort Confederation for a job as cashier at the bank in St. Stephens, wrote to let her know that "there would be a sale of African Negroes who had been seized and tried out for three years, and that in general they were good sorts" (February 29 [sic], 1821). She urged Girard to send her the money that she absolutely needed "for the purchase of Negroes who will be sold by the sheriff at St. Stephens in late February. They have never been as good a buy as right now, and it is absolutely in our interest to acquire some. Please procure for me as soon as possible the means to have Negroes, we cannot do without them" (December 14, 1820). Might the influx of slaves explain the fall of prices?

Stephen Girard took the necessary steps to give Victoire George credit at the St. Stephens bank. Informed by his letter, which arrived from Philadelphia in late February 1821, she immediately sent her son Achille to St. Stephens. Thanks to Girard's signature and assistance from Gaines, whose reputation was excellent, the bank freed the three thousand dollars that permitted her to buy several slaves in March (March 2, 1821). Victoire informed Girard that Gaines had been kind enough to advise her son and help him "do very good business in this area," for he had already been offered a profit, but the need of hands was too great to resell (April 13, 1821). The two men continued to help her purchase more slaves that she asked for that autumn to work in cotton fields (November 8 and 14, 1821).

For lack of funds, Lajonie was never able to carry out his plan to breed blacks and sell them. Like most American colonists, he had to content himself with a small number of slaves. In Mobile, in August 1819, after talking so much about it, he was finally for the first time in a position "to buy a mulatto woman with her two children; the elder is six years old. They want 5000 francs for them and that is not expensive. I do not yet know whether I shall conclude the purchase. You must remember that soon we will not have any more workmen and that you undoubtedly do not want me to work my fingers to the bone all my life" (August 29, 1819). The deal was not concluded. It was only when his wife arrived that he took on a little ten-year-old black girl, whom he paid twenty francs per month, to make up for the absence of his own children, who could have carried out the same tasks (January 2, 1821). In June 1822, when he went to the river port to get some vines, Lajonie bought his first slave, an eighteen-year-old, for the sum of 560 dollars: "I am very pleased with him; he is most useful at present when the heat is excessive" (June 13, 1822). Through at least 1823 he had only the one: "My Negro is sufficient to help me with everything possible" (December 20, 1823). It is known that he later bought other slaves whom he had to sell when he returned to France.

Lajonie adapted fairly easily to a human environment that was totally new for a man who had recently emigrated from France. A close neighbor of Indians, in daily contact with American farmers so different from those of southwestern France, and in perfect agreement with the slavery system, he assimilated people and practices as he was discovering them. His feelings of fraternity for the native peoples did not prevent him from wanting their lands, nor did his awareness of white cruelty prevent him from having slaves. Disconcerted by an immoral capitalist mentality, in the end it was with the Anglo-Americans, whose attitudes toward the natives and blacks he adopted, that his relations were the most difficult. From this point of view, Lajonie's case can assuredly be extrapolated to the entire group of French colonists.

14

A Model Colonist

I now consider myself as nothing but an American farmer. . . . I am the first in this region to have shown the Americans the fruit with which one makes the claret that they are rather fond of. . . . I willingly gave up the title of first farmer of the colony in order to always merit that of good payer on whom one can count.

—Jacques Lajonie, Aigleville, June 24, 1820, August 15, 1820, January 2, 1821

The nature of the relations between the French colonists and those who shared their environment does not sufficiently explain the success of some and the failure of most. The Indians, pushed west of the Tombigbee, were peaceful, and the slaves docile. Some Anglo-American farmers were certainly unfriendly, taking advantage of the naïveté of the French, occupying their lands, defaulting on payments, but this was not the rule: as neighbors, they got along, struck up friendships, and intermarried. Could the French colonists, for the most part Catholics, have lacked the famed "Protestant work ethic" of the English, German, and Swiss, which enabled them to carry out their agricultural or commercial ventures so successfully? Lajonie was a Catholic by baptism, a Huguenot at heart, and an atheist by conviction.

Most of the Colonial Society members who went to Alabama abandoned their allotments shortly after settling on them, discouraged by obstacles and harsh natural conditions. During his twelve years in the colony, Lajonie was not spared. But pride, and his desire to return home with enough assets to start over, enabled him to succeed. He was a true American farmer, both in his own mind and in the minds of the colonists who saw him at work, admiring his exceptional strength of will and integrity. How many did as much as he for the sake of the colony?

FROM DAY TO DAY

A pioneer planter supposedly had no time to write,[1] but all evidence is to the contrary. Work, illness, and the countless vicissitudes of everyday life for the émigrés in Alabama never prevented them from communicating with each other in the United States or with people in France.

CORRESPONDENCE ISSUES

Lajonie's correspondence and the evidence from other letter-writing exiles show that most of them frequently took up the pen: Victoire George writing to Stephen Girard, Desnoëttes to Clauzel, Mangon to Lajonie, the younger Lallemand to his wife, Pénières to his brother. . . . Between the two continents, the most difficult thing was not writing and sending mail but waiting for an answer, as the movement of letters was as long and uncertain in one direction as in the other—navigation on the Mobile and Tombigbee, transit through the St. Stephens post office and sea-ports, the ocean crossing, frequent losses necessitating duplicate and triplicate cop-ies. From the banks of the Dordogne to the White Bluffs via Bordeaux, New Or-leans, and Mobile, two and a half months was the minimum, and twice that time for return mail.[2] It is understandable that in the solitude and isolation of wilder-ness, two hundred miles from Mobile, which was itself just beginning to develop and was periodically made even more isolated by deaths from yellow fever, main-taining contact, asking for material or moral support, and giving and receiving news was more important to the colonists than to their addressees in France or in other, less isolated parts of America.

During the period before his wife arrived, Lajonie was constantly asking for news from France and complaining that he did not get enough. Before leaving the East Coast on his long journey to the South, he wrote to his family: "I should like to tell you that it is rare to see one of your letters completely filled. Tell me, is it for lack of subjects or for fear of boring me? . . . I want details down to the largest cab-bage of your garden. . . . Everything around you interests me. As for me, I am ea-ger to be in my new possessions. What descriptions will I not be able to give you!" (April–May 1817).

Six letters from Philadelphia and Baltimore, announcing his departure for New Orleans, went unanswered: "I found consolation for the difficulties of my jour-ney by persuading myself that several letters were awaiting me. . . . What was my fate at my arrival? No letters" (June 1817). This discrepancy between France and America was an endless source of dissatisfaction for Lajonie, ever hungry for mail: "A letter every week, however small it may be, will suit me better than a thick par-cel every six months" (November 1817); "I want a letter by each ship that leaves France" (June 21, 1818). His wife, sister, and brother-in-law were unpardonably lazy: "What, is it so difficult to scribble on a sheet of paper? There are so many things that interest me, so many thoughts that I should like to know and you do not condescend to inform the poor fugitive" (May 13, 1818).

Lajonie, on the other hand, countered his family's neglect with unfailing regu-larity, taking every opportunity to write them while in New Orleans—that is, whenever a ship sailed (July 28, 1817). Writing was a priority. Even when he was overworked, Lajonie devoted time to his family and friends: "When writing to a friend, there is never enough paper" (June 1, 1818), for, he added in a later letter, "it is foolish unconcern to live in an *extraordinary* country without talking to one's friends about it" (June 7, 1818). His curiosity about the new things surrounding him explains his enthusiasm and his avowed ability to describe them, but less his urgent need of the aid he requested. He justified the slowdown in his letter writing after arriving in Demopolis by complaining about the inconvenience of writing on a trunk in a tent (October 12, 1817). Considering his situation as a camper and the work to be accomplished, it may seem surprising that he sacrificed time to correspondence, but it was this very precariousness and the sheer size of his task that obliged him to hurry and write so as to hasten the arrival of much-needed supplies. At the start of 1818 he had still not received a letter in Demopolis: "What might you be doing? In spite of our occupations, which certainly surpass yours, in spite of our ills, I have found the time to write to you nine times," he complained bitterly (January 19, 1818). Six months later: "Has there ever been such an irregular correspondence between people who love each other so? Never has there been such a thing" (July 4, 1818).

On October 1, 1818, Lajonie reviewed a year's exchange of letters: "We ought to have two kinds of correspondence, one commercial and the other personal. You have been scrupulous neither in the one nor in the other." He attributed this to laziness, even though, in defense of Le Soulat, few ships sailed between Bordeaux and the ports on the Gulf of Mexico. It was therefore difficult for his family to gratify him with a little letter every two weeks rather than a large parcel every few months, a wish he kept repeating. And as the years went by, letters from Le Soulat were less frequent. August 3, 1820: "Have you deprived me of letters long enough? . . . Where are your hearts?"; August 15: "Wicked friends, you have undoubtedly sworn not to write us any more. . . . What I should not give now to receive from you letters like those that I have sometimes received! What remedies would they not bring for my accursed melancholy!" Lajonie's laments had no effect. What was worse, until 1825 the letters came ever further apart, from three, six, or seven months to even a year: "For three months, my dear friends, I have not received a single letter from you" (October 9, 1822); "Why are you not writing us anymore? . . . It is only too true that since the month of June we have received no letter from you" (February 1, 1823); "I have finally heard from you, my dear friends, dated December 4, 1824. Do you know that we had not heard in nearly a year" (July 2, 1825). Thus it remained during the last years until 1829. Lajonie, for his part, continued to write, but less often, because, unlike the Taupiers, who had fourteen servants and whose chief correspondent was Lajonie himself, he had many friends to write to and more and more to do: "If you are not ashamed to tell the people who have written me that a vineyard worker, a farmer, a cowherd and carter rolled into one, who combines with the occupations of his condition, with his commercial affairs, his foodstuff to sell, his supplies to buy, is excusable, if he cannot answer all the letters he receives" (March 14, 1820). Or again: "I am forgivable . . . for not writing. In the winter, we take the axe in hand, in the spring, the plow and in the summer the spade" (November 3, 1820).

Lajonie could find no excuses for the inhabitants of Le Soulat, but when they in turn criticized him for his neglect, he found a great many excuses for himself:

moral weariness, fatigue, fevers, a hernia, work necessary for survival, the price of postage. "I have not written you a single line for several months," he admitted on December 17, 1826. Between his relatives' infrequent letters and his own writing only every now and again, he puzzled over the situation: "How is it that, loving each other the same as six years ago, we seem to forget each other?" (March 20, 1824). Lajonie was certainly the more excusable in waiting for rain to sit down and write.

Home, Sweet Home

Neither Lajonie nor his companions camped out for very long in Demopolis. They received small plots for presettlement near the town. The goal was to get a roof up and plant a few acres. The first harvest would offset supply problems and high food prices.

Nauzelbine. Lajonie's sixteen-acre plot, number 12, was, like its neighbors, long and narrow. Perpendicular to the Black Warrior, it crossed a large prairie bordering the river, about a mile northeast of Demopolis (see map 12). Ten days after arriving, Lajonie described how the work was going: "Every day we work on our house, which is 16 feet wide and 28 long. For the moment, it will have only two rooms, but it is arranged for four, with a corridor in the center. It is 60 paces from the river and is situated on a little elevation some 50 feet above the ordinary level of the river. The Touqualousa in front of our door is about 80 paces wide" (October 22, 1817).

The houses that the French colonists built were modeled on the log cabin, whose rustic design had long proven its worth and was not about to disappear. It was an integral part of the American landscape, particularly that of the frontier, where pioneers, urgently needing a permanent shelter, put first things first.[3] No great skills were needed: a woodcutter's ax, an adze, and a carpenter's saw, along with a good sense of observation for its basic architecture were sufficient. Construction was easy and free. Raw materials were plentiful on-site: stones picked up in the fields for foundations, clay taken from the soil for cob, and lumber in profusion— red cedar, oak, and conifer—for the logs. Put together quickly with perishable materials, the log cabin was to precede the construction of a real house, also made of logs, when the colonist's farm was firmly established and his economic situation had improved. Unlike the brick houses that appeared later in Demopolis—Bluff Hall, built in 1832, is the oldest—these crude log cabins, ephemeral in essence, did not resist time's erosion. But we can get a good idea of them from contemporary descriptions like that of William L. Adams, a Treasury Department agent visiting Aigleville in 1826:

> The owner or some one on his or her account, built on the allotment a log cabin of a common height for such kind of buildings, hewed down inside and out, covered with a good board or shingle roof, laid with a plank or puncheon floor, with a log chimney, and made quite comfortable for a building

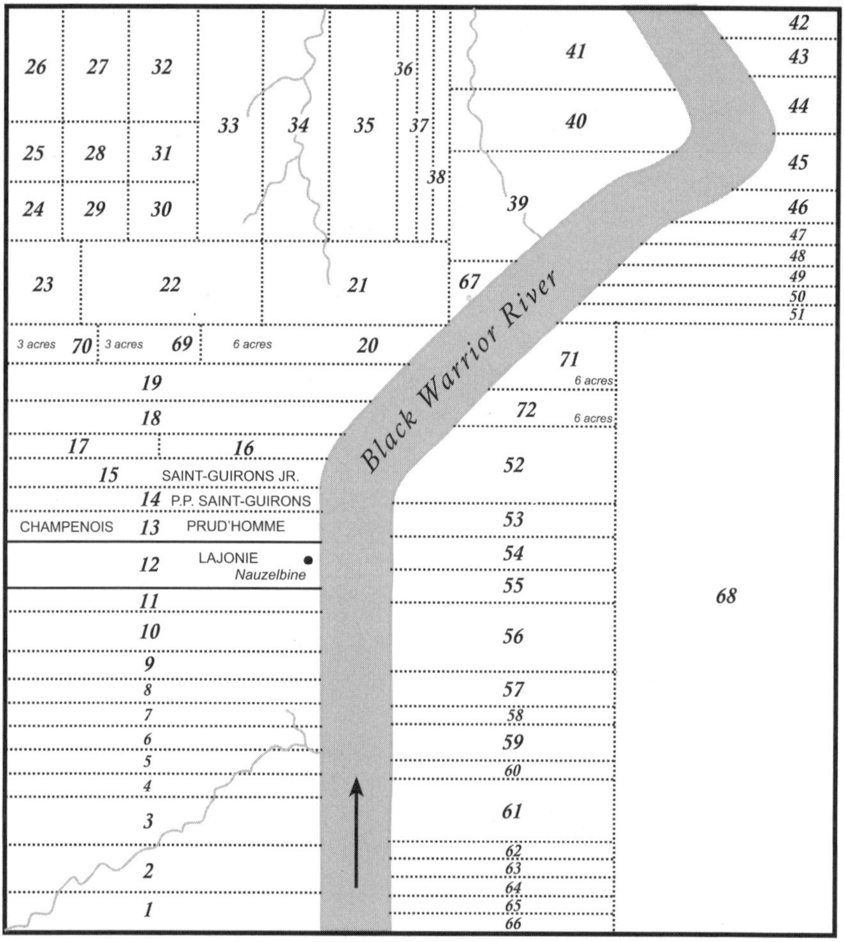

Map 12. "Nauzelbine," Township 18, Range 3 East (ca. 1820).

of the cabin kind. The smallest size cabin which I examined was 16 by 18 feet on the inside; and the largest 19 by 23 feet. Every building had enclosed about it from one to five acres of land, and cost the owner from 85 to 150 dollars, varying in price according to the size of the cabin and the quantity of land cleared and enclosed.[4]

Give or take a few feet, Lajonie's log cabin must have looked like this. Like the Americans, he had wanted to do things simply: "On their properties they do not seek to procure for themselves any amenities; a wooden shed is sufficient for the entire family. If we do not imitate them, it is impossible for us to compete" (January

10, 1818). With the indentured men who constituted his entire family at the time, there were eight people in a 448-square-foot building. Lajonie had decided to live as the local people did, but in accordance with his enterprising character, he soon aimed higher, as shown by the very accurate plan and sketch he sent to Le Soulat exactly four months after his arrival:

> In spite of our infinitely occasional little vicissitudes, we have nonetheless managed to have a central building whose layout promises us eight roomlets, of which four will be eight feet square and four, seven feet by four feet in width. Add to that a corridor in the middle, which will open onto a gallery that will go around the entire house. Fifteen paces from the house and to the right is the workshop. At the same distance, on the same line and toward the left is the kitchen. These three little buildings are linked in the rear by a balustrade of thick stakes, burned at the base, which forms the last of the railings of the gallery of the house. The farmyard is closed off by this balustrade, the two wing buildings, a continuation of stones meeting the balustrade at right angles, and by an oak board fence that forms the front of the garden. The garden is surrounded following the shape of the terrain, so that standing in the front door of the house one sees first the yard that is twenty feet wide and fifty long, below it the garden (five boards) whose shape is original. Here is the plan. (February 12, 1818)

He had not lost his onomastic proclivities. Lajonie called his property "Nauzelbine": Nauze for the brook flowing below his family house of Lapeyre in Gensac, and Elbine for Marie-Elbine, his daughter. On October 1, 1818, the house was being modified in view of family arrivals:

> I am directing the work on a fine wing of the house which, for the time being will serve as a storage place for our wood and other things. If you come we will make of it three 12 foot square bedrooms. We can take pride in being the best lodged and in having turned our land to the best account. In ten years, our residence will be the Tivoli of Aigleville. If you do not come, I may even make of it a building of some establishment, for example a secondary school. It would be perfectly situated. The other wing will be built later, so that the house will have approximately the following façade overlooking the river. The main house will have a gallery (it is double) and has a 4-foot corridor in the center. In the back and in squares, we shall have to add two buildings to serve as stables, etc. You know the layout of the garden and the front yard.

Despite the pride its owner took in it, the most famous place of the colony was not Nauzelbine but Lefebvre-Desnoëttes' log cabin, located on the Black Warrior not far from Aigleville, at the point where the river curves the farthest south. Its exact site was determined in the 1930s by James Whitfield, and in 1974 a historical marker was erected on the grounds of the cement factory whose present area covers that of Desnoëttes' plantation. Rustic though it was, this cabin was nonetheless a sanctuary where those nostalgic for the Empire could meditate. In it, Desnoëttes is said to have displayed a collection of sabers and pistols taken from the enemy, and the interior walls were hung with French flags. A bronze bust of the Emperor adorned the center of this exotic branch of the Invalides of Paris, perhaps the first Napoleonic museum.

With the exception of this lair of remembrance, the other dwellings must initially have been very much alike, since the first colonists had come to Demopolis without furniture. According to Lajonie, who was not the poorest, his house was lacking in comfort, but this was of secondary importance: "Fougnet and I, so as not to lose time that is precious for agriculture, have not spent anything on sleeping arrangements. The floor is our bed and our mattress is our only resource to rest our weary limbs" (April 27, 1818). Lajonie long continued this practice. On September 3, 1819, for the edification of his children, he informed them that "their papa sleeps every night on a mattress lying on the floor, that he makes his bed every evening and that every morning he rolls his wool blanket up in his old mattress and sets it in a corner of his bedroom, rolled up in this way." Élise Fougnet had the right to "a soft bed" (April 27, 1818), and when Dorothée Lajonie was to join her husband, a bed was prepared in the room that the young woman had to give up. This monastic sleeping arrangement, understandable in a region where all decent people were ruined if they deviated from what was "*strictly necessary*" (April 27, 1818), was a relatively minor inconvenience compared with the far more important daily worries, such as food. Nauzelbine remained for Lajonie his "sweet home."

Squirrel or Maize. The colonists had to adapt to their lot as pioneers very quickly by sleeping on the ground and, before the first harvest, eating what nature offered rather than foodstuffs that were imported or bought at high prices from local tradesmen. "In America the newcomers enrich those who are already here. I pay 40 francs for one hundred pounds of wheat flour, 30 francs for Spanish wheat or maize, and a sack of sweet potatoes costs us 15 francs. We can find no other vegetables to buy. Each planter has only his own supply. Wine is too expensive. Whiskey and rum are 50 sols a bottle. Meat is 15 sols a pound," lamented Lajonie on October 28, 1817, just after arriving in Demopolis. Two months later, the situation had worsened: "Everything is extremely expensive here: 50 francs for a sack of Spanish wheat, 80 francs for a barrel of flour, the same for rice, 100 francs a barrel of salt meat, 150 a barrel of pork, 5 sols a pound of fat, 25 sols for fresh beef, add to that that we have almost no food left and that we have only very little money" (December 17, 1817). These high prices were due to a demand that far exceeded supply, since trade networks as well as farming in the region were only in their infancy: "In areas that are being settled, provisions are at the highest possible price; in others that are already settled, one lives for almost nothing," concluded Lajonie when he was in New Orleans, where on May 14, 1818, he compared prices: "One hundred ears of maize are 10 francs in Aigleville; here one gets a full barrel for 7 francs, etc., etc."

The colonists turned to resources at hand locally to get around the obstacle of ruinously expensive supplies, but this was only a stopgap solution. Time spent hunting and fishing, the obvious answer, was taken from the all-important activities of clearing the land and planting. Moreover, they could not fish as much as they wanted for lack of nets that Lajonie repeatedly asked his French relatives to send (and that he sold for a good price when he got them), nor hunt game that had become scarce—at best, a roast squirrel supplemented their meal. It is true that the Choctaws regularly furnished chickens, turkeys, and legs of venison, but these could only be additions to the colonists' diet, since the Indians themselves were victims of food shortages caused by the arrival of the whites and their slaves on their former territory. Lajonie spoke of dire straits:

> We have much to endure before we have more or less all the foodstuff necessary for our household. For two months we have been living on Spanish wheat (maize) with which we make our bread and soup with rice. With our thrift we manage never to lack meat, in spite of the high prices. Game is lacking in the region, which is most bothersome. We lack nets to catch fish, and, oh extreme misfortune, money takes to its heels. We eat, now listen to this, we eat, I say 35 francs a month in maize, and yet that is the cheapest food. A barrel of crackers costing 50 francs here and used in the same way as maize lasts only five days. A barrel of flour, costing 75 francs here and serving the same use lasts us two weeks. (February 12, 1818)

To get other foodstuffs and forget the omnipresent taste of maize, they had to travel or send for things from France. Failing to find Irish potatoes and white beans near the falls of the Black Warrior, Pierre Pascal Saint-Guirons, Lajonie's neighbor at Nauzelbine, had to go all the way up to Huntsville in northern Alabama (March 6, 1818). From New Orleans, Quessart forwarded the supplies that arrived from Bordeaux for Lajonie, but most of these were destined for sale, not consumption. Among other problems, that of supplies discouraged the least motivated colonists and hastened their flight from the colony. Those who persevered saw the situation improve with successive harvests and the economic development of the region. While rice, and especially maize, which Fougnet complained of eating "in soup twice a day" (June 6, 1818), remained the basic staples, their menus began to diversify to include potatoes, sweet potatoes, beans, pumpkins, squash, and so forth, as well as meats: poultry, beef, and pork that most of them were now raising. The French did not die of hunger, nor did they lapse into the rustic simplicity of local eating habits. They never gave up bread making or drinking wine shipped from France, and made an effort to keep up the tradition of good meals: "I still enjoy my food," wrote Lajonie in November 1817. If difficulties had persisted, the survivors would not have turned to cotton rather than food crops after planting and growing the grapevines and olive trees stipulated by their contract. Food and shelter were certainly a concern, but one that could be resolved, unlike other worries, like illness, for which there was often no remedy.

Illness, loneliness, and the absence of dear ones were probably the most distressing in the litany of personal trials suffered by the colonists. To reassure himself as much as his family when yellow fever was killing people around him, Lajonie had, during his stay in New Orleans, repeatedly sent them heartening messages. The region was healthy, and the fevers everyone was so upset about did not deserve their fatal reputation. Until he arrived in Demopolis, Lajonie, along with Quessart, continued to vaunt the healthiness of the South and Alabama's climate, the best in the United States: "Colds, drafts, chills, pleurisy, and fevers are unknown here: Fougnet and his sister had them, but that was due to our long stay on the river" (November 1817). But because he could now be contradicted by the testimony of his companions and because he himself had fallen victim, Lajonie soon no longer hid the physical woes that overwhelmed them as they did the other colonists at grips with the same ills. The first they were faced with, and the only one, insisted Lajonie, though most annoying, were skin afflictions, pimples, and boils of all kinds. In November 1817 Lajonie said that his buttocks had been bothering him for six days because of a "kind of red rash" that itched mercilessly. It was, for that matter, with boils in the same place that he had made his inaugural (and therefore disagreeable) journey from Mobile to Demopolis, as had Pierre Dupuis, whose were enormous. The sulfurous and nitrous waters they drank were said to be what caused the skin disorders, the only ones known here, Lajonie repeated. It did not take him long to change his mind, since their fevers and "infirmities," as he called them, only increased (January 10, 1818).

At about the same time, Lajonie began to suffer terribly from a felon, an infection on the middle finger of his right hand that handicapped him so much that for three months he could neither work nor write (December 4, 1817). When the infected joint fell off, the finger finally began to heal. Fougnet, already suffering from fever, came down with the same thing: "It would seem that hands that are not accustomed to work are made to pay tribute," concluded Lajonie philosophically (January 19, 1818), but also playing doctor and attributing the loss of his finger joint to the impurity of his blood, impaired by scabies (February 12, 1818). After Labrousse and Mangon, Lajonie had in turn contracted it, and local medical science was of no help, claimed Saint-Guirons: "The doctors here are not familiar with citrine ointment. They have their scabies patients swallow sulfur flour and then rub them with mercury. That is how we treat our horses or our dogs" (March 6, 1818). Several weeks later, Lajonie was finally able to announce: "In two weeks I shall no longer be the prince of scabies" (April 3, 1818).[5]

But fevers, boils, felons, and scabies were nothing compared to the sword of Damocles held over them by yellow fever—not in the colony, a healthy region where no one had even had a cold, he claimed, but in the ports of the Gulf of Mexico, where it raged as never before during the summer of 1819. In August, Lajonie was in Mobile, surrounded by the dead and dying, exposed to the ravages of malignant and putrid fevers. He buried two of his friends—Puech (Pénières's nephew) and Charles Desaifres—preceded by many persons he did not know and followed by a steady stream of others of all kinds. Aside from himself, only Louis Edmond Bourlon, General Clauzel's nephew, Charles Batré, and a doctor had not caught the illness that was "nicely consuming the Americans" (August 13, 1819). Dorothée Lajonie, who had arrived in New Orleans on July 30, 1819, came down with the fever, but she recovered, Jean Quessart wrote reassuringly: "She fell ill here, had five attacks of fever of which three were very strong, but is, thank God, well recovered and is today as acclimated as those who have had yellow fever, because this illness thinned her blood, which is the only protection against the illness which sometimes wreaks havoc here, but which this year does not as yet appear to be very bad. So set your minds at ease concerning her" (August 20, 1819).

Lajonie was waiting for his wife in Mobile. Did he fall ill due to conjugal solidarity? "Mobile's diabolical fever finally tired of protecting me; one day it decided to send me to bed and to make me wait for my wife in this pleasant position. It had undoubtedly foreseen my wife's arrival, for three days after my first attack Dorothée arrived toward eleven o'clock in the evening, just at the moment when the fever was entertaining me the most. I assure you that our embraces were hot," he wrote on October 1, 1819. Lajonie regretted that his family had let Dorothée leave when she would arrive during the worst season. They had both been extremely fortunate to survive the fever that had devastated a large part of the United States. News received the following month convinced them of this:

Monsieur Quessart has just written me that never had illnesses been as devastating as this year and Batré has just announced to me the death of the rest of my acquaintances in Mobile. Monsieur Fournier, who arrived a fortnight ago, told us more of poor Mobile. He assured us that corpses lay unburied on the various streets, that he himself had seen four without going very far into the city. What a calamity! Death did not spare all the butchers nor all the carpenters, so that people died of hunger before dying of the fever. When I was there, a coffin for a poor man sold for 100 francs. Later one coffin was used for a dozen and finally, according to Monsieur Fournier, they were not used at all any more. In one year, Mobile lost half of its former population and all the foreigners who had settled there, except Batré and Devangeux. (November 22, 1819)

Lajonie attributed this disaster among the population to the greed of Americans when populating a city: "Half of the houses are built on piles and no one bothered to fill up the excavations beneath them which are full of miry water, and a thousand foul things that the last hurricane brought there gave off a pestilential odor that wrought the havoc to which I almost fell victim" (November 22, 1819).

It is certain that far from the seaports, where the fever was centered, the French colony was more protected, as Lajonie said: "Never has there been a healthier region. All the ill people who came here recovered in very little time, except one who no doubt had, internally, an excessively strong germ of the fatal illness" (November 22, 1819). But Demopolis was not miraculously protected, and the increasing frequency of exchanges between Mobile and the colony via the Tombigbee brought the illness upstream. No one, including Lajonie, had recognized the relationship between mosquitoes and yellow fever. Of all the scourges that afflicted the colonists, mosquitoes were mentioned the least, and then only indirectly, with an urgent request for mosquito netting. But from March to October there must have been swarms of them, a permanent annoyance and a danger far greater than the bears and wolves that attacked cattle but rarely people. We have no list of the French colonists struck by epidemics of smallpox, dysentery, and fevers, but it can be asserted that almost all were severely afflicted, and likely several times. After surviving the fever in Mobile in 1819, Lajonie contracted "a nice and good malignant and putrid fever" the next year that he said he cured with camphor and vesicants (July 15, 1821).

Many died. In 1825 Lajonie attributed the death of his friend and associate Pierre Fougnet to the fact that he refused preventative remedies and his medical help. Lajonie had no faith in the expensive and ineffective local doctors and thought of himself as a disciple of Asclepius, the Greek god of medicine and healing. This assurance came from a comprehensive knowledge of the works of Louis Leroy, the only man he thought possessed the art of completely curing human illnesses and whose methods were supposedly working miracles in the region (March 20, 1824; July 2, 1825).[6]

On the advice of Dr. Laclaverie of Bordeaux,[7] a blacksmith from Castillon named Gilet had brought the good news to New Orleans and to Quessart, who still could not get over it: "I thank God for having shown me his method, which gave me back my health and that of my family, and of many others who, like I, have learned to value him. Consequently I have given up all other remedies forever. I have used it in twenty different cases, on weak, strong, old and young subjects and always with success" (April 20, 1824). Quessart had begun the treatment in January 1824, when he thought he had less than six months to live. On March 30, 1826, he was on his 550th dose. Getting better and better, he had his entire household benefit from it: "Seventeen people in my household have used it, for chronic as well as recent ailments, and for various cases such as chronic asthma, measles, smallpox, pneumonia, hemorrhage, deep-rooted mouth ulcers, fever, venereal diseases, diarrhea, always successfully." One of his friends had saved his five-month-old only son with eleven vomitories and one medicine in six days; a young woman with a case of black jaundice that had resisted four doctors for three years recovered her natural skin color and freshness after ninety-five doses. Quessart could give a whole list of cures that resulted from Leroy's "reasoned method," in spite of all that his detractors might say (March 30, 1826). But Quessart himself gave them cause to speak ill of it, since he died shortly thereafter, on July 14, 1827, in his Burgundy Street house.

Quessart had had time to give Leroy's works to Lajonie, who became an ardent advocate. "Thanks to his belief" (March 12, 1826), he enjoyed good health and had helped others. He had convinced Mangon, who had treated himself successfully for two years, and had rid his own wife of a very painful growth on her breast in twenty-five days of treatment. Above all, he saved the life of Antoine Fougnet, who, with the same symptoms as his older brother, refused the bleeding that the doctor prescribed and had faith in his prescriptions, the secret of which was taking jalap:[8] vesicants on his legs for five days, two mercury medicines, 200 to 300 grains of jalap, and several grains of emetic. After weeks of fever, the illness was conquered. Ten days later the convalescent was on his feet again, back at his normal life, but he was to continue the treatment, as Lajonie diagnosed that his lungs had been "affected by masturbation" (July 2, 1825). On the strength of his results, Lajonie proselytized in the French colony. Captain Raoul wrote him concerning Dr. Leroy's works: "Read, reread, if you do not want me to outdo you. I have finally seen the light of truth and our charlatans are still walking in darkness" (March 20, 1824).

The colonists endured genuine physical suffering, which, for the military men, was added to that of years of war. The sum of their injuries made a glorious total but was not well suited to such a harsh conversion into agriculture. For Desnoëttes, who had been wounded by a pistol shot in 1808 and pierced twice by bayonets in the battle of Brienne in 1814, or Lajonie, whose arm had been shattered by a bullet, work in the fields and forests was not the ideal treatment. After two and a half years in Alabama, Lajonie complained of lacking strength, of not having "enough energy to work in all seasons," particularly in the summer because of the heat, and of not being "able to bear up alone under the fatigue of the hard farm work" (June 24 and August 15, 1820).

Nevertheless, these hardened men doubtless tolerated physical pain far better than they did the moral sufferings of exile. While Lajonie did not hide his ills, he did not actually complain, preferring to minimize them or attribute them to circumstances. Loneliness, homesickness, and the absence of his family, on the other hand, wore down his strength of character: "I cannot seal this letter without telling you that I feel very lonely" (April 27, 1818), he wrote after sixteen months of exile. "My solitude troubles me at times," he wrote on another occasion (April 7, 1818). Also at this time, Élise Fougnet was homesick and wept over the separation from her loved ones. Unlike the Domingan refugees, many of the new émigrés were living without their families, whom they hesitated to send for as they were uncertain of their fate. During the first years, Lajonie's letters were filled with these questions to his family at Le Soulat: Should you come? When? How? Where? It will also be recalled how traumatic his Atlantic crossing had been, and he feared it for his loved ones. At the same time, he greatly missed his daughter, Elbine, and his son, Léopanno, whom he had never seen. Their upbringing, in which he could not have a hand, was all the more worrisome because their aunt and uncle, who were in charge of it, had no children, but too many servants who might spoil them. Lefebvre-Desnoëttes was in a similar situation. His daughter, Charlotte Lévinie, was born too late in 1816 for her father to be able to see her.

To the absence of children was added that of his wife. After a long resistance, Lajonie finally gave in to her request and asked Dorothée to come: "I should like us to be reunited; desire it also, and soon we shall be, and be happy." For the first time he speaks of great discouragement and even mentions the idea of suicide: "Why so many sorrows in this world when one can leave it so easily?" (October 1, 1818). In 1819 Dorothée was twenty-five years old, with two children, aged four and two and a half, and had been separated from her husband for those two and a half years. Besides her natural desire to be with the man she loved, the jealousy that Lajonie had imprudently aroused from America drove her to him. During his stay in Baltimore he had sung the praises of the beautiful, charming, musically talented Eliza Latour, all the while claiming to be resisting temptation, which made him all the more suspect: Lajonie was a dapper man and a smooth talker. "It is true that these pleasures are innocent and there is nothing about them that should trouble your peace of mind," he said reassuringly, adding that in the United States love outside marriage was considered unworthy of the human race (March 1817; January 20, 1817). But the damage was done; he had said too much. He was suspected at best of little misdeeds, at worst of fickleness. His wife wrote him of her worries and suspicions, which showed "delirium rather than affection," protested the persecuted husband (September 1817).

At Nauzelbine, Élise Fougnet, about the same age as Dorothée Lajonie, succeeded Eliza of Baltimore. Rather than leave her unmarried state as a lodger at Lajonie's, she pouted when offers of marriage came in: "She has refused two or three more or less suitable matches. Mademoiselle's pretensions are a bit too lofty. I should fear, if she were not so young" (October 1, 1818). Lajonie deplored the fact that she turned down Juan Rico Vidal, forty-six, a false Spanish general but a true Capuchin of the order of St. Francis of Assisi and a former liberal member of the Cádiz Cortes, whom the absolutist restoration of Ferdinand VII in 1814 had driven into exile: "His properties have been confiscated. Nonetheless he has something, and friends. Mademoiselle laughed and bantered for awhile at this gentleman's indirect suggestions. I felt it necessary to put an end to these communications. The gentleman made his request in due form and Mademoiselle refused because he was over forty and wore a wig. I reproached her for her flirtatious ways, a few tears resulted and everyone went home. In fact, this gentleman is a decent fellow, has learning and is hard working, which says a great deal. Her brother and I would have liked it, but Mademoiselle has loftier pretensions" (April 1, 1819).

Pierre Fougnet shared with Lajonie the concern caused by this feminine presence and was worried "about keeping an unmarried young lady, or rather her reputation," wrote Lajonie, after Lise turned down Labrousse's marriage offer. He had offered her "his hand and his labor," but she had "found it quite extraordinary that he should dare to do such a thing" (November 3, 1820). Despite the interest he took in Lise's married future, Lajonie occasionally felt it necessary to mention in a jokingly embarrassed way that they sometimes found themselves alone together.

Tormented by doubts, Dorothée left her children at Le Soulat and sailed from Bordeaux to New Orleans, where she arrived on July 30, 1819. She joined her husband in Mobile in early September, and by the end of the month they were safe and sound in Aigleville. Four months later, Quessart wrote to Le Soulat that Dorothée "did not drink in vain the prolific water of the Mississippi River, and that she is well on the way to seeing her family increase" (February 7, 1820). On July 31, 1820, exactly nine months after they arrived at the White Bluffs, Jeanne-Elisabeth, called Helena or Ninon, was born at Nauzelbine. Her birth was registered in Gensac[9] on the testimony of a certificate of May 15, 1833, in Mobile: before Basile Meslier, notary public, five former colonists, Jean Jérôme Cluis, Achille George, Simon Chaudron, and Claude and Auguste Payen, stated that "in the early days of the settlement of the French émigrés in the State of Alabama that had been granted them by the government of the United States, they were totally deprived of a church and a public records office to declare and register birth certificates or any other official documents."[10]

It seems that rather few exiles who left for the United States alone were joined by their wives and remained faithful to them. Accustomed to living far from home, officers compartmentalized their family lives and their sexual lives on military campaigns. They did the same while in exile. Lajonie, who had not been impervious to the charms of Eliza Latour in Baltimore, became the amused reporter of the colony's gossip. General Lefebvre-Desnouëttes was living with "a friend of his wife" (April 3, 1818), Léontine Desportes, the countess's former lady's companion. We recall that they had left Le Havre for America, but their ship, caught in a storm, had put into port in England, where the women are supposed to have quarreled. The countess turned back, but Léontine continued her journey to the United States, where her father was said to be employed in the Murat Company.[11] According to Consul Pétry, she lived across the Delaware River in New Jersey in the home of a poor fisherman whom she paid ten francs a week, doing unskilled, heavy labor to survive: an ugly woman of forty, scatterbrained and very excitable, who came to Philadelphia two or three times a week. She supposedly refused offers of help from Lallemand and Desnoëttes, whom she claimed to have saved. Not for long, however, since in November 1817 she left with the latter for the Tombigbee colony via Mobile, where a 120-acre allotment and a husband awaited her. What was the nature of her relationship with Desnoëttes? It was he who drew up the certificate of her marriage, on May 18, 1818, in Demopolis, to François Violle, a doctor at the Champ d'Asile and former honor guard, twenty-five years old like Léontine herself, whom the consul had aged prematurely.

The regicide Pénières was also living "with his chambermaid," who had followed him into exile and had been solacing him, said Lajonie, "for I know not how many years" (April 3, 1818). Unlike that of Desnoëttes, his marriage had been in shambles since, ten years earlier, he had threatened his wife with "100 kicks in the ass" if she did not immediately leave their home in Corrèze.[12]

General Clauzel was living with Constance Ogé, the widow Démérest, whose affairs he was managing and whom he called "My dear cousin" in his correspondence.[13] Owning the 240 acres of allotment 302, she was a neighbor of Edmond Bourlon, Clauzel's nephew. Finally, Lajonie closed his review with Nicolas Simon Parmantier. Having abandoned his family in France, he "was cuckolding an old naval captain surnamed Brise-Raison with whom he [was] living" (April 3, 1818).

It can be seen from these few examples that the essentially male society of the French exiles saw to satisfying its desire for women, who were in short supply. Little is known of the colonists' relations with black women, but they existed. In 1819, concluding a letter to General Clauzel in Mobile, Lefebvre-Desnoëttes let it be understood that Clauzel had a child with a black woman: "I am delighted with the increase of your family, come on, admit it and do not look elsewhere for parents for this poor child. Madame Démérest knows as well as I that you have always had a predilection for the mother's color."[14] General Rigau, as will be seen, also seems to have had an affair, and a daughter, with a woman of color in Louisiana.

Little by little, more women came to the colony, and women of easy virtue flocked in as the number of potential clients increased: "A great number of colonists are arriving. Already more than twenty ladies have arrived. Houses of ill repute are already springing up; they are everywhere" wrote Lajonie on April 1, 1820. This realistic picture of the young colony contrasts with the idyllic Currier-and-Ives-type vision of a happy, family-based society. It contradicts the historiography that delights in highlighting the incongruity of French elegance in a frontier wilderness: balls, suppers, silk gowns, dress uniforms, music, and refinement. Lajonie certainly did see this: "The beginnings were charming: balls, evening gatherings, good dinners after which they made fun of your friend who beat his wife to make her spin and burn thick logs; upon hearing that expression, thick logs, there were great peals of laughter from the kind guests" (December 20, 1823).

But these pleasures of the colony's beginnings did not last, and they never made up for the ills assailing them from all sides. "It is impossible to be happy in our position," wrote Pierre Fougnet on June 20, 1818, summarizing the general feeling. And on July 4 Lajonie added: "Here my only pleasure is the hope that I cannot ruin myself." Like others, Fougnet and Lajonie worked at this by diversifying their economic activity before embarking on the cultivation of grapevines and olive trees.

THE WHEEL OF MISFORTUNE

Lajonie had entered upon a military career by inclination and abandoned it with regret. Farming did not attract him. When he resigned, he did not plan to trade in his battle mount for a draft horse, but to live on his investments and wait and see. When he left France, he had to consider with his family at Le Soulat how they could help him support himself in the United States. They agreed that together they should try their luck at trade.

THE LUCK OF TRADE

While still in the Gironde estuary, Lajonie had closed his first letter as an exile by underlining that he was "full of the idea of trade" (November 4, 1816). In the next, written from the other side of the Atlantic, he still believed in his idea, convinced that the continent welcoming him was the promised land of trade, vital in a country where "merchant" and "American" were synonymous. He was conscious of the problems to be overcome: learn the language, get together funds, rent a warehouse, and create a network of contacts in the country's interior and trusted agents in France. But he did not doubt, nor did his sister Nauzille, "that all we had to do to get rich was to buy" (early January 1817).

His enthusiasm waned quickly in the face of unscrupulous businessmen. There were too many obstacles in this career to be able to pursue it honorably, and not enough business for the man who wanted to remain above reproach. The little exposure he had to business had already taught him that three-quarters of the merchants had gotten rich only at the expense of good people, victims of their dishonesty (March 13, 1817). This discovery had disgusted him and made him realize that he would not measure up in the United States, a feeling confirmed by the fact that he had to sell at a loss the jewelry that Latour sent with him to New Orleans. But his arrival in the pioneer world of the South convinced him anew that trade went hand in hand with agriculture, which he was increasingly determined to engage in on the banks of the Tombigbee: "There are few newly established planters who do not have a storehouse of objects indispensable for existence. Their trade is with the savages or with the planters who do not have the means to build a storehouse" (July 15, 1817).

Two weeks later he confirmed his decision to be a planter, farmer, and merchant in Alabama. He planned to set up a little storehouse in Demopolis for his personal use and to have another in St. Stephens or Mobile, managed by Fougnet and his sister Lise. Lajonie could supply the Fougnets free of charge from the product of his harvests, and the Taupiers would do the same from Bordeaux with the merchandise Lajonie would advise them to send to Mobile. Until this direct line was inaugurated, Quessart would be their necessary agent in New Orleans. They still had to determine what kind of merchandise was suitable for trade.

Wines and Dry Merchandise. Given his origins and the absence of local production, Lajonie did not hesitate as to what to import first and foremost from France: wine, brandy, and liqueurs (early January 1817). The wine was to be in bottles, cases, or casks, old or new, preferably red, but also a small amount of white, dry, or a bit of sparkling like that of Le Soulat, never sweet, because it would have been worth nothing in Alabama. Americans liked hard liquor and full-bodied wines. They liked brandy even more than wine. During the first eight months of his stay in America, Lajonie put wine at the top of the list of things to send, but his variations on the subject reveal his ignorance of the market. Noticing at first that wines were outrageously expensive in his new homeland, he then changed his mind, writing that France furnished the United States with more wine than was wanted and that, in the end, the price of wines was not high and even lower than in France, where they had come from. White wine arriving in Baltimore from Bordeaux was said to have been sent right back for lack of consumers: "The Americans are extremely sober at present," he noted in March 1817. In reality, as we have seen with the Dufours in Indiana, the taste for home-brewed alcohol won out over wine. But Lajonie did not become discouraged and persisted in thinking that he would succeed in this trade if he could offer attractive prices.

In October 1817, following his advice, his family sent brandies and thirty-two casks of wine from Bordeaux on the *Marie-Thérèse,* a shipment that Quessart received in New Orleans in February 1818. Quessart immediately sold the brandy for $2.75 a gallon, totaling $547 (2,737 francs), but expenses and customs duty reduced the net profit to 40 or 45 francs (March 13, 1818). The wine did scarcely better. During the crossing, it had fermented a great deal. Five casks had to be emptied, and two others, one white and one red "that were neither wine nor vinegar," were distilled into twenty jugs of good brandy. Quessart sent twenty casks to Mobile, where Lajonie had his warehouse. The wine survived this new journey, but as it was early April, Lajonie feared what heat would do to it. He sold it for 275 francs per cask and 35 sols per bottle, or 30 sols if the buyer took ten dozen (April 7, 1818), which was closer to its value in France than the 100 percent profit he had been counting on. A month later, he was in despair because his wine did not seem to suit local tastes: what "you need here is the red wine bought from the farmers in Médoc; it is black as ink." Besides, prices were not holding up and costs were exorbitant: "All the merchants who shipped here this year have generally lost. Old Aubrion wine, three or four years old, is 500 francs a cask here; it costs more than that in Bordeaux. Duty is about 350 francs per barrel brought by a foreign ship" (May 6, 1818). The result of his sales in Mobile is unknown, but Lajonie was obliged to transport casks up to Aigleville, where there were even fewer consumers. On October 1, 1818, he admitted that he had been unable to sell a single cask in six months. Did the French colonists have so little money? As for the Americans, they did not take the plunge, but stuck to rum, whiskey, and corn liquor. Lajonie may have expected to do better with dry merchandise, natural products, and manufactured items.

He stressed the importance of sending foods produced locally: dried fruits, including prunes, figs, pears, and peaches folded into fig leaves; jars of large whole peaches in syrup; grape jam, preserves made with good wine and other more ordinary ones, but with a nice label; lentils and pea beans. He also asked for seeds (fruit stones and pips of all kinds, sainfoin and alfalfa seeds), and homemade rose and jasmine oils and lavender water. But they could not expect nearly as much profit from these items as from manufactured items like arms and textiles.

Lajonie requested four-foot carbines with bayonets, but above all shotguns, preferably double-barreled, not very long, and very large-bored, from Tulle in Corrèze. To avoid customs duties, which did not apply to personal effects and tools, each indentured man was to come to America with a new shotgun. His family was to send tablecloths, lace and batiste, cottons and canvas from Brittany, and wide blue cloth from Montauban. They were to negotiate with local artisans for articles of clothing: shoemakers of the region for all kinds of shoes and soft Souvaroff-style boots, named after the Russian marshal; straw hat makers for good hats with rather large brims for men and ordinary ones with black linings for women; tailors for shirts, three-quarters of them rather nice with coarse batiste jabots, waistcoats, cotton and drugget trousers, and so forth.

To these items that Lajonie thought would sell on the American market he added what he called trinkets—without a pejorative connotation—that would also be desirable: large, silver, English-style watches and French silver; a huge number of risqué or naughty used books, works in Spanish, pamphlets on important people like Bonaparte and the members of his family, prints, paintings, and caricatures. Lajonie insisted on the need of novelty to pique the curiosity of buyers, but this was not enough. Arriving on the *Marie-Thérèse* along with the wine and brandy, these items did not sell as Lajonie had hoped. His determination to engage in trade was badly shaken.

In fact, each problem had its explanation. The candied grapes had fermented during the journey because they had not been cooked long enough. The lighters would have sold well if they had not been damaged; not even 25 percent were in good condition. The straw hats would have brought 50 percent the preceding year, but by flooding the market the English had made them fall to next to nothing. The shoes would have brought in just as much if they had not been too small: Americans had large feet. Lajonie was still hopeful about the shotguns, which would sell and bring in a profit because there had been no duty on them. But everything was in jeopardy, for the country was flooded with merchandise going for a pittance, and the slow outlook for sales would add to storage costs in Mobile. Lajonie was not equal to the task in a field he did not know, requiring talents he did not have, as he admitted on April 7, 1818: "Honor makes me as timid in business affairs as it made me daring in the army."

For fear of bankruptcy, Lajonie was not ready to resort to credit, neither to granting it nor to asking for it: "This is certainly a capital flaw in a merchant since more than a million tradesmen in the United States work only on the funds of others. . . . Here I could have more than 50,000 francs in merchandise very reasonably, with four months of credit, but I lack the courage." Lajonie valued security above all else and committed his money only to a sure thing: "Let us be prudent. Experience teaches me something new every day. I appear to know so little about business that the most stupid people rush to tell me what I often know better than they. It is impossible to gather too much information about trade" (May 10, 1818). Even before taking final stock of the situation, Lajonie admitted that his commercial speculations had failed. They should go on to other things: "Ah! Let us no longer speak of our poor business; we have too much competition to succeed; it is better to limit ourselves to raising cotton and bringing in vines if it is possible, than to scatter our capital here and there without real profit. I have been convinced of this idea for a long time but the charms of agriculture made me forget that I had already given you advice. So in the name of God, if it is not too late, send me only money" (May 6, 1818).

He would use the money to buy land on which to grow cotton and vines. Lajonie therefore combined his activity as planter and winegrower with that of landholder. But in this field, his investments were no more successful than his commercial ventures.

Land Speculation

Lajonie owned two sites in the colony: Nauzelbine, northeast of Demopolis, a 16-acre allotment that he received when he arrived in October 1817, subject to flooding in the winter when the river's water was high; and a 480-acre allotment in Township 20, awarded in the Philadelphia drawing, also in October 1817. He had been lucky in the drawing. This allotment, nicely situated, was "superb and very healthy," half in prairie with excellent soil, the other half in black oak (August 16, 1819). There was no lack of water, as was the case in many places in the colony, for there were springs. Believing that vines could be grown there, Lajonie named his large allotment "Wine Spring," the source of wine.[15]

Lajonie did not love money for its own sake. He felt he had to justify his desire to become rich by insisting he had nothing else to do in Alabama, and wished to leave something to his children. Moreover, exiled due to judicial error, he very much wanted to return to France with his head held high, preceded by a reputation of honorable success. This is how his venture into real estate must be understood, rather than as a desire to profit from lands bought under advantageous conditions: in April 1818 he regretted that some had "speculated on what was supposed to be sacred and fraternal" (April 7). Lajonie aimed to enlarge his holdings and also to speculate, but never against the colony, of which he was one of the most resolute members: "If we have money before us, it will be easy to increase our territorial possessions very reasonably, either by buying from the government itself or from members of the Society who do not wish to establish themselves here," he wrote on April 27. Because of the commercial activity in which he intended to engage in Mobile, it was around this up-and-coming town, an American door to the ocean since 1813, that he first centered his plans: "Before this time, the acre, in many places in which cotton can be grown, went for 25 sols. Sites in cities went for one hundred gourdes (five hundred francs). Today the same land and the same sites are worth from two to 60 gourdes the acre and the sites from one thousand to ten thousand gourdes (50,000 francs). In the eight months since I first came through this city, sites have doubled. Right now my lack of money keeps me from buying a piece of land for one thousand gourdes; in four years I should need four thousand to have it" (June 7, 1818).

Lajonie did not have the means. He did not acquire land in Mobile, but rented a storehouse where he could sell some of the merchandise that arrived from Bordeaux in 1818 and made his first purchases in conjunction with his commercial plans in the colony. The site of Demopolis, cleared and occupied by mistake and then given up by the French colonists, had been bought from the government by the Company of the Town of Demopolis. Lajonie bought a share and became the owner of two "town lots" on the White Bluffs, on each of which he put up a store-house or house—according to Lajonie himself, nothing more than barns. There he stored the merchandise that did not sell in Mobile and the products of his harvests, like maize; then attempted to rent them out to pay the company for the drafts he had signed. At the time, people were betting on Demopolis, and the banks of the Tombigbee were prime property, wrote Quessart on June 16, 1819, citing a letter from Lajonie dated April informing him that lands had sold at White Bluff for fifty-one piasters per acre whereas four years previously they would not have been worth more than two. The price fever subsided within a few months' time, and on January 2, 1821, Lajonie recalled that period when "the rage for buying govern-ment land and particularly lots in the new towns flared up in all the men living in the United States (particularly the Americans). Lots in all the developing towns sold at wild prices, with some in Demopolis selling for 5,500 francs."

At that time he had been renting out one of his houses for 750 francs per year. But due to the slump the town found itself in 1821, he was not able to rent out his buildings and despaired of everything but paying what he owed: "My speculations on town lots in Demopolis will bring me absolutely nothing and . . . I shall be most fortunate if my share in the company can pay for half of the lands and the lots that I have bought from said company." Not honoring his drafts worried him the most: "Nothing on earth will from now on be able to get me to speculate again until my drafts (which are in the hands of the Demopolis Company) are paid" (January 2, 1821). The lots in Demopolis were good but expensive. If the company agents agreed to lower prices, sales might pick up again, but in any case it would not be a good deal (September 17, 1821). His plan for a port on the Black Warrior, above Arcola, was no better.

Claret-Ville. In the first half of 1819, Lajonie and two associates, each own-ing a third, bought from the government 445 acres of land on a scenic bend of the Black Warrior beyond the boundary of the French colony, about three hundred yards west of Township 19, Section 98. As landing was possible there, the site had been partly developed by Jason Candy, the first storekeeper of the town of Troy (Greensboro), and even of Greene County,[16] and so was called Candy Bluff. La-jonie, whose goal at the time was to leave Aigleville, saw only advantages in this move. Rent from his houses in town and his small allotment of Nauzelbine, along with the money he received from Le Soulat, would allow him to build a ware-house there and found Claret-Ville or Harmonie-Ville. The port was good, the surrounding lands excellent and beginning to be settled, and it was at an equal dis-tance from his properties: first, the town lots in Demopolis and the small allot-ment of Nauzelbine to the southwest; second, the large allotment six miles north in Township 20; and third, a 120-acre allotment in the same township, about four leagues away in the northernmost part of the colony, easily reached from Claret-Ville in less than three hours on foot, with a possible stop to rest at the large allotment, which the road passed through (August 16, 1819). But Lajonie abandoned the Claret-Ville plan, preferring to sell his lands for a profit. The first sale, an auction, failed for lack of bidders: "Our sales of Harmonie did not work; the inhabitants are in the bag," he lamented on November 3, 1820, before regaining hope thanks to an attractive drawing scheme organized shortly thereafter: "There are 234 lots at 50 francs a ticket. Each ticket is a winner. There are six town lots reserved. If each of them is filled, there will remain the fractional lots that we shall sell in private sales. I have already sold 600 francs in tickets in one day. The drawing is to take place next May 1st. Batré is in charge of selling tickets in Mobile and Monsieur Quessart in New Orleans. My co-associates are already shouting victory. I do not yet dare to rejoice" (January 2, 1821).

Optimistically announced in January, the ticket sale was still not completed in April, and nothing indicates that the drawing took place in May, as the ticket takers were slow in paying: "My holding in Candy Bluff, town of Harmonie . . . would easily have grown if all the shareholders of our town had paid; we would have sold 10,000 francs worth of deeds and more, but they are so right that I should do the same thing in their place. They have too many examples to pay thoughtlessly; several have, however, paid, but this money is not sufficient to liquidate the entire debt of 750 gourdes owed to the government" (September 17, 1821).

In a letter to Le Soulat dated February 3, 1822, Quessart said that the death of one of Lajonie's associates had led him to take measures so that the United States would not repossess its lands for non-payment: "Because they made a good deal at the time and he would rather become owner of it all by paying for it than to miss this deal by not paying, which justifies what is said of him: of all the inhabitants of Tombigbee, Monsieur Lajonie is the one who best knows his business." Months went by. In October, despite his bad speculations, Lajonie was still hopeful: "Our town of Harmonie has yet to prosper" (October 9). But it did not prosper. Harmonie never materialized, and Candy Bluff remained a simple boat landing, later known as Candy's Landing.

In August 1821 Lajonie attributed his speculating failures to a lack of both liquid assets and settlers: "Everything is going downhill in this country. Money is appallingly rare, land is going for nothing, cows also and my speculation on land and town lots singularly defies imagining. I am not alone" (August 1). One year later the situation had worsened. In Demopolis itself, town lots were losing value "every day" because no one lived on the surrounding French lands (October 9, 1822). Lajonie considered this the chief cause of his business misfortunes. Having dropped his commercial ventures, he did not persevere in his real estate dreams, considering himself fortunate that he had not invested too much and thus not lost too much. He now devoted himself to managing his allotments in the colony.

Nauzelbine, Wine Spring, and Sainte-Hélène. On June 14, 1818, while struggling to sell his Bordeaux wine in his Mobile storehouse, Lajonie acquired a quarter share, that is, 120 acres of a large allotment, eight acres of land near Demopolis and half a town lot in exchange for five casks at 225 francs each plus a barrel of rum at 150 francs, for a total of 1,275 francs. Lajonie was satisfied with this acquisition, which doubled his chances of success and anchored him more firmly in the colony: "It gives us at least two opportunities. We are almost sure not to be badly placed everywhere" (June 21, 1818). Pierre Jean Champenois, the seller, divided his 240-acre allotment between Lajonie and Basile Meslier, who took the other half.[17] Allotment 266 was located in Township 20, at the northern boundary of the colony. Lajonie called it Sainte-Hélène. His prestigious but absent neighbor was the younger General Lallemand.

Lajonie now owned three principal allotments in the colony: Nauzelbine (18 acres),[18] Wine Spring (480 acres), and Sainte-Hélène (120 acres). Unable to live on all three at once and sure that he would not settle at Candy Bluff, he had to make choices: sell Nauzelbine, rent out Wine Spring to a tenant farmer, and settle at Sainte-Hélène.

In March 1821 he sold Nauzelbine, "the masterpiece of [his] first exploits" (December 1, 1821), that is, the land and all its cabins, for "the modest sum" of about 750 gourdes, or 3,800 francs (April 12, 1821). He sold it on credit, with half to be paid in one year, the other in two. Lajonie trusted the two well-endorsed notes he had received, but both the buyer and the endorser, "these crafty Americans," suffered losses as soon as the transaction was concluded and offered him cows as payment. "They will be quite expensive if I accept," commented Lajonie, who refused the proposal, which may have been a mistake because the affair went from bad to worse: "The consequences of the sale of Nauzelbine are as harmful to me as his divorce might have been for the Corporal who has died" (December 1, 1821).[19] Lajonie teasingly blamed Nauzille, annoyed that her brother had separated himself from a property that was like "a part of herself." Did she want it to return to the family circle? Well, her wish was granted. Learning that his debtors were continuing their bad business deals and planning to leave for Mexico without paying the first draft, Lajonie went to Demopolis on November 26, 1821, to repossess his property. The occupants agreed to return it to him on January 15, 1822, along with an indemnity of 175 gourdes payable on the following March 1. Lajonie, who had only his "own two hands," was annoyed at finding himself "with two farms to look after" (December 1, 1821). By October 1822, he was thinking of taking the matter to court, since he had still not received one cent.

Lajonie unwillingly took back his first allotment, and enlarged it, also unwillingly, by six acres taken from the neighboring allotment of the younger Saint-Guirons, who had returned to France without paying a debt. This was the only way he could get his money back. Nauzelbine was now a 24-acre farm with no one to work it. What is more, the farm was no longer the fine place that Lajonie had made it. His buyer had left after doing "a frightful amount of damage both in the surroundings and in the house" and destroying a pretty little brick bridge in the woods (January 25, 1822).

He rented Nauzelbine as was to Nicolas Raoul, who paid him a year's rent of three hundred francs in advance, but in the form of a mare and ten piasters, which did not exactly please Lajonie, who needed cash. Raoul remained there several months, during which he did "useless work for 170 bushels of maize that he will have a great deal of difficulty getting," predicted Lajonie. Attracted by other adventures and perhaps discouraged by the high water that flooded his land in 1822, Raoul sublet Nauzelbine to Americans before Lajonie took it over again, lamenting its continuing deterioration: "Poor Nauzelbine has changed a great deal. Now it is nothing more than a farm, pure and simple, whereas in my day it was a little country house" (October 9, 1822). As a result of arrangements between Lajonie and Labrousse, who had been running a refreshment stand in Demopolis since 1821, the latter received half of Nauzelbine (December 17, 1826). As for the other half, Lajonie could not sell it in 1826 for the six gourdes per acre he was asking. It was a small farm subject to flooding, but Lajonie always cared about it for sentimental reasons: it was his first American land, which he cleared and planted before the others, on which he had built his first house, and where Ninon was born.

His relationship with the large allotment was different, since he had never settled there and because the Fougnets farmed it well. Fougnet Sr. had received the forty-acre reserved allotment P, but did not occupy it, since he remained at Nauzelbine for a long time with his sister Lise. He was a hardworking, capable man in whom Lajonie had complete confidence. On November 1, 1819, they became partners.[20] Lajonie put at Fougnet's disposal the 480 acres of land, which included "two or three hundred *journaux* of prairie" (November 22, 1819);[21] twenty-five head of cattle, nineteen ewes, twenty hogs and sows, two pair of yoked oxen and three harnessed horses; a cart and five plows, tools (axes, spades, scythes, carpenter's tools, a saw), a maize mill, kitchen and household utensils; one year of food for his family;[22] two men for six months, Martial and Ragon, and two others for two years, Dupuy and Mangon. Fougnet, in return, was to farm the land and oversee the partnership's business. The nine-year contract could be broken by Lajonie after five. Upon its termination after nine or five years, the lands, animals, and goods, their products and interest, were to be divided equally between the two partners.

One month after signing the contract in Demopolis, Lajonie took Fougnet to the large allotment. Fougnet fulfilled his task admirably and built two houses of red cedar and two large sheds for the horses and oxen "on a fine knoll from which one can see for a distance of more than two leagues" (February 2, 1820). The partnership dissolved when he died in 1825, his young brother Antoine and sister Lise remaining on the large allotment. In 1826, seventy acres under cultivation on this large allotment were in the best possible condition, but the presence of "Lime Stone Water" gave little hope of getting more than four gourdes an acre, lamented Lajonie (August 1, 1826).

With Nauzelbine sold, bought back, rented out, taken back, and partially resold, and Wine Spring farmed by a tenant, Lajonie could occupy and farm Sainte-Hélène, which he acquired in June 1818. Only after settling Fougnet and selling Nauzelbine did Lajonie, in March and April 1821, move to Sainte-Hélène with his family: "I am on the 120 acres that I have in section 4. See your map, at one and one half leagues from Fougnet (north). My lands are very fine and sandy. I have four or five springs, but at 150 paces from the house and around it. There is a main road 80 paces from the house. I am on a road between two towns. One is the county seat and the other a town in the interior of the country" (April 12, 1821).

The first town was Erie, the seat of Greene County since 1819, located to the west on the Black Warrior; the second, to the east, was Troy, the first name of Greensboro. In spite of their quality and location, Lajonie's lands would not, he said, make him rich: "Land is going for nothing. Near me there are 120 acres, 30 of which are cleared, that one could have for 3,000 francs. The government is selling its land for one and ¼ gourdes per acre" (April 12, 1821). Despite everything, he got to work and by December 20, 1823, he was pleased: "My plantation is small but in good condition and well situated. I owe nothing to anyone and have all the conveniences of life, at least those that one can procure in this region." By August 1, 1826, his plantation was envied: "My farm is so fine that Adolphe Batré, passing through on his way to the North, wrote to his brother from Tuscaloosa to encourage him to buy it. I am asking 12 dollars an acre: 120 acres, one year of credit with security = 100 acres of good land that can be cultivated, between Erie and Greensborough. . . . I have 40 acres under cultivation in the best condition. Good water."

On February 3, 1822, Quessart praised Lajonie but deplored the fact that he was working too hard, for this did not seem necessary: "According to what I have heard from those who know him and who know his situation, I believe that he does not need to give himself so much trouble, he should simply see to harvesting enough to meet his expenses; his large properties will increase in value without work as the population in the area increases." In reality, Lajonie's properties had value only because of thankless hard work, as Quessart himself, who never saw the lands he owned in the colony, had to acknowledge. On October 26, 1819, he had acquired (in Township 20, Section 132) 60 acres of land—that is, one-eighth of the 480-acre allotment belonging to Samuel Vorhees.[23] A corner of Quessart's allotment happened to border Lajonie's large allotment. Lajonie gave him a lukewarm description of the acquisition made in his name: good land, water, but accessible only from wells; a healthy region, vast prairies, but located on either side of the allotment, which was "almost all woods." Since good, sandy lands with springs were selling for two gourdes, Lajonie did not think he could sell his easily for more than one gourde per acre in the next two years. Consequently, he dissuaded Quessart from making speculative land purchases: "So if you are buying only to resell, it is not worth the trouble unless you were offered the 60 acres for some thirty gourdes. If on the other hand you are buying in order to have the imposed conditions fulfilled and with the idea that perhaps you might live there yourself one day, then it is worth from 60 to 80 gourdes, but if you want to sell, you must wait to get 80 to 100 gourdes for it, unless you are in need of money, which I do not suppose to be the case with all your Negroes" (July 15, 1821). Quessart listened to Lajonie, made no more investments in the colony, and in September 1821 even gave his sixty acres as a gift to Ninon Lajonie, his godchild (September 18).

Unlike Quessart, Charles Batré and his brother Adolphe did live in the colony, but, chiefly interested in their business in Mobile, neither of them put in as much effort as Lajonie. The Philadelphia drawing had given Charles 120 acres in Township 20, Section 113, allotment 280, and he was thus a neighbor of Sainte-Hélène. In 1822 the two brothers acquired the 160-acre allotment 105 of Township 18, Section 42, ceded by Joseph Moynier,[24] but Adolphe's interest in Sainte-Hélène, near his brother's property and well developed by Lajonie, did not lead to a purchase.

In the end, Lajonie's efforts to improve and promote his properties were not really rewarded by the hoped-for capital gains. The French colony's lands had value above all because of the hard work devoted to them and the harvests they produced. Therefore, after admitting that neither commercial ventures nor land speculation would succeed in making him wealthy, Lajonie devoted himself entirely to what he had come for: farming, winegrowing, and, incidentally, olive trees.

THE CHARMS OF AGRICULTURE

Developing vineyards and olive groves was a legal obligation for Lajonie and his friends, but it was never essential or exclusive: new and demanding, its success was remote and its return uncertain. It most particularly did not exempt them from recourse to the traditional agricultural sectors of stock raising and growing food-producing plants, or to others, more speculative but tried and proven in the South, such as cotton and wood.

Cows, Calves, and Pigs. Despite all his setbacks, Lajonie never rejected the principle of speculation that had fascinated him since arriving in the United States. Singing the praises of a country that was ready to grant the French emigrants a bit of its land, this is what he anticipated he would do: "All domesticated animals reproduce easily here and are an investment," he wrote in his fifth letter (January 1817). Fifteen months later, listing all the sorts of speculation open to him in the South, he judged that the best was pigs and cattle, whose nourishment costs no more than "the care to watch that they do not wander from the settlement" (April 27, 1818). To these farm animals he soon added horses. He held to these three species and, despite obstacles more numerous than foreseen, persevered on this course: "I have put all my energies into the animals, the only work that suits me," he stated on June 24, 1820, after three years of stock raising at Nauzelbine. Typically counting his chickens before they hatched, Lajonie had earlier announced his plan to one of his friends—"Supposing I had ten sows and multiply by four years" (June 1, 1818)—but little by little he had carried it out. On August 3, 1820, he gave a first listing of his livestock: "I now have here, of my own, two fine carriage horses, two mares with one colt, 19 head of cattle, cows and calves, 65 hogs and soon more I hope. I have several pregnant sows." Lajonie devoted the most care to the hogs: in early January 1821, he had 111 of them and three sows ready to give birth. On the prairies of the large allotment, Fougnet was doing likewise, raising cows, horses, pigs, ewes, and sheep. Speculation on animals, whose number increased with regularity, could be profitable, if not for losses, thievery, illness, and poor sales.

Nauzelbine was not fenced. Back after a few days spent with Fougnet on the large allotment, Lajonie found all his pigs gone. It took him ten days to find his "deserters," three leagues away (August 3, 1820). This was time he could have spent on his other occupations, but he was happy to have recovered his hogs, for he often lost them to predators and sometimes to dogs. More isolated, Fougnet suffered more from the depredations of the bears and wolves that were devouring his hogs every day, wrote Lajonie on August 3, 1820. Their livestock also tempted thieves. Lajonie cites the case of a certain Menoui (or Menou), a man from northern Italy who claimed to be French and was living more or less like the Indians after getting into some sort of trouble, except that instead of hunting as they did, the "scoundrel" was eating cows, calves, and pigs. Caught at their camp surrounded by large pieces of meat, his wife confessed that they had eaten or butchered eighteen cows belonging to a single American farmer (January 2, 1821). In the spring, when his cows returned from the woods to the prairies, Lajonie worried about how the count would come out. When they were not disappearing into the stomachs of predators or thieves, the animals died of illnesses, often epidemic, lamented Lajonie: "Last year we were unfortunate as concerns horses. This year, it is the ewes: [the Fougnets, on the large allotment] have lost more than two dozen to sickness. The rest are doing well. I personally lost many hogs. It seems that my new home does not suit them very well. This year I am losing more than 100 gourdes in hogs and almost all my enthusiasm. Last year I dreamed of nothing but hogs; this year I no longer dream of anything. Experience has wrought terrible havoc in my European mind" (July 15, 1821).

Lajonie had left Nauzelbine at the head of a "fine band of hogs" (November 3, 1820); after several months at Sainte-Hélène, he had only about twenty left. Some had been butchered and preserved as salt pork, as the going rate of "four sols a pound" was too low for Lajonie, who was hoping for a better price (January 2, 1821). Others had succumbed to illness. The region, he said, was bad for hogs and he constantly had to feed them maize. So he let Fougnet raise about one hundred magnificent hogs in very good condition, and a herd of fine cattle on the large allotment (December 1, 1821). He himself expanded his "cavalry" of some ten horses as well as his herd of cows. In this regard, in 1826, after much thought, he was able to state with certainty that cattle raising was more profitable than tilling the soil; however, if one knew how to go about both, the two went well together (August 1, 1826). Lajonie and Fougnet had both in hand.

At Nauzelbine, cutting the grass, pulling up cane—a reed with a strong, flexible stem growing in such dense stands that the region around Demopolis came to be called "The Canebrake"—plowing, and planting followed each other in rapid succession. It was important to become self-sufficient as quickly as possible so as not to have to pay high prices for food. They planted maize or Spanish wheat in the largest field because it was the staple of human and animal nourishment, and also a product for barter, if not profit. They planted near the house a fine garden that, as the years passed, included more and more varieties (cereals, legumes, tubers, and gourds): buckwheat, French potatoes, sweet potatoes, peas, haricot beans, red kidney beans, artichokes, carrots, tomatoes, pimento, parsnip, pumpkins, and melons, which grew very well in Alabama, unlike onions and cabbages, according to La-jonie. Spading preceded plowing with draft horses: in March 1818 Lajonie ordered plows from the blacksmith near the falls of the Black Warrior through his neigh-bor Saint-Guirons (March 6, 1818); by June 6, he and Fougnet were about to plow with draft horses, and the next summer he mentioned oxen harnessed French-style (July 21, 1819). But before that, he was hoping to get 120,000 ears of maize for his first harvest in June 1818, and good results for his other crops. Things did not always go smoothly, due to weather rather than the nature of the soil, which had grown maize for a long time and did well by the seeds imported from France.

The spring of 1818 was exceptionally hot and dry, surprising even the Ameri-cans, according to Fougnet. Lajonie had not warned him that his shirts and trou-sers would be wet "from head to foot," except by the rain he had told him would fall every two or three days (June 6, 1818). By early June, the maize was not too good. If it did not rain in the next few days, some would be too dry to harvest, which was already the case in Poculot's field. But several dewy mornings rapidly revived it and gave new hope to Fougnet, who concluded that it actually took as long to ripen as in France (June 13, 1818). The French colonists had to get used to a climate dif-ferent from what they had expected and learn by experience, as Fougnet found to his cost on July 21, 1819: "Would you believe that two days after pulling up our potatoes, six bushels were entirely rotten and they are continuing to rot. Thinking that leaving them in the sun would preserve them, we ruined them, for it is only the great heat that started them fermenting." When the heat was not threatening to roast everything, the flooding of the Black Warrior, covering the fields of Nau-zelbine, threatened to rot the plants.

The farmers' efforts were eventually rewarded, and Lajonie was happy with his first real harvest, that of 1819, consisting of 500 sacks of maize, 150 sacks of sweet potatoes, and 8 sacks of beans of all kinds. This success came too late, he com-plained on November 22: "If we had had this harvest last year, it would easily have brought us 15,000 francs, but the price of foodstuff has decreased by more than half. The colonists who are arriving were so fearful of having to pay for supplies at last year's prices that they are all coming with a great deal of food." Lajonie did not become discouraged. In 1820 he had another "very fine" harvest and Fougnet a "quite good" one, despite its being the first planting on what had been prairie land (August 3 and June 24, 1820). With fifty acres of "extremely well-worked" land on the large allotment (January 2, 1821), Fougnet planned to harvest four hundred sacks of maize the second year and Lajonie up to twelve hundred bushels for his third at Nauzelbine; but Lajonie sold his property and Fougnet had only two hun-dred sacks of maize and an equal amount of sweet potatoes (December 1, 1821). These disappointing harvests were not as worrisome as the persistent economic slump. This time it was no longer caused by low prices, but by a lack of money that also affected land purchase. On January 2, 1821, Lajonie wrote of poverty: "My maize is selling slowly and on credit besides, my potatoes likewise, our eggs likewise, our salt pork likewise, in short no one has any money. We make trades with the merchants who have merchandise that we want, and at that we are very happy that they give us preference. We have decreased our expenses; we use roasted maize instead of coffee and rarely have rum. I often go barefoot and use very little expensive clothing."

At Sainte-Hélène, while still favoring maize, Lajonie diversified his crops and began to concentrate on others. For instance, he highly valued dry or "caro" rice, "very productive," growing without irrigation in soft, light soil (March 20, 1824). He thought the same of sweet potatoes, "the bread of the poor and the rich," doing well in bad soil and harvested in September (June 13, 1822). The first person to bring sweet potatoes and upland rice to Dordogne should have the region's eternal gratitude (March 20, 1824). From his homeland, Lajonie had imported peaches and figs, which he produced in abundance: in August 1826 he had harvested twenty-five wagonloads of peaches and had more than one hundred fig trees bringing in a great yield (August 1). But while everything was going better, there were always difficulties. Squirrels, partridges, ravens, and woodpeckers feasted on seeds and seedlings, harvests sometimes remained unsold, and the work was all the more unrewarding in that it left them no respite.

These difficulties never made him renounce his call to farming, as others had made him give up speculating. Lajonie did not feel comfortable in money matters, where honesty was a handicap. In agriculture, on the other hand, he took pride in being "a vineyard worker, farmer, cowherd and carter rolled into one" (March 14, 1820) and asserted that farmers were the class that should be scorned the least (April 12, 1821). For his sister, attended by her fourteen servants, he described a typical day on his plantation of Sainte-Hélène on August 1, 1821:

> I must still get up every day before the sun rises (no matter what the weather) in order, when we have no little Negress, to carry water and wood, light the fire, care for the sick animals and see to it that the others can get their food, and if there is something to do with the horses, care for them if they are here or else go find them in the woods in dew to above my knees (they have bells around their necks); milk the cows and work in the field until breakfast, until dinner the same work, plus fresh water for dinner, after dinner the same work or spading or plowing, at sunset unharness the horses and care for them, milk the cows, bring the flocks into a pen to fertilize it (we use it as a field the next year), take care of the hogs and carry wood and water, all that out of doors. The pigs have only the trees for shelter and the cows are also milked outdoors.
>
> After that, as a diversion during the day, we run after a cow, sows or a mare that have given birth in the woods. We go to the neighbors to ask them to come help hold down calves, hogs or foals that we want to castrate, all this in due season. We must also take the mares to the stallion (25 francs apiece for one season), we must sometimes look for an entire month for stray

hogs, horses or cows. For almost the past two months I have had more than twenty big hogs and sows and a big bull which, incidentally, led us (Fougnet and me) on a fruitless chase for two entire evenings in the woods and plowed fields.

Care of the animals took up a large part of Lajonie's time, to the detriment of his crops, most particularly commercial crops like cotton, which he did not decide to grow until relatively late.

Cotton and Oak Boards. Cotton always had its place in Lajonie's speculative daydreams, as he was aware of its importance in the South and in American foreign trade. During his stay in New Orleans, he had written with envy of planters who had discovered "the secret of making up to 300,000 francs in revenue, either from sugar or from cotton and maize" (July 20, 1817), and of merchants who, losing money in a market flooded with merchandise, made it up on cotton (August 23, 1817). Money brought a return of up to "50% if invested in cotton, maize and potatoes" (April 27, 1818); these crops, whose acclimatization in Alabama was less risky than that of grapevines, were the ones in which to invest. In May 1818 Lajonie embarked on his customary calculations: "Cotton is at 30 sols; an acre can give 300 pounds. A good Black can, with a horse, cultivate 20 acres. Six thousand francs that this Negro will bring you." Lajonie was urged to turn to growing cotton, which most planters were growing at the time—first by Fougnet, acknowledging that it would be better to plant cotton than be a merchant (June 20, 1818); then by Latour, in Baltimore, asking him about the advantages and conditions of growing it (August 29, 1820); and finally by Quessart, who, on December 8, 1820, asked him the decisive question: "Why do you not plant cotton, which would bring in more than maize and potatoes?"

Shortly thereafter, but before he planted any, Lajonie spoke of exporting it, along with beef, pork, and flour (January 2, 1821), but he put it off. Having become cautious, he was wary of export conditions, fearing that duties would eat up his profits. Besides, the prospect of exporting cotton to consignees in Bordeaux was not very attractive; connections between the French port and those of Mobile and New Orleans were limited to just a few round trips per year, as Quessart lamented on June 8, 1822: "It grieves me that everything or almost everything goes to Le Havre de Grâce and that Bordeaux receives very little of our cotton; Rouen and Paris offer a great market for this item. If the capitalists of Bordeaux and the surrounding area wanted, they could also have factories and in this way give life to their region, for labor is as cheap in Périgord and Saintonge and in the Limousin as it is in Normandy and those regions have rivers that would be very useful for that. I am not sorry to see Le Havre de Grâce prosper, on the contrary, but I should also like to see Bordeaux keep the rank it had among trading cities."

Lajonie planted cotton at Sainte-Hélène for the first time in 1823, but very little, given that he had more urgent things to do and lacked help. Trying again in 1824, he was able to sell at a good price, since the economy was looking up and he took full advantage of it. These results and advice from Le Soulat encouraged him, and he planted a great deal of cotton in March 1825, but much too much for his strength. His maize suffered on this account and so did his cotton, which he did not plant well, as he wrote from Sainte-Hélène on July 2, 1825: "A fortnight ago my cotton was as clean as a French garden. Today the weeds are eight inches high and not an inch of earth is visible. I use a plow with a moldboard and without a landside. The weeds near the cotton plant are indestructible, but the drawback is that I bury half of the low branches, consequently those with the most flowers. I think that I planted my cotton too low. If it had been higher, the lower branches would not be as exposed. I must be patient for another year."

To Lajonie's inexperience were added vexations from which his harvest suffered: the unavailability of Antoine Fougnet, attacked by fevers, and of two horses, sick or injured; the proliferation of crabgrass (July 2, 1825); and in June, worms causing great damage to his cotton. Despite all this, he would see his efforts rewarded, promised Quessart: "If Tombigbee cotton, that this year was worth up to 25 sols a pound here, holds at only 15 sols for a few years, there is no doubt that our friends will make a fortune on the settlements that they have created. They have had a great deal of difficulty, but today they are comfortably off. How many Frenchmen who were in the same place and abandoned it would be happy like they if they had had their courage and their industry, instead of which they are in poverty" (August 10, 1825).

In January 1826, even if he reported a "rather fine cotton harvest" for Lajonie, Quessart was more restrained because the rise of the previous year had been short-lived (January 24). Cotton did not live up to Lajonie's hopes, since on July 2, 1825, when his crop was not doing as well as he wished, he inquired about one last possible little speculation on *merrain*, oak wood split into narrow boards for barrel making, which he considered exporting to Bordeaux: he asked about the shipping price, customs duties, and how many thousand *merrains* a ship could carry. It is not known whether Lajonie himself went into this operation, but in 1826 Mangon took ten thousand *merrains* and two to three thousand feet of cedar down the river to Mobile on his raft (March 12, 1826). He was to continue this activity for several decades.

Despite their efforts, competence, and conscientiousness, Lajonie and his associates never succeeded in making capital out of the land that Congress had conceded them. Farming it allowed them to eat without ever lacking anything, but what they did not need was traded more often than sold. Lajonie and Fougnet were exemplary colonists; what is more, they defended the smallholders whom land grabbers had in their sights: "In all modesty, the weak of the colony need me," wrote Lajonie (January 2, 1821). One can imagine the situation of the others, and understand why most left. The success of the vine and the olive for which they had come could have prolonged the colony's life expectancy. On the contrary, it was cut short by their failure.

When the site for the colony was chosen in Philadelphia, not a single voice was raised to question its suitability for winegrowing. On the contrary, it was compared to southern France, where sunshine, the slope of the hillsides, and the quality of the soils produced such fine grapes. No one seemed to realize that between the future vineyards of Demopolis and their elders in the Bordeaux country there lay two thousand years of history and thirteen degrees of latitude that made all the difference in the world. There were no experienced winegrowers among the society members, not one of whom, additionally, was familiar with the site. Alabama as a land of vineyards was a theoretical view, but some believed in it, Lajonie first and foremost.

WINE SPRING

While still at Nauzelbine, before moving to his property of Sainte-Hélène, Lajonie had, as we have seen, settled the Fougnets on the 480 acres of the large allotment, granted in the drawing and to which he thought he should, "politically, give the name *Win's Spreen* (spring of wine)," adding:

> These two English nouns go together well and appeal to the ear and to the palate. That is what we need. I am beginning to believe that it is sometimes necessary to sacrifice the present for the future. The pronunciation of the word is about that of *ouin sprinn*, the second *inn* like *in* in Latin and the first like the nasal French *in*. Without wine, farewell to fortune. My lands will be something for my children, and these wines some comfort. Our lands will triple in twelve years, quintuple in sixteen, and so forth, and we, and we, my poor friends, and we, oh! can fret all our lives and not enjoy them for a single moment. Farewell. Farewell. (April 1, 1820)

In April 1820—that is, three years after passage of the law of March 3, 1817—the wine he would produce remained for Lajonie the gold mine that would make his fortune and that of his children. At this time it was only a firm conviction, since not a single bunch of grapes had as yet grown on any grapevine. Lajonie had not come to Demopolis in October 1817 without first thinking about how he would meet his obligation to plant vines. He did not see it as restrictive, unlike most of the other society members. Shortly after arriving in Philadelphia, he had announced that on the lands he wished to purchase, blacks were to be slaves, French was to be on a par with English, and it would be possible to grow grapevines. Lands and slaves required money; grapevines—"very rare in the United States" (January 1817)—required vigorous young vines, specialized tools and equipment, and skilled workers. From January to October 1817, this is what he regularly asked his family to send.

To attract people, Lajonie knew that they had to "make the peaceful wine growers want to come," otherwise they would not move readily. Knowing that his letters were read beyond his immediate family circle, Lajonie added innocently on March 13, 1817: "Ah! If they knew how fertile this soil is, if they knew how rapidly the fortune of landowners increases." Lajonie was thinking of entire families, like the Charlots of Gensac, "three men and two women. They would be a great asset for the planting of vineyards which could be done most successfully" (February 1817). But capable, trained, single men who were easier to convince were welcome. Very young ones too, like fourteen-year-old Antoine Fougnet. Lajonie insisted that while waiting to leave he learn English and Spanish and work for a distiller with whom his uncle Taupier could place him (February 1817). Antoine and his brother Jean lived on the farm of Saurel, in the parish of Coubeyrac, where their parents were farmers and winegrowers, as were those of Jean Mangon, born in a village of the same name, surrounded by vineyards.[25] By April, Jean Fougnet had apparently promised to send ten workmen, a number Lajonie then thought sufficient, before hiring others (June 27, 1817). The important thing was that they come with their tools and, those leaving after December 15, with young vine plants.

Did Lajonie know of experiments in winegrowing in the United States by Legaux in Pennsylvania or Dufour in Indiana? It is certain, on the other hand, since he had seen some on the banks of the Ohio, that he knew there were indigenous grapevines, which he never thought of using. The vines he would transplant in Alabama could come only from southwestern France. With the exception of *enrageat*,[26] he said nothing about types, accepting young vines of all sorts that would produce either white or red wine (June 27, 1817). But Lajonie was particular about time and packaging for shipping: everything was to be numbered and the young vines put into old barrels well filled with a sort of clayey earth mixed with a bit of water in which their roots would be immersed. At first Lajonie asked for just enough to plant "a *journal* of thick vines." For now, that would do (June 27, 1817).

The young vines were to be sent in winter, while dormant, and arrive in the United States in early spring. Boarding the *Marie-Thérèse* in Pauillac on November 26, 1817, farmers Antoine Fougnet and Pierre Ragon arrived in New Orleans on February 16, 1818.[27] Jean Quessart wrote to Le Soulat that the cargo of the *Marie-Thérèse* had been brought ashore and that the two men had left for Demopolis via Mobile on March 9, on a schooner laden with supplies, barrels of wine, brandy, and young vines. He had checked the condition of the barrels when they were unloaded and found that the young vines had kept well (March 13, 1818). This first shipment probably came from Gensac and Juillac or their environs.

Lajonie left the elder Fougnet on his allotment and went down to Mobile, where he met young Fougnet and attempted, until July, to sell the wine and a portion of the merchandise. But competition was stiff and business poor. It would be better "to limit ourselves to raising cotton and bringing in vines if it is possible, than to scatter our capital here and there without real profit," Lajonie wrote on May 6, 1818, from New Orleans. So it was Ragon who escorted the young vines, which the elder Fougnet planted at Nauzelbine in April. On June 6 Fougnet informed Lajonie, now back in Mobile, that they still had "a few young vine shoots, but shall lose quite a few." On July 14, 1818, the anniversary of the French arrival at the White Bluffs, he was delighted that everything but the vines was growing nicely on their lands, with the exception of forty-five superb young plants (a variety that made red wine); but Fougnet did not expect to have more than fifty like this. The young vines were only about a year and a half old and would not bear fruit until they were three, that is, not before the summer of 1820.

That year Lajonie received a second shipment of young vines from Le Soulat, transplanting them at Nauzelbine on March 13, 1820. He wrote that he now had "about one hundred fifty young ones and forty old ones," adding, "I believe that they will ruin us" (March 14, 1820). There we have in a nutshell the problem with winegrowing. Three years of investment and effort necessarily preceded picking the first grapes. Already tired after staking two hundred vines, Lajonie was concerned about several of his old plants that had not sprouted and about the threat of rising water: the end of the winter had been very rainy and the river was at their doorstep (March 21, 1820). But since they had not sprouted, he at least did not have to fear frost. Alabama, where it could be very hot as early as May, was not spared late frosts that could kill grapevines and olive trees until the end of March or even April. Flooding could also have unfortunate consequences: in the winter and spring the Black Warrior and Tombigbee often overflowed their banks and submerged broad stretches of the countryside. Nauzelbine, situated on low land, was not safe from flooding, but in 1820 this did not occur. August 15, 1820, was a red-letter day: that was the day Lajonie picked the two fine bunches of grapes and took them to the election office as his prime qualifications. He prided himself on being the first to show the Americans the fruit used to make wine.

After fulfilling their commitment to plant grapevines so quickly, Lajonie and Fougnet harvested their fruit in the shortest possible time given the conditions of settling in, preparing the soil, receiving and transplanting the young vines, and the vines' natural cycle. This promising outcome spurred Lajonie to continue. That same year, he bought from another society member, Pierre Hurtel, for twenty-five sols apiece, one hundred more young vine plants which most probably came from the Bordeaux region where Hurtel had lived after returning from Saint-Domingue. On January 2, 1821, Lajonie noted that the vines had almost all rooted. Furthermore, while until then he had been rather evasive, he now made remarks about consumer tastes and the types of vine that should be sent him: Americans "prefer white wine that is very dry and heady. Send me white *enrageat,* sauvignon, muscadet and a few other types that sprout very late. The young vines will have to leave before December 15th."

Early in 1821 Lajonie complained of the lack of money in the colony, but he had hopes for the future, for life was easy when one had slaves, and wealth would come if the vines succeeded. It was at this time that he left his small allotment of Nauzelbine in Aigleville for half allotment 266 purchased from Champenois in Section 108, in the northern part of the colony. His neighbor in the section that bordered his on the south was Charles Batré, owner of the 120-acre allotment 280: "Batré is here planting grapevines. He was sent some from Bordeaux and they arrived in good condition," he wrote from Sainte-Hélène on April 12, 1821.

Lajonie was absorbed in settling in, clearing the land, and starting the crops on his new plantation, but in the same letter he also expressed concern that Le Soulat no longer wrote about sending young vines. (He took the opportunity to remind his family that he was also supposed to plant olive trees and requested a few.) Did he receive some earlier or get some in another way? It was not until a year later that he wrote Le Soulat of the arrival of "a superb shipment of vines" that he had gone to pick up at the river port:

> I received them only a fortnight ago. One barrel that I think is all red had entirely sprouted, the other that I think is all white and ¾ *enrageat* had not. Everything has been planted and everything is sprouting. Ah! Except for the *enrageat* that is late in taking hold, I hope, because it is quite green. I sold 75 plants at 12½ sols apiece, which will pay part of my freight. From New Orleans to here I paid 105 francs in freight or costs. I do not know how much Monsieur Quessart has paid. You packed the vines very well. However one barrel seemed damper than the other. We have enough for the present. (June 13, 1822)

Three and a half months later, Lajonie sent an update on this shipment. The barrel of white vines had probably not received the same care, since the two plots of young vines had died. Although they were green when they arrived, they had not sprouted as the red vines had. This was probably due to a difference in packing: "So all you must do is remember whether the sand of the latter [barrel] was damp and the other not, so that if necessary you can act with near certainty from now on" (October 9, 1822). On the field, Lajonie had become a true winegrower, capable of giving advice to his family at Le Soulat, surrounded as they were by vines: "If by chance you are having vines planted, anywhere except on a steep hillside. Plant them at 8 feet wide by five in distance in poor soil, six in good. With a good horse, I shall show you how one can work" (June 13, 1822).

On August 1, 1826, Lajonie said he had two acres of vineyards on his 120-acre allotment. That is, after nine years in the colony, he had met one of the conditions prescribed by the law of March 3, 1817, but had not done more—or rather, had not been able to do better. It was not for lack of trying, together with Fougnet: in March and April 1821 they had planted eight hundred young vines from the Bordeaux region that a young man had given them provided he got half of them the following year (April 24, 1821). But while announcing this great effort, Fougnet feared that "grapevines will not do very well in this country." On October 9, 1822, Lajonie wrote Le Soulat confirming his friend's fears: "The vines that we transplanted at Fougnet's gave a few bunches of grapes this year, but the birds devoured them before they were entirely mature. I fear that for several years it will be difficult to avoid the damage of these winged creatures."

The year 1820 had been the year of hope, materialized by two beautiful bunches of grapes exhibited with pride. It had also been that of the questions that Latour and Quessart had successively put to Lajonie. Latour, seeking information for a young Frenchman, asked him on August 29, 1820, whether those who had planted vines and olive trees thought they would succeed, whether the soil was suited for them, whether the colonists were forced or obliged to grow these trees, and finally, whether planting cotton was not more advantageous, whether it sold quickly, and at what price? Quessart wrote from New Orleans on December 8, 1820, also inquiring whether the soil was suitable for grapevines and whether it might not be better to plant cotton (rather than maize and potatoes), but most important, he asked *the* critical question: "I ask you again, did you make wine or did you merely eat grapes?"

Nothing in Lajonie's letters permits us to answer this question definitely, but most likely birds ate more grapes than he himself did, and he never produced or sold any wine, otherwise he would have boasted of it. Young vine plants are mentioned frequently in his letters, bunches of grapes rarely, and grape harvests never. This is what he wrote from Sainte-Hélène on July 2, 1825: "I started this farm four years ago and already have peaches and figs in abundance and two large baskets of grapes. I have only about twenty plants that are producing." Lajonie's winegrowing efforts fulfilled the contract with the government and gave him legal ownership, but it did not in any way make him wealthy or encourage him to continue. Competent, courageous, resolute, and honest, Lajonie was the very model of the French colonist who ought to succeed. What could be the fate of the others, particularly those, in the majority, who did not possess all these qualities? And, consequently, what was the fate of the vine and the olive trees, and thus of the Franco-Alabamian lands coveted by Americans?

V

Choice of a World

15

Rebirth in America

My dear friend . . . everything you say about French affairs makes one shudder so that I am happy to be here. Yesterday we had the big election for president. There has been a lot of disorderliness, and there were fights in several places, but when it is known who was elected, whoever he may be, it will be all over. We have settled here for good. . . . On January 1st I am going to open a café and a grocery. A lot of barges loaded with cattle, merrains and all kind of wood stop here. The place is doing really well. I am going to succeed, with my wife and my two children.

—Pierre Mangon Sr., New Orleans, November 8, 1848

Jacques Lajonie, heroic farmer and winegrower, was not the exception in the French colony. For varying lengths of time, in their own way and with the means at their disposal, others also put a lot into its development: his associates Mangon and Fougnet, General Lefebvre-Desnoëttes, Roudet the nurseryman from Isère, and Domingan refugees George, Ravesies, Stollenwerck, and Chapron. But they, and others who persevered in their efforts, did not acquire a following. When they came, the budding planters almost all learned from the failure of vines and olives and parted with their allotments without delay. They sold them either to their compatriots or to increasingly insistent Americans, whom these crops would put in compliance with the law so that the land would be theirs permanently, not make them money, a privilege reserved for cotton. However, this concerned only a few members of the Colonial Society, survivors of the small contingent that came, even for a little while, as far as the White Bluffs. The others remained on the East Coast, settled in Louisiana where French was spoken, pushed on into the continent's interior as far as Missouri, or bypassed it at midcentury to go to California. Only ten years after its founding, the colony had lost the core of its French identity. Twenty years later, only traces remained.

THE COLONY TEN YEARS LATER

Without the papers of the Tombigbee Company in Demopolis, which were destroyed by fire, it is impossible to arrive at a precise count of the French present between 1817 and 1827 and, consequently, to calculate the colony's population density, but reference points can be set up. The second half of 1817 saw the successive arrivals of the three groups, led respectively by Parmantier, Lajonie, and Generals Desnoëttes and Clauzel, probably not more than eighty to one hundred persons in all. During the following year, the colony continued to receive a regular stream of individual and collective new arrivals, some of them from the Champ d'Asile. In February and March 1818, a petition from the French company, requesting the appointment of a justice of the peace in Marengo County, showed that there were in Aigleville more than 150 individuals to whom Congress had granted lands, and as many Americans in the immediate neighborhood.[1] In December the company informed the secretary of the treasury that thirty-four grantees, and seven others with reserve allotments, had settled on their land with their families.[2] Three years later, in December 1821, a petition from Lefebvre-Desnoëttes indicates that the high point of occupancy had most likely been reached, with 81 householders settled on their allotments at the head of a population of 327 persons.[3] The decline probably began the following year.

The colony was thus much more a transit zone than a place to take root, even if counting those who passed through without leaving a trace is much less certain than counting those who attempted to stay and left their names on notarized administrative deeds and legal documents in the newly created Marengo and Greene Counties: public records, rental, sale or transfer of lands, powers of attorney, administration of estates, slave transactions, and so forth. Rather than go into their tedious detail, we will look at two documents that set out clearly the reality of the French presence ten years later: the list of customers at the Bierne sale in Demopolis in 1826, and the Adams report of 1827 on the occupancy and development of the colony.

THE GROUP OF FIFTEEN

Captain François Bierne had arrived in Alabama in about 1820.[4] A survivor of the Russian winter, he ironically met his death on July 25, 1826, coming up the Tombigbee on the steamboat *Herald.* A month earlier, a last letter written from Mobile to his wife in France mentioned the store selling European merchandise that he had in Demopolis.[5] Achille George was executor of his estate.[6] This did not include land in the colony—which Bierne had intended to acquire[7]—but only the contents of his general store and his personal effects, whose sale, from October 19 to October 25, brought in seven thousand dollars.[8] The inventory of thousands of items had attracted dozens of buyers, whose listing amounts to a census of the colony's lifeblood. Almost all were Americans, except the Swiss Charles Breton, the German Gotlieb Breitling, and fifteen men of French origin: Pierre Édouard Chaudron, François Louis Constantin, Clément Ducoutumany, Alexandre Follin, Alexandre Fournier, Théodore Guesnard, Jean Hurtel, Mathieu Labrousse, Jacques Lajonie, Pierre Mangon, François Martin, Jules Martinière, Frédéric Ravesies, Louis Auguste Stollenwerck, and François Violle.

This group certainly does not include all the French people still in the colony, but its number and composition suggest a few remarks. Of the original society members, the Frenchmen banished after 1815, active in Philadelphia and Washington to obtain a grant, and then settling in Demopolis, only one remains: Lajonie. Fougnet, Mangon, and Labrousse cannot really be compared to him, as they were merely holders of reserve allotments who had come at his request, like Constantin, who had emigrated with his uncle at the age of fourteen. Neither Guesnard nor Ducoutumany was on the original list, having joined the colony through marriage: the first to Marie Angélique Zéline, daughter of the elder Jean Baptiste Herpin, in Philadelphia in 1814; the second to Louise, daughter of Claude Payen, in Alabama. The military complexion of the colony had never been clearcut, since in 1817 half of the society members who were officers had preferred the adventure of the Champ d'Asile to working their allotments[9] and only a quarter—about fifteen veterans—had lived for a time in Demopolis. Neither Vandamme nor Grouchy gave a moment's thought to honoring the colony with their presence; at best the marshal sent his son Victor to alleviate his boredom. Clauzel seems to have accompanied Desnoëttes to the White Bluffs on their first journey,[10] but, quickly returning to Mobile, he responded to none of the urgent and reassuring invitations of his friend and associate: "You must not be afraid of being uncomfortable; I have a good house and I live well."[11] The younger Lallemand enjoyed Desnoëttes' hospitality, but he came very late and undertook nothing on his allotment. Raoul tried. Of the Napoleonic stars, only Desnoëttes put any effort into his role of soldier-farmer, but he was always determined to take advantage of the first opportunity to return to France. With his departure, the colony lost a bit of its soul. The Grand Army's campaign on the Tombigbee was soon reduced to a few underlings like François Violle from Auvergne, who had enlisted in the Imperial Guard in the waning days of the Empire.[12] He emigrated in 1817, was a physician at the Champ d'Asile, and then went to Demopolis, where, on May 22, 1818, he married Léontine Desportes, the former chambermaid of Countess Desnoëttes, whose husband drew up their marriage certificate.

Contrary to a historiography according the lead roles to military men, the predominance of civilians and overrepresentation of society members who had long been in America stands out: eight refugees from Saint-Domingue (Chaudron, Follin, Fournier, Guesnard, Hurtel, Martinière, Ravesies, and Stollenwerck) and François Martin. After emigrating from Marseille in 1806 with his older brother, Joseph Martin-Picquet, Martin was a confectioner in Philadelphia before the brothers separated, the elder going to Kentucky and the younger to Alabama, where he had cleared his 480-acre allotment, which was frequently flooded by the Black Warrior.[13] Finally, it should be noted that this group of fifteen was not even the standard-bearer of France in the colony, since a number of them were American citizens, either by birth, like Edward Chaudron and Alexander Fournier in Philadelphia, or John Hurtel in New York; or through naturalization in Philadelphia, like Frederick Ravesies in 1803 and Francis Martin in 1808.

The data from this reference group are confirmed by the results of the survey ordered by Secretary of the Treasury Richard Rush. On September 22, 1826, he had asked William L. Adams of Tuscaloosa to check on the progress made by the French emigrants in respecting the clauses of the contract signed with the government, and had informed Frédéric Ravesies, the agent of the Tombeckbee Association, of this. After fleeing from Saint-Domingue to Bordeaux, his father's native city, Ravesies had gone with him to the United States in 1796,[14] settling in Philadelphia, where he was associated until 1814 with the Gareschés, agents of the Dupont and Bauduy powder plant in Wilmington,[15] and then with Curcier. Widowed in 1815, he married Domingan refugee Adèle David, née de Sevré, the widow of an officer of the Saint-Domingue expeditionary corps whom she had married in Philadelphia in 1804.[16] Ravesies played an important role in the organization of the Colonial Society, then acquired many allotments quickly gotten rid of by their purchasers, before reaching Alabama with his family in 1820. A large landowner in the colony, he lived on the banks of the Black Warrior in Arcola, on the plantation since known as Hatch Place, where his son Paul was born in 1825. Rush contacted Ravesies because he was the association's agent, a position that reflected the local supremacy of the Domingan refugees. Indeed, before analyzing the survey data in terms of occupancy and cultivation of the soil, it should be recalled that the Frenchmen cited there had already been present at the Bierne sale or, once again, came for the most part from the Domingan community: Bayol, Brugières, Chapron, Descoins-Belair, Fournier, Gallard, George, Herpin, Noël, Pfister, Teisseire, Teterel, and so forth. The Adams report confirms their takeover of the colony as well as the disappearance of the exiles for whom it had initially been created, the soldiers and winegrowers.

Lajonie and Violle are joined by only two other junior officers—though their American titles suggest otherwise: *Colonel* Cluis and *Major* Descourt. Lieutenant Jean Jérôme Cluis held the highest rank of the four, but he had long since left the cavalry for his position as secretary to the minister of police, and then as an ephemeral gendarme captain during the Hundred Days. From 1804 to 1808, Descourt was a mere fortification guard at Ostende, then a civil service officer on state ships, and finally a corporal in the sappers of the German duchy of Juliers.[17] What position did he hold when in 1815 he decided to go to the United States?[18] In 1817 he was suspected of gathering "the most striking individuals" of the French community in exile for reprehensible reasons in his rooming house on Courtland Street in New York.[19] His wife assisted him in his questionable activities by taking to France "the letters and packages of all of Bonaparte's partisans."[20] After this long stay on the East Coast, Cluis and Descourt went to Aigleville with their families.[21] Their signatures appear on real estate transactions of the 1820s, either for themselves or on behalf of society members not present in the colony. Descourt settled and planted vines on half of his 240-acre allotment 191, transferring the other half to Cluis; rather than cultivate his own, which was twice as large, the latter had preferred to associate with Descourt. But both eventually gave up working the land and left the colony to become innkeepers in Greensboro. Unlike Lajonie, they were not part of the viticultural inner circle, but in any event, of those French colonists who might have been, very few distinguished themselves by their motivation and agricultural results, with the notable exception of Jean Claude Roudet, known as Corneille junior,[22] whom Rafe Blaufarb humorously classifies with the few "eccentrics whose commitment to unorthodox crops set them apart from their cotton-minded peers."[23]

THE LAST WINEGROWER

In 1831 a local newspaper gave Roudet the title of best French winegrower of Alabama but, so as not to discourage vocations, refrained from adding that he was the last winegrower worthy of the name.[24] He was not a nurseryman by training, nor was his caterer father, but his grandfather and ancestors before him had been. Twelve years in America had given him the opportunity to demonstrate his expertise. He had left Bordeaux in September 1818 with Marie Thérèse Melizet (called "Miette," a Louisiana Creole he had married in Grenoble ten years earlier), their two sons, and several other society members.[25] Roudet emigrated with the plan of settling in Alabama, where his allotment no. 4 awaited him, located near those of his Izère friends Antoine Vial and Antoine Bugey, who had gone to America before him. Two weeks after his arrival in Philadelphia on November 2, he and his brother-in-law François Melizet acquired another allotment, no. 302 (240 acres), belonging to Constance Démérest, represented in the sale by General Clauzel.[26] In 1819 Roudet was on-site; by November he was already on the commission of the Association of French Emigrants of Aigleville with Michel Mestayer and Jacques Lajonie.[27] With them and Lefebvre-Desnoëttes, he was one of the first to grow vines in Aigleville, probably on his small allotment rather than on allotment 4, which he sold in 1823 for three hundred dollars to Frédéric Ravesies, who planted vines there.

Roudet's beginnings were difficult. After ten years of work, his explanations were similar to Ravesies' explaining the Adams report and Lajonie's in his letters. They all agreed. The first problem was the comparatively high death rate of young vines, both during the journey and afterward, regardless of whether they arrived after the best season for transplanting or were delivered on time and in good condition. For example, allotment 149, whose grantees, Michel Le Bouteillier and his wife (who married Juan Rico Vidal, Lise Fougnet's bewigged suitor, after Le Bouteillier's death in 1819), belonged to the Parmantier group, received three shipments of young vines that all died because they had arrived too late, so that in 1826 the allotment still had no vines. But most often the consignments were shipped successfully. Ravesies pointed to the regular fall in price per vine as proof of the great number that arrived: 25 cents at the beginning, then 12.5 cents, and finally 6.25 cents in 1826.[28] But even when they were properly transplanted and well cared for, these vines never came up to the colonists' expectations. At worst, the young vines died as soon as they were planted. At best, they took root and grew for two or three years until they bore a few bunches of grapes for eating, not enough to make wine. After that the vines lost much of their vigor. Roudet thought this failure was due to the émigrés' incompetence and the climate: any careful observer could have foreseen what would happen to these French vines in such an unfavorable latitude, where they lost all their value. Furthermore, the soil had never been worked, and Ravesies claimed that old soils gave the best results.

The two experts agreed that the vines could not adapt to their host country—in his adopted Indiana, Jean Jacques Dufour also gave his point of view[29]—but they differed in the rest of their analysis. Ravesies was pessimistic, but realistic. The art of winegrowing went back to the dawn of time. In Europe it had taken centuries of experimentation to train winegrowers, improve growing methods, develop soils, find good vines, and so forth. It might take seventy years to determine what type of vine would succeed in the Alabama soil and under the Alabama sun, not a mere seven. This assertion was all the more justified because, in addition to concerns about transportation, climate, and soil, there was a shortage of skilled workers, that is, people possessing or inheriting a proven technique formed over generations. This could not be the case with Anglo-Americans, who quickly far outnumbered the French, working in the colony as landowners, agents, renters, and agricultural laborers.

In 1831 Roudet was presented as the only successful winegrower, first of all because he was one of the few truly competent ones, and second because he abandoned the French vines with which he had started and substituted, among others, the Madeira vine that François Violle is said to have introduced in Demopolis. Roudet states that, after returning from Galveston to New Orleans, Violle received from a merchant for Desnoëttes four lots of authentic young vines unloaded from a ship arriving from the island of Madeira; this took place in 1818. According to Roudet, who seems to have confused his dates, it was in the spring of 1821 that Violle delivered them to the general, who gave half of them to Roudet.[30] By spring or summer 1822, three lots had died, but the fourth had survived transplanting. Roudet got it to thrive so well that in the autumn of 1824 he produced a demijohn of Alabama wine, which Colonel McKee, passing through the colony, wished to taste. Roudet sent three bottles of French Alabama wine to him in Washington, probably the only ones ever to make the journey. Jefferson had drunk Dufour's Swiss wine in 1803; twenty-two years later, President John Quincy Adams and the members of his cabinet drank Roudet's wine. According to McKee, they liked the wine from the Madeira vines, despite the fact that it was young, whereas that from the French vines had turned vinegary.

Roudet's allotment near Demopolis was too small to develop his vineyard and tree nurseries. In 1824 or 1825 he went to Greene County, where he transplanted all his vines and the fruit trees he had been able to bring to a larger area. On his new lands, never plowed, he planted about one hundred young vines when the season was already advanced. Roudet was also concerned that he had to begin with a limited stock; going from one hundred plants to several thousand would require time. He took it. In 1828, after three years of effort, his vineyard had expanded to five acres, and was still growing in 1831. That, he concluded in the newspaper article, was the true story of the introduction of Madeira in the French colony. He contradicted one of his neighbors who claimed that it was not a Madeira, but a wild vine found in Louisiana. This winegrower also had a fine vineyard, admitted

Roudet, but he had not chosen his young vines well, particularly the Schuylkill muscadine, whose extraordinary qualities of robustness, growth, and resistance to hot March days were spoiled by serious flaws: an inelegant, bitter-tasting grape and a yield that was barely enough for bottling. In a new country, the winegrower's first duty was to experiment repeatedly in order to select the vines best adapted to local natural conditions, otherwise success was doubtful. Besides muscadine, Roudet mentioned other possible varieties, Isabella, Chasselas, and especially Persian, but he preferred Madeira for its luxuriance, productivity, and harvest time in early September, after the hot days of August.

These considerations are the last known concerning the cultivation of the vine in the French colony of Alabama. It can be deduced from them that in the early 1830s there were still vineyards, but most likely Americanized, small, with low yields and no commercial markets. In his introduction to the article, a reporter had presented Roudet as the only person who had managed to endow his adoptive country with a new source of riches, adding, however, that only future generations would benefit from this.

Roudet himself got no profit from his work. In 1837 the death of his elder son, Pierre Corneille,[31] and the fire that destroyed his home drove him from Greene County to Mobile, where his death on September 13, 1839, left a widow and a son, Jean Baptiste. His death certainly tolled the knell of Alabama's viticultural hopes, born of the founding of the Colonial Society of French Emigrants twenty years earlier. But his very departure had further reduced, if possible, the presence of the French in a colony eroded since its creation by the Americanization of its lands and occupants, contested only by the Domingans, though they were now closer to the United States than to France, with which their ties were growing slack. One of them was nonetheless the originator of the first French attempt to save a situation that was very quickly seen as compromised.

Officially Recognized Landowners

We have seen how the French colonists' situation had quickly deteriorated and the rate at which departures had increased, but also that a certain number, like Lajonie, did not give up and sought to diversify their means of subsistence in order to supplement their uncertain chief agricultural enterprise. Inexperience and natural disasters (frost, drought, a tornado, storms, rains, flooding) had hit the colony hard during its first years, resulting in poor harvests. In a region lacking everything and just beginning to be developed, on the other hand, it was possible to find work in services and trade. Nicolas Raoul and Lucien Ensfelder[32] each ran a ferry to make up for the absence of bridges in a colony subject to flooding, and Lajonie hauled freight to St. Stephens and Mobile on his barge, the *Bonaparte;* Labrousse had a refreshment stand in Demopolis; Bierne had the best-stocked store, but before him Lajonie and Lefebvre-Desnoëttes had tried selling their own products and items imported from France, with as little success as they had business sense. Desnoëttes' correspondence with Clauzel, his associate and supplier in Mobile, shows they were better at managing a division than at selling herrings, hats, or any other kind of merchandise. Desnoëttes, not really interested and furthermore too busy extending and improving his estate, had given the job of managing his store to Ensfelder, who ran up one loss after another and, as revealed by an inspection in January 1819, let rats eat a great many items not yet paid for.

Discouraged by the results of their agricultural endeavors and disappointed in the profits of their supplementary activities, the colonists who had wagered on success on the Tombigbee found their praiseworthy efforts unrewarded. Now, if these pioneers were to leave, the colony would inevitably die out, particularly because of the principle that tied the society members jointly to one another. According to the terms of the law of March 3, 1817, the settlers who had, either in person or via an agent, worked their allotments and planted vines and olive trees would receive the deeds to their property, but only when the entire amount for the grant of the four townships at two dollars an acre had been paid. In other words, individual contributions became meaningful only as part of the collective effort. Why slave away on an allotment that would in the end be difficult to obtain if others were totally uninterested in theirs, either not coming, or abandoning it very quickly, or not hiring anyone to occupy it? This injustice had to be brought to an end to by seeking a revision of the contract made official by the law, and converting a domain held in common into individual private properties.

Early in 1820, or perhaps even sooner, Jean Marie Chapron and other grantees petitioned Congress that each emigrant might receive title to his property separately as soon as the contractual conditions had been met and the government had been paid for his proportionate share in the colony. Dissociating, in writing, the lot of the most deserving from that of the members in general would reward and increase their motivation, cultivate the best allotments—a quarter of the colony's lands being of very poor quality—and thus increase their value on the real estate market. But the petitioners' interest was not that of the treasury secretary, who felt that his country had everything to lose: the colony's good lands, while the bad ones would go back into the public domain; and, above all, the opportunity to become self-sufficient in wine production.[33] The chief purpose of the grant was not that a small number of allotments be planted in vines and olive trees but that it be entirely settled by people skilled in growing them, at least the vines, as a prelude to rapid geographic expansion. In addition, Crawford pointed out, the land had been granted not to individuals, but to an association which had knowingly accepted a contract that he, moreover, judged to be generous enough that it did not have to be made more flexible. For these reasons, he was opposed to any modification, which, besides, he did not have the power to make in the place of the legislators. In any case, his argument would seem to prove, three years after the fact, that the lands had been granted to the French emigrants for genuine economic reasons (to establish a domestic viticultural industry) rather than strategic ones (for which there was no longer any reason, with the annexation of the Floridas).

The association returned to the attack. On December 8, 1821, in Demopolis, General Lefebvre-Desnoëttes put his fame to use by signing a new petition that the colony's agent, Charles Villar, addressed to the secretary of the treasury, accompanied by a letter dated the twelfth that was more detailed than the petition itself. Both asked for repeal of the principle of solidarity, justifying it with old and new arguments: the difficult beginnings, the initial error in location, the 25 percent of unworkable lands, the fall in price per acre, the need for communications in a region that had scarcely been explored, the constant trouble with flooding, illness, the deaths of twenty-three co-associates, an investment estimated at $160,000, the negligence or unwillingness of some members, and the determination of many Americans to finish off the colony. Despite all this, 81 society members and their families (that is, 327 persons—many of whom had only just enough to get by) wished to continue their efforts on the Tombigbee that were already productive: eleven hundred acres had been worked by them personally, and fifteen hundred by their renters. Although this number would have been tripled if not for the losses during maritime transport or arrival out of season, ten thousand young vine plants were in full growth. How could the French colonists better assure the American government of their intention to respect their contract? But, wrote Charles Villar, "the mere idea of solidarity slackens our energy, and we dont [*sic*] presume to be blameable for respectfully begging to answer each of us for our own facts. Moreover, the repeal of that solidarity, in relieving our anxiety, would be a sufficient proof for the squatters, who are daily trespassing on our land, that the government do not consider us as mere intruders, as they believe."[34]

Crawford reacted positively to this approach, influenced less by Desnoëttes (whom he met in Washington in March 1822 when the general was more concerned with his return to France than with the fate of the colony he had just left) than by an urgent situation that had not existed at the time of the first petition. The first article of the contract entered into on January 8, 1819, by his department and the French Emigrant Association, stipulated that within three years of that date, that is, by January 8, 1822, each allotment of the four townships was to be occupied, either by the members themselves or by someone else acting for them.[35] On March 18, 1822, ten weeks after the three-year deadline had expired, Crawford informed the Senate that a considerable number of members had already lost their right because no settlement had been made on their respective allotments by the preceding January 8. Now, considering the law of March 3, 1817, which tied the granting of individual property titles to respect of the contract and complete payment for the four townships, it was becoming clear that, because of the loss of interest of some of the members, those who had fulfilled the contract's conditions— settlement and vine cultivation on their allotment—would not be able to acquire title to the land they had improved. Such an eventuality would wrong, and even ruin, those emigrants who had made such great efforts, "not only to avail themselves of the benevolent intentions of the government, but to make the only return for such an act of munificence, which could be received by the donors."[36]

It was only after bitter debate that Congress followed Crawford's recommendation and, on April 26, 1822, voted to repeal the solidarity clause. Thus the society members who had respected the contract's clauses could individually obtain title to their allotments, and their heirs as well. But proof was still needed that within seven years of January 8, 1819 (so before January 8, 1826), at least one acre of vines had been planted in each quarter section—that is, a total of 576 acres—and five hundred olive trees in the colony as a whole. That is why President Adams ordered his treasury secretary to send a special agent to the Tombigbee to check on the progress made. As we have seen, Rush entrusted this inspection mission to William L. Adams.

THE ADAMS REPORT

On November 1, 1826, in Tuscaloosa, Adams received his instructions from the secretary, along with a list of the grantees, their allotment numbers, and a map of the townships. He was expected to do a complete, objective assessment, but illness obliged him to defer the start of the project and request the aid of Frédéric Ravesies, the association's agent, whose position as judge and party to the contract could not guarantee impartiality. Adams arrived in Aigleville in late November and immediately got to work, but illness struck again and kept him in bed for six weeks. So it was not until February 1827 that he could finally report to the secretary of the treasury on a mission whose limitations he was forced to admit. Assuring Rush that his report contained "no errors that will materially injure the rights of either of the contracting parties," he implied that it was inaccurate and insufficient, being not "so full to every point as may be wished for."[37] Paying homage to Ravesies in his final report "for the constant and friendly aid which he rendered [him] in this work," he acknowledged a collaboration that was too close to be trustworthy.[38]

That is how Richard Rush understood the matter when he read the report and Ravesies' enclosed letter, of which he acknowledged receipt on May 17, 1827.[39] While he said that on the whole he was satisfied with the information gathered and transmitted, he insisted on more details for each allotment: how many acres had been cleared and planted, how many acres had been planted in vines, and how many olive trees had been planted. Rush waited months for answers to his questions, because the persons to whom he wrote needed coaxing. Not Adams, his direct correspondent, who had meanwhile died, but Charles D. Conner and Frédéric Ravesies, the secretary and the agent of the association, respectively, who took their time with his letters that reached Aigleville in the course of the summer. Conner failed to answer because of travels to North Carolina; as for Ravesies, he had already left for Philadelphia, where, on July 30, the treasury secretary sent him a duplicate of his original letter of June 19. Returning in late November—that is, about four months after receiving the letter in Philadelphia—Ravesies first asked the executors of Adams's estate what instructions he had received from Rush on May 17, and then set about following them, sending his report to Washington on January 16, 1828; Rush received it a month later.[40] Was this nine-month delay entirely due to the above-mentioned circumstances, or were the settlers reluctant to correct the impressionism of the Adams report with the arithmetical realism of the additional one? Whatever the case may be, Rush chose the first option. When he transmitted the Adams report to the president of the Senate on December 21, 1827, he attributed the delay to its author's illness, and when he passed on Ravesies' report on February 14, 1828, he implicitly attributed this additional delay to Adams's death, which had obliged him to turn to the association's agent for help.

Published by order of the Senate on December 24, 1827, the first report—twenty-five pages plus directions for use from Adams and a statement from Ravesies—gave an inventory of the condition of the colony. Each of the 346 principal allotments and the 40 reserve allotments was more or less fully described, according to the amount of information obtained and the situation observed. Was there someone living on the allotment? Was it the original grantee himself, his agent, his renter, a squatter, or some new French or American owner? Had there been a harvest? Had vines been planted? An optimistic analysis of the report shows that 220 allotments had someone living on them and 160 had been planted in vines, a result by no means insignificant, given the difficulties encountered. But on the downside, the same report also shows that 156 large and small allotments, 40 percent of the total, had not been worked at all, nor therefore produced anything. (On fourteen other allotments, no one had settled, but there had been planting, most often of vines, to respect the clauses of the contract.) Moreover, the report clearly establishes the fact that few of the original society members, even if they came, settled on their allotments and improved them themselves; at best, they had someone do it for them. Finally, it was obvious that Americans were gaining a foothold: Thomas Newman, a Philadelphia merchant, alone owned several thousand acres.

Realizing that this report would cause the treasury secretary to have doubts concerning the colony's real progress, Ravesies had thought it necessary to draw up a statement[41] deploring the poor results and listing the natural and human reasons for this, reasons that Lajonie before him had detailed to his family in letter after letter: the fact that the region was a hostile wilderness when they arrived, the unequal quality of allotments received in the drawing, the lack of drinking water, the continuous invasion of squatters, the lack of farmers, the initial necessity to plant food crops rather than vines and olive trees, the difficulties of travel, the prohibitive price of supplies, physical and financial exhaustion, discouragement, illness, and death. The picture was true to life, but Ravesies had deliberately darkened it and minimized the deficiencies of the French. Lajonie and a few others had shown that with hard work and the will to succeed, obstacles could be overcome and the region's resources be put to good account.

In the final analysis, the results of the Adams report, as explained by Ravesies, could not possibly satisfy Rush: on one hand, the fact that the columns were filled in, allotment by allotment, gave the impression of satisfactory occupancy of the colony; on the other, the statement of the difficulties encountered raised the fear that all was not going well. Because he did not know what to think, Rush requested an additional report that would give figures and thus be more rigorous. Back from Philadelphia, Ravesies, as said above, got hold of the original instructions for Adams and, using the individual reports that had until then been submitted to the executive committee of the emigrant association, for the first time gave a precise, if not absolutely accurate, idea of the amount of land under cultivation in the colony: in January 1827, that is, after nine and a half years of effort, 7,651 acres had been cleared, including 302 acres planted in vines and 484 olive trees.[42] How should these data be interpreted?

The number of acres cleared covered 8.2 percent of the total area of 92,000 acres, but 11.8 percent of the useful agricultural surface, which was about 65,000 acres, after subtraction of lands unfit for cultivation, roads, streams, urban expropriation, and buildings. The result, while far from exceptional, was improvable, a prospect less certain when it came to vines and olive trees. The proportion of acres in vineyards was less than 4 percent of the cleared area, and their distribution over 160 allotments made for an average of fewer than two acres per allotment, with most having only one. As for the olive trees, their number did not reach the threshold of the five hundred legally required; two hundred of them were concentrated in one place and the rest scattered over some thirty allotments. It was obvious that the settlers were planting vines and olive trees only to abide by their contract, the passport they needed to receive their individual property titles from the government. Even if they did only the minimum, can they really be accused of deception? In the difficult situation in which they found themselves they had quickly realized that vines could not succeed in Alabama, and olive trees even less. They could not produce enough quality fruit to give impetus to a domestic industry, create a market of consumers, and bring wealth to its promoters. The cases of Lajonie and George show how planters had diversified their activities and basically pinned their hopes of prosperity on sure values like cotton, to the detriment of other products as delicate as vines and olive trees. There is, however, a single counterexample: the attempt of a few settlers to go into silkworm breeding, like Pastor Gibert's half a century earlier on his Silk Hope plantation in North Carolina.

After years of apathy, beginning in 1825, Congress and state legislatures passed several resolutions and laws encouraging its development and the planting of mulberry trees, the only food of the *Bombyx mori*, the silkworm. Specialized periodicals flourished, promising it a great future, thanks particularly to the recent importation of a kind of mulberry, the *Morus multicaulus*, considered to have many desirable qualities due to its Asian origin.[43] In the 1830s a huge craze prompted American farmers and speculators to want to cover the countryside with mulberry trees and set up cocoon nurseries. Jean Marie Chapron was caught up in the fad. Living on his Alabama estate in the fall, he spent the rest of the year with his family in Philadelphia. It seems to have been there, in the epicenter of this interest in silk, that he became fascinated by this sector of activity, which had become a veritable mania in the Northeast, he wrote his agent, John McRae, on May 25, 1839.[44] Having purchased fourteen hundred acres in the French colony, McRae was living on the Black Warrior in Arcola—founded by his father-in-law Ravesies—when Chapron suggested they add silk to cotton and develop the silk industry. McRae already had mulberry trees and silkworms under way. If he hired skilled workers and succeeded in making silk, Chapron proposed joining him along with Parson

Hatch, another planter of a great number of mulberry trees, who was returning to Alabama after stocking up at Philip Syng Physick's nurseries in Germantown, among the country's largest. The project came to nothing, and Chapron did not get heavily into sericulture, except through James Martin, the manager and overseer of the blacks on his plantation, probably located less than a mile from present-day Gallion, Marengo County.[45] Assuring him that silk would be as successful in America as in Europe,[46] he sent him mulberry cuttings and *Bombyx mori* eggs, got him a subscription to the *Silk Grower*,[47] and plied him with advice, but Martin probably did not devote much of his time to this project. The general trend in favor of silk slowed very quickly due to a series of natural and financial problems. On August 7, 1841, Chapron mentioned sericulture for the last time with these conclusive lines: "I dont [*sic*] know whether it is too late for worms. As you do not say any thing about them I flatter myself that their reign is over."[48] Cotton had never been displaced, but its position as the dominant cash crop was reaffirmed.

THE COLONISTS AFTER THE COLONY

Silk, wine, or olives—we have noted several times that none of these potentially lucrative crops enriched or permanently kept a single colonist on the French grant, and this, as we have seen, despite the efforts of a few valiant pioneers. Three principal options now presented themselves to these casualties of experimental agriculture: return home if political or economic conditions permitted, depending on whether they had left Europe by edict or choice; stay where they were but move into a new activity, either within the colony itself or in the broader framework of the state of Alabama; or start over elsewhere in the United States, preferably in the South, but not exclusively there. The final chapter will deal with the option of returning to France, so this one will instead accompany a few of those society members who continued their American venture—that is, adopted a language and culture totally foreign to them, had children, and passed down their names. These will in most cases be French people brought in by the post-1815 migratory wave rather than Domingan refugees, old immigrants and new citizens whose rebirth in America had begun when they arrived in the 1790s—and who, moreover, have been the subjects of excellent studies.

One thing should be stressed at the outset: the society members who remained and integrated into American life were, with rare exceptions, anonymous civilians or soldiers not tempted by the prospect of a return, and sometimes even illiterate. Among the celebrities in exile, Joseph Bonaparte and Joseph Lakanal did indeed live for many years in the Northeast and on the Gulf of Mexico, respectively, but they eventually returned to die on their native continent, the first in Italy, the second in France.

Former Convention member Garnier de Saintes had drowned in 1817 as he was preparing to go down the Mississippi. His colleague Pénières-Delzors, Demopolis's first mayor, had been more fortunate: he was appointed subagent of Indian affairs in Florida by the federal government in March 1821, but in August he died of yellow fever in St. Augustine, on the Atlantic coast. This same illness had, in September 1820, carried off General Rigau in New Orleans, exhausted at sixty and recovering with difficulty from the Champ d'Asile venture. A brief stay with Lefebvre-Desnoëttes had not encouraged the younger General Lallemand to settle on the French grant; he died in Bordentown in September 1823. After a stay in Louisiana, Charles Villar, the society's second president, is said to have died in 1840—in his seventies, in Texas, and in dire poverty.[49] We close this mortuary review with the two discoverers of the site of White Bluff on which Demopolis was built. Benoît Marguerite Poculot, from Lyon, drowned drunk in the Tombigbee in the spring of 1818 at the age of forty.[50] He did not live to enjoy his 120 acres, but his uncle Villar—said to have owed him quite a bit of money—got $160 for them. Nicolas Simon Parmantier, the Breton secretary-explorer, appears for the last time in 1831 in Pensacola, Florida, where he was active as justice of the peace and interpreter at almost sixty years of age. He is the only society member of any importance about whose death nothing is known.

These were all mature men, or even advanced in age when the French colony was born. An already full life restricted their chance or desire to prolong it in a context that was also difficult for them. The situation was different for younger men, whose lives lay before them.

The Alabama Survivors

Of the Tombigbee colonists who populated the southern United States, very few settled on the territory of the four townships or in its immediate neighborhood. Among those who did stay were Domingan refugees Jean Marie Chapron, Jules L'Amitié Martinière (whose son, Julius A. Martiniere, a Yale Law School graduate, was elected mayor of Demopolis in 1873),[51] and the Stollenwerck brothers, as well as the following émigrés who will serve as case studies: Mathieu Labrousse, Jean and Pierre Mangon, François Martin, and François Louis Constantin.

Unable to find employment in his craft of dyer in New Orleans, Mathieu Labrousse followed Lajonie to Demopolis, where, rather than farm his reserve allotment (which the Adams report noted as unimproved in 1827), he worked for Lajonie for three years. After this, in late October 1820, eager to set up his own business at the age of twenty, he left to run a little refreshment stand and sell bread in Demopolis.[52] Six years later he was still there, or at least nearby, since he is on the list of buyers at the Bierne sale. He married, and "by the Indian rights of his latest wife" (which implies her origin and that she was not the first) he received "a Bluff" on the site of old Fort Tombeckbé, now called Jones Bluff.[53] The place was particularly well situated, right near the new town of Livingston and a large road that, once finished, would go to the state of Mississippi. As the road was to cross the Tombigbee River at this point (a bridge is now there), Jean Mangon saw the interest of purchasing land on the other bank, in a region in full development where buildings were going up fast and "crowds are rushing in."[54] In 1832 Mangon had a small business there that he feared he would not manage to keep, for he sold only on credit: in six months he had made a thousand dollars in sales but had taken in only one hundred. In 1833 he and Labrousse seem to have decided to run a ferry that they hoped would "in some time" bring in something like one hundred dollars per year.[55] In 1837, despairing of finding "a good woman" to marry in the region, Mangon temporarily returned home to France. As for Mathew Labrousse, he found himself forsaken at the age of forty: almost all his compatriots had died or left.

In 1841, however, Labrousse married (a third marriage?) Mathilda Martin Latham, a young American widow of French descent, the daughter of society member François Martin from Marseille, naturalized as Francis Martin, who had gone into business as a shopkeeper in Philadelphia, where Mathilda was born in 1816.[56] While his brother had chosen to go into business in Louisville—probably after the death, in 1822, of their father, from whom they inherited the 120-acre allotment 82—Francis Martin went to live on his own allotment, but found that it was frequently flooded. So he took a lease on allotment 44 (320 acres), which the original grantee, Emile Voerster, a German infantry major, had sold to Lefebvre-Desnoëttes, who had planted it in vines before leaving the colony. In 1825 he bought a small allotment from Johannes Bütche for fifty-two dollars, and in 1834 he purchased Antoine Latapie's 160-acre allotment 25 for twice that amount. Moreover, he was probably the Martin, "a Frenchman and a good carpenter," who together with Pierre Mangon built a steam mill on the river, expected to be very profitable, starting in July 1832, most likely the date it was put into operation.[57] By November 1833 "the steam mill [was] attracting many people who always bring a bit of money," wrote Mangon.[58] It is not know how his business prospered, but only how his life ended. Shortly after their daughter remarried, his wife left him to go live with the Labrousse couple on Jones Bluff.[59] On November 5 or 6, 1842, at nearly seventy years of age, Francis Martin died of indigestion for which he had refused the help of doctors.[60] His oldest son, Marius Martin, and his son-in-law Mathieu Labrousse were present at his death. He was buried in Demopolis in the loop formed by the Black Warrior—the "Bend," as it was then called—where in 1818 and 1819 the French had cleared their first farms side by side. Labrousse saw to administering the estate, consisting chiefly of land, to the great displeasure of his mother-in-law, who claimed he did not defend her interests.[61]

Labrousse appears for the last time in Greene County in 1850, listed as farmer on the census rolls, along with his wife, Mathilda, and their six-year-old son, Napoleon B. LeBruce—an interesting example of onomastic syncretism, combining the Americanized family name with the flagship first name of nineteenth-century French history. In 1861 Napoleon was living at Tishabee (Bragg's Bluff), west of Forkland. Was this also his parents' new address? At seventeen this grandson of a Marseille cooper on his mother's side and a Gironde shoemaker on his father's enlisted as a private in the Confederate army, Eighth Alabama Cavalry Regiment, Company D, known as "Ball-Hatch."[62]

Forkland was on the road from Demopolis to Eutaw; it was in this town that François Louis Constantin lived for twenty-five years. The natural son of a war hero, a noncommissioned officer who acknowledged him when he married his mother in Lorient in 1810, he arrived in New Orleans as an adolescent with his maternal uncle, Jean Joseph Baron. But according to family tradition, probably embellished, the uncle died very early in Alabama on his way to the French colony, and his nephew was deprived of the twenty thousand dollars he was to inherit. So in association with a man named Deschoules he acquired a forty-acre reserve allotment that remained unimproved, for, still according to family tradition, he accepted General Lefebvre-Desnoëttes' offer to put him up, perhaps out of consideration for his father's service record. He survived his host's departure, since in 1826 he was one of the buyers at the Bierne sale, and the following year, his name changed to Constantine, he married a very young woman from North Carolina, Clementine Cornelia Hamlett. Preceding his compatriot Labrousse, he paid homage to the Empire by naming the last three of their eleven children Joséphine, Hortense, and Murat Ney. When his first child, Dominique Francis, was born in 1829, the couple was living in Erie, at the time the seat of Greene County and a rapidly expanding river port. What a distance he traveled in a decade: likely starting out as a farmhand, then a farmer and merchant, he had become the highly respected Dr. Francis L. Constantine.

It is said that his father would have liked Constantine to go to the École polytechnique, perhaps because of his native intelligence rather than his school performance: a letter to his friend Lajonie reveals a spelling nearly as phonetic as Jean Mangon's.[63] Was it through contact with Lajonie, who took such pride in his medical knowledge, that Constantine developed his vocation? He is said to have practiced medicine successfully, first for his family, following the "Thomsonian system," which was at the time claimed to be effective against malaria. He then produced anti-bilious pills and sold them in Alabama, Louisiana, and finally Mississippi, where he settled in 1834. There, in Noxubee County, he acquired nearly fifteen hundred acres of land and ran a mule-driven "peck mill" (sawmill) near Shuqualak. Business prospered. By 1860 he owned forty-two slaves; his eldest son had five.[64] But while keeping his Mississippi holdings, Constantine had returned to Alabama in the late 1840s, settling in Eutaw, where he bought a lot on which he had a house built; it survived until 1973.[65] The Civil War broke out. At the age of sixty he was too old to fight, unlike his three sons who committed to the southern cause: William Lewis Constantine in the Twelfth Mississippi Cavalry Regiment and Dominique Francis and Francis Louis Constantine both in the Forty-first Mississippi Infantry Regiment. The latter was wounded in the Battle of Stone River (Tennessee) and taken prisoner.[66]

When the war was lost, Constantine left Eutaw for the northern part of the state. Near Blount Springs he purchased as a summer home a property he named Lorient after his native city. In 1873 he settled in the growing nearby town of Birmingham, where he made large real estate investments. In 1877 he and his wife celebrated their golden wedding anniversary there, with their seven living children, twenty-four grandchildren, and one great-grandson. Moving to Atlanta soon afterward, Constantine died there on May 21, 1891, at the age of eighty-eight. Reverend Patton of the Methodist church, which Constantine had joined in 1838, read the eulogy of a "remarkable man" who had spent most of his life in the United States. His widow died in Atlanta four years later. Their many descendants received and handed down like relics the objects testifying to his origins.

Quartermaster Sergeant Dominique Constantin, the French grandfather, having distinguished himself by his bravery at Arcola, Gouvernolo, and Hohenlinden, was rewarded by the First Consul with a saber of honor and then the Cross of the Legion of Honor on September 16, 1802. After their recipient's death, the saber and cross passed to his sister Terre Sainte Amélie Nathalie Constantin, then to her daughter in Plouay (Morbihan), who left them to Francis Louis Constantine, his American grandson. In 2010 they were in the possession of Henry Phillips Constantine of Clearwater, Florida, one of many descendants.

Constantine had lived in Alabama for half a century, most of this time on or near the French townships, but died in Georgia. Pierre Mangon, alias Peter Mangone, was buried near the Presbyterian church in Jefferson, Marengo County, in the family cemetery, which has since been destroyed by a farmer who is said to have thrown the tombstones into a well and planted cotton in the area he had thus cleared.[67] Appearing for the last time in the 1880 census in Spring Hill, Marengo County, at the age of seventy, Mangon thus appears to have died before 1890. He survived his wife, Jemina Red from Kentucky, whom he had married in Greene County in 1840 and with whom he had six daughters and six sons. John and Franck Mangon, the oldest sons, served the Confederacy in the Forty-third Alabama Infantry Regiment, Company A, in 1862 stationed at Fort Morgan near Gulf Shores.[68] In 1880, two single daughters and three grandchildren were living with Peter Mangone. He now had a farm, after engaging in a variety of activities since arriving in Mobile in 1829 with his older brother Jean Mangon. Unable to carry out his plan to set up as a baker, in 1832 he became a co-owner with Francis Martin of a steam mill they had built on the Black Warrior. The following year, according to his older brother, Pierre Mangon was in business for himself with a nephew. He was happy in this country and was earning money. In 1850 Peter Mangone was working as a wheelwright near Jefferson, in Township 17, Range 2 East, just south of Township 18, where the French colony had its beginnings. By 1860 his personal estate came to six thousand dollars. His descendants kept the name "Mangone," used before Peter's death.[69] There are still people with this name in Memphis, Tennessee.

Mangone is the last example of French colonists who lived and put down roots in the Alabama counties that partially covered the French colony. Many more preferred to withdraw from an unrewarding situation and begin anew in cities where economic activity was infinitely more promising, particularly in the two large ports on the Gulf of Mexico, Mobile and New Orleans.

Yellow fever had decimated the French in Mobile during the early years of their immigration, and the danger continued to ravage the population with the same deadly regularity. This was not, however, reason enough to discourage the Tombigbee colonists. They were far less isolated in this rapidly growing seaport, home to a French community represented by a consular agent answering to the consul in New Orleans. From 1825 to 1836 this office was held by Adolphe Batré and, briefly in 1829, his older brother Charles. Under the corporate name of C. and A. Batre and Company, their firm of cotton brokers was located on Planters Alley in 1837.[70]

The Batré brothers had soon left Demopolis, where Charles's presence is confirmed in 1818 and Adolphe's in 1820. They acquired many allotments in the colony, either from the outset or by purchasing them progressively from other society members. Interested in the project, they planted vines. Adolphe in particular worked the land he had acquired near Lajonie's Sainte-Hélène plantation. But this was only a very small part of the Batré brothers' business, and their economic future in Alabama did not depend on it. They rapidly developed a highly complex network of contacts with many society members, and marriage was for both of them a means to becoming established. On December 25, 1824, Charles married Adèle Macré, the widow of a society member who had died in Philadelphia in 1817 after the allotments had been shared out. The mother of two children from this first marriage, she bore six to Charles Batré, who died in 1838 in Havana at the age of forty-four, assured of descendants. His body was brought back to Mobile and buried in Magnolia Cemetery.

Adolphe Batré survived Charles by half a century in Mobile, where he lived out his American life and was married twice: in 1823 to Sylvanie Chaudron, the daughter of Jean Simon Chaudron, and in 1828 to his sister-in-law by marriage, Jane Tombarelle, widow of Paul Émile Chaudron, Jean Simon's late son. Several of the children born of this second marriage died very young,[71] but the others also assured Adolphe of descendants.

Mobile's old French cemetery, the Old Church Street Graveyard, contains the burial sites of the Batrés and Chaudrons, among them that of patriarch Jean Simon Chaudron.[72] Already elderly when he arrived in the French colony, Chaudron had soon tired of the rustic country life he had praised so highly in the columns of the *Abeille Américaine.* According to the Adams report, he took possession of his 480-acre allotment 276 and planted vines, but he did not prolong the venture. A buyer named Ira Carlton acquired his allotment, and Chaudron finished his long life in Mobile, where he died on October 28, 1846, his eighty-eighth birthday. The city did not forget him. The next day the local newspaper paid him homage in verse;[73] in 1996, on the 150th anniversary of the death of the man who had proclaimed "I am a Mobilian," the Mobile Museum of Art exhibited his portrait painted around 1806 by Rembrandt Peale and kept since then by Chaudron's descendants.[74]

Five rows from the burial plot of Chaudron, his wife (née Stollenwerck), and several of their children is the gravestone of Catherine Victoire George (née Le Grand de Boislandry), another outstanding figure among the Domingan refugees who had come from Philadelphia to Alabama. Preceding Chaudron, she died just before reaching the age of eighty, after living in her home on Spring Hill overlooking Mobile from about 1827 to 1843. Like Chaudron's descendants, those of Victoire George handed down her pastel self-portrait through succeeding generations, from her son Edward (buried beside his mother) and daughter-in-law Mary Potter to Robert Hunter, and then Catherine McPhilipps, his niece (who still had it as of 2010).[75] Other French colonists buried in this cemetery are the Pfister, Meslier, Herpin, Payen, Ravesies, and Cluis families. The last of these will serve as another case study.

Since 1817, Jean Jérôme Cluis' American destiny had been tied to that of his friend Descourt. After working the same allotment together, each of them owning half, they succeeded one another as innkeeper of the Eagle Hotel in Greensboro. In November 1830 Descourt was managing it when the French and Americans, led by Honoré Bayol and Frédéric Ravesies, celebrated the July Revolution and fall of Charles X there.[76] Cluis had been the first to leave the town and had settled in Mobile, where his stay was short. He died on October 15, 1833, at the age of sixty. In France he left a lawful wife and three daughters, who inquired about an inheritance that they were told was not very large and in any case not available, since he had a common-law wife and two natural children in Alabama.[77] The size of the estate is not known, but it enabled the Mobile heirs to survive and remain in the city. Émilie Mézières Cluis died there in 1884, in her nineties, at the home of one of her granddaughters, while her two children, Émilie Éléonore (1820–60) and Frédéric Victor (1815–64), preceded her in death some twenty years earlier. Émilie Éléonore's 1836 marriage to States Gist Deas of Charleston, and Frédéric's 1851 remarriage to Mary Ellen Weir produced a great many descendants who scattered throughout the United States into the twenty-first century. The case of François Louis Constantine is a nice illustration of personal success. That of Jean Jérôme Cluis rather highlights the success and integration of his descendants, as shown over three generations: (1) his son Frédéric Victor Cluis, a wholesale grocer, tobacco merchant, real estate speculator in Mobile, lieutenant in the Twenty-sixth Alabama Infantry Regiment, prisoner of war, demobilized at the rank of captain; (2) his grandson Victor Manning Cluis (1862–1925), an insurance agent and wood and property dealer in Birmingham in the 1880s who then moved to Arkansas, Illinois, and finally Georgia, where he died; (3) and his great-grandson Frederic Victor Cluis (1893–1951), who returned to Mobile after enlisting in Atlanta on June 5, 1917, during World War I. The family's century in the United States had come full circle.

The French Diaspora

Numerous though they were in Mobile, there were even more founding Colonial Society members and their descendants in New Orleans and the Louisiana parishes, from the bayous of the Mississippi estuary to Baton Rouge and Natchez in the north, and even farther upstream to Missouri. They share the same characteristics: integration into the proslavery world and defense of the southern cause in the Civil War. This is understandable in the emigrants arriving after 1815: they found themselves rubbing shoulders with immigrants and Creoles already living there before the French Revolution or after that of Saint-Domingue or after their expulsion from Cuba in 1809. The economic interests of all these groups merged, and they forged family ties. The meeting between Lajonie and chocolate maker Jean Quessart Sr. illustrates this. A Domingan refugee who had come from Cuba to New Orleans, Quessart had acquired a sixty-acre allotment in the Tombigbee colony; never setting foot there, he gave it to his goddaughter Ninon Lajonie. He died childless in his Burgundy Street house on July 14, 1827, at the age of fifty-seven.[78] The following summer, his widow and his brother, who had joined him nine years earlier, died in short succession of yellow fever.[79] The epidemic was responsible for the early deaths of even more society members in New Orleans than in Mobile. Between 1818 and 1822 these included Joseph Truc, José Maria Balbuena de Sotomayor, Claude Joseph Lefeuvre, Antoine Bugey, Antoine Vial, Antoine Rigau, André Pagnerre, Abel Farcy, and César Payen. The journey continued more or less successfully for many others, a few of whom will serve as case studies (conscious of the sometimes overwhelming abundance of data): a dignitary and three imperial officers, a merchant of the Domingan community, and a Bordeaux merchant and his brother.

As mentioned earlier, Chevalier Jean Baptiste Jame, the Bonapartes' right-hand man and manager of the Bank of France, had inexplicably fled from Paris on Christmas Eve 1813. Five years later, on May 16, 1818, neither widowed nor divorced, he married the Creole Françoise Sigur, widow Marion, who was then living in a settlement called Plaquemine, about six leagues from Baton Rouge.[80] The wedding took place in a church in Saint Gabriel, Iberville Parish, with Jame using the pseudonym "James de Bellièvre." The couple settled in Baton Rouge, and Bellièvre acquired land there, worked by the eight farm laborers among his thirteen slaves.[81] He appeared to be well settled and affluent: an authorization of April 1819 shows that he asked an individual to receive on his behalf in New Orleans the sum of $4,358 in bills of exchange drawn in Louisville one year previously on Peter Paul Francis DeGrand, a Boston merchant of French origin.[82] In 1820 the College of Baton Rouge, headed by Reverend M. Martial, moved into Bellièvre's house; annual room and board was three hundred dollars.[83] It is not known what became of this college. In any event, his life in Baton Rouge persuaded him not to leave it for the Tombigbee. In 1825 he had Alexandre Léonard Descourt, a society member still living there, sell his 240-acre allotment to an American, John Cocke,[84] and the following year he regularly placed advertisements in the local *Weekly Messenger* looking to hire servants or craftsmen by the day or month.[85] He is said to then have invested in a steamship enterprise that failed.[86] He died in Baton Rouge on June 19, 1833, at the age of sixty-six, supposedly in "the saddest misfortune."[87] His lawful French wife, Caroline Jame née Boscary de Villeplaine, renounced her rights to his estate but demanded that by duty to herself as his widow and to his children, his true name be officially and legally substituted for that of James de Bellièvre on American legal documents.

The case of Alphonse Louis Delaroderie, said to have been a cavalry captain in Murat's army in Naples, is the opposite of that of Jame, despite the fact that both died in Baton Rouge. When mentioned earlier, he was in Baltimore during the spring of 1817 with Athanase Garnier, the son of former Convention member Garnier de Saintes, whom they were going to meet in New Albany, Indiana, on the banks of the Ohio, an indispensable stop on the way to the Tombigbee. At the elder Garnier's, Delaroderie made the acquaintance of the family living nearby: the mother, Angélique Aimée Thibaudeau, widow Audibert, and her four children, among whom Thérèse Aimée, the eldest, had just turned eighteen.[88] All belonging as they did to the Colonial Society, did they set out on the Ohio together? The Garniers probably drowned before reaching the Mississippi; the Audiberts and Delaroderie joined forces, and rather than venture onto the river below New Madrid, they preferred the seismic risks of the city that an earthquake had destroyed some years earlier, as Lajonie had noted when he passed through.

There, in late 1817, Delaroderie, thirty-three, married Thérèse Aimée Audibert, fifteen years younger than he. They had at least five children, four of whom survived to adulthood.[89] Alfred Alphonse must have been the first born on the land that his father had purchased for a few dollars several miles south of town,[90] near Point Pleasant, a port that a Frenchman from Quebec, François Lesieur, had just founded on the west bank of the river, building a hotel and store.[91] In November 1820, Delaroderie bought a neighboring piece of land with a house, into which he moved. Obviously, at this time he no longer intended—if he ever had—to leave the place and go down to Alabama: five months earlier, he had sold his 240-acre Tombigbee allotment to Joseph Barbaroux.[92] In 1821 he confirmed his local settlement in purchasing five lots in New Madrid itself, where he had a house built,[93] but above all in becoming the long-term deputy clerk of the town's circuit court and working to make it the county seat in the newly created state of Missouri. Said to have descended from old Périgord nobility, Delaroderie was an educated man who played the pianoforte and spoke and wrote English perfectly.[94] This facilitated his integration into an area that was mostly Anglophone but also reinforced his ties with a large Francophone group, which often called on him, particularly to administer wills.[95]

As his business had over the years required journeys to Louisiana, in his mid-sixties Delaroderie thought of moving permanently to Baton Rouge and put his properties up for sale. Three years later, in the summer of 1850, having found buyers for most of them, he went down the river to Baton Rouge with his family and twelve slaves; they are listed there on the census in September. On October 10, he and his oldest son, Alfred, signed a nine-year contract creating the firm Alphonse Delaroderie and Son: to a first steam sawmill that they had apparently been operating for a long time in New Madrid, they added a second in Baton Rouge. The business prospered. When the contract expired, the associates signed another for the next nine years.[96] By 1860 the Delaroderie family had changed: two daughters had died, and in 1851, in Baton Rouge, the youngest son, Rodolphe Gabriel, had married Henriette GrandPre, the daughter of a former Louisiana governor. At their home on St. Louis Street, the parents lived with their oldest son, Alfred, now a widower, and his sons, Alphonse Joseph and Timoléon. With fourteen slaves, real estate valued at more than thirteen thousand dollars, and a personal estate of ten thousand dollars, the household was living comfortably. Then the Civil War broke out.

It is not known whether the Delaroderies' possessions and property suffered when Baton Rouge was occupied by Union forces and partly destroyed. It is known, however, that Alfred and his son Alphonse Joseph fought for the South: the former, in the Louisiana militia, was a prisoner paroled at Vicksburg in 1863; the latter served in an infantry company.

On October 25, 1864, six months before Lee's surrender, Alphonse Delaroderie, the grandfather, died in Baton Rouge at the age of eighty and was buried in the city's Catholic St. Joseph Cemetery. His life had run from the French Revolution to the American Civil War. His widow continued to live with her oldest son, her grandson Timoléon, and seven black servants, probably former slaves who had stayed with them; after her death in 1876 she was buried beside her husband. In the half century they had lived in the country that had welcomed them, Alphonse Delaroderie and Thérèse Audibert had combined longevity, success, respectability, and posterity, surviving into the twenty-first century, principally through their youngest son, Rodolphe. Widowed in 1871, he was remarried the following year to Susan Robertson Schoenbrodt, widow Mueller. Their male line successively produced four Rodolph Audibert Delaroderies in Baton Rouge from 1873 to 1951. The American Delaroderies bear their name proudly and proclaim their French origin on their website.[97]

A native of Burgundy like Jame, a military man like Delaroderie, Achille Chapotin had preceded them in death. He was buried in New Orleans on January 31, 1831, the day after his death at the age of thirty-six, perhaps in his house on Tchoupilas Street, Faubourg Lacourse. A second lieutenant artillery cadet at odds with the army before going into exile, he had successfully changed direction in New Orleans, practicing law like his father in Auxerre before him. In 1856, a certain Rosalie de Chappotin, living in Paris and bold as brass, said that she was related to him and could thus claim an inheritance worth a million, which, in any case, the inventory after Chapotin's death would not have led anyone to expect.[98] Had he been rich, his fortune would have gone to his widow, Adèle Guillotte (whom he had married in 1821) and their three sons,[99] among them Charles, later a bank officer at the Bank of America, and Henry, a bookkeeper, both listed in the census on the eve of the Civil War as living in Jefferson, on the east bank of the Mississippi. Enlisting in Captain Guy Dreux's Louisiana cavalry unit in Mobile in 1863, prisoners paroled in Meridian, Mississippi, once freed they returned to New Orleans where, in 1869, they helped found the military-inspired Southern Historical Society.[100] The brothers had at least five daughters who carried on their line. Achille died too young.

Chapotin had done better at ensuring his posterity than at enjoying his new American life. Born like him in 1795, but in Lot-et-Garonne, Charles Louis Coquillon, an honor guard during the Empire, lived half a century longer. "Obliged to go to the United States after the Toulouse affair,"[101] he seems to have joined his father and uncle, both Domingan Tombigbee Society members, in 1817 in Au gusta, Georgia, where he applied for American citizenship the following year.[102] It is unlikely that any of them ever saw the 240-acre allotment awaiting them in Alabama, which, moreover, long remained unfarmed. Young Coquillon went to New Orleans, where on June 26, 1820, he married Éléonore Chavenet, a Creole from Saint-Domingue who was related to him and had a daughter from a first marriage, whom Coquillon adopted before having four other children, two in New Orleans in 1823 and 1824 and the others in 1830 and 1832 in Mandeville, St. Tammany Parish, where he had moved in 1826. The 1880 census indicates that fifty years later the family was still there: Louis Coquillon Sr., its head, widowed fourteen years previously,[103] old and retired, was living with his daughter Victoria, fifty, single, and unemployed, and his son Louis Coquillon Jr., forty-eight, a married bookkeeper, plus a laundress and a maid. Coquillon had just recently retired. Until a few years earlier he had been postmaster in Mandeville. He had known better days before the Civil War, which had been disastrous for him: "Before the emancipation of the slaves, before that, I had means, of which I lost 25 thousand piasters [dollars] in black skins, plus at the same time, in my absence, my house took fire and burned up along with everything it contained. I did not save a thing, about 5 thousand piasters, so within just a few days, I found myself without a cent."[104] His loss and the freeing of his slaves must have been heartbreaking: his practice of slavery, etched into his Domingan genes, went back to his arrival in New Orleans.[105] This, at any rate, is how, in 1870, he justified to the New Orleans consul his request for a pension relating to the St. Helena Medal he had received ten years earlier as a veteran of the Napoleonic Wars. In 1882 the Ministry of Foreign Affairs granted him seventy francs in aid. This was during the Third Republic. Coquillon died on December 12 of that year, at the age of eighty-seven, and was buried in the cemetery of his town of Mandeville, where he had lived for more than half a century.[106]

Charles Cavaroc shared with Louis Coquillon the fact of being born in France (in Bordeaux ca. 1788) of a father who had been a colonist in Saint-Domingue and emigrated to the United States.[107] Among other Domingan merchants in economic difficulty in Philadelphia in 1816, he found himself insolvent and had to put what he owned into the hands of Nicolas Parmantier and an American, who were to settle his debts, particularly to Léonard Englebert, a future Tombigbee shareholder. Cavaroc himself became one, probably convinced by Parmantier that he could start over in Alabama. But while he did actually go south, it was without his allotment, which he had already sold in 1818. He went directly to Louisiana.[108]

In 1822, in New Orleans, Cavaroc had a draper's shop on Chartres Street in partnership with a man named François Baulos; then after 1830, under only his own name, a clothing and dry goods store on the Levee.[109] Two years earlier, his marriage to Marie Irma Roy in New Orleans had given him Pierre Charles Cavaroc, who succeeded him in business after his death at midcentury.[110] This son was in the 1860s president of the New Orleans National Banking Association while heading the firm C. Cavaroc and Company, wine merchants and importers, at 111 Exchange Place.[111] When his bank failed in 1873, so did C. Cavaroc and Son,[112] but the firm that succeeded it, P. A. Cavaroc, importer of wines and liquors, particularly known for finding a market for California wines in New Orleans, still existed in 1900 at 220 Carondelet Street.[113] Charles Cavaroc had not experienced the winegrowing venture of the French in Alabama, but half a century later his descendants were helping to promote domestic viticulture.

In New Orleans the name Cavaroc spanned decades. That of the Plantevignes was also long known in Louisiana, rather than in Alabama, where Jean Jacques Justin Plantevigne, also from Bordeaux, did not go to engage in the viticultural activity suggested by his name any more than did his younger brother Antoine Plantevigne. Unlike Cavaroc, however, he kept his 160-acre allotment and, while he did not farm it himself, he entrusted its care to his agent, Louis August Stollenwerck, who was living there and working it in 1834.[114] Moreover, he maintained contact with many society members, starting with the Canonges from Saint-Domingue, one of whom, Jean François, a linguist and judge on the New Orleans Criminal Court, was his brother-in-law.[115] It was Plantevigne who received at his home the shipment of Lefebvre-Desnoëttes' trunks when he was returning to Europe via Philadelphia.[116]

The Plantevigne brothers, in their thirties, integrated into Louisiana society as soon as they arrived, following the classic pattern. They married Creoles, the elder in 1817 in New Orleans, the younger in 1819 in Pointe Coupée, having together at least seven children;[117] set up in business in New Orleans at various addresses;[118] and purchased slaves.[119] Much later, a James Plantevigne enlisted in Pointe Coupée's Confederate artillery.[120] While it is easy to follow the paths of the brothers during the founding decade of their immigration, after this there are only occasional milestones. It is known that in 1826 Jean Jacques Justin Plantevigne was living in Mexico, apparently as a merchant associated with Pierre Laborde, a former French consular agent in Veracruz.[121] Was his return to New Orleans in 1828 temporary, and did he go back to Veracruz? In 1856 the City Directory lists a "J. J. Plantevignes [*sic*]," 187 St. Ann Street, and in 1861, associated with C. Soniat as weighers at 11 St. Louis Street. As for Antoine Plantevigne, he remained in Pointe Coupée, where he died three years after his wife, in 1844, at the age of fifty-five; he was buried in the cemetery of St. Francis Church. His descendants may include James, mentioned above in the Confederate army, and do include two married children, one of whom, Pierre Édouard, listed in the 1860 census as a schoolteacher, transmitted the name Plantevigne to a son born two years earlier in Avoyelles Parish.[122] And what about John Joseph Plantevigne, a Creole born in Pointe Coupee Parish and who, in 1907, was one of the first men of color to be ordained to the Catholic priesthood in the United States?[123]

Last on a list that could go on for a long time are the Fourniers. In their case (unlike the other émigrés') I became acquainted with the descendants before knowing the story of their ancestors: Warren J. Fournier Sr. (whose house in Metairie near New Orleans was destroyed by Hurricane Katrina in 2005) and his son, Warren J. Fournier II, a psychiatrist. We met in June 2001, in Gironde, in the Foy region on Lajonie's land. I did not know at the time that Lajonie had in Alabama been the neighbor, friend, and later correspondent of Alexander Fournier, their earliest Franco-American ancestor, who used the title "Colonel" with his name. He was born on Christmas Day 1796 in Philadelphia, where his parents— Alexandre Fournier, a silversmith originally from Tours, and the daughter of a Le Havre merchant—had chosen to live after fleeing Port-au-Prince.[124] When his father died in 1819, Alexander Fournier went to the French colony in Alabama. In Greene County he worked his plantation of Orange Grove, where his wife, Adèle Bouttes d'Estival, a Domingan Creole, bore him six children. Fournier, a model colonist, engaged in numerous transactions with French and American colonists. The 1850 census lists him as a merchant in Marengo County—in Demopolis he had a store on South Market Street (now Main Street)—with a wife, three children, and five slaves. He died in Demopolis on the last day of 1867 at the age of seventy-one and was buried four days later in the Mobile cemetery. His widow lived another fourteen years. Lorensky Alexander Fournier, the second-to-last of their six children, was Warren J. Fournier Sr.'s great-grandfather.[125]

Thus a great many members of the Tombigbee Society, similar to the cases cited, populated the southern United States, from Florida to Louisiana and on to Texas. Relatively few of them tried their luck in other states, like Jean Mangon, Lajonie's friend. He had twice left Demopolis for France, first in 1829 and again in 1837. Ten years later he returned to Louisiana permanently with Elisabeth Perret, whom he had married in Gironde, and their two children, Berthe and Charles, born in 1843 and 1845. He set up a business of *merrains*, oak staves, on the Mississippi, first in Lafayette, upriver from New Orleans, where he also intended to open a café and grocery store. In 1850 he was in Troy, Tennessee, where he had rented twenty-five acres of good land; in 1855 he was in Sterling, Arkansas, an unhealthy area but one abounding in white oak, good for his business as a dealer in wood; in 1860, in the same place, he is listed as a stave maker. His family grew from two to five children. Life was difficult, he often complained in letters to Lajonie, but he never returned to France, and his children married and had descendants. His oldest son, Charles Mangon, fought for the Union in the Second Ohio Infantry Regiment, while his cousins defended the Confederacy.[126] Then he moved to Nebraska, where he went from arboriculture to politics and married Margaret Anna Habig, the daughter of German émigrés who had arrived there some years earlier. Multinational roots and many moves passed on a taste for adventure to one of his most extraordinary descendants, Bertha Evelyn Mangon Thomson (1889–1950). A medical doctor, she and her husband, sponsored by the United Christian Missionary Society of the Christian Church, went to British India in 1915, where, until 1923, she was in charge of a dispensary and mission hospital in the Mahoba district while her husband, a minister, worked as a missionary.[127] A century earlier, her ancestor, an uneducated farmer who had answered Lajonie's call, was plowing the earth of the French Tombigbee colony and planting vines and olive trees.

Members of the Colonial Society who did not go to Alabama and thus never saw the region in which they owned (or had owned) land did not necessarily remain forever on the country's East Coast. Thus the Bellemères, sons of a hatmaker from the area around Troyes, who had become hairdressers in New York, left in 1850 to practice their trade in San Francisco, where they lived out their lives. Auguste Noël Follin, born in Charleston in 1827, the grandson of Auguste Firmin Follin and Marie Jeanne Mélanie Noël, two society families, also left to seek his fortune in California. There he met the dancer and courtesan Lola Montez, who took him along on a tour to Australia, India, and China from which he never returned.

The Colonial Society did without a doubt not attain its goal of settling French colonists in a natural wilderness to be tamed and covered with vines and olive trees in order to ensure their happiness and rebirth far from the corruption of cities. But in not fulfilling this purpose, these same immigrants, who had the will to succeed in a new existence in America, did not all fail. Many began anew, founded, and built, differently and elsewhere. But it is also true that many returned home.

16

Return to the Homeland

Our generals are getting ready to leave in a year, especially Grouchy, Clauzel if it is possible. Vandamme and Lefebvre will not stay either. However the latter is constantly busy on his large allotment, has many people with him and is having a lot of work done. They say that if his wife agrees, he will not go back to France for a long time.

—Jacques Lajonie, Aigleville, March 2, 1819

Beloved France!
Sweet homeland!
May your sons thus all see you again!
At last here I stand,
And upon your strand,
On bended knee I give thanks to heav'n.
I embrace you, oh dearest land!
God! How an exile must sigh.
But now I'll be able to die.
All hail to my homeland!

—Pierre-Jean de Béranger, "Return to the Homeland," August 1819

Of 352 observable cases, 86 members of the Colonial Society—that is, one-fourth of them—left America to return to Europe permanently, in ways determined by the conditions of their departure: motivated by economic considerations or political harassment, those who had left voluntarily could return whenever they wanted; on the other hand, those condemned to death or banished, victims of the proscription edicts, were entirely dependent upon royal goodwill. The return of some other fugitives, such as Colonel Douarche, who was pursued by creditors, or Lajonie, sought for murder, was linked to the decision of a court of justice. Of the many returns, the first ones examined here are those cases of the thirteen exiles who were the most at odds with the king: six of the seven who enjoyed his clemency returned during the Restoration, while Desnoëttes, the seventh, perished in a shipwreck; two others returned during the July Monarchy; four died in the United

States. The chapter will conclude with several more ordinary situations, among them that of Jacques Lajonie.

THE KING'S PARDON

In May 1818, upon his return from a year in America with his father, Alphonse de Grouchy wrote him that he found France completely changed:[1] reason and liberal ideas had replaced terror and partisan feeling. Public opinion wanted liberty and was asking that the banished be recalled. The government was ready to satisfy it, but how could it extricate itself from the position into which it had been placed by the double rigor of the edicts of proscription? A law that would wipe out the lists was out of the question, for this would be hard to justify to the majority of the deputies; but partial and gradual returns could be considered, beginning with those banished by article 2 of the July edict, some of whom had supposedly already been selected. Furthermore, the promise of the allied occupation forces to soon leave the country could probably favor a recall.[2] Although hoped for since early 1817, the first returns were nonetheless slow in coming, as was royal pardon itself, finally granted on an individual basis. But it was indeed public opinion that hastened the decision to put an end to the exiles of many outlawed men, in America and elsewhere.

The Pressure of Opinion

Opponents of the Bourbon regime, whether Bonapartist, liberal, or revolutionary, first seized upon the affair of the Champ d'Asile to affirm their solidarity with the Texas adventurers and repudiate the government that had exiled them. They found expression in the *Minerve française,* a newspaper founded by spokesmen of the Liberal Party[3] that quickly became the principal voice of public opinion, particularly of those favoring a constitution.[4] Benjamin Constant was one of its most noteworthy editors. The others, and their contributors, had had trouble with the authorities because of their imperial past and their association with the *Nain Jaune:* its founder, Cauchois-Lemaire, was in exile in Brussels, and Étienne, editor in chief of the *Journal de l'Empire,* was outlawed in 1816 and expelled from the Académie française, of which Étienne de Jouy, imprisoned for a month for something he wrote, was also a member.[5] Their colleagues in the Académie were also harassed: Aignan, the Emperor's former aide for ceremonies and later secretary-general of the Paris prefecture, and Arnault, minister of public instruction during the Hundred Days and, in 1816, expelled from the Institut to which Bonaparte had had him appointed seventeen years earlier.

"*Alms for Misfortune.*"[6] These men, whom the Restoration harassed, expelled, or banished for their opinions, opened their newspaper to a subject dear to their hearts: exile. In February 1818 the *Minerve* published two ballads on this theme: one, "L'exilé," was by opposition songwriter Béranger. Singing the praises of the Empire's outlawed men, he lent them these words in a new ode to the Champ d'Asile (August 1818): "We are Frenchmen; Take pity on our glory." Poet Béraud answered in echo: "Noble remains of the field of honor, Fertilize the Field of Refuge [Champ d'Asile]!"[7]

This poetic-patriotic celebration was the opening salvo of a fervent national humanitarian campaign in favor of the colony that even survived its failure. The subject of reports and written testimonies, put into verse, music, and pictures, all tinged with romanticism, the colony became an extraordinary fad. The *Minerve* praised the merits of the Lallemands and "several intelligent officers" who had chosen the Texan wilderness to shelter their comrades in misfortune, supposedly deprived by speculation of the lands Congress had granted them: seven-eighths of the settlers and landowners on the Tombigbee were thus said to be Americans, while they were supposed to be exclusively French.[8] The extensive treatment given the elder Lallemand's responsibility in previous chapters will not be repeated here. The disaster was not public knowledge; he was at this time the unfortunate and virtuous hero whose proclamation of the Champ d'Asile was published in the *Minerve,* which encouraged its readers to join him.[9] The undertaking was noble and generous; in the midst of forests and privations, the French were clearing the land to endow it with the benefits of civilization. Free men were encouraged to come to their aid.

Former prefect Félix Desportes, himself exiled in the German states, had the idea of opening a subscription for them and asked the writers of the *Minerve* to publicize it.[10] He set an example with the sum of three hundred francs, followed, between August 1818 and May 1819, by hundreds and then thousands of compassionate people of all ages, social classes, and means. The newspaper grouped the donors by city, published their names, occupations, and the amount of their offerings, whose total came to ninety-five thousand francs.[11] On the lists, café owners were side-by-side with attorneys, unknown folk with celebrities—Lieutenant General Gérard, deputies Dupont de l'Eure and Caumartin, the lawyer Manuel, Talma and Mademoiselle Mars, members of the Théâtre-Français. . . . Some said they had no connection with the refugees, while others were related to them: Jean Mathieu Bujac of Bordeaux; Charles Lefebvre-Desnoëttes and stockbroker J.-B. Laffitte of Paris, respectively the father and brother-in-law of the exiled general; Eugénie Rigau, daughter of the Champ d'Asile's second-in-command. The donations were both individual and collective, from Masonic lodges, school classes, or groups of "good Frenchmen" celebrating the departure of the allies. A rare few said they were devoted to the king or the Charter, or even called themselves Ultras, while the vast majority of others composed a picture of anti-Bourbon France with veterans of the imperial army and the ex-Guard—officers discharged, on inactive duty, half pay, or retired—as the most numerous and visible figures. The subscription was for them a legal way to be noticed and counted, and show the government that their exiled compatriots had the sympathy and support of public opinion unanimous in its condemnation of the reaction of 1815: France itself had become a true Field of Refuge for the French. This published demonstration of strength should not, however, reduce the fund-raising to a simple pretext. When the Texas failure became known, some suggested that the aid, intended for founding the Champ d'Asile, be transferred to the Tombigbee colony. But the objection was raised that since the latter owed its existence to the American government and its prosperity to the experienced men at its head, this colony had less need of money than of unity and perseverance.[12] It was therefore decided to carry through the initial project by adapting it to circumstances, that is, by sharing out among the individuals, according to their rank, situations, or needs, right where they were, in the United States or in France, the funds they would have received collectively.[13]

Petitions. Without being called such, the donor lists published by the *Minerve* constituted unofficial petitions, reinforced during this same period by official petitions, both individual and especially collective, put forward on behalf of other categories of banished men. Early in 1819 a management committee in Paris drew up a model petition and sent it to the large cities of France, from where it was distributed throughout the departments.[14] At least twenty-seven petitions requesting the recall of those banished by the amnesty law[15] were submitted for examination to a committee of the Chamber of Deputies, which wondered: Should they be sent to the government or be put on the assembly's agenda? The latter case might result in a discussion dangerous for the ministers, ill at ease with such a subject. On May 17, 1819, the committee's reporter took the floor to read an indictment of the exiled and banished men and concluded that the king alone could, in his wisdom, decide their fate.[16] This did not avoid a debate between the orators of the opposition—Lafayette, Constant, and Caumartin declaiming that "forgetting the past had become a duty" when this past was no longer a danger to anyone[17]—and the partisans of the Bourbons, who found the petitions offensive and refused pardon to their enemies.[18]

The minister of justice announced the government's position. Bonapartes and regicides were *forever* excluded from a recall, except for reasons of advanced age or infirmity for the latter. As for the individuals "temporarily exiled"—particularly by virtue of article 5 of the amnesty law, which concerned the vast majority of them—the will of the king was to recall those who would openly promise loyalty to him and submission to his government.[19]

The Most Important Recalls

Former minister Regnaud de Saint-Jean d'Angély, living in New York, was the first to receive an authorization to return for health reasons. He claimed to be very ill.[20] Consul d'Espinville did not deny this, pointing out that for some time his words and actions had shown very great "mental derangement."[21] In June 1817 he was hospitalized and his affairs placed under guardianship,[22] before repatriation was considered. While contemporaries confirm his insanity, a letter from Regnaud to Stephen Girard shows perfect good sense on the eve, he said, of being called back to France by "unexpected and urgent" circumstances.[23]

On July 19 he boarded the *Neptune* for Antwerp with a woman and two servants.[24] The king of Prussia permitted him to stay in Aix-la-Chapelle, where he joined his wife, mother-in-law, and son, but he evaded the surveillance he was under, upsetting the French consuls in Antwerp and Amsterdam.[25] From Belgium, where he settled, Regnaud on June 27, 1818, sent a petition to the Chambers in which he complained of the rigors of exile inflicted upon outlawed Frenchmen. On February 10, 1819, the king authorized him to return, due to the state of his health. His son brought him the authorization, and the diplomatic representative in Brussels issued him a passport for Paris. He arrived on March 10 and died the next day.[26] He was fifty-seven years old. On his monument in Père Lachaise cemetery we read that on one and the same day he had seen the end of "his woes, his exile and his life."

Pierre-François Réal, his former colleague on the Council of State, was less anxious to return home and die. He had gone to the United States without his wife, with whom a police report said he was on bad terms; however, on October 22, 1818, she and her daughter Eulalie, the Baroness Lacuée de Saint-Just, asked the king to recall him, minimizing the reasons for his banishment.[27] Their request was accompanied by a petition signed by the usual members of the opposition—Lafayette, Kellerman, Dupont de l'Eure, Caumartin, Keratry, Manuel, Boissy d'Anglas—but it was also followed by attestations from individual royalists to whom Réal had been helpful, such as the Count and the Duke of Polignac, the Marquis of La Guiche, the Marquis of Dreux-Brézé, and the Duke of Céreste.

On May 26, 1819, the king expressed his will to recall only those who would openly promise loyalty to him and submission to his government. Réal and Dirat were among these men. The minister of foreign affairs quickly authorized Hyde de Neuville to issue them the passports they would need to return.[28] On July 25 the ambassador answered that he had informed them but that they did not plan to leave yet.[29]

The certificate restoring his rights, Réal confided to Joseph, brought back all his memories and nostalgia for "the state of splendor, glory and happiness in which France had once found herself" under Bonaparte. The fear of seeing a country deprived of its luster may have contributed to his prolonging his American stay by eight more years, but surely it was less a factor than his financial investments to develop agriculture on his farm in Cape Vincent on the Black River in New York State. Intending to settle in the area, General Henri Lallemand was there on November 13, 1821, when he wrote his wife that Réal's farm was "very nicely situated, at the junction of Lake Ontario and the St. Lawrence River, good land and a fine house."[30]

In 1826 the death of Marguerite Agnès Pérignon, his wife, prompted Réal to return, giving up his agricultural work and occasional forays into mechanics and applied chemistry with Claude Charles Pichon, his companion and former member of the Colonial Society.[31] Leaving in May for New York, his daughter brought them back a year later on the *Edward Quesnel*, arriving in Le Havre on May 29, 1827.[32] Réal likely used this delay to dispose of his property. According to a police report he was said to have lost most of his fortune on "bad speculations" in the United States,[33] perhaps based on a claim made by Réal himself that he had no money and needed to return there so he could pay a creditor[34]—who was fantasizing about 80,000 livres in private means that he supposedly had there. He lived out the remainder of his life in Paris, studying chemistry with Pichon, visiting Cuvier, whom he delighted with "a few seeds from America,"[35] and browsing in bookstores.[36] The regime that had cost him eleven years of exile fell. Réal advised the new police prefect,[37] then withdrew from public affairs, and died on May 6, 1834.[38] He was given military honors at Père Lachaise cemetery, and General Bernard, recalled from the United States by Louis-Philippe, was a pallbearer.[39]

When Hyde de Neuville informed Dirat that a passport awaited him, he replied on July 28, 1819, that he would come to Washington to pick up the document and take advantage of the first ship sailing to France, if he could. But he was said to be as penniless after three and a half years in America, where he had piled up debts, as he had been when leaving France, and his wife's embroidery lessons in Philadelphia had not gotten them back on their feet.[40] Since the government had in 1816 promised him a sum of three thousand francs, he requested it to settle his debts, along with a bit more to pay for his return or else passage on a government ship. Would the ambassador please forward to the throne the plea of an unhappy but faithful subject whose conduct in exile must have been well enough known "to make him worthy of aid and allow him the prompt enjoyment of His Majesty's kindness."[41]

Hyde confirmed that Dirat had indeed conducted himself well and that, whatever error he may have committed in the past, his desire to amend his life and live peacefully in France was sincere. He deserved the king's clemency. The ministries consulted as to "how to fulfill the hopes that may have been given" Dirat when he left France.[42] It is not known who paid. On December 1, in New York, Dirat, his wife, their son,[43] and their maid took passage on the *Bonne-Henriette*, which reached Le Havre on January 6, 1820.[44] He reappeared during the July Monarchy and was said to be a police captain in Bordeaux in 1832, and later in Toulouse.[45] He died around 1860.

Regnaud, Réal, and Dirat were civilians. They had been banished, as was General Vandamme. Generals Grouchy and Clauzel had fled. These were the only three generals to return.

"*Never again to see my Normandy retreat.*"[46] Marshal Grouchy's letters from exile show that he had a very difficult time adjusting to his banishment from France and probably longed more than anyone for the king's pardon. Unlike the Lallemands, Desnoëttes, and Rigau, who knew why they had been condemned and did not question it, Grouchy said he did not understand the fate that had befallen him, and which, until the very end, he had not at all expected.[47] He spoke of severity and injustice. There was thus no question of *settling down* in exile, but rather of doing everything possible to shorten it and come back. Even before he fled in November 1815, he told his wife that he thought only of returning, and told his son that he had no plans whatsoever to remain in America. His American stay only reinforced his determination, and he regularly repeated what motivated it.

He wanted to see his homeland again, and those dear to him whose absence was unbearable, and relaunch his sons' careers; to flee a climate incompatible with his rheumatism and injuries; to recover his fortune and get away from the excessive expenses of life in America; and to forget the customs of its inhabitants, overly influenced by England. Grouchy pleaded: "In the name of Heaven, do not let me die in this dreary country . . . , on this foreign soil." Better to shorten an existence "doomed to too much bitterness, privation and misfortune" for it to be borne long:[48] "My life is a torment worse than death."[49]

Grouchy got his family into battle order—his wife, his sister, the Marquise of Condorcet, his elder daughter, the Marquise of Ormesson, and his faithful Alphonse. They were to contact any important person who might intercede for him with the king: the Duke of Richelieu, his ministers, marshals, the Duke of Angoulême, William I, the Emperor of Austria, ex-Empress Marie-Louise. They should all be told that he was ready to go to Belgium, Bavaria, or Austria rather than continue to vegetate in a "monotonous and pallid" existence.[50] The king should know that if he agreed to commute his sentence to detention in a fortress he would joyfully give himself up;[51] if he permitted him to return, he promised to live on his estate, as submissive to the royal government as he had been to the preceding one, to remain there and end his "stormy career" in "the pleasures of rural life."[52]

His tense relations with the French diplomatic corps, criticizing him for using the title of marshal and for his socialite's life, had finally calmed. Grouchy's conduct, and that of his elder son, at the time of the 1818 explosion at the Dupont de Nemours powder factory in Eleutherian Mills seems to have been a decisive factor. But Grouchy had quickly shown himself in a more favorable light, by distancing himself from his conspiratorial comrades-in-arms with whom he felt he had been unfairly categorized:[53] as mentioned earlier, he was reluctant to attend Henri Lallemand's wedding and exasperated at seeing his name in the papers on that occasion. Grouchy and his sons had indeed subscribed to the Tombigbee venture, but without getting involved in the colony—Victor had gone there to occupy his time—and not at all in the Champ d'Asile. Diplomatic reports state that Grouchy was unaware that it was being planned[54] and had not wanted to be involved in earlier Mexican ventures: "Be that as it may, there is no longer any talk of Grouchy's or Clauzel's coming: here [in New Orleans], in the eyes of their friends, they are nothing but renegades to their party who are now trying to get back into the good graces of the government by attempting to justify themselves. . . . At least that is what they are being accused of by Lefebvre and Lallemand [the younger], who are honest enough to admit that they have well deserved their fate."[55]

With the exception of Clauzel, whom he took in and called his friend, Grouchy's relations with the exiled generals did not improve. Discussing writings in which he justified his conduct at Waterloo, he said he had no reason to go easy on Vandamme, who had been so "detestable" toward him on the field. In the same letter, he regretted that a year earlier he had not closed General Lallemand's mouth as he had wished, that is, sword in hand.[56]

When his hope of returning to France began to take shape, Grouchy made every effort, in the United States, never to get "on hostile terms with [his] government," he wrote in August 1819.[57] Already the previous year, during a dinner where a spy posing as a Bonapartist had tried to sound him out, he claimed he had spoken highly of the Bourbons.[58] The following year, on the occasion of the "deplorable" assassination of the Duke of Berry[59] (February 13, 1820), he showed how much this had grieved him: on April 13 in Philadelphia he attended a funeral service in the duke's memory.[60] But by this time, Grouchy, who, as a final concession, had just substituted the title of count for that of marshal, had fallen back into line.

On September 22, 1819, the Ministry of Foreign Affairs had informed Hyde de Neuville that Grouchy's conduct had persuaded the king to grant him a passport for the Netherlands,[61] but after his long exile the beneficiary was not satisfied, he said, by this act of half-justice. He preferred to remain in the United States and resume negotiations to succeed the French ambassador, who was scheduled to return.[62] At worst, he would vegetate in Philadelphia, where he had "part of life's _comforts_," rather than go settle "in furnished rooms in Brussels."[63]

But on February 14, 1820, Hyde de Neuville, who was still the ambassador, informed him that, by an edict of the preceding November 24, the king had authorized his return. He added: "I need not tell you that this decision of royal kindness is due chiefly to Mgr. the Duke of Angoulême's repeated, kind entreaties."[64] Grouchy knew this, thanks to his elder son, who had paid the duke many visits. Among those who had helped reopen his country's doors to him, he particularly thanked Hyde for the healing action of his paternal hands upon "wounds that were still bleeding after five years."[65] Hyde had sincerely wished for his return, long convinced that "by giving France back one of her best generals, the king would win forever the loyalty of a subject faithful to legitimate power, and along with him a family that should, in all respects, belong to the Bourbon monarchy."[66]

Restored to the titles, rank, and honors that had been his before March 20, 1815, Lieutenant General Grouchy set sail from Philadelphia in May 1820 on the _Océan_, the same ship that had brought his sons three years earlier, and arrived in Le Havre in June. Avoiding a passage through Belgium, Grouchy effected the honorable return that had been the indispensable condition of his leaving the United States. Granted retirement in 1824, he recovered his marshal's rank in 1831 and that of peer of France the following year. He died in Saint-Étienne in 1847, after returning from a trip to Italy.

Each of these exiles—Regnaud, Réal, Dirat, and Grouchy—had wronged the Bourbons, but none had acted in a way that made all hope of a pardon futile, nor had any of them been sentenced. They belonged to the amendable category of exiles from which Joseph Bonaparte and regicides Pénières and Lakanal were permanently excluded. Of the other French celebrities in the United States, six generals remained: five condemned to death and Vandamme, who was not, but whose unpopularity was as good as a sentence. Nonetheless, he was the first to return, and one of the three to see their homeland again, by the edict of December 1, 1819, authorizing the return of the individuals listed in article 2 of the edict of July 24, 1815, who were still abroad. By this time Vandamme was already in Belgium.

After his noisy speeches during the Colonial Society meetings in the fall of

1817, the hotheaded general seems to have been excluded from the plans to divert the Tombigbee enterprise to the benefit of the Champ d'Asile, according to Consul Pétry, who stated that he knew nothing about it.[67] Vandamme went neither to Alabama nor to Texas, but to the country near Philadelphia, where he dropped from public view. As in 1815, when she had implored pardon for her husband, on August 22, 1818, Sophie Vandamme begged that he be recalled and, bemoaning his absence, asked for protection for herself and her son that France had never refused the unfortunate.[68] She pleaded that her husband be allowed to join her in Belgium, where he had a few companions in misfortune, particularly in Ghent, where her family lived. In March 1819 the police minister informed his colleague in the Ministry of Foreign Affairs that since information on Vandamme in the United States was favorable, he was granting the countess's request.[69] Armed with this assurance, she informed her husband that he might benefit from a general recall on the occasion of the king's coronation. On April 24, 1819, the consul in Philadelphia announced to Hyde de Neuville that Vandamme had earlier received a passport for France from the government and that he had just left Philadelphia for New York, from where he was to sail.[70] But it would seem that Vandamme, based only on his wife's letter, anticipated the issuance of the passport that had not yet been sent him.[71] So it was without authorization that he took passage on the *Marcus* on May 8 or 9, 1819;[72] he arrived in Le Havre a month later with a certain Elizabeth Rieffel, with whom the prefect of Seine-Inférieur thought he was intimate.[73] The local authorities retained him until he was issued a passport for Ghent via Cassel on June 14.[74] The edict of December 1, 1819, permitted him to return to France. On December 20, at the Antwerp consulate, he swore an oath of loyalty to the king and obedience to the constitutional Charter and the laws of the kingdom.[75]

Placed on reserve and then retired in 1825, General Vandamme died in Cassel, in his château called La Frégate, on July 15, 1830.[76] There are said to be relics of his American exile: in order to make a living, impoverished French officers in America had decided, to the best of their ability, to manufacture furniture, which Vandamme bought, "paying top price to be able to help out his former brothers in arms while avoiding humiliating them by giving them charity."[77] The family is supposed to have brought it back to France and then kept it in a house they owned in Flanders.

Despite his bad reputation and his participation in the battle of Waterloo, Vandamme had done less to be forgiven of than the other generals. Louis XVIII had refused him audience in 1814 and excluded him from the new military organization: thus he had not betrayed, and even less had he rebelled. This was not the case with Lefebvre-Desnoëttes, nor the Lallemand brothers, instigators of a revolt at La Fère, nor their accomplice Rigau, nor Clauzel, who had driven the Duchess of Angoulême from Bordeaux and kept the tricolor flag flying over the city well after Napoleon's second abdication.

Of the five generals condemned to death, Clauzel was the only one who had not plotted.[78] He could consequently mount a defense against the accusations brought against him at a trial that he wanted reviewed. Already in May 1817 Hyde de Neuville stated that, like Grouchy, Clauzel was openly announcing that he wished to return.[79] It was only in October 1818, when he was in Baltimore with Desnoëttes, that Clauzel asked the consul whether their conduct in the United States, known to the king, would be enough grounds to provide hope for his clemency and whether he would forward to Hyde in Washington their "petitions to the king for pardon."[80] The consul dodged the first question and in answer to the second said that on his own authority he could only register their declaration to give themselves up to be taken to France. The ambassador authorized the consul to receive all the documents the generals might present and send them to him: on October 23, 1818, their petitions were in his hands. The joint steps taken by these two outlawed men who were preparing to come see him troubled him.[81] On one hand, the letter of the official instructions ruled out all contact with those sentenced in absentia; on the other, its spirit softened their severity. But Clauzel fell ill and got him out of his predicament by canceling his journey, pinning his hopes on a petition to the king in which he asked for justice rather than pardon, since he was not guilty.[82]

The king did not see it this way. A year later he had not changed his mind. In September 1819 the minister of foreign affairs informed Hyde that, except for Grouchy, the king did not feel that the time for leniency had arrived for these other Frenchmen exiled in America, Desnoëttes and Clauzel.[83] The latter then decided to go to Belgium in order to be closer to France, whose doors he hoped would quickly be reopened to him. On May 23, 1820, the consul in New York issued him a passport, which, after a good crossing on the *Douglas*, he had stamped a month later in Antwerp.[84] Clauzel seemed eager to submit to the French government and petition the king to review his case, stated the consul,[85] who could decide nothing without the consent of his superiors.[86] Everything went very quickly. Marshals Marmont and Davout intervened in favor of his rehabilitation and the king put an end to the legal proceedings against Clauzel. Back in Paris on July 30, he was reinstated in the army in August, but put on reserve duty with the general staff. With no hope of a command, he returned to his estate of Le Secourieux near Hauterive (Haute-Garonne), where he remained under constant police surveillance. The king's pardon had its limits.[87]

Clauzel had to wait until 1829 to make his official return, as deputy of the Ardennes (a position he retained until his death), and until the establishment of the July Monarchy to resume his career: commander in chief of the army of Africa in August 1830; marshal of France in July 1831; and governor-general of the French possessions in North Africa from 1835 to February 1837, when he was relieved of his position after failing to take Constantine. Based on his experience in America, where he had admired the success of white colonization and sensed the extraordinarily rapid development that European settlement would generate on the new continent, Clauzel "believed in the possible extension of France in Algeria and in the cohabitation of the French and the Arabs, an ambitious dream of liberal, progressive minds at that time," wrote a descendant.[88] Along these same lines, perhaps inspired more by his Alabama stay than by the custom of the ancient Romans, Clauzel suggested establishing colonies of demobilized veterans in strategic Algerian locations.

Pardoned by the Bourbons, put back in the saddle by Louis-Philippe, Clauzel rounded out fifty years of military life by devoting his last efforts to the man to whom he owed his fame and who had included him in his will.[89] In 1840 he chaired the Chamber of Deputies' commission for the transfer of Napoleon's ashes from St. Helena to Paris. He died on April 12, 1842, at Le Secourieux. Since 1858 his body has lain in the cemetery next to the commune of Mirepoix, where he was born.

United in their American exile, Clauzel and Lefebvre-Desnoëttes acted together to seek their return, but each in his own way, according to his personality and his own history. Clauzel, who said he was innocent, asked for redress; Desnoëttes asked for pardon because he acknowledged his wrongs and was not seeking excuses.

Over a period of three years, the relations between the royalist ambassador and the Bonapartist general show the developments that permitted the two camps to draw closer to one another. Hyde de Neuville laid the groundwork for Desnoëttes' first "secret" visit by suggesting to the minister that his example would certainly be quite advantageous in a country where the Lallemands were constantly bustling about "to gain henchmen and drag unfortunate Frenchmen into the abyss."[90] Hyde got ready to receive him without promising him anything, to speak to him frankly, and to see how far he would be willing to serve as an example to enlighten a crowd of Frenchmen whom two or three ambitious men were urging on to error: "You have constantly recommended to me in the name of His Majesty, Your Excellency, to do everything possible to bring these lost children back and divert them from all dangerous and criminal undertakings. Nothing could be more useful in this than to persuade certain men to isolate themselves from all intrigues and to point them out," he concluded.

On December 11, 1818, General Lefebvre-Desnoëttes came to see the ambassador, accompanied by the senator from Louisiana, Eligius Fromentin, a former French priest who had emigrated to America during the Revolution. A long conversation took place in an atmosphere of reciprocal trust. Hyde treated him tactfully, he said, but did not gloss over the difficulties he would have to overcome and reminded him of the seriousness of his offense; as for Desnoëttes, he did not cast blame on others. Of course he had been deceived, but as a general officer he could not plead his youth or refuse the duty to accept his responsibility, alone and without complaint. He was now living at peace. While he may have known of intrigues, like Lakanal's Napoleonic Confederation, he had never become involved in them, and agreed to do all he could to set straight those of his compatriots whom madmen might still seek to lead astray. Desnoëttes also offered some insights that convinced the ambassador that there was no longer anything to fear from the French exiles in the United States. He believed in Desnoëttes' sincerity; indeed, in a certain candor that made him exclaim: "Amnesty for all, if all can bear in their souls the same conviction of the honesty of their repentance."[91]

The quality of the relations between the diplomatic corps and generals like Clauzel and Desnoëttes cast a new light on things. Whereas in 1816 the ambassador and his consuls took offense at the welcome the United States gave the French exiles, two years later their dispatches to the minister were free of indignant comments. On December 28, 1818, Consul Pétry sent to Paris a clipping from a recent Philadelphia newspaper stating that public opinion expected of the king's kindness a general amnesty that would benefit these distinguished generals, the honor of France, and also that of the century.[92] President Monroe and his secretary of war were convinced of it, as were most of the enlightened people who met them in Washington. The article made amnesty a test for the Bourbons: since the allies had left France, it was presumed they had recovered their freedom of action and were therefore able to choose between a harsh or a generous policy.

The government backpedaled by first pardoning the least compromised exiles, and took its time absolving the others. In September 1819, neither Clauzel's return nor Desnoëttes' was on the program. While the former was able to go back to France the following year, this was not so for the latter, who despaired in Demopolis on November 1, 1821:

> Death is a thousand times preferable to my dreary, hopeless existence here. If I am not granted a recall, I shall go present myself to my judges, not in the hope of clearing my name, having no reason to give, but to put an end to all my ills at once. I must admit that I no longer have the courage to endure such a long exile; I feel that I shall find enough of it to die like a Frenchman should die. I shall leave for Europe in four or five months and, one way or another, I shall go to live or die in France. You know what kind of life I have led in this country; except for a journey to Washington, I have not left my fields. In distancing myself from cities, I followed my opinion rather than my inclinations; unfortunate as I was, I had to flee people's pity or curiosity. I wore myself to the bone with hard field work in a burning climate so as to kill my mind's activity. Now I have destroyed my strength and health and remain alone with my thoughts.[93]

Desnoëttes implored the ambassador to transmit to the king the petition enclosed with his letter, of which the ambassador acknowledged receipt on the following December 1: "I have received your letter dated November 1. I hasten to inform you of it. I shall do for you what I would do for my own brother. The noble, touching, and in all respects proper overview that you have sent me no longer permits me to see in you the soldier who strayed; as of now you can be in my opinion only the brave warrior to whom I owe my esteem, the Frenchman to whom I owe my support. You are unhappy; I too have been so. Always count on me as I count on you. I shall write immediately and shall inform you without the least delay of the result of what I undertake."[94]

What the ambassador promised to undertake was part of a process begun three years earlier, whose latest step went back to a letter to the king dated May 3, 1821. Six months had gone by without Hyde's knowing what effect it may have had, but Desnoëttes had received reassuring letters from France: the matter of his return had been raised in the king's Council, and the king himself had promised to deal with it; the Duke of Richelieu had given his wife an audience and recommended a transit via Belgium, from where a return would be negotiable. On February 1, 1822, writing from New Orleans, Desnoëttes, feeling that his deliverance was near, again promised not to become involved in anything in France—"that happened to me only once to my misfortune"[95]—and to abandon all desire of ambition, honor, or fortune for a calm life with his family. Sometime in February he left for the East Coast. On March 19 he was in Washington. The ambassador received him and suggested he delay his departure for Belgium by a few weeks; as he himself was due to return to France soon, he could more easily be useful to him there. Desnoëttes agreed and left Washington on March 24, without having the opportunity to thank Hyde again or get the certificate proving that he had made his statement of repentance.[96] But on March 30 Desnoëttes wrote him from Philadelphia that he had received many letters from France, the most recent, dated December 24, revealing that Richelieu had asked his minister in Holland to receive him. Desnoëttes credited this success to his family's intervention and to all that Hyde had repeatedly said and written about him.[97]

After six years of exile, the knowledge that he could return home increased Desnoëttes' impatience, and he could not bear to remain in the United States for even two more weeks. Hastening his departure, he boarded the American liner *Albion* on April 1, bound for Liverpool, from where he planned to continue to the Continent, probably to Amsterdam.[98] On board were twenty-four crewmen and twenty-nine passengers, among them Marguerite Garnier-Thiollière, the wife of Charles Garnier, a member of the Colonial Society, and their son Charles Joseph, who was almost seven years old.

"A kind of fatality has always seemed to dog him since his arrival in the New World," Hyde de Neuville wrote concerning Lefebvre-Desnoëttes more than a month and a half after his death in the *Albion*'s shipwreck.[99] On April 22, 1822, the ship was caught up in a storm off the southern coast of Ireland, near the headland known as the Old Head of Kinsale, at a place called Garret's Town. It was smashed against the rocks: only eight people, including two passengers, were saved. The bodies of Marguerite Garnier-Thiollière and a few others were found and buried in the Templetrine cemetery, but that of young Garnier, whose death by drowning was confirmed by two surviving sailors, was never found, nor was that of Lefebvre-Desnoëttes.[100] Hyde, who had worked for his return and believed in the sincerity of his redemption, was deeply grieved by his death, as he had been by Moreau's nine years earlier: "Poor wretch! So it is only too true that the day he left camp he plunged into the abyss. . . . God did not permit him to see his homeland again! Let us bow to the decrees of Providence and hope that he who had so nobly made reparation for his error, and who nonetheless had suffered so much, is now enjoying the bliss that no shipwreck can destroy."[101]

Desnoëttes died in dramatic circumstances before reaching the age of fifty, a victim of fate since arriving in America, said Hyde. Actually, fate had dogged him since his capture in Spain in 1808, the prelude to his detention in England, the retreat from Russia, the lost French campaign, the northern conspiracy, the defeat at Waterloo, his death sentence, and exile. Countess Desnoëttes became the guardian of remembrance. In memory of her husband lost at sea, she had a monument in the shape of a sugarloaf erected on the heights of Sainte-Adresse, to serve as a seamark to navigators in sight of the headland of la Hève at the entry of the port of Le Havre.[102] In 1837 she wrote to the minister of war to express her astonishment that her husband's name was not on the Arc de Triomphe "among those of his brothers in arms." She protested against this undoubtedly involuntary oversight: "The honor of being inscribed upon the monument devoted to the military glory of France will not be refused the courageous and devoted general whom the Emperor remembered on his deathbed."[103] Her request was heard: the name LEFÈVRE-DESNte [sic] is inscribed on column 31, west pillar of the Arc de Triomphe de l'Étoile.

Preserving her husband's memory did not prevent the countess from attempting to safeguard her own and her daughter's interests, linked to their American inheritance. The general must have left a fortune in property on the banks of the Tombigbee that had to be converted into hard cash. The countess had the impression that he had purchased and developed hundreds of acres of land, on which he had made costly improvements, and that he had owned slaves. Twenty letters, written between 1833 and 1840, relate to Lefebvre-Desnoëttes' estate in America and therefore to the sale of his "considerable property in the province of Alabama."[104] Their contents offer a perfect example of the complexity of the problem to be solved at a distance of thousands of miles and the impossibility of achieving results. Too many people—French émigrés Nicolas Raoul and John Hurtel, then American attorneys Francis S. Lyon and George N. Stewart—had been involved in the management, administration, and transfer of the lands and the circulation of the money. More than ten years after Desnoëttes' death, and despite the best efforts of the consul in New Orleans and Batré, his agent in Mobile, understanding the ins and outs and seeing what could be saved was a tricky business. They managed to learn that Raoul received the equivalent of four to five thousand francs from the first sales, which were presumably still in his possession in Paris, and that in 1832 the sale of a piece of property had brought in another two thousand dollars, collected by John Hurtel for the general's heirs. As to the income from the sale of the slaves, Hurtel said it was used to pay debts. In 1834 the countess said that her husband "had acquired lands on the Tombecbi [sic], he had put a lot of money into them," but that she "never got a sol out of it."[105] It is true that there were 422 acres of land still unsold, but the considerable expense involved in disposing of them and paying back taxes would take up all the money of the sale. In short, if she wanted to get anything, the countess would do well to hire an attorney, except that she should always remember that out of 1,000 Americans, 999 were rascals, even if they were lawyers, the New Orleans consul pointed out.[106] In 1840, having concluded that it would cost more than it would bring in, the consulate's lawyer decided not to pursue Countess Desnoëttes' complaint. So the general's legacy came not from the Tombigbee but from the provisions of Napoleon's will, which were executed during the Second Empire.[107]

Lefebvre-Desnoëttes would have returned to France after a brief Belgian purgatory, if fate had not been against him. His participation in the northern conspiracy did thus not put him beyond the pale as far as the king was concerned, because in the United States he had made amends and had partially redeemed himself by his exemplary conduct. This was not the case with the Lallemand brothers, whose Texan escapades had made headlines, or, to a lesser extent, with old Rigau, once more led astray into a bad bargain.

THE KING'S VETO

In May 1819, after the petitions had been sent to the Chamber of Deputies, the government had left the door of return ajar to some but closed to others: to officers compromised by their risky venture in America, and to the Bonapartes and regicides with whom the dispute was too serious. Regicides Garnier de Saintes and Pénières and Generals Rigau and the younger Lallemand died in exile. Ex-king Joseph, the elder General Lallemand, and regicide Lakanal returned to France during the July Monarchy.

DEATHS IN EXILE

Garnier de Saintes was the first to die, drowned with his son in the Ohio in the shipwreck of a paddle steamer, probably between Louisville and New Madrid in late 1817.[108]

His former Convention colleague Pénières, a lawyer and Demopolis's first mayor, survived him by four years, during which he was successively a farmer, jurist, carpenter, architect, road contractor, and cook. As none of these occupations had made him rich, a letter from William Lee in Washington persuaded Pénières to leave Alabama. He shook, he said, the dust of his "rags" onto the heads of the Demopolis colonists, sold a half lot in the town, facing Market Street, to Jean François Duchemin for one hundred dollars,[109] and gave Parmantier power of attorney. He left in August 1820 at the earliest. The next month, due to "a thousand accidents," he claimed to be buried "in a foot of snow more than 80 miles from Washington."[110] Lee, with whom he met in the capital, would have liked him to stay there and live a gentleman's life, but Pénières did not want to owe anyone anything, even if it meant "selling onions or sausages."[111] The only person to whom he would have liked to owe something was the king, from whom he had hoped to find a recall letter upon arriving in Washington. While in 1817 Pénières would not have dreamed of returning to the country that had rejected him and whose regime he abhorred, by 1820 homesickness had done its work: "the most beautiful day will be the one on which I shall clasp you in my arms and go from yours to those of my dear family and our good friends," he wrote to his brother.[112] Without a hope of returning, he accepted the position of subagent of Indian affairs in Florida, to which Monroe appointed him on March 31, 1821.[113] He was to travel around this huge, unhealthy region to inventory its natural and human resources. Flattered at being selected and immersed in his dreams of an idyllic society, Pénières had an idealistic view of his job, but the task of prospecting and organizing was too great for one man. After making contact with Seminole and Creek chiefs, whose language he did not know, his health gave way. To recover and to swear an oath of naturalization, he went to St. Augustine on the Atlantic coast.[114] On October 4, 1821, he died of yellow fever.[115] He was buried in the local cemetery. Pénières left nothing in America,[116] but he did leave property in Corrèze that his heirs squabbled over in an interminable series of legal proceedings brought to an end by a judgment of the Court of Limoges in 1849.

The same fever had killed Rigau a year earlier, in New Orleans on September 4, 1820.[117] He was sixty-two years old. He was buried in St. Louis cemetery. The French consul reported that a few days before his death Rigau expressed to him "once again his regret and sorrow for the part he had played in the disastrous event of 1815."[118] In 1818, back from the Champ d'Asile, on bad terms with Lallemand and "entirely separated from him" because of how badly Lallemand had treated him, Rigau had visited the consul to ask for monetary aid and tell him of his desire to come back into favor.[119] To this end, Rigau addressed a petition to the king, ready to submit and serve him, confident in his kindness toward an aging man with many family obligations, whose reputation was intact and whose life was above reproach. Moved to pity by his lot, Hyde de Neuville authorized the consul to help him, with as modest a sum as possible; official compassion had its limits, those of public funds, and Hyde had to explain this liberality to the minister.[120] In January 1820 Narcisse Rigau, living in St. Martinsville, Attakapas County, with Antonia and their father, informed Baroness Rigau that the general was again going to New Orleans to see the consul: he was keeping up his spirits and thought he could be back in France by spring or fall.[121] Rigau died before being recalled.

Their respective allotments of 480 and 160 acres on the Tombigbee, sold for a dollar an acre before they left for Galveston, had made neither the father nor the son wealthy.[122] Worn out after their return from Texas, they had settled in St. Martinsville, where Narcisse had set up a prosperous business employing two men who received board and 150 francs a month. His father, who went back and forth between St. Martinsville and New Orleans, was financially dependent on him. At his death his accounts were chiefly on the debit side, including expenses for his illness and funeral and a large sum of money owed a woman of color, doubtless intimately involved with him.[123] She was "holding hostage" all the general's personal effects—his papers, certificates, warrants, and so forth—which she had resolved not to return until the debt was completely paid. For this, Narcisse had to request a contribution from Mme Rigau in France. The general did not leave his heirs much. In 1829 his widow brought a lawsuit to obtain a pension; in 1854, Antonia Rigau, the daughter of his first wife, was involved in a case relative to her father that was then pending in the Council of State.[124] Was this due to the decision of Charles Kestner, the husband of Eugénie Rigau, Antonia's half sister, to refuse to accept from Napoléon III, guilty of tearing up the constitution, the 100,000 francs Napoleon had left Rigau?[125]

General Henri Lallemand's only daughter did not have such reservations regarding the Emperor's nephew. In the petition she addressed to him, she reminded him of her father's death sentence, exile, death (of illness) on September 14, 1823, in Bordentown,[126] and his close ties to the family of Joseph Bonaparte, who had surrounded him with esteem and affection to his dying moments. During his exile, she continued, the general had made such "great" sacrifices "to help the Frenchmen who had come to the *Champ d'Asile*," that he had left nothing to his widow and daughter. Now, married and the mother of three young children, she found herself in a difficult personal situation that impelled her to request a position for her husband, Count Green de Saint-Marsault,[127] and payment of the arrears due her father as commander of the Legion of Honor.

One must wonder about the dire straits of Caroline Lallemand, Stephen Girard's great-niece, who had married into a very old French family and whose brother-in-law was the prefect of Seine-et-Oise, a man devoted to the Second Empire.[128] But it is true that her father had not profited from his years of exile any more than had his friends, and this was not due to his improbable aid to the victims of the Texan disaster. Unlike Rigau and dozens of other officers, he had not even sold his holdings in the Tombigbee settlement: his original 480-acre allotment plus another of 160 acres, purchased in 1819 from Amédée Martin for $326. This is the land that his widow, Harriet Lallemand, and his daughter had inherited and that, in 1824, Achille George was asked to recover in their name.[129] Three years later, the smaller allotment was not under cultivation; on the larger an agent had planted grapevines to comply with the terms of the law of March 3, 1817. When they were later sold, they brought little profit.

Succeeding those of Desnoëttes and Rigau, Henri Lallemand's death made his older brother the sole survivor of the plotting generals exiled in America. But Charles Lallemand did not take advantage of this to obtain the pardon of the king, whose ears must have shuddered at hearing the names of the places where he had attracted attention between 1815 and 1818: La Fère, Waterloo, Rochefort, Smyrna, Philadelphia, Galveston, and then in Europe, after leaving the United States. Louis-Philippe's accession to the throne was for Lallemand and a few others the final resource that permitted them to return to France after long years of exile.

RESURRECTIONS

Knowing his case to be hopeless, Lallemand never followed up on the slight chance he may have had to return to the king's favor and repeatedly courted bad press, even after the failure of the Champ d'Asile. And yet, back in New Orleans, he gave the impression of wanting to settle down: in late 1818 he joined the lodge Triple bienfaisance, applied for American citizenship,[130] and in 1819 purchased a house in nearby Métairie and a farm at Bayou Saint Jean, with animals and slaves.[131] But money problems ruined his calling as a respectable planter, and many considered him a scoundrel. Louis de Fériet complained that he had not been paid for the farm he had sold him;[132] Louis Lauret regretted endorsing a sizeable sum, which he was obliged to pay instead of Lallemand, who failed to keep his word;[133] Rigau and other refugees accused him of failing to complete the task, given him by a committee, of drawing up a list of the colonists at the Champ d'Asile, and of thus preventing the distribution of the fifteen thousand dollars that the *Minerve* had already sent to Philadelphia and New Orleans in 1819.[134] In May 1820 the pro- and anti-Lallemand camps battled each other in newspaper articles until a list of seventy-eight refugees was published and Lallemand relinquished to Rigau the portion due him.[135] Was this belated demonstration of altruism his way of clearing himself of the suspicion that he had wanted to lay hands on the entire amount to wipe out his debts?

Lallemand had nothing left to hope for in the United States, where his reputation had fizzled out, or in Mexico, where he no longer had the means to operate. He decided to let matters drop. He liquidated his holdings and sold his lands in the Tombigbee colony for almost three thousand dollars to Charles Amédée Hurtel, a young man of twenty-two, the son of the Peter Hurtel mentioned elsewhere in this account.[136] In June 1821 he left New Orleans, supposedly for Havana and then the kingdom of Naples,[137] but by the end of the year he was in Philadelphia.[138] On June 10, 1822, he boarded the *Ulysse* in New York, bound for Bristol.[139] According to the consul, who thought Lallemand was on a brig for Nantes, his goal was to "place himself at the head of an armed party in France against the Bourbons."[140] But he went to Spain, where the constitutionalist movement was attracting foreign liberals thirsting for adventure, like the former military men of the Grand Army. In 1820 a revolution had again given the upper hand to the constitution over royal absolutism, a thing that the Europe of the Holy Alliance could not tolerate: in 1823 the French army invaded Spain and restored the ancien régime.

Like other officers, Lallemand tried to oppose this, but in vain. Refusing Colonel Fabvier's offer to lead various groups of Frenchmen who could have united under his name, in the spring of 1823 he preferred to recruit on his own, in Lisbon, a corps that is said to have been very small.[141] The following July he was seen in Malaga commanding cavalry, then in Cádiz, and finally in Gibraltar, which he left for England in early August. In December 1824 he went from London to the Netherlands, where he was suspected of wanting to rally malcontents and carry out some base political maneuver. It was also said that in Brussels, where his wife happened to be in July 1825, he lived for a time in poverty. Back in England, he was this time interested in freeing Greece from the Ottoman yoke. In 1826 the Greek Committee of London entrusted him with $150,000 to purchase two frigates for its nationalist cause in the United States, but only one ship was delivered because of fraudulent squandering of part of the funds. Lallemand was once again involved in a muddled affair.[142] In the end, his only prospect was to remain in New York, where he opened a school.

The July Revolution put an end to his exile and restored him to active duty in January 1831, after more than fifteen years without employment or a salary. Lallemand subsequently held high positions—in the cavalry and at the head of territorial military divisions—and various other posts in the high command. Following a chaotic career in exile, from frauds to fiascos, this was a miraculous recovery. At the pinnacle of his career, a lieutenant general and peer of France, he died in Paris on March 9, 1839. He was buried in Père Lachaise cemetery, where Joseph Lakanal joined him six years later.

Of all the great exiled figures, Lakanal was the last to return from the United States, in 1837. When he came back to the Institut, to the section of the Academy of Moral and Political Science, the oldest of his former colleagues were stunned, as if at the sight of a ghost: twenty-two years had gone by since his departure. It will be recalled how vague his situation had been at the time with regard to exile. The authorities had tried hard to link him to the Hundred Days, claim that he was a relapsed regicide, and consequently banish him; but they had not hesitated to expel him from the Institut, deprive him of his retirement pension, and dismiss him from his position of inspector general of the metric system. Lakanal summed up this ambiguity in one sentence: "I was not exiled from France, but I was obliged to leave and must never return."[143] This mutual rejection, at the height of the reaction, eventually subsided: in 1825 the state recognized his rights and awarded him a pension; two years earlier, Lakanal supposedly preferred to remain French rather than become an American citizen, as required to remain at the head of the University of Louisiana.

Moving south from Kentucky, Lakanal had indeed become president of Orleans College in May 1822 before being forced to resign in July 1823 in favor of his son-in-law Lucien Charvet, a graduate of the École polytechnique.[144] After this he went to the Tombigbee colony, where he had a large allotment, but, discovering that "this establishment, worthy of a prosperous fate, had fallen to pieces bit by bit, by fault of the colonists," he went to "acquire a possession on the romantic banks of Mobile Bay," where he worked his slaves, raised cotton and then sugarcane, and increased the size of his plantation.[145]

When he heard the news of the "admirable" July Revolution, he thought he would very soon see again his "beautiful" France, freed from "the most inept, cowardly, odious" government that had ever existed, but he was not in a rush. Lakanal liked the United States, a country he claimed to know thoroughly, understanding the customs and outlook of its inhabitants and speaking the language fluently. During his exile he had gathered the materials he needed to write a comprehensive book on this country, wanting to offer a new view of it to the French.[146] He liked Mobile, a busy, pleasant town that was expanding rapidly, where he claimed to be happy, dividing his time between his books, a laboratory of experimental physics, a botanical garden, and the management of his estate. Lakanal was fond of his daughters, who had settled in the country.[147] Even if he had wanted to leave quickly, the need to sell his properties prevented this. It would take time, for in this region, as Lakanal knew, important business matters were rarely cash transactions.[148] The result of this was that, in the autumn of 1830, conscious both of his ties and his local assets, Lakanal applied to the Ministry of Foreign Affairs for a diplomatic or consular position in America: he could thus stay where he was and serve his country. After this proposal came to nothing, two events brought him back to France: his 1834 reinstatement into the Academy of Moral and Political Science, which permitted him to return "through the door of honor";[149] and the death of his wife in 1836. He soon had the French consulate in New Orleans issue him a passport, and on June 14 he sold to Adolphe Batré his 480 Tombigbee acres for five hundred dollars. Because a sudden departure would have lost him 200,000 francs, he delayed leaving New Orleans for another year, and arrived in Bordeaux in September 1837.

A few weeks later he moved to the Paris region, to the country house of Labillardière, his retreat while revising the three volumes of his *Séjour d'un membre de l'Institut de France aux États-Unis, pendant vingt-deux ans* (An Institute of France Member's Twenty-two-Year Stay in the United States). Lakanal expected great things of the publication of this work about a country that the French did not know, meticulously documented but as accessible as a novel. Completed around 1840, the manuscript mysteriously disappeared at its author's death and was therefore never published. While Lakanal was disappointed in his hope of seeing his great opus in print, the end of his life offered him cause for satisfaction: in 1839 his mistress (whom he married in 1842, at the age of eighty)[150] bore him a son, Joseph Hippolyte; and the Academy, into which he had been reinstated and of which he was the most senior member, elected him vice-president for 1844 and president for the year following.

He died on February 14, 1845.[151] Of the exiles of his rank he was the last to die, after living the longest in America. With him, the review of French celebrities in America closes. Although posterity has not recorded their names, or only as footnotes, other exiles, both military and civilian, would merit as much consideration. A few remarkable career paths deserve mention.

Following in the footsteps of their ex-minister or general officer fathers, several former Colonial Society members went on to fine careers. An aide-de-camp to the Emperor after Waterloo, cavalry Major Auguste Regnaud de Saint-Jean d'Angély had been demoted by the Restoration and listed as resigned when he returned from New York in 1816. In 1825 he participated in the Greek struggle for independence, before being readmitted to the army under Charles X and promoted by his successors: Napoleon III, in particular, who awarded him the Grand Cross of the Legion of Honor, made him commander of the Imperial Guard, and finally marshal of France in 1859. A week after his death on February 1, 1870, his body was placed in the crypt of Saint-Louis des Invalides.

Before also ending his career with a flourish, Nicolas Raoul, like Regnaud, participated in a movement of national liberation, but in Central America. Reaching Demopolis in 1821, Raoul and his mistress, Teresa (Alvora) Sinibaldi, managed only to lead a hand-to-mouth existence in the colony, farming Nauzelbine, which Lajonie had rented them, as best they could, and making ends meet by running the ferry across French Creek and selling its passengers homemade ginger cakes and crepes. In 1822 they sold their 120-acre allotment, with buildings and improvements, to James Pickens for $360.[152] They did not go to Mobile until 1824, where, on November 29, they married. After this they sailed for South America, where the liberators were hiring. In January 1825 Raoul was in Cartagena, Colombia, where he was encouraged to vaunt his military experience to the recently created Federation of the Republics of Guatemala, Salvador, Honduras, Nicaragua, and Costa Rica. He accepted the federation's offer of a contract as colonel, inspector of artillery and engineering, and in June went to Guatemala City. He played an active role on the region's political and military scene, but Arcé, the president of the federal government, was hostile to him and had him incarcerated. Freed, Raoul went to El Salvador, which was trying to gain independence from Guatemala, served as head of its army, and defeated the federal troops at Milingo. In 1828 he could then return to Guatemala, where, while retaining his military title, he managed his nopal plantation (nopal is a cactus on which lives the cochineal insect used in a scarlet red dye) and, probably after a divorce, married the Guatemalan widow Maria Dolores Vidaurre on April 10, 1832.[153]

Two and a half months earlier, Raoul had written to Soult, the new war minister, putting his fate into the hands "of a hero who knows the worth of an old Waterloo soldier."[154] A year later he was in Paris, where he notified the minister of his arrival on June 30, 1833: "After seventeen years of an exile that was always rigorous and sometimes cruel, I saw the birth of the fine day on which I could rest on the hope of once again serving France under her Flag, object of all my devotion."[155] His return to active duty was followed by an appointment as sub-director of artillery in Douai on July 31. In 1845, promoted to the rank of brigadier, he was put in charge of the Artillery School of La Fère, where thirty years earlier the northern conspiracy had failed. He died in Paris in 1850, leaving a nine-year-old daughter, born of a third marriage to a widow from a socially prominent family.[156]

Unequally treated by the Restoration, the Grouchy sons' careers opened up again along with their father's, whom Louis-Philippe once more made a marshal. A lieutenant general in 1842, Alphonse, the elder, was elected senator ten years later; Victor, the younger, a brigadier in 1847, finished his career as inspector general of cavalry in Africa in 1858.

As mentioned earlier, Clauzel, governor-general of the French possessions in North Africa, established a link between the colonization of America and that of Algeria.[157] Did crossing the Atlantic inspire others, even in very indirect ways, to cross the Mediterranean?

Michel Combe, a major in the First Regiment of the Imperial Guard's foot grenadiers at Waterloo, paid for this by thirteen years of exile in New York, where, in 1823, he married Eliza Walker of Utica, divorced from French officer Lallemand de Villehaut, and the wealthy heiress of Colonel Benjamin Walker, a former aide-de-camp to the Prussian general Baron von Steuben and later to General

Washington. Combe returned to France in 1830. Promoted to colonel the following year and sent to Algeria in 1835 at the head of the Forty-seventh Infantry Regiment, he was mortally wounded on October 13, 1837, during the assault on Constantine. Painter Horace Vernet immortalized the hero, his native town of Feurs erected a statue to him and, until her death in Versailles in 1850, his widow received an annual pension of two thousand francs as a "national recompense."[158]

Other former Colonial Society members also ended their days in Algeria, where they had gone to take part in the colonizing effort, which was civilian as well as military, as was the case with the Colonna d'Ornanos, who returned to France permanently via Le Havre in the summer of 1831.[159] The father, Barthélémy, who had sold his 320 Tombigbee acres in 1818[160] but had traveled back and forth between America and Europe several times over more than ten years (suspected in 1820 of taking letters to the Bonaparte family in Rome and of intriguing to serve them),[161] took up again in Algiers the position of magistrates' civil court judge that he had held in Ajaccio before his exile, and died there in 1855; his son Émile, an officer in the Second Light Infantry Battalion of Africa, fought in Mexico before being sent to Algeria, where, in 1867, he died of double pneumonia in Mersel-Kébir near Oran, having reached the rank of lieutenant colonel in the foreign regiment.[162] We also have the example of several families that had intermarried in the United States. Jean François Roland de Bussy, nephew of Pierre François Réal and secretary-general of police headquarters during the Hundred Days, had in 1816 joined his uncle, with his wife and their three children. In 1818, in New York, the eldest, Marie Françoise, had married Alexandre Ambroise Germond, a Paris merchant, and Marie Françoise Julie, the youngest, had married Camille Arnaud, a merchant from Lyon. None of them ever went to Alabama any more than had Colonna. Roland de Bussy sold his allotment, at cost, to Peter Hurtel; Arnaud and Germond sold theirs to Thomas Newman and Victoire George, respectively. In April 1821 they all boarded the American ship *Stephania* in New York, bound for Antwerp, from where they planned to go to Paris.[163] They were together again in 1830, having followed the French expeditionary corps to Algeria, where they settled permanently. There, Roland de Bussy was in turn a police captain, presiding judge of the magistrates' court, justice of the peace, presiding judge of the criminal court, and member of the board of governors of the Regency in Algiers.[164] His death in Algiers in June 1858 was an occasion of public mourning, and the city council erected a monument in homage "to his merit and his virtues";[165] his son Jean carried on during the Second Empire as prefecture councilor, deputy mayor, and later mayor of Algiers, where a street long bore his name.[166] His son-in-law Germond died in 1840 in Algiers, where he was a deputy police captain and city registrar; Arnaud, his other son-in-law, was the last to die, in 1859 at La Sénia, south of Oran. These decades in Algeria had succeeded four years in New York.

For Father Mathieu Bernard Anduze, the reverse was true. After two decades in Louisiana, he did not survive three months at the foot of the Atlas Mountains, where he had come to bring Christ to the indigenous peoples. The astonishing story of this priest, whose correspondence is in the archives of Notre Dame University (Indiana), remains to be written. Born in Rodez (Aveyron) in 1795, Anduze had in 1816 emigrated to the United States, where he was ordained. From 1824 to at least 1844 he ministered to the New Orleans diocese at St. Louis Cathedral and as priest in Iberville, where he founded the College of St. Gabriel. To his talents of preacher, theologian, and educator must be added that of negotiator, used on two occasions: from 1827 to 1830, in Spain, England, and Italy, where he met with very influential political and religious figures; and after the capture of Veracruz by Baudin's French squadron in 1839, in Mexico and Texas, at the time two independent and adjoining republics. Why did he, at the age of fifty, exchange one continent for another where the material and spiritual conditions of life would be far more difficult? A priest who went to visit him in Blida, where he had been parish priest for a short time, reported that his death in 1847 was "terrifying."[167]

These few portraits of society members whose lives after their return from America were out of the ordinary are perhaps no more interesting than others, much less brilliant or exotic. The destiny of Jeannet-Oudin, known as "One-Arm," comes to mind, a cousin of Danton and major actor at the Champ d'Asile who died almost indigent in his native town of Arcis-sur-Aube in 1828; or that of Charles Haraneder (a cousin of the Catholic historian and essayist Frédéric Ozanam), whose Creole wife, Emerante Bonin, brought back from Louisiana, died in Paris in 1826 after giving birth to their second daughter, who in turn died at twenty-six, a nun at the convent of the Soeurs de la Retraite in Aix-en-Provence. But rather than add more examples to counterbalance those of the celebrities, this chapter will conclude with an account of the reinsertion into French life of Jacques Lajonie, the symbolic figure of this American adventure.

LAJONIE AFTER EXILE

In 1826, ten years after charges had been brought against him, Lajonie learned that the legal action against him had lapsed. The doors of France were once again open to him, but it was out of the question to sell cheaply lands that had appreciated thanks to his care, just to get home more quickly. On December 17, 1826, he itemized his sales.

Despite having to grant fourteen months of credit, the Sainte-Hélène property, 120 acres with 40 under cultivation, sold very well: 9.55 gourdes per acre—that is, five times the purchase price. He also sold everything he could do without: cows, horses, and tools. The sale brought in a total of 1,500 gourdes. While his wooden buildings were in poor condition, the value of what he had kept—two slaves, two horses, a few cows and hogs, tools, and household utensils—came to 1,200 gourdes, making a total of 2,700 gourdes or dollars. It is unclear whether the 700 gourdes from the sale of seventy head of cattle is included in this sum or should be added to it. He had not yet by this date been able to find a buyer for his 480-acre allotment at two dollars per acre. He feared that he would not be able to sell it before leaving or that he would have to rent it out until the government demanded the payment of two dollars per acre in 1831.

Compared to the majority of situations, Lajonie's was financially quite enviable. His work had borne fruit. All the same, he was bitter because the colony for which he had fought seemed destined to disappear: "One must come to this country to get an idea of the changes that take place within very few years in the value of property. Seven years ago, Demopolis was prospering and five years ago it declined; three years ago there were only four families and now there are eight inhabited houses. They say that it will pick up again. It is possible, but I have lost 2,000 francs there and the interest I have in it is not worth thinking about. The little 24-acre Demopolis farm that I called Nauzelbine, of which Labrousse has half as a consequence of arrangements, could not be sold for six gourdes an acre."[168]

All things considered, how much did Lajonie's property, improvements, slaves, cattle, and crops bring in? The only certain information that we have is an invoice for the insurance he purchased before leaving the United States. He insured a sum of 4,000 in cash at .75 percent and another of 600 at 1.5 percent, making 4,600—but were these dollars or francs? Nor do we know how much the Fougnets had when they left. In 1826 Lajonie wrote that they would be reduced to a sum of 8,000 francs. The earnings likely did not tally with Lajonie's dreams of fortune, but neither he nor the Fougnets was returning to France penniless.

A MATTER OF HONOR

In late March 1829, six passengers—Jacques, Dorothée, and Ninon Lajonie, Antoine and Lise Fougnet, and Jean Mangon—boarded the *Brighton* in Mobile, bound for Bordeaux, which they reached in May 1829. Nine-year-old American Ninon discovered France; her parents were rediscovering their homeland after twelve and a half years of absence for one and ten for the other. One can imagine their excitement as they approached Juillac, and the tears and embraces at the château of Le Soulat between a sister and her older siblings, a father and his son, seeing each other for the first time, a mother and her children, ten years older than when she had left. The Taupier-Letages gave the reunited family a warm welcome and took them in while Lajonie reaccustomed himself to his country and took care of his most urgent affairs. They were to move into the Lapeyre house a bit later.

Clearing his name of the murder that had caused his exile had been the first step toward rehabilitation; his reinstatement in the Order of the Legion of Honor was the second.[169] On August 2, 1829, Lajonie was in Paris, where he requested an audience with the order's grand chancellor in order to recover his legionnaire's certificate and his eligibility for the pension of which he had been deprived since 1812. The collection of original certificates and documents submitted to the secretariat of the grand chancellery and the on-site assistance of an officer, Lerminier, a former comrade in the Seventh Dragoons, had led him to hope for a rapid conclusion, but his years of exile had made him forget the convolutions of French bureaucracy: it took correspondence and recourse to the most highly placed individuals in the order's chancellery before Lerminier was able to announce to Lajonie, in January 1830, that he would receive arrears through 1828 and his annual pension for 1829. But with Lerminier gone on the African expedition to punish the Dey of Algiers, the office sank back into lethargy and Lajonie had to wait several months for the first installment of his arrears and the issuance of his certificate.

This recognition of his rights permitted him to wear once again his medal of the Legion of Honor and to be "proud of his family's esteem," as his friend Lerminier wrote him. Lerminier was not worried about him, despite the distress and adversity that had wounded him: "But you are not alone: your wife and three children are well aware of your misfortunes and will endeavor to make you forget them, I am sure; nonetheless I believe that your chief concern is to give your children the good education that is the basis of a good future; you are ready to make every sacrifice to manage this, and they will appreciate and respond to this. It is wonderful: you will live on hope and privation."[170]

His personal honor restored, Lajonie could indeed see to his children, whose upbringing had been in the care of their uncle and aunt Taupier, suspected by Lajonie of pampering them, as they had no children of their own. The wish to give them the best education had won out over that of keeping them near him after being deprived of them for so long.

In 1830 Marie-Elbine was sent to a private school in Bordeaux, with the payments managed by Étienne Laroque and Cie, the bank handling Lajonie's accounts. This financial effort did not last long, since in May 1832 Elbine, seventeen, married Louis François Eugène Gentillot, twenty-six, a landowner and the mayor of Vayres. Both died very young: Elbine at the chateau of Le Soulat on July 16, 1841, and her husband the following year, leaving no children.

Lajonie took Léopanno to Paris in November 1830. A boarding student at the Collège Royal Saint-Louis, he studied under renowned teachers. His progress seems to have been so rapid that in 1834, after finishing the class of rhetoric, he devoted himself to mathematics in which he excelled, intending to take the entrance examination of the École polytechnique. He was following in the footsteps of his father, a graduate of the special imperial military school of Fontainebleau. "It is, as you see, a new way to attain the goal you wish for. I foresee that your son will give you a great deal of satisfaction and I am not afraid to congratulate you on it in advance," wrote Pierre Mathieu Mannoury, who managed Léopanno's education and boarding expenses for Lajonie.[171] This native of Champagne, in business in Paris, was a former Tombigbee colonist who returned from New Orleans to Le Havre in 1824[172] after giving Lajonie power of attorney to sell his allotment 69.[173] He always referred to himself as his very devoted friend Mannoury.

Léopanno did not continue his Parisian studies beyond the spring of 1835. He was then eighteen years old. He returned to Gironde and lived at Le Soulat with Jean Pierre Taupier, who adopted him legally on April 16, 1844. When his adoptive father died on July 25, 1849, at the age of eighty, Léopanno, unemployed, inherited most of his estate and became a wealthy man.[174] He died in Gensac on September 1, 1875, a childless bachelor landowner. He was not yet sixty years old.

Ninon, born in Demopolis in 1820, remained with her parents at Lapeyre. The only one to survive her father—she died in Gensac in 1911 at the age of ninety-one—she was also the only one to give him descendants: two daughters born of her marriage in 1851 to her cousin Jean-Jacques Lajonie, the steward of the estate of La Siguenie in Saint-Seurin-de-Prats (Dordogne).[175]

His honor and role as father restored, Lajonie turned to reestablishing himself economically and socially on lands where he was to live for another fifty years. His return from a long, far-away, and unjust exile enhanced the local reputation he had had before his departure. The inhabitants who had signed petitions in his favor in 1816 had followed all his storybook adventures thanks to his letters to Le Soulat and the testimony of émigrés who returned earlier, either permanently, like Monfrant in 1825, or temporarily, like Adolphe Batré in 1828. Lajonie was returning with his head held high, preceded by the news of his role in the Colonial Society of French Emigrants in Philadelphia, his socializing with the important outlawed men, his handshake with President Madison, his journey on the Ohio and the Mississippi, and his pioneering exploits in Alabama. Although he did not bring back a fortune, Lajonie returned with an example of courage and perseverance in a particularly difficult and distressing context.

A Notable

The Revolution of July 1830 and the accession of Louis-Philippe were for Lajonie the opportunity to erase fifteen years of a regime that had done him such harm and to obtain a promotion. Elected that year as commander of the national guard of the chief town of his canton, he was able, in his correspondence with the United States, to use the title "colonel," greatly coveted by the Napoleonic officers during their exile as well as by the Americans.[176] After holding this position for several years and serving on Gensac's municipal council, Lajonie had retired from public life when the Revolution of 1848 awakened his republican instincts. In March he became president of the democratic central committee of Gensac; in 1849, after supporting Louis-Napoleon Bonaparte's candidacy for the presidency of the Republic, he was elected commander of the two companies of the Gensac canton battalion:

> In taking command of the battalion of the canton of Gensac by virtue of the rank to which you have promoted me by the wishes of the majority, and after being confirmed at this rank on the 13th of this month by the superior authorities, my first act as a citizen must be a direct appeal to your civic-mindedness to arrive at a frank and loyal conciliation uniting under the same flag men who may have diverse opinions and beliefs, but who should all work together toward the same goal—the good of all—equality, fraternity and liberty with order. . . .
>
> Citizens of the towns of Gensac, Flaujagues, Juillac, Pessac and Coubeyrac who are members of the battalion that I have been asked to organize, rally around me, through your officers and your non-commissioned officers,

so that together we may work to acquire the two strengths about which I spoke to you above. Thus armed, the national guard will no longer be a problem for the authorities. Long live the Republic.[177]

At the age of sixty, Lajonie dominated the social hierarchy of his canton: on March 14, 1849, the Gensac parish priest presented "his respects to Monsieur the Commander of the National Guard and asked that he do him the honor of coming to dine with him and Monsignor the Archbishop, on Monday at noon."[178]

His papers abound with personal writings and speeches that compose his credo: "I am a republican, by birth I believe. I am a republican by long-standing conviction, reasoned and strongly reinforced by experience. Finally, I am a republican by gratitude." He proclaimed that he was "excessively a friend of democracy with a single president," but that he did not like excess. It is therefore not certain that he favored the forced passage from the Second Republic to the Second Empire. Lajonie retired from public affairs to see exclusively to improving his estate.

After his return from America and a few months spent at Le Soulat, Lajonie had moved back to the estate of Lapeyre at Gensac, where he was born, and where his mother lived until shortly before her death in Juillac in 1833. The overseas agricultural expert became a landowner, farmer, and winegrower back home.

Various inventories of his possessions given to insurance companies—La Rurale, against hail, or La Paternelle, against fire, lightning, and later, gas explosions—give an accurate idea of his activities and of the increase in the value of his property, insured over a period of twenty years: 36,700 francs (1852); 40,100 francs (1861); 43,100 francs (1868); 55,050 francs (1872). Taking the 1868 inventory as a point of reference, we see two principal elements: 25 percent of the insured property concerns the house of Lapeyre, its outbuildings and furniture, signs of a certain affluence;[179] 42 percent concerns winemaking activity (storehouses, a vat, wine, and winemaking furnishings); and the rest is divided between agricultural buildings (a barn, dovecote, aviary, hired hand's house, and outbuildings), all kinds of cattle, fodder, straw, grains and grasses on one hand, and on the other, the buildings of La Chaulette, a part of Dorothée (Noguey-Maransin) Lajonie's inheritance. Located in the commune of Thoumeyragues, this twenty-seven-acre smallholding was planted in wheat and grapevines. Lajonie's accounts for the 1840s indicate the sale of dozens of casks of red and white wine to local merchants, among them Beylot and Cie of Libourne.

Lajonie increased his holdings in spite of himself, thanks to provisions in the wills of his children and wife who predeceased him. Without going into the details of complicated inheritances, we can note that at Elbine's death in 1841 he found himself coheir of half the principal manor house of the Robert estate in Juillac, a wedding gift to Elbine and Eugène Gentillot from Taupier-Letage, and half of the quarter of the Lapeyre property that was part of the settlement of her estate.[180] Dorothée's death on May 9, 1860, at their daughter Ninon and son-in-law Lajonie's in Saint-Seurin-de-Prats (Dordogne), gave him principally a life interest in the smallholding of La Chaulette, which changed nothing, since he already had this by his marriage.[181] But in 1875 Léopanno's death brought him a bit more than thirty-seven acres at Robert and Muset in Juillac, their value estimated at 66,640 francs. He thus owned the parcels of Casse-Praufit, Gabarey, Pey-de-Roque, and so forth, whose young vine plants had sixty years earlier made the transatlantic journey to their improbable transplantation in Alabama. Le Robert and Le Muset remained in the Coustou family, descended from Jacques Lajonie through one of Ninon's daughters, until the last quarter of the twentieth century. In 1985 the Robert manor house and its twenty-seven adjoining acres were promoted to "Château la Roberterie," the name of a red, white, and rosé wine whose excellent quality is due to the joint talents of its new proprietors, winegrowers Marie-Thérèse and Alfred Pantarrotto, children of Italian immigrants from the area around Venice.

Born just before the Revolution, injured at Wagram, hunted under the Bourbons, exiled in Alabama, returned to his country, republican and patriot ("I become fanatical at the very mention of the word fatherland")[182] during the Revolutions of 1830 and 1848, Lajonie had an existence as turbulent as his times. His long life did not spare him the suffering of seeing the death of his loved ones, or of experiencing the humiliation of defeat in the Franco-Prussian War of 1871. The black hair and beard of his youth had long become white when he died in his home of Lapeyre on August 22, 1878, at the age of ninety-one. He was buried in the neighboring cemetery of Gensac, where his body lies to this day.

Conclusion

The second Restoration's wave of migration carried many French passengers across the Atlantic, but it was not the surge driven by the American dream often portrayed by historiography, deserving neither the strength ascribed to it nor the recurrent qualifier of Bonapartist. Quite a few military and civilian adversaries of the Bourbons were admittedly exiled for rallying to the Emperor during the Hundred Days, but a great many others preferred to emigrate, temporarily or permanently, rather than be subjected to a regime they despised or judged incapable of securing their future. The Bonapartist officer fleeing his country because of his conduct or ideas is the tree that hides the forest of merchants and workers guided first and foremost by economic choices.

The encounter, in 1816, of this particular exodus with the preceding exodus of the refugees of Saint-Domingue reinforced the heterogeneous character of French immigration to American soil, but it did not hinder the creation of a manufacturing and agricultural Colonial Society of French Emigrants in Philadelphia. During the "era of good feelings" it was a fine idea to want to gather together men differing in origin and in the reasons for and length of their presence in America, but enduring identical sufferings as refugees or exiles in a foreign land. Inspired by the symbolism of a return to nature, the members' plan aimed to find on the frontier virgin land whose soil and climate were similar to those of southern France, that is, suitable, after being cleared and enriched, for growing grapevines and olive trees.

There was broad consensus on this plan. The French diplomatic corps, frightened by the notion that a plot was being hatched in the United States to deliver the prisoner of St. Helena or create a Napoleonic Confederation, was pleased to see the partisans of the Emperor and his brother Joseph, living in New Jersey, move into the distant interior of the country and undertake something useful. As for the federal government, it was fulfilling a duty of hospitality toward exiles, both famous and anonymous, who were unhappy to be here and had no prospects other than to populate a territory that had just been conquered from the Indians in a geostrategically sensitive region and introduce new crops profitable to the national economy. On March 3, 1817, Congress passed a law granting the émigrés 92,000 acres of land at financially advantageous conditions, provided they plant a small portion of it in grapevines and olive trees. But the vote was not unanimous. Some congressmen objected to granting the land, arguing that there was a risk of speculation in this Mississippi Territory that was being opened to American farmers. Others grumbled that it was a gift. It was up to the society members to sweep away these objections by their behavior, knowing that the French in general had a reputation for not being able to get along, exporting and continuing their internal quarrels abroad, or even lacking the pioneer combativeness of Anglo-Americans. But the Colonial Society missed the opportunity to refute them both as an organization and in the implementation of its plans on-site.

Difficult though they were, the journeys taking the French settlers from the East Coast to the banks of the Tombigbee and the Black Warrior in Alabama did not cause any to give up: some died on the way, but the majority of those who left arrived safe and sound. These, however, were never more than a portion of the hundreds of members concerned; many never moved, refusing, in the end, to substitute a hypothetically better life in the depths of the forest for the comforts of urban civilization. They ceded their lands or, if they kept them, did not bother to improve them. This is an important explanation of the French failure. Others were not set in Alabama either, but in Philadelphia, where the society's members tore each other to pieces when the time came to draw allotments in the autumn of 1817, and then in 1818 in Texas, where the elder General Lallemand led so many officers to their ruin in the unfortunate Champ d'Asile affair.

The defection of part of the society certainly thwarted its success, but so did a host of other, even more damaging reasons, which those who suffered because of them—ex-lieutenant Jacques Lajonie, writing to his family during his twelve-year stay, or Domingan refugee Frédéric Ravesies, reporting to the American secretary of the treasury in 1827—had no trouble listing. Both men put forward natural, structural, and human causes, interacting to such an extent that it is impossible to dissociate them.

The officers, merchants, and craftsmen who chiefly composed the society had a very limited knowledge of the art and practice of agriculture, while the number of farmers and winegrowers was very small. This basic deficiency, almost crippling in itself, was worsened by the conditions faced by the first colonists when they arrived in a wild, uncultivated region where nothing had been prepared for their settlement and where consequently everything centered first of all on survival: clearing brush, pulling up stumps, plowing and sowing to have food, and building for shelter and storage were absolute priorities that delayed the planting of grapevines and olive trees for commercial markets. The immediate need to plant food crops was due to the sometimes prohibitive cost of basic foods like meat and maize. Significant transport difficulties justified these prices, as no road worthy of the word went to the colony.

After a few months, the first colonists, exhausted by their work of clearing the site of White Bluff on the river and building their own temporary settlements on their small allotments near Demopolis, learned that the configuration and location of the townships that the association's secretary had given them were erroneous, since Parmantier had not scrupulously respected the contract's stipulations. In his defense, it must be said that the French had preceded the Land Office agents into a region that had not yet been surveyed or mapped. Had they arrived six months later, the situation would have been completely different.

When they had to abandon the newly cleared site of White Bluff to start all over again next door in Aigleville and then go to the large allotments that had been assigned them in Philadelphia, many colonists no longer had the financial means or physical strength to go on. Inflammations and fevers of all sorts attacked bodies weakened by fatigue and privation. Illness spared no one, and many died. Those who remained and attempted to live on their large allotments often found their efforts unrewarded.

Laid out arbitrarily, far from the actual site, the division of the townships had respected geometry, but with no idea of the quality of the allotments. This was very uneven. Many of the allotments, entirely wooded, had so little open land that it was futile to try to settle there. Others, lacking drinking water, required the digging of very deep wells with no guarantee of success. Still others, especially along the Black Warrior, were subject to flooding over a large part of their area. Finally, no allotment was safe from the devastation wrought by tornadoes or the voracity of bears and wolves that decimated the livestock.

However, the colony was not merely a vast, uninhabitable, and unproductive zone. Many plots were usable but, by this very fact, had attracted squatters who were not only unwilling to give up what they considered the right of the first occupant, but hostile or threatening toward anyone who wanted to dispossess them, even if it was a judge. Protected by the law but unfamiliar with the language and the customs of their new country, could the French assert themselves against Americans accustomed to harsh, precarious frontier life? Unprepared for the world that awaited them, they had arrived too soon in a region without infrastructures; but they had also arrived too late, since the portion of the colony where settlement and farming were unproblematic was for the most part illegally occupied.

In addition to these general conditions, there were specific ones equally unfavorable to growing the grapevines and olive trees that were the contractual reason for the grant to the émigrés. Their lack of foresight is an essential cause of their failure. How could they have compared the southern regions of the United States with those of France, separated as they were by thirteen degrees of latitude, and think that their respective soils would be identical before even being able to verify this? The French, totally lacking in specific knowledge, are responsible for wanting to acclimate foreign vine and olive stock in an environment that did not suit them, while they could have used domestic vines, as did Roudet or as the Swiss had done in Indiana. But even if they had succeeded in producing enough wine to be able to market it, it is doubtful that the venture could have continued. The Americans, who had little by little supplanted the French in their colony, had no reason to continue planting grapevines and olive trees once their landownership had been registered with the government. Cotton was the American farmers' best bet; they could lose their shirts on an uncertain crop, demanding skilled workers, foreign to their traditions, and not answering the needs of local consumers who liked hard liquor or, at least, good wines imported from Europe. The six vineyards that exist in Alabama two centuries later, like Perdido Vineyards near the Florida border, confirm the very strange flavor of wine made from native muscadine grapes.

The harsh living conditions endured by the French settlers in the Alabama countryside and the fact that they found it impossible to get grapevines and olive trees to grow properly are not enough to explain why most of them left so quickly, and almost all within fifteen years. The exiles, the most famous of whom—like Lefebvre-Desnouëttes, who was very much involved in the colony—had never lost hope of obtaining the king's pardon and consequently of an early return to France. Lajonie, more committed than anyone, had put off his departure only because he wanted to go back to his family with his head held high, after selling all or part of his lands and leaving the remainder of his business in the hands of trusted persons: the money he took back to France was merely compensation for the sacrifices he had made for twelve years. The end of an unjust exile and homesickness are two major reasons for return, not fortunes resulting from the sale of lands, agricultural produce, or cattle; even less from barrels of Alabama French wine. A study of wills shows that the grantees' heirs did not benefit from their American property deeds either. In the end, only the refugees from Saint-Domingue, whose legitimacy had been challenged in September 1817, stayed on the territory of the former French colony. On the eve of the 2017 bicentennial of its founding, only a few of their descendants remain on this territory: in Greensboro, Colonel Ramsey, descended from the Noël and Hurtel families, whose house in the woods features a magnificent Empire console table; and in Demopolis, John Cox Webb and his sister Louise Webb Reynolds, from the Chapron and Nidelet families, whose homes contain quite a few objects, jewelry, and furniture mostly dating from the Domingan period.

Deprived very early of its French founders and their sons, and, even more noticeable, of French surnames, the city of Demopolis was rapidly cut from its first roots, after almost disappearing in the 1820s when there remained only a few houses and the French gravestones that have since disappeared. The city was reborn from its ashes in the following decade thanks to American planters, many of whom had quickly bought the French lands, and experienced its glory days before the Civil War.[1] It was inhabited by an elite, enjoying the pleasures of music, travel, and leisure, men like Francis S. Lyon. This lawyer, politician, planter, and in 1860 owner of two hundred slaves was, it is said, the man who did the most for the life and well-being of Demopolis in the nineteenth century: a street bears his name, a stained-glass window of the Episcopalian church is dedicated to him,[2] and the Honorable Francis Strother Lyon Prize is awarded every year for efforts made for the historic preservation of Marengo County. Lillian Hellman referred to this cotton aristocracy of the Deep South in lines she gave one of the characters of *The Little Foxes*, supposedly set in Demopolis: "Ah they were great days for these people . . . They had the best of everything. Cloth from Paris, trips to Europe, horses you can't raise any more, niggers to lift their fingers."[3] Demopolis soon had an opera house and then a theater, where boxer Jack Dempsey and magician Harry Houdini gave public performances, before the opening in the 1930s of a large movie theater, the Marengo, the only survivor a century later of this cultural life. As a sort of exception, Demopolis was distinctive in its difference and could also have inspired what Carl Carmer wrote about a little neighboring town ennobled by cotton: "There are only two places in the world where one may live a civilized existence . . . Paris and Uniontown."[4]

Cotton has now disappeared. An economy based on farm-raised catfish, wood, cattle, and soybeans has replaced it. Uniontown has sunk into unemployment, drugs, and insecurity. Demopolis has lost its singularity; it has become one of those small American cities that have interesting things to show but that no one knows about because people pass it by. Life has inexorably flowed from its historic center toward U.S. Route 80, stretching along the southern edge of town, attracting most of the local and through traffic, lined with businesses and services. A victim of "Walmartization," the concept of downtown as a business center has lost its significance. Around the town square are the establishments that have always been there: the city hall, post office, newspaper office, lawyers' and insurance offices, gunsmiths . . . After 5:00 P.M., when Washington and Walnut Streets have closed their doors, their dark red, ocher, or white brick facades look out only at magnolias, pecan trees, and red cedars lined up in front of them or on the opposite sidewalk. Atop his pedestal at a corner of the square, in full sunlight facing south, the bored Confederate soldier leans on his gun. The only things to distract him from his drowsiness are the rhythmic vibrations imposed on the town by the Southern Railway diesel engine as it passes through, or the tugs pushing their barges on the Tombigbee River. The municipality, the chamber of commerce, and various associations are attempting to revitalize the downtown area and bring back shoppers and visitors. A return to the city's origins with the promotion of its past and its geography could be a solution.

Since its creation in 1819, a virtual historic thoroughfare has run along the top of the limestone wall bordering the river: Arch Street, named after its curve and its homologue in Philadelphia, the city from which the French colonists came.[5] The creation of a bicycle and pedestrian path should attract walkers, combine exercise, nature, and culture, and act as a footbridge linking other historic sites maintained in Demopolis by the Marengo County Historical Society: Bluff Hall, Lyon Hall, Gaineswood, and the large museum yet to be built. This place would bring together everything that has contributed to the making of Demopolis and its region, from American Indian settlements to the reign of cotton—including the French interlude.

Arrowheads and unused cotton warehouses bear witness to their respective eras, while at first sight nothing remains from the founding years: only three large bricks thought to be French-made because of their unusual size,[6] and traces in place-names: Desnouettes Street (pronounced DEZ-ou-nettes), Demopolis's most winding street, homage to the general who was the leader of the French immigrants; dusty Arcola Road, running through the cement plant that has been taking stone out of the ex-colony's subsoil for a century; and just before the town, near the highway from Selma, houses reached via a Napolean (*sic*) Drive and a French Creek Circle. At both of the town's entrances, imposing stone markers announce to those arriving that Demopolis, "the city of the people," was founded in 1817, before Alabama became a state. But by whom? By Greeks, according to the name's etymology; by Italians, if one knows where Arcola and Marengo (the name of its county) are located.

Demopolis does not deny its French ancestry, but unlike Vevay, founded at the same time in Indiana by Swiss winegrowers, which has remained faithful to its Swiss roots,[7] it has very little concern with France. Forty years after the celebration of its 150th anniversary, do all its inhabitants know of Demopolis's origins, particularly its African American inhabitants, implicitly not involved in the town's cultural activities and plans? In Demopolis as elsewhere in the Deep South, progress in material civilization has often been more rapid than the human progress praised by the French ambassador in 1967. Blacks and whites daily leave their respective neighborhoods, from social ghettos to Victorian homes, cross the color line, greet one another and work together, but they do not often mingle socially, and they usually worship separately. This divide diminishes as one goes up the social ladder but, conversely, strengthens to the detriment of society's poorest people living in the rusted trailers of the American dream, mobile homes on their concrete blocks having ironically become the permanent lodgings of those unable to move to areas where there are jobs.[8] Their enslaved ancestors, whose owners were French, helped build Demopolis, but their contributions are rarely mentioned.[9] In the Robertson Bank Company, customers are greeted by a large mural fresco depicting the town and the men of its beginnings: American Indians, pioneers, a dashing imperial general—but no blacks. The bank is located on a corner of the central square, Confederate Park, where in 1910 the town put up its monument to fallen southern soldiers, for many the true foundation stone of Demopolis.[10] This book recounts a chapter of their history, but the full story of the slaves in Demopolis has yet to be written.[11]

Better than a monument raised to the French pioneers or historical markers with necessarily short summaries, the planned museum mentioned above could broaden the treatment of Demopolis's history, no longer systematically limiting it to the Civil War, but going back to Philadelphia and the creation in 1817 of a society of emigrants desiring a new life. It would be an opportunity to move Demopolis from the pedestal of a lost war toward the renewal epitomized by these émigrés who contributed, if only by their appearance on the list of society members, to the birth of the city and its county.

Tracing the history of the Vine and Olive Colony, as it is now called, opens a window on an episode of French emigration to the United States. The majority of society members, whether or not they came to the banks of the Tombigbee, did not return to France. Aside from those who died of yellow fever, the others found in their adopted country the motivation and means for a new life. Work, acquiring fluency in English, marriage, and citizenship kept them in the United States, where they continued, first they and then their children, to migrate from one state to another. Those who came to Alabama spread throughout the South, whose interests they embraced and defended when the time came. They continued to reside in Alabama itself, especially in Mobile, settled in Louisiana from New Orleans to Baton Rouge, went up the river and then fanned out into Arkansas, Tennessee, Kentucky, Missouri, and Nebraska . . . Most of those who did not come to Alabama remained for a while in Philadelphia, where they had purchased a share in the colony, but the nation's expansion offered opportunities that some took up, trying their luck in Texas or California.

There are still a great many descendants of the French Tombigbee émigrés, but compiling a register of them is an impossible task owing to the size of the country and the disappearance or Americanization of their original surnames. Despite everything, I have been able to find several dozen across the country, proud, though they no longer bear their ancestor's name—Champenois, Cluis, Constantine, Delaroderie, Dufour, Duval, Follin, Fournier, Gardien, Génin, Mangon, Martin, Ravesies, Stollenwerck, Tulasne—at least to assert their French roots. The vine and the olive symbolized their ancestors' rebirth: today they owe them their lives.

Appendix

Appendix key: Biographical data on 460 Vine and Olive colonists

ABBREVIATIONS

NAME
1. Given name(s)   Last and first names + titles

BIRTH
2. Ori   Origins (geographical roots): see country/origin abbreviations

3. Place   birth place: City, State (Country). Note: All states are
                    abbreviated with their 2-letter standard.

4. Date   birth date: see variations under DATES category.

DEATH
5. Place   place of death: 2-letter U.S. state abbreviations used

6. Date   date of death: see variations under DATES category.

MARRIAGE
7. Place   place of marriage: 2-letter U.S. state abbreviations used

8. Date   date of marriage: see variations under DATES category.
                    Y means this person was married but the date is unknown;
                    C means cohabitation without marriage; W means widowed.

9. Spo   spouse's origin: see country/origin abbreviations

10. Desc   Descendants' place of residence: see country/origin abbreviations

OCCUPATION
11. Occ.   Exiles' job, trade, profession or occupation ca. 1815

EMIGRATION
12. Date   date of departure from home country: see variations under
                    DATES category.

13. Acc	Accompaniment: fam means left with family; serv means left with servants; col means left with other exiles.
14. Departure	European port of departure
15. Ship	name of ship to U.S.
16. Arrival	American port of arrival
SET	Settlement
17. Place	place of settlement
NAT	Naturalization
18. Am	date of American naturalization
RET	Return
19. Fr	date of return to France

Country/Origin abbreviations are as follows:

Bel	Belgium
CdA	Champ d'Asile
Cre	Creole
Du	Dutch
Elb	Elba
En	England
Fr	France
Fr/StD	French born in Saint Domingue
Ir	Ireland
Ita	Italy
Ge	Germany
Gua	Guatemala
LA	Louisiana
Mex	Mexico
Ne	Netherlands
Pol	Poland
Por	Portugal
Pru	Prussia
Scot	Scotland
Sp	Spain
Sw	Switzerland
Am	United States of America

Place Abbreviations:

Ark.Post	Arkansas Post, Louisiana Territory
Balt	Baltimore, MD
Bat Rge	Baton Rouge, LA
Bchs-d-Rhô	Bouches-du-Rhône
Bostn	Boston, MA
Charlstn	Charleston, SC
Cte-Mtme	Charente-Maritime
Ht-Rhn	Haut-Rhin
Hte-Grne	Haute-Garonne
Hte-Marne	Haute-Marne
Hte-Vne	Haute-Vienne
Htl	Hôtel
Hts-Pyr	Hautes-Pyrénées
Hts-Sne	Hautes-de-Seine
Ille-Vil	Ille-et-Vilaine
Ital-Sav	Italian-Savoy
La Croix-Bqts	La Croix-des-Bouquets
Loire-Atl	Loire-Atlantique
Lot-Grne	Lot-et-Garonne
Noflk	Norfolk, VA
NoLA	New Orleans, LA
Phila	Philadelphia, PA
Pyr-Atls	Pyrénées-Atlantiques
Sne-Marne	Seine-et-Marne
Sne-Mtme	Seine-Maritime
St Aug	St. Augustine, FL
St-Do	Saint-Domingue
D.C.	Washington, D.C.

DATES

/date	event occurred before the listed date
date/	event occurred after the listed date
italicized date	the event occurred circa the listed date

TABLE

Blank box	No information found
Box with dash inside	No information available

TABLE 1A

01	NAME	BIRTH			DEATH		MARRIAGE			
	1	**2**	**3**	**4**	**5**	**6**	**7**	**8**	**9**	**10**
	Given name(s)	ori	place	date	place	date	place	date	spo	desc
001	**ACHARD** Victor	Fr	LaGarde, Var	1784	Am	/1829	Var (Fr) / AL	1811 / 1822	Fr / Sw	
002	**AIMÉ** François Gabriel Valcour	Fr	LA	1798	LA	1867	LA	1819	Cre	Am
003	**ALLAIN** Joseph	Fr			LA?					
004	**ALLARD** Henri	Fr	France		AL	/1821				
005	**ALLOUARD** Jean Pierre Fructidor	Fr	Isère (Fr)	1793			LA	1821	Cre	Am
006	**ANDUZE** Mathieu Bernard	Fr	Rodez, Aveyron (Fr)	1795	Algeria	1847	—	—	—	—
007	**ANGELI** Hyacinthe de	Fr	Corsica (Fr)	1786	PA	1850/	Phila	/1829	En	Am
008	**ANTOINE** Claude–Charles	Fr	Moselle (Fr)		Algeria?					
009	**ARNAUD** Camille	Fr	Lyon, Rhône (Fr)	1787			NY	1818	Fr	Fr
010	**ASTOLPHI** Laurent	Ita	Italy							
011	**AUDIBERT** Angélique Aimée (widow)	Fr	StGeorge-de-Montaigu, Vendée	1771	MO	1834/	Nantes	1794	Fr	Am
012	**AUZÉ** Charles	Fr	Bchs-d-Rhô (Fr)		PR	1823	Balt	1808	Fr	Am
013	**AUZÉ** Joseph (brother)	Fr	Bchs-d-Rhô (Fr)	1784	SC	1838	Phila	1825	Fr	Am
014	**AZAN** Joseph	Fr	Marseille, Bchs-d-Rhô (Fr)	1775			France	/1802	Fr	Am
015	**BACLÉ** Alexandre, elder son	Fr	France							
016	**BADARAQUE** Joseph Thomas	Fr	Marseille, Bchs-d-Rhô (Fr)	1769	Phila	1834		/1805		Am
017	**BAILLY** Michel	Fr	Autun, Saône-et-Loire (Fr)		France	1787	France	/1815	Fr	Fr
018	**BAIZEAU**	Fr	Western France	1784	NoLA	1819				
019	**BALBUENA DE SOTOMAYOR** José Maria	Sp	Toledo (Sp)				—	—	—	—
020	**BALTARD** Prosper	Fr	Paris	1796	Hts-Sne (Fr)	1862				
021	**BARBAROUX** Joseph	Fr	Marseille, Bchs-d-Rhô (Fr)		KY?		Phila	1821	Fr	Am
022	**BARBE** Antoine	Fr	France							
023	**BARIÉ** Louis	Fr	France	1782	GA?	1850/	Savannah, GA	1809	Fr/StD	Am
024	**BARRAUD** Auguste	Fr	Saintes, Cte-Mtme (Fr)	1796						

TABLE 1B

02	11 OCCUPATION occ. ca. 1815	EMIGRATION 12 date	13 acc	14 departure	15 ship	16 arrival	17 SET place	18 NAT Am	19 RET Fr
001	naval administration	1815/	—	—	—	—	AL	—	—
002	LA planter						LA		—
003		1816					AL		—
004		1815/					AL		—
005	soldier	1816		Bordeaux (Fr)			CdA; LA		1844
006	catholic priest	1816	—	Leghorn (Ita)	*Ann Maria (Am)*	Phila	LA	1823	—
007	captain?	1817		Nantes (Fr)			PA		
008		1815					LA		
009	merchant	1817		Bordeaux (Fr)	*Gen. Jackson (Am)*	NY	NY		1821
010	soldier or confectioner	1816	col	Leghorn (Ita)	*Nymphe (Fr)*	Phila	PA		—
011	landowner	/1804	fam; col	Nantes (Fr)		NY	IN; MO		—
012	ship's captain	/1815		St-Do		NY	MD; PA		
013	merchant	/1808					AL; GA		
014	grocer	1815/					PA	1808	
015	soldier?	/1800					PA		
016	merchant	1817		St-Do?		Phila	PA	1803	
017	tailor	1817	fam; col	Bordeaux (Fr)	*Hunter (Am)*	Phila	CdA; PA		1819
018							—		
019	officer for King Joseph (Sp)						CdA; LA		
020	architect	1816		LeHavre (Fr)			AL		1820
021	tradesman	/1810		St-Do?		Phila	PA; KY		—
022	tradesman	1815/					LA		
023	merchant	/1809		St-Do			GA		—
024	guard of honor	1817					CdA; LA		1820

01	NAME	BIRTH			DEATH		MARRIAGE			
	1 Given name(s)	2 ori	3 place	4 date	5 place	6 date	7 place	8 date	9 spo	10 desc
025	BATRÉ Charles	Fr	Bordeaux, Gironde (Fr)	1794	Havana, Cuba	1838	Mobile, AL	1824	Fr	Am
026	BATRÉ Simon Adolphe (brother)	Fr	Bordeaux, Gironde (Fr)	1800	Mobile, AL	1870/	Mobile, AL	1823	Fr/StD	Am
027	BAUMIER César	Fr	France							
028	BAUZAN Joseph	Fr	Bchs-d-Rhô (Fr)							
029	BAYOL Honoré	Fr	Marseille, Bchs-d-Rhô (Fr)		AL	1835	Phila	/1808	Fr/StD	Am
030	BELAIR Louis DESCOINS-	Fr	Cap Français, St-Do	*1785*	Am	1834/				none
031	BELANGÉ Martial-Denis	Fr	StGermain-en-Laye, Yvelines	1780	France	1854				Am
032	BELLEMÈRE Edme-François	Fr	Romilly-sur-Seine, Aube (Fr)	1767	NY	1839	France	/1800	Fr	Am
033	BELLEMÈRE Jean-Baptiste son	Fr	Paris	1793	Phila	1842	Phila	1822	Am	Am
034	BELLIÈVRE Jean-Jacques de aka JAME Jean-Baptiste	Fr	Chagny, Saône-et-Loire (Fr)	1767	Bat Rge	1833	Paris	1810	Fr	Fr
035	BERGASSE Nicolas Cadet	Fr	Lyon, Rhône (Fr)	*1766*	Am	/1833	LA	1818	Cre	none
036	BERNARD Henri	Sw	Switzerland?				Phila			Am
037	BESSON Louis-Antoine	Fr	Phila	*1798*	Mexico City	1832				
038	BEYLARD Jean Jr.	Fr	Gironde (Fr)?				Phila	/1821	Am	Am
039	BEYLE Joseph	Fr	Sassenage, Isère (Fr)	1771	Phila	1832	Phila	1797	Fr/StD	none
							Phila	1804	Fr/StD	
040	BILLINGTON John	En	Am?		Am					
041	BINSSE DE SAINT-VICTOR Louis François de Paule	Fr	Cap Français, St-Do	*1773*	NY	1844	Nantes	1795	Fr/StD	Am
042	BISTOS Jean-Baptiste	Fr	Southwest France				NY	1812	Fr	Am
043	BLANCON Dominique	Fr	Southwest France	*1795*			Albany, NY	1824	Am	
044	BLANDIN Jean	Fr	France							
045	BLAQUEROLLE	Fr								

02 / 11	OCCUPATION	EMIGRATION					17	NAT 18	RET 19
	occ. ca. 1815	12 date	13 acc	14 departure	15 ship	16 arrival	SET place	Am	Fr
025	merchant	1816	col	Bordeaux (Fr)	*William (Am)*	Phila	AL	—	—
026	merchant	1817					AL	—	—
027	soldier	1815/					CdA; LA	—	—
028	soldier?	1815/					NY		—
029	merchant	/1808		St-Do		Phila	AL; PA	1808	—
030	bookseller	1796		St-Do		Phila	PA		—
031	naval administration	1815		LeHavre (Fr)	*Argo (Am)*	Phila	PA	—	1822
032	stockingmaker	1816		LeHavre (Fr)	*Argo (Am)*	Phila	PA	1817	—
033	artisan	1816		LeHavre (Fr)		Phila	PA	1834	—
034	head tax collector	1815/	—				LA	—	—
035	merchant	/1805					PA	—	—
036	grocer	/1815					PA		
037	merchant	—	—		—	—	PA	Am	—
038	tradesman	/1815				Phila	PA	1824	1828
039	merchant	/1797		St-Do		Phila	PA	—	—
040						Phila	AL	Am	—
041	miniaturist painter	1803	fam	St-Do			PA		
042	confectioner	1816		Bordeaux (Fr)		NoLA	LA; NY	1834	—
043		1817		Bordeaux (Fr)			PA		
044		1815/		France			PA		
045		1815/					NY	—	

01	1	2	BIRTH 3	4	DEATH 5	6	MARRIAGE 7	8	9	10
NAME	Given name(s)	ori	place	date	place	date	place	date	spo	desc
046	BOGY Ignace	Fr	Ark. Post	1793	Am		AR	1823	Cre	Am
047	BOISLANDRY Eugénie LE GRAND DE, husb. **Chieusse**	Fr	Phila	1795	Mobile, AL	/1868	Mobile, AL	1824	Fr	Am
048	BOITEAU François	Fr	France		AL	/1832				Am
049	BONNAUD	Fr			AL?	/1821				Am
050	BONNEAU Charles	Fr	Am					/1821	Cre	Am
051	BONNOT Gaspard	Fr	Lyon, Rhône (Fr)		AL	1832		C	Fr	Am
052	BORDAS Élie	Fr	France	1776			France	/1801	Fr	Am
053	BOULAND Étienne-Vincent	Fr	Paris	1784	NY	1863	NY	1815	Fr	Am
054	BOURDICHON Barthélémy	Fr	Razac, Dordogne (Fr)	1794						—
055	BOURLON Louis-Edmond	Fr	Port-au-Prince, St-Do	1795	Mobile, AL	1819				—
056	BOUTIÈRE François-Gaspard **de**	Fr	Rhône (Fr)		AL	1827				—
057	BOUTIÈRE Jean-Claude Benoît **de**	Fr	Rhône (Fr)		AL	1823	Am	1815/	Fr	Am
058	BRAUD Jacques	Fr	St-Do		LA	1838	Am	/1821	Fr	Am
059	BRECHEMIN Claude Louis	Fr	Paris	1787	Phila	1858/	Phila	1823	Scot	Am
060	BRINGIER Paul-Louis	Fr	St. James Parish, LA	1780/	Am	1860	LA	/1829	Cre	Am
061	BROWN Samuel	Ir	Augusta Co, VA	1769	AL	1830	Natchez, MS	1808	Am	Am
062	BRUGIÈRE François Annet Charles	Fr	Bussy-Varache, Hte-Vne (Fr)	1774	NY	1837	Phila	1803	Fr/StD	Am
063	BUGEY Antoine	Fr	Corenc, Isère (Fr)	1794	LA	1819				—
064	BUJAC Alfred	Fr	Castelmoron, Lot-Grne (Fr)	1782	Phila	1830/	Phila	1812	Fr	Am
065	BUJAC Mathieu-Jules (brother)	Fr	Castelmoron, Lot-Grne (Fr)	1790	Am		Balt	1823	Fr	Am
066	BULLIARD Étienne	Fr	Besançon, Doubs (Fr)	1786	LA	1834	LA	1821	Cre	Am
067	BÜRCKLÉ Emanuel	Ge	Wurtemberg (Ge)	1796						—
068	BURGUÉS Jean-Bernard	Fr	Southwest France	1790	France		AL	1822	Fr	—
069	BUTAUD Isaac	Fr	La Rochelle, Cte-Mtme (Fr)	1780	AL	1821	Phila	1815	Fr/StD	Am
070	BUTAUD Victorie, née George	Fr	Port-au-Prince, St-Do	1792	Phila	1827	Phila	1826	Fr	none

02	OCCUPATION 11	EMIGRATION 12	13	14	15	16	SET 17	NAT 18	RET 19
	occ. ca. 1815	date	acc	departure	ship	arrival	place	Am	Fr
046	tradesman	—	—	—	—	—	AR	Am	—
047	none	—	—	—	—	—	AL;PA		—
048		1816		Nantes (Fr)?			AL		
049		1815/	—			NY	AL	—	—
050					—		MO	—	—
051	merchant	1815/					NY;Mex		—
052	merchant	/1815	—	LeHavre (Fr)		NY	NY		—
053	merchant	1817		Bordeaux (Fr)	Rebecca (Am)	NoLA	AL	Am	—
054	carpenter	1817		France			AL		—
055	tradesman						AL		—
056	tradesman?						AL		—
057	tradesman?	1816	fam; col	France			AL		—
058	shopkeeper	1816	col	Bordeaux (Fr)	Susquehanna (Am)	Phila	PA;LA		—
059	jeweler	1815	col	LaRochelle (Fr)	Alex. Paulowitsch (Am)	Phila	PA		—
060	topographic engineer						LA	Am	—
061	physician						AL	Am	—
062	merchant	/1803		St-Do		Phila	PA; NY	1803	—
063	law student	1817	col	Marseille (Fr)	Sachem (Am)	Phila	PA; LA		—
064	merchant	/1803		St-Do		Phila	PA		—
065	master of studies	/1806		St-Do		Phila	PA		—
066	health officer	1817					CdA; PA; LA		—
067	merchant	/1813			Susquehanna (Am)	Phila	NY	1818	—
068	merchant	1816	col	Bordeaux (Fr)		Phila	AL		/1833
069	engraver	/1810		France		Phila	AL; PA		—
070	none	1794	fam; col	St-Do		Phila	AL; PA		—

	NAME	BIRTH			DEATH		MARRIAGE			
	1	2	3	4	5	6	7	8	9	10
01	Given name(s)	ori	place	date	place	date	place	date	spo	desc
071	CAILLEBAUX Guillaume	Fr	Oléron, Cte-Mtme (Fr)	1766	Phila	1831				—
072	CAMPARDON Jean-Baptiste	Fr	France	1786			Mexico	/1835	Fr	Mex
073	CANOBIO François	Ita	Nice, Comté de Nice	1790						
074	CANONGE Pierre-Auguste	Fr	Jérémie, St-Do	1791	AL	1821				
075	CARRÉ Jean-Thomas	Fr	Normandie (Fr)	1744	Phila	1829	St-Do	1779	Cre	Am
076	CASTAN Étienne	Fr	France		Am	/1834	St-Do	/1795	Fr	Am
077	CAVAROC Charles M.	Fr	Bordeaux, Gironde (Fr)	1788	NoLA	1852	NoLA	1828	Fr	Am
078	CHAMPENOIS Pierre Jean Isaac	Fr	Beaugency, Loiret (Fr)	1769	AL	1842	Phila	1811	Am	Am
							Mobile, AL	1817	Am	Am
079	CHAPOTIN Achille	Fr	Auxerre, Yonne (Fr)	1794	NoLA	1831	NoLA	1821	Fr	Am
080	CHAPPON Alexandre-Alphonse	Fr	Meaux, Sne-Marne (Fr)	1793	France	1793	France	1818/	Fr	Fr
081	CHAPRON Jean-Marie	Fr	Port-de-Paix, St-Do	1787	AL	1868	Phila	1814	Fr/StD	Am
082	CHARASSIN Jean-Philibert	Fr	Dijon, Côte d'Or (Fr)	1786	France					
083	CHARRETON-RASPILLER Joseph-Louis	Fr	Givors, Rhône (Fr)	1793	NoLA	/1832		Y	Fr	
084	CHASSÉRIAU Benoît	Fr	La Rochelle, Cte-Mtme (Fr)	1782	Puerto Rico	1844	St-Do	1804	Fr/StD	Fr
085	CHAUDRON Jean-Simon	Fr	Vignery, Hte-Marne (Fr)	1758	Mobile, AL	1846	St-Do	1790	Fr/StD	Am
086	CHAUDRON Pierre-Édouard	Fr	Cap Français, St-Do	1792	AL	1836	—		—	—
087	CHAUVEAU Charles	Fr	France							
088	CIRODE Louis-Guillaume-Marie	Fr	Nantes, Loire-Atl (Fr)	1791	Am	1840/	Lexington, KY	1818	Am	Am
089	CLAUZEL Bertrand, General Count	Fr	Mirepoix, Ariège (Fr)	1772	Lot-Grne	1842	NY	1804	Fr/StD	Fr
090	CLUIS Jean-Jérôme (known as Colonel)	Fr	La Châtre, Indre (Fr)	1773	Mobile, AL	1833	LaChâtre (Fr)	1794	Fr	Fr
							Paris	1815	Fr	Am
091	COLOMEL	Fr								
092	COLONNA D'ORNANO Barthélémy	Fr	Cognocoli-Monticchi, Corsica	1778	Algeria	1855	Leghorn (Ita)	1819	Ita	Fr
093	COMBE Michel, Colonel	Fr	Feurs, Loire (Fr)	1787	Algeria	1837	NJ	1823	Am	none

02	OCCUPATION	EMIGRATION			15	16	SET 17	NAT 18	RET 19
	11	12	13	14	ship	arrival	place	Am	Fr
	occ. ca. 1815	date	acc	departure					
071	baker	/1805				Phila	PA		
072	tailor	1817	fam	LeHavre (Fr)		Phila	CdA; PA; LA		—
073	health officer	1815/				Phila	CdA; LA	Am?	—
074	tradesman	/1815		St-Do		Phila	AL; PA	Am	—
075	language teacher	1794	fam	St-Do		Phila	PA		—
076	clockmaker-jeweler	/1800	fam	St-Do		Phila	PA		—
077	tradesman	/1815		St-Do		Phila	PA; LA		—
078	tradesman	1801	—	Nantes (Fr)		NY	PA		
079	artillery officer cadet	1815	fam	Bordeaux (Fr)	South Carolina (Am)	Phila	AL,	1816	—
080	merchant	1819		LeHavre (Fr)	Providence (Fr)	NoLA	AL; LA		1818/
081	merchant	1816	fam; col	LeHavre (Fr)		NY	MD	1808	—
082	captain	1800		St-Do	La Jeune-Henriette (Fr)	Charlstn	AL; PA		1821
083	merchant	1816		Antwerp (Bel)		NY	CdA; LA		
084	planter	1815/		LeHavre (Fr)			LA		—
085	gold/silversmith	1793	fam; col	St-Do	Charming Betsy (Am)	Phila	AL; PA	1805	—
086	tradesman	1793	fam; col	St-Do	Charming Betsy (Am)	Phila	AL; PA	Am	—
087	soldier?	1815/		St-Do		Phila	PA		
088	tawer	1815/		Nantes (Fr)	Unknown (Am)	NY	KY; LA		
089	lieutenant general	1815	—	LaRochelle (Fr)	Medora (Am)	NY	AL; PA		1820
090	secrty. for minstr. of police	1816	fam	LeHavre (Fr)	Shakespeare (Am)	NY	AL; NY		
091							—		
092	judge	1817	serv; col	Leghorn (Ita)	Saunders (Am)	Phila	PA; NY		1831
093	infantry major	1816		LeHavre (Fr)		NY	NY		1830

01	1	2	3	4	5	6	7	8	9	10
	NAME	BIRTH			DEATH		MARRIAGE			
	Given name(s)	ori	place	date	place	date	place	date	spo	desc
094	COMBES Germain	Fr	Hte-Grne (Fr)	1748	Phila	1828				
095	COMBES Jean Vincent	Fr	Hte-Grne (Fr)							
096	CONDÉ Charles	Fr	France	1778	Phila	1835		/1800	Fr	Am
097	CONSTANTIN François-Louis	Fr	Lorient, Morbihan (Fr)	1802	Atlanta, GA	1891	AL	1827	Am	Am
098	CONTARDI Louis	Ita	Italy							
099	CONTE Honoré	Fr	Marseille, Bchs-du-Rhô (Fr)		France					
100	CONTE Marius	Fr	Marseille, Bchs-du-Rhô (Fr)		France					
101	COQUILLON Jean-Charles	Fr	Lot-Grne (Fr)	1750	LA?	1826			Fr/StD	Fr
102	COQUILLON François-Barthélémy (brother)	Fr	Lot-Grne (Fr)	1748	NC	1827			Fr/StD	Fr
103	CORSO François	Ita	Mondovi, Piedmont (Fr)	1792	LA	1865	—	—	—	—
104	COUSIN Louis-David	Fr	France	1784				/1820	Fr	
105	CUCHET Jean-François	Fr	Jura (Fr)		Am	1821				
106	CURCIER André	Fr	Bordeaux, Gironde (Fr)	1776	PA	1850/				
107	DAGNEAUX Françoise, Widow, née Guessy	Fr	Quebec (Ca)	1750	AL	1848	St-Do	/1777	Fr/StD	Am
108	DALMAZEAU Jean	Fr					PA	1792/	Fr/StD	
109	DAREMBERT	Fr	Seine (Fr)?							
110	DAVID Étienne H.	Fr	France							
111	DAVID Louise Adèle Gertrude, née de Sevré	Fr	Port-au-Prince, St-Do	1790			Phila	1804	Fr	Am
112	DAVIS Laurence A.	Fr	Saint-Malo, Ille-Vil (Fr)	1789	Mobile, AL	1858	Phila	1807/	Fr/StD	Am
113	DEBROSSE Charles	Fr	France		Am	/1827				
114	DELACROIX LOUVRAIS René-François	Fr	Saint-Servan, Ille-Vil (Fr)	1785	France					

02 OCCUPATION

11 occ. ca. 1815	12 date	13 acc	14 departure	15 ship	16 arrival	17 SET place	18 NAT Am	19 RET Fr
094 physician/pharmacist	/1800				Phila	PA		
095 —	1816	fam; col	Bordeaux (Fr)	James Murdock (Am)	Phila	PA	—	—
096 hairdresser/perfumer	1816	fam; col	Bordeaux (Fr)	James Murdock (Am)	Phila	PA	—	—
097 none	/1810	fam	Nantes (Fr)	Petit-Paul (Fr)	NoLA	AL; GA	—	—
098 soldier?	1817				Phila	CdA; LA	—	/1824
099 tradesman	1815/				Phila	PA	—	—
100 tradesman	1815/				Phila	PA	—	—
101 planter	/1815					LA	—	—
102 planter	/1815					CA	—	—
103 soldier	1815/		LeHavre (Fr)		NY	CdA; LA	—	1820
104 cutler	1816					AL	—	—
105 shopkeeper	1816					—		—
106 merchant	1795		Bordeaux (Fr)		Phila	PA	1803	—
107 none	/1805		St-Do		Balt	AL; NY	—	—
108								
109								
110 shoemaker?	1817	fam	Nantes (Fr)?		Phila	PA		
111 none	/1798	fam	St-Do	Hunter (Am)	Phila	AL; PA		—
112 tradesman	1816		St-Do	Eagle (Am)	Phila	PA		1828
113 lawyer	1815/					AL; PA; FL		—
114 merchant marine officer	1816	—	Brest (Fr)	Caravane (Fr)	Balt	CdA; NY		1818

01	1 Given name(s)	2 ori	3 place	4 date	5 place	6 date	7 place	8 date	9 spo	10 desc
		BIRTH			**DEATH**		**MARRIAGE**			
115	DELAPORTE Louis	Fr	France	*1797*	Bat Rge	1864	MO	1817	Fr	Am
116	DELARODERIE Alphonse Louis	Fr	Dordogne (Fr)	*1784*	AL	1824	Bordeaux (Fr)	1804	Fr	Am
117	DELAUNAY Joséphine, née **Verrier,** widow **de** Boutière	Fr	LaCroix-Bqts, St-Do	*1779*						
118	DELAUNAY Pierre, Junior	Fr	Phila	1808	Am		Am	1818	Fr	Am
119	DELAUNAY Dr	Fr		*1821*						
120	DELPIT	Fr								
121	DÉMÉREST Widow, née **Ogé** Anne Thérèse Constance	Fr	LaCroix-Bqts, St-Do	1791	France		France	/1815	Fr/StD	none
122	DEPREST René	Fr	Châtellerault, Vienne (Fr)	1788	Vaux (Vienne)	1829	Colombia	*1824*	Sp	Fr
123	DEPREST Zacharie (brother)	Fr	Châtellerault, Vienne (Fr)	1793	Chât (Vienne)	1868	Chatellerault	1813	Fr	Fr
124	DEROURE Guillaume	Fr	France							—
125	DESAIFRES Charles	Fr	Ardèche (Fr)		AL	/1819	—	—	—	
126	DESCAVES Adèle Marc Louis	Fr	Dememary, Dutch Guyana	1787			France	1811	Fr	Fr
127	DESCHAMPS François-Marc	Fr	France				France	/1815	Fr	Fr
128	DESCHOULLES Victor	Fr	Rhône (Fr)	1796/	Mobile, AL	/1821	—	—	—	—
129	DESCOURT Alexandre-Nicolas	Fr	Paris	1781	NoLA	1833	France	/1815	Fr	Am
130	DESFOUCH Charles	Fr								
131	DESMARE Nicolas Alphonse	Fr	Caudebec-en-Caux, Sne-Mtme	1792	NoLA	1853	NoLA	1817	Cre	Am
132	DESORMES Charles	Fr	France							
133	DESPLANS Samuel	Fr	said to be Sardinian							
134	DESPORTES Léontine Marie Elizabeth, husb. Violle	Fr	France	1794	LA	1853	AL	1818	Fr	none
135	DEVANJEU Charles **BROSSIER**	Fr	Indre (Fr)	1790/	Mobile, AL	1826				
136	DIRAT Louis-Marie	Fr	Nérac, Lot-Grne (Fr)	1774	Paris	1860	France	1806/	Fr	Fr

02 11 occ. ca. 1815	EMIGRATION						NAT 18 Am	RET 19 Fr
	12 date	13 acc	14 departure	15 ship	16 arrival	17 place		
115 tradesman	1815/		France			AL; PA; LA	1819	Peru
116 cavalry captain	1815/		?			MO; LA	?	—
117 none	?					AL		
118	—	—	—	—	—	AL; PA	Am	—
119						AL		—
120	/1815							1820
121 landowner	1815	—	LeHavre (Fr)	Terrier (Am)	NY	AL	—	
122 clockmaker	1816		Bordeaux (Fr)		NY	PA		
123 clockmaker	?		?		?	—		
124	1817	col	LeHavre (Fr)?		NY	NY; PA		—
125	1815/					AL		—
126 army employee (1813)	/1809; 1816		French Guyana?; LeHavre (Fr)		NY; Balt	MD	Am	
127 soldier	1817	fam; col	Antwerp (Bel)	Wm. P. Johnson (Am)	Phila	PA; LA		1818
128	1816	fam; col	Marseille (Fr)?			AL; LA		—
129 soldier	1815	fam	LeHavre (Fr)		NY	AL; NY		
130	1815/					PA		
131 guard of honor	1816		LeHavre (Fr)	Minerve (Fr)	NoLA	LA		—
132 merchant	1815/					CdA		
133 soldier?	1817		Bordeaux (Fr)			CdA; LA		
134 maid	1816		LeHavre (Fr)			AL; CdA; LA		—
135 tradesman	1815/					AL		
136 sub-prefect	1816	fam; col	Bordeaux (Fr)	Magnet (Am)	Phila	PA		1819

01	NAME	BIRTH			DEATH		MARRIAGE			
	1 Given name(s)	2 ori	3 place	4 date	5 place	6 date	7 place	8 date	9 spo	10 desc
137	**DOR** Marius	Fr								Fr
138	**DOUARCHE** Pierre, Colonel	Fr	Bessan, Hérault (Fr)	1769	Jamaica	1819	France	/1815	Fr	Fr
139	**DROUËT** Pierre	Fr	Southwest France					/1815	Fr	
140	**DUBARRY** Jean Baptiste Marie	Fr	Antist, Hts-Pyr (Fr)	*1764*	Phila	1830	Phila / Phila	*/1797* / 1803	Ge / Fr	Am
141	**DUBOCQ** Guillaume	Fr	Bordeaux, Gironde (Fr)	1772	Phila	1847	St-Do	1799	Fr/StD	Am
142	**DUCHEMIN** Jean François	Fr	France							Am
143	**DUCOING** Pierre Sr.	Fr	Bordeaux, Gironde (Fr)	1762	StThomas	1821	Bordeux	1807	Fr	Am
144	**DUCOMMUN** Joseph	Sw	Bergues StVinox, Nord (Fr)	1776			Nantes	/1803	Fr	Fr
145	**DUCOUTUMANY** Clément	Fr	Vijon, Indre (Fr)	1785	Mobile, AL	1837	AL	/1820	Fr	none
146	**DUFOUR** Jean-Jacques	Sw	Vevay (Sw)	1763	Vevay, IN	1827	Vevay, IN	/1796	Sw	Am
147	**DUFOUR** Daniel-Vincent	Sw	Vevay (Sw)	1789	Vevay, IN	1842/	KY	1825	Am	Am
148	**DUFOUR** Jean-François	Sw	Vevay (Sw)	1783	Vevay, IN	1850	KY	1806	Am	Am
149	**DUMAS** Antoine	Fr	France							Am
150	**DUMESNIL** Antoine	Fr	Paris	*1772*	KY	1833	MA	*1800*	Am	Am
151	**DUPONT** Pierre-Charles	Fr	France	*1768*						
152	**DUPOUY** Nicolas-Alexandre	Fr	Southwest France					Y		
153	**DUPUY** Pierre	Fr	StFoy-la-Grande, Gironde	1792	AL	1820	—			
154	**DURAND** Jean-Baptiste	Fr	France					—	—	—
155	**DURIVE** François Marie **MARTIN**	Fr	Bergerac, Dordogne (Fr)	*/1785*	AL	1834	Bergerac (Fr)		Fr	Fr
156	**DUTERTE** Mathurin	Fr	France		Phila	1825				
157	**DUVAL** Jacques Séraphin	Fr	Mantes-la-Jolie, Yvelines	1766	Phila	1842	Phila	1796	Am	Am
158	**EMERY** Louis Jr.	Fr	France	*1786*	Phila	1860/	Phila	/1810	Am	Am
159	**ENGLEBERT** Léonard	Fr	France	*/1770*	Phila	1843	Phila	1799	Am	Am
160	**ENSFELDER** Lucien	Fr	Strasbourg, Bas-Rhin (Fr)	*1794*					Fr	Fr
161	**ESTRIBAUD** Charles	Fr	Southwest France							
162	**FAGOT** André	Fr	Kaskasias, IL	*/1768*	Am		Ark Post	1798	Cre	Am

02	OCCUPATION	EMIGRATION					SET	NAT	RET
	11	12	13	14	15	16	17	18	19
	occ. ca. 1815	date	acc	departure	ship	arrival	place	Am	Fr
137	artisan	1815/					PA	—	—
138	colonel	1817		Bordeaux (Fr)			CdA;PA	—	—
139	tradesman	/1815		St-Do		Phila	PA;VA	—	—
140	merchant	1793		St-Do		Phila	PA	1796	—
141	merchant	1803	fam	St-Do		Phila	PA;KY	1804	—
142		?						—	—
143	merchant	/1797		St-Do		Phila	PA	1797	—
144	pharmacist/physician	1815		LeHavre (Fr)		NY	PA	—	—
145	tradesman	1816/		LeHavre (Fr)		NY	AL	—	—
146	winegrower	1796	—	LeHavre (Fr)	Sally (Am)	Phila	KY;IN	—	—
147	military cadet	1818	—				IN	1823	—
148	winegrower	1801	fam	LaRochelle (Fr)	Voodsop (Fr)	Noflk	KY;IN	—	1819
149		?	—	St-Do?			LA	—	—
150	merchant	1791	—	St-Do		Phila	MA;KY	—	—
151	merchant	1815		LeHavre (Fr)		NY	PA	—	—
152	wine merchant	/1804				Phila	PA	—	—
153	husbandman	1817	col	Bordeaux (Fr)	Rebecca (Am)	NoLA	AL	—	—
154	tax collector	1815/					CdA;PA	—	—
155	soldier?	1815/					AL	—	—
156	hairdresser	/1815				Phila	PA	—	—
157	merchant	1791	—	StHelena	Unknown (Am)	Phila	PA	1798	—
158	hatter	/1802				Phila	PA	1807	—
159	tradesman	/1799		St-Do		Phila	PA	1806	—
160	soldier	1815/					AL;LA	—	1819
161	merchant	1816		Bordeaux (Fr)			LA	—	—
162	tradesman	—	—	—	—	—	AR	—	—

01	NAME	BIRTH		DEATH		MARRIAGE				
	1 Given name(s)	2 ori	3 place	4 date	5 place	6 date	7 place	8 date	9 spo	10 desc
163	FALLOT Eugène Hyacinthe	Fr	Paris	*1794*	TX	1818	—	—	—	—
164	FANCHON Honoré	Fr	France		NoLA	1822				
165	FARCY Abel Charles Marie	Fr	Montamy, Calvados (Fr)	*1782*				Y		
166	FARROUILH André	Fr	St-Do	*1781/*	Am?			Y		
167	FAURÈS Placide–Laurent	Fr	Phila	*1794*						
168	FIRMIN B. Charles	Fr	France?							
169	FISCHER	Fr	France?							
170	FOLLIN Jean Charles Auguste	Fr	Cap Français, St-Do	*1778*	Charlstn	/1839	Charlstn	1805	Cre	Am
171	FOLLIN Mathieu Firmin	Fr	Cap Français, St-Do	*1783*	Charlstn	/1851	Charlstn	1806	Cre	Am
172	FOLLIN Firmin Auguste	Fr	Cap Français, St-Do		AL	1828	Charlstn		Fr/StD	Am
173	FOLLIN George	Fr	France	*/1794*	Phila	1849				
174	FONTANGES Pierre Frédéric	Fr	Port-au-Prince, St-D	*1790*	PA	1860/	Balt	1826	Fr	Am
175	FORMENTO Felice Maria Bartolo	Ita	Bagnolo, Piedmont (Ita)	*1790*	Italy	1890	NoLA	/1837	Cre	Am
176	FORNI Fabius, Colonel	Ita	Valenze, Piedmont (Ita)	*1776*						
177	FOUASCHE Pierre	Fr	Sne-Mtme (Fr)							
178	FOUGNET Pierre Sr.	Fr	Coubeyrac, Gironde (Fr)	*1796*	AL	1825	—	—	—	—
179	FOUGNET Antoine (brother)	Fr	Coubeyrac, Gironde (Fr)	*1802*	Gironde	1860	—	—	—	—
180	FOURESTIER Élie B.	Fr	St-Do?		Phila	1851	Phila		Fr/StD	Am
181	FOURNIER Alexandre	Fr	Phila	*1796*	AL	1867	AL	/1823	Fr/StD	Am
182	FOURNIER Honoré François	Fr	France		Phila	1844				
183	FRÉDÉRIC Louis–Auguste	Fr	France				Am			
184	FRENAYE Anne Gilbert Marc Antoine	Fr	Rivière Froide, St-Do	*1783*	Phila	1873	Phila	1812	Fr/StD	—
185	FRENAYE Pierre (brother)	Fr	Gannat, Allier (Fr)	*1789*	Phila	1861	Phila	/1819	Am	Am
186	FUX Jean-Louis	Fr	Eastern France	*1780/*	MS	1873	LA	1826	Cre	Am
187	GAINES George Strother	En	SC	*1784*		1873	AL	1812	Am	Am
188	GALABERT Louis Jacques, Col.	Fr	Castelnaudary, Aude (Fr)	*1773*	Paris	1841	France		Fr	—

02	OCCUPATION 11	EMIGRATION					SET 17	NAT 18	RET 19
	occ. ca. 1815	12 date	13 acc	14 departure	15 ship	16 arrival	place	Am	Fr
163	soldier	1817	col	LeHavre (Fr)		NY	CdA	—	—
164	soldier	1815/					CdA; LA	—	—
165	accountant	1815/					CdA; LA		
166	merchant	/1815		St-Do		Phila	PA	1815	
167	physician	?		—	—		PA	Am	
168		?					LA?		
169	soldier?	1815/					CdA; LA		
170	merchant	/1800		St-Do		NY	SC	1820	—
171	merchant	/1805		St-Do			SC	1805	—
172	veterinarian	/1799		St-Do			AL; SC		—
173	merchant	?		France		Phila	PA	1817	—
174	merchant	/1801		St-Do		Phila	PA		—
175	health officer	1817	col	LeHavre (Fr)	Elisabeth (Fr)	NoLA	CdA; LA		/1860
176	lieutenant colonel	1816					CdA; LA		
177		1815/					PA		
178	husbandman	1817	fam; col	Bordeaux (Fr)	Rebecca (Am)	NoLA	AL		—
179	husbandman	1817	col	Bordeaux (Fr)	Marie–Thérèse (Fr)	NoLA	AL		1829
180	distiller	/1809		St-Do?		Phila	PA	1809	—
181	tradesman	—		—	—		AL; PA	Am	—
182	tradesman	/1805		St-Do			PA	1805	—
183	tradesman	1815/					AL; PA; MS	—	—
184	tradesman	1804	fam; col	St-Do	via Jamaica	Balt	PA	1808	—
185	French teacher	1804	fam; col	St-Do	via Jamaica	Balt	AL; PA		—
186	tradesman	1815/				Phila	CdA; LA		—
187	captain; lawyer	—		—	—		AL	Am	—
188	infantry major	1817		Liverpool (En)	—		CdA; LA	—	1820

| 01 | NAME | BIRTH | | | DEATH | | MARRIAGE | | | |
	1 Given name(s)	2 ori	3 place	4 date	5 place	6 date	7 place	8 date	9 spo	10 desc
189	**GALLARD** Pierre Louis	Fr	St-Do	1761	AL	/1829				none
190	**GARDIEN** Joseph Étienne	Fr	Boissy-sous-StYon, Essonne	1788	AL	1831	LaRochelle (Fr)	1825	Fr/StD	Am
191	**GARESCHÉ LA POTERIE** Jean-Pierre	Fr	St-Do	1780	StLouis, MO	1861	Phila	1813	Fr/StD	Am
192	**GARESCHÉ MAISONNEUVE** Vital Marie	Fr	St-Do	1780	Havana, Cuba	1844	Wilmington, DE	1809	Fr/StD	Am
193	**GARNIER DE SAINTES** Jacques	Fr	Saintes, Cte-Mtme (Fr)	1755	Am	1818	Charente (Fr)	/1789	Fr	Fr
194	**GARNIER** Simon Henry Athanase	Fr	Saintes, Cte-Mtme (Fr)	1792	Am	1818	—	—	—	—
195	**GARNIER** Charles, known as **GARNIER THIOLLIÈRE**	Fr	Pérouges, Ain (Fr)	1779			Paris	1816	Fr	Fr
196	**GASQUET** François Bonaventure	Fr	Montfort/Argens, Var (Fr)	/1790	Guatemala	1832/	NoLA	1832	Fr/StD	Fr/StD
197	**GATTI** Anne Antoine	Ita	Mandello, Piedmont (Ita)	1762	Marne (Fr)	1828	Var (Fr)	/1810	Fr	Fr
198	**GAUNY** Nicolas	Fr	Verdun, Meuse (Fr)	1771	StThomas	1835	Marne (Fr)	1822	Fr	Fr
199	**GÉNIN** Charles François	Fr	Mirecourt, Vosges (Fr)	/1790	LA	1819	France	/1810	Fr	Fr
200	**GEORGE** Pierre Édouard Côme	Fr	Louvres Parisis, Val-d'Oise (Fr)	1769	Phila	1829	LA	1825	Cre	Am
201	**GEORGE** Catherine Victoire, née **Le Grand de Boislandry**	Fr	L'Aigle, Orne (Fr)	1770	Mobile, AL	1843	L'Aigle (Fr)	1791	Fr/StD	Am
202	**GEORGE** Edward Jr.	Fr	Phila	1795	Mobile, AL	1841	Phila		Am	Am
203	**GEORGE** Félix Achille	Fr	Phila	1800	Mobile, AL	1844	AL	1822	Fr/StD	Am
204	**GÉRARD DE VOUAILLES** Hyacinthe	Fr	Neuf-Brisach, Ht-Rhn (Fr)	1789	France	1789				
205	**GERMAIN** Henri	Fr	Séez-en-Tar, Ital-Sav (Fr)	1791	France	1860/	Bat Rge	1829	Fr	Am
206	**GERMOND** Alexandre Ambroise	Fr	Paris	1791	France		NY	1818	Fr	
207	**GIGON** Martial Isaïe	Fr	Gensac, Gironde (Fr)	1804	AL	1825	—	—	—	—

02	OCCUPATION	EMIGRATION					SET	NAT	RET
	11	12	13	14	15	16	17	18	19
	occ. ca.1815	date	acc	departure	ship	arrival	place	Am	Fr
189	jeweler	/1808		St-Do		Phila	AL; PA	1808	—
190	lieutenant	1828	—	LaRochelle (Fr)	*Nile (Am)*	NY	AL; PA	—	—
191	merchant	1795	fam; col	St-Do	*via La Rochelle*	Phila	PA; DE	1805	—
192	merchant	1795	fam; col	St-Do	*via La Rochelle*	Phila	DE; MO	1804	.
193	Chamber member	1816	fam; col	Antwerp (Bel)	*Prince of Orange (Am)*	Phila	PA; IN	—	—
194	cavalry lieutenant	1816	fam; col	Antwerp (Bel)	*Prince of Orange (Am)*	Phila	PA	—	—
195	merchant	1816	fam	LeHavre (Fr)		NY	LA	—	—
196	officer	1816			*Gen. Jackson (Am)*	NY	KY; LA	—	—
197	captain	1817	col	Leghorn (Ita)		Phila	PA	—	/1824
198	armorer/gunsmith	1816		LeHavre (Fr)		Phila	PA	—	—
199	dealer in laces	1817	col	LeHavre (Fr)		NY	LA	—	—
200	supercargo	1794	fam; col	St-Do		Phila	AL; PA	1817	—
201	headmistress, wm's boarding house	1794	fam; col	St-Do		Phila	AL; PA	—	—
202	ship's captain	—	—	—	—	—	AL; PA	Am	—
203	merchant	—	—	—	—	—	AL; PA	Am	1818
204	second lieutenant in infantry	1816	—	Antwerp (Bel)	*Manufactor (Am)*	Phila	CA?	—	1818
205	artillery lieutenant	1817	col	Antwerp (Bel)	*Concordia (Bel)*	Phila	CdA; LA	—	1860/
206	merchant	1815		LeHavre (Fr)		NY	NY	—	1821
207	husbandman	1817	col	Bordeaux (Fr)	*Rebecca (Am)*	NoLA	AL	—	—

01	NAME	BIRTH			DEATH		MARRIAGE			
	1	2	3	4	5	6	7	8	9	10
	Given name(s)	ori	place	date	place	date	place	date	spo	desc
208	GILBAL Antoine	Fr	Moulins, Allier (Fr)	1785						
209	GILBERT Jean Hypolite	Fr	Versailles, Yvelines (Fr)	1790						
210	GODAT Charles	Fr	Doubs (Fr)	*1780*	Am?		Am	1815/	Fr	Am
211	GODEMAR Jean Baptiste	Fr	Guadeloupe?							
212	GODON Rose Victoire née **Brun**	Fr	France	1770/			France	/1807	Fr	Am
213	GOUIRAN Joseph Michel	Fr	Marseille, Bchs-du-Rhô (Fr)	1786			Phila	1824	Fr	
214	GRATZ Hyman	Ge	Phila	1776	Phila	1857	Phila	/1810	Am	Am
215	GRÉGOIRE Étienne	Fr	France							
216	GRILLET François Joseph	Fr	Doubs (Fr)?	*1796*	Am	1836/	PA	1817	Fr	Am
217	GRONING Caspar	Ge	Germany	1776	Am	1836	Am		Am	Am
218	GROUCHY Emmanuel de, Count	Fr	Paris	*1766*	StÉtienne (Fr)	1847	France	1785	Fr	Fr
219	GROUCHY Alphonse de (son)	Fr	Condécourt, Val-d'Oise (Fr)	1789	Paris	1864	France	1822	Fr	Fr
220	GROUCHY Victor de (brother)	Fr	Condécourt, Val-d'Oise (Fr)	*1796*	Paris	1864	—	—	—	—
221	GRUCHET Louis Adrien, Colonel	Fr	Besançon, Doubs (Fr)	1786	Am?		Phila	1818	Fr	
222	GUBERT Joseph Hilaire	Fr	Draguignan, Var (Fr)	1778	Phila	1836	Cuba	1805	Fr	Am
223	GUIBERT Henry	Fr	France				KY	1819	Am	Am
224	GUILLAULT Jean	Fr								
225	GUILLOT Pierre	Fr	France		LA?					
226	HAEZ Jean	Fr	Lyon, Rhône (Fr)	*1765*						
227	HALMA Anselme	Fr	Sedan, Ardennes (Fr)							
228	HAMEL Victor	Fr	Normandy (Fr)		France	1821				
229	HARANEDER Charles	Fr	Paris	1799	Phila		LA	1820	Cre	Fr
230	HAVARD	Fr			Brussels (Bel)	/1848				
231	HERPIN Jean Baptiste André	Fr	Port-au-Prince, St-Do	1795	Mobile, AL	1873	Phila			
232	HIMELY Henry Barthélémy	Sw	La Neuveville (Sw)	*/1795*				/1829	Fr	
233	HUMBERT Jean Jacques Étienne	Fr	Paris	1780	Am?					Am

02	OCCUPATION	EMIGRATION					SET	NAT	RET
	11	12	13	14	15	16	17	18	19
	occ. ca. 1815	date	acc	departure	ship	arrival	place	Am	Fr
208	second lieutenant	1817				NY	CdA; SC		
209	student; merchant	1816		LeHavre (Fr)			NY		—
210	clock box maker	1816	fam				KY		
211		1815/					AL		
212	shopkeeper	1807	fam	France		Bostn	PA		
213	jeweler	1815/				Phila	PA	1822	—
214	merchant	—	—	—		—	PA; KY	Am	
215	soldier?	1815/					PA; LA		
216	officer	1817		Bordeaux (Fr)			CdA; PA		
217	soldier?/inventor	/1815		Bremen (Ge)?			CdA; SC; PA	Am	—
218	lieutenant general	1815	—	Guernsey (En)	Two Brothers (Am)	Balt	PA; DE		1820
219	cavalry colonel	1817	fam; col	LeHavre (Fr)	Ocean (Am)	NY	PA; DE	—	1818
220	cavalry lieutenant	1817	fam; col	LeHavre (Fr)	Ocean (Am)	NY	AL; PA; DE	—	1818
221	cavalry captain	1817	col	Leghorn (Ita)	Saunders (Am)	Phila	CdA; PA; LA	—	1819
222	porcelain dealer	1809	fam	Cuba	Mary (Am)	Phila	PA	1835	
223	dancing master	1815/					KY		
224							AL		
225	artillery officer	1817	col	Leghorn (Ita)	Saunders (Am)	Phila	CdA; LA	—	—
226		1816					AL		
227	painter	1816	col	Bordeaux (Fr)	Bainbridge (Am)	Phila	PA		1820
228	confectioner	/1808				Phila	PA	1808	—
229	merchant's clerk	1817	col	Leghorn (Ita)	Saunders (Am)	Phila	CdA; LA	—	/1826
230		?							
231	tradesman	/1815		St-Do		Phila	AL; PA	Am	—
232	merchant	1797	fam	Hamburg (Ge)	Active (En)	Charlstn	PA; SC	Am	—
233	captain	1816	col	Antwerp (Bel)	La Jeune-Henriette (Fr)	Charlstn	NY	—	—

01 NAME	BIRTH			DEATH		MARRIAGE			
1	2	3	4	5	6	7	8	9	10
Given name(s)	ori	place	date	place	date	place	date	spo	desc
234 HURTEL Pierre Hyacinthe Baptiste	Fr	Brain, Ille-Vil (Fr)	1770	AL	1824	St-Do	1793	Cre	Am
235 HURTEL Jean François	Fr	Brain, Ille-Vil (Fr)	1790	Ille-Vil (Fr)	1856	Bordeaux (Fr)	1818	Fr	Am
236 HURTEL Jean (son & nephew)	Fr	New York, NY	1798	Mobile, AL	1877	Ille-Vil (Fr)	1840	Fr	
237 ILARI Benedetto (Benoît)	Ita	Italy				AL	1821	Fr/StD	Am
238 JACKSON Samuel	En	Am?		Am?					
239 JAMET Joseph Victor	Fr	Rouen, Sne-Mtme (Fr)	1777	France	1840				
240 JEANDREAU Jean	Fr	Cambes, Gironde (Fr)	1772	Mobile, AL	1822	St-Do	1795	Cre	Am
241 JEANNET Louis René, Colonel	Fr	Arcis/Aube, Aube (Fr)				France		Fr	Fr
242 JEANNET Joseph (nephew)	Fr	Troyes, Aube (Fr)	1789	France	/1852				
243 JEANNET-OUDIN Georges Nicolas, aka "le Manchot"	Fr	Arcis/Aube, Aube (Fr)	1762	Arcis/Aube	1828	Arcis/Aube (Fr)	1788	Fr	Fr
						Arcis/Aube (Fr)	1807	Fr	Fr
244 JENIM Édouard	Fr								
245 JOGAN Antonin	Fr	Gironde (Fr)	1784	Am	/1822	France	/1815	Fr	
246 JORDAN Ambroise, Colonel	Pol	Poland	*1769*	TX	1818				
247 JOUNY Louis Michel	Fr	France		Am?					
248 KELLER Jonas	Sw	Switzerland		Am					
249 KIMBAL	En	Am?							
250 KNAPPE Jacques Philippe	Fr	Poitiers, Vienne (Fr)	1797			Phila	*1819*	Fr	Am
251 LABROUSSE Mathieu	Fr	Gensac, Gironde (Fr)	1800	AL?		AL	1841	Fr	Am
252 LACOMBE Pierre	Fr	Bordeaux, Gironde (Fr)	*1759*	Phila	1833	Phila	1798	Fr/StD	Am
						Phila	1818	Fr/StD	Am
						France		Fr	
253 LADURELLE M. François Auguste	Fr	Ardennes (Fr)	*1787*			France?	*/1823*	Fr	
254 LAGAY Paul	Fr	France		NoLA	1850	AL	*/1828*	Fr	Am
255 LAJONIE LAPEYRE Jacques	Fr	Gensac, Gironde (Fr)	1787	Gironde (Fr)	1878	Gironde (Fr)	1814	Fr	Fr
256 LAJONIE Jeanne Élisabeth Hélène	Fr	Demopolis, AL	1820	Gironde (Fr)	1911	Gironde (Fr)	1851	Fr	Fr

02	OCCUPATION 11	EMIGRATION 12	13	14	15	16	SET 17	NAT 18	RET 19
	occ. ca.1815	date	acc	departure	ship	arrival	place		
234	tradesman	1794	fam	St-Do		NY	AL;NY	Am	Fr
235	printer	1815/				NY	NY	—	Fr
236	tradesman	—	—	—		—	AL;PA	Am	—
237	soldier?	1817	col	Leghorn (Ita)	Gen.Jackson (Am)	Phila	PA		
238							PA		
239	sub-prefect	1816	fam	LeHavre (Fr)		NY	NY		—
240	tradesman	/1805		St-Do		Phila	AL;PA		—
241	officer	1817	fam; col	Antwerp (Bel)	Wm. P.Johnson (Am)	Phila	CdA;LA	—	1820/
242	lieutenant	1817	fam; col	Antwerp (Bel)	Wm. P.Johnson (Am)	Phila	CdA;LA	—	1835
243	sub-prefect	1817	fam	Setubal (Por)	Free Ocean (Am)	Phila	CdA;LA	—	/1828
244			col	Bordeaux (Fr)	Susquehanna (Am)	Phila	PA		—
245							—		—
246	captain	1816	col	Amsterdam (Ne)	Amazon (Am)	Phila	CdA;PA	—	—
247	locksmith	1817	col	Bordeaux (Fr)?			PA;NY	—	
248	mechanic	1816	col	Bordeaux (Fr)	Laguira (Am)	NY	PA;NY	1816	—
249								Am	
250	bottler	1816	col	Bordeaux (Fr)	Susquehanna (Am)	Phila	PA		—
251	dyer	1817	col	Bordeaux (Fr)	Rebecca (Am)	NoLA	AL		—
252	merchant	/1798	col	St-Do		Phila	PA	1798	—
253	lieutenant (Spain)	1816	col	Bordeaux (Fr)	Susquehanna (Am)	Phila	PA		—
254	office worker	1815/					AL;LA		—
255	farmer	1816	col	Bordeaux (Fr)	James Murdock (Am)	Phila	AL;PA;LA	—	1829
256	none	—	—	—		—	AL	—	1829

01	NAME	BIRTH			DEATH		MARRIAGE			
	1 Given name(s)	2 ori	3 place	4 date	5 place	6 date	7 place	8 date	9 spo	10 desc
257	LAKANAL Joseph	Fr	Serres, Ariège (Fr)	1762	Paris	1845	France	/1796	Fr	Am
258	LALLEMAND the elder Charles François Antoine	Fr	Metz, Moselle (Fr)	1774	Paris	1839	NY	1804	Fr/StD	none
259	LALLEMAND Henri Dominique	Fr	Metz, Moselle (Fr)	1777	NJ	1823	Phila	1817	Am	Fr
260	LANDEVIN François	Fr	France							
261	LAPEYRE Jean	Fr	La Réole, Gironde (Fr)	1764			St-Do	*1784*	Fr	Fr
262	LAPEYRE Jean-Baptiste	Fr	France		TX	1818				
263	LATAPIE Antoine	Fr	Lourdes, Hts-Pyr (Fr)	1787	Am	1848/	TN	1825	Fr	Am
264	LATAPIE	Fr	Southwest France							
265	LAURENT Clément	Fr	Cap Français, St-Do	1777	NoLA	1844	NoLA	1820	Fr/StD	Am
266	LAURENT Maurice (brother)	Fr	Cap Français, St-Do	1777	Htl Invalides	1852/	France	1819	Fr	Fr
267	LAURET Jean Louis Barthélémy	Fr	Baigts, Pyr-Atls (Fr)	1795	Am	/1846	NoLA	1819	Cre	Am
268	LAVAUD Jean François Sully	Fr	St-Do							
269	LE BOUTEILLIER Michel	Fr	France	1766	AL	1819	Phila	1806	Am	none
270	LE CAMPION François	Fr	France?	1744	Phila	1825	VA	1789	Am	Am
271	LECLERC Joseph P.	Fr	France	*1780*			France	/1810	Fr	Am
272	LECOQ DUMARSELAY André	Fr	Nantes, Loire-Atl (Fr)	1754	Louisville, KY	/1827	Nantes (Fr)	1795	Fr	Fr
273	LEFEBVRE-DESNOËTTES Charles Jean, General Count	Fr	Paris	1773	Ireland (at sea)	1822	Paris	1805	Fr	Fr
274	LEFEUVRE Claude Joseph	Fr	Nantes, Loire-Atl (Fr)	1789	NoLA	1819	Nantes	1809	Fr	none
275	LE FRANÇOIS Jacques	Fr	Rouen, Sne-Mtme (Fr)	*1787*						
276	LE FRANÇOIS Tougnet (brother)	Fr	Rouen, Sne-Mtme (Fr)							
277	LEGRAS Gilbert	Fr	France?		Am?					Am
278	LE GRIS BELISLE Basile	Fr	Clécy, Calvados (Fr)	1754	LeHavre (Fr)	1830	Calvados (Fr)	*/1800*	Fr	Fr
279	LE MAIGNEN Pierre Paul	Fr	La Chapelle-Hareng, Eure	*1797*			Am	*1816*	Fr	none

02 OCCUPATION	EMIGRATION					SET	NAT	RET
11	12	13	14	15	16	17	18	19
occ. ca.1815	date	acc	departure	ship	arrival	place	Am	Fr
257 weights & measures inspector	1816	fam; col	LeHavre (Fr)	*Eugene (Am)*	NY	AL; KY	Am	1837
258 brigadier	1817	—	Smyrna	*Triton (Fr)*	Bostn	CdA; LA; PA	—	1821
259 brigadier	1816	—	Hamburg (Ge)	*Flora (Fr)*	Phila	AL; CdA; LA; PA	—	—
260						—	—	—
261 tradesman	1793		Bordeaux (Fr)		Phila	PA	1795	—
262 infantry captain	1815/				NY	CdA; NY		—
263 cavalry second lieutenant	1815/		France			TN; KY		—
264						CdA		
265 ship's captain	1811	—	Cadiz (Sp)	*Edward&Charles (Am)*	Phila	LA	—	—
266 officer	1817	col	Gibraltar (Sp)	*Savannah Packet (Am)*	Phila	CdA; MO; LA	—	/1825
267 guard of honor	1815/				NY	CdA; LA	—	—
268 tradesman	/1794		St-Do		Phila	PA		—
269 ship's captain	1793		St-Do?		Phila	AL; PA	1798	—
270 tradesman	/1815				Phila	VA; PA	Am	—
271 tradesman	1815/					PA		
272 landowner, private income	1815	fam; col	Nantes (Fr)	*Tennessee (Am)*	Phila	KY	1816	—
273 cavalry lieutenant general	1816	—	Antwerp (Bel)	*Zoé (Fr)*	Phila	AL; PA; LA	—	1822
274 practitioner (physician)	1815/	fam	Nantes (Fr)?			PA; LA	—	—
275 merchant	1817		Amsterdam (Ne)			AL		
276 merchant	1817		Amsterdam (Ne)			AL		
277						AL; LA		
278 merchant	1800		France		Balt	MO; LA	1821	
	1816		LeHavre (Fr)?		?	OH	1826	1829
279 shopkeeper	1816	col	LeHavre (Fr)	*Deux-Frères (Fr)*	NY	PA	1826	

01	NAME	BIRTH			DEATH		MARRIAGE			10
	1	2	3	4	5	6	7	8	9	desc
	Given name(s)	ori	place	date	place	date	place	date	spo	
280	LEMEUSNIER Félix Antoine	Fr	Western France		TX	1818				
281	LEMEUSNIER Jean Joseph brother	Fr	Western France	1790	Mobile, AL	1835	Am?			Am
282	LÉPINE Jean François	Fr	France							
283	LE ROUYER François	Fr	France		TX	1818				
284	LESUEUR Charles Alexandre	Fr	Le Havre, Sne-Mtme (Fr)	1778	LeHavre (Fr)	1846	—	—		—
285	L'HUILLIER MANSUIS Jean	Fr	France				France	/1815	Fr	
286	LUCIANI Pascal M.	Fr	Corsica (Fr)	1791	AL	1853	PA	1815/	En	Am
287	MACRÉ Jean-Marie	Fr	Southwest France	/1785	Phila	1817	France	/1815	Fr	Am
288	MAILLET Henry Pierre Alexandre	Fr	France							
289	MAHÉ Théobald Vincent	Fr	Morlaix, (Finistère)	1792						
290	MALCZEWSKI Constantin Paul	Pol	Warsaw (Pol)	1795	Mexico City		Mexico City	/1824	Mex	Mex
291	MANE Honoré	Fr	Var (Fr)	/1790						
292	MANFREDY Mathieu Ferdinand	Fr	Southeastern France	1792	Havana, Cuba	1823				
293	MANGON Pierre the elder, known as Jean	Fr	Pineuilh, Gironde (Fr)	1799	AR?	/1861	Gironde (Fr)	1840	Fr	Am
294	MANGON Pierre Jr.	Fr	Pineuilh, Gironde (Fr)	1810	AL	1880/	AL	1840	Am	Am
295	MANNOURY Pierre Mathieu	Fr	Buxeuil, Aube (Fr)	1787	France					
296	MARCHAND Louis Pierre Joseph	Fr	St-Do?	1777					Fr/StD	
297	MARIANO Pompée M. A., aka Count de Lestadas, Colonel	Ita	Italy	1784	D.C.	1821				
298	MARTIN Amédée François	Fr	Paris	1793						
299	MARTIN Joseph (father)	Fr	Marseille, Bchs-du-Rhô	1737	Phila	1822	Marseille (Fr)	/1768	Fr	Am
300	MARTIN PICQUET Louis Joseph François Marie (elder son)	Fr	Bastia, Corsica (Fr)	1771	Am		Marseille (Fr)	1796	Fr	Am

02	OCCUPATION 11 occ. ca. 1815	EMIGRATION 12 date	13 acc	14 departure	15 ship	16 arrival	SET 17 place	NAT 18 Am	RET 19 Fr
280	soldier						CdA; NY	—	—
281	soldier	1817	col	Antwerp (Bel)	*Wm. P. Johnson (Am)*	Phila	AL; NY	—	—
282	soldier	1818		LeHavre (Fr)	*Ceres (Am)*		AL; NY; LA	—	1819
283	soldier	1815/					CdA; PA	—	—
284	naturalist artist	1816		LeHavre (Fr)		NY	PA	—	1837
285	musician	1816	fam				PA; LA	—	—
286	non-commissioned officer	1815/				Phila	PA	1817	—
287		1816	fam; col	Bordeaux (Fr)	*William (Am)*	Phila	PA	—	—
288	soldier?	1815/					CdA; PA		
289	guard of honor	1817							
290	officer	1815/		Bordeaux (Fr)	*Magnet (Am)*	Phila	CdA; PA; LA	—	1820
291	surgeon	1816	col	Bordeaux (Fr)			CdA; PA; LA	—	—
292	soldier	1816		LeHavre (Fr)			CdA; LA	—	—
293	husbandman	1817	col	Bordeaux (Fr)	*Rebecca (Am)*	NoLA	AL		1829
		1829	fam; col	Bordeaux (Fr)	*Waltham (Am)*	NoLA	AL		1838
		1848	fam	Bordeaux (Fr)	*Victoria (Am)*	NoLA	TN; AR		—
294	husbandman	1829	fam; col	Bordeaux (Fr)	*Waltham (Am)*	NoLA	AL		—
295	merchant's clerk	1817		LeHavre (Fr)			AL; LA		—
	tradesman							—	
296	tradesman	/1808		St-Do		Charlstn	GA; SC	—	—
297	finance minister of Parma	1815/					PA; KY; DC	—	—
298	engineer	1817		Antwerp (Bel)			PA		—
299	cooper	1807	fam	Marseille (Fr)	*Louisiana (Am)*	Phila	PA		
300	tradesman	1806	fam; col	Bordeaux (Fr)	*Charleston Packet (Am)*	Phila	PA		—

01	NAME	BIRTH		DEATH			MARRIAGE			
	1 Given name(s)	2 ori	3 place	4 date	5 place	6 date	7 place	8 date	9 spo	10 desc
301	MARTIN François	Fr	Bastia, Corsica (Fr)	1774	AL	1842	Am?			Am
302	MARTIN DU COLOMBIER Joseph (father)	Fr	Port Margot, St-Do	1761	Phila	1846	St-Do	1783	Cre	Am
303	MARTIN DU COLOMBIER Prosper (son)	Fr	St-Do	*1790*	Phila	1851	Phila	1805/ *1820*	Am	Am
304	MARTINET Pierre Louis	Fr	St-Do	1790	NY	/1835	NY	1826	Fr/StD	Am
305	MARTINIÈRE Jules Marie l'Amitié CHASSELOUP DE LA	Fr	Port-au-Prince, St-Do	1793	AL		Phila	/1821	Fr/StD	Am
306	MATHIEU Joseph A.	Fr	France	*1756*	Phila	1831				Am
307	MAYER				Am	/1821				
308	MELIZET François	Fr	NoLA	*1785*	Phila	1823	Phila?		Fr	Am
309	MENOU Dieudonné	Ita	Piedmont (Ita)							
310	MERLE Étienne	Fr	France							
311	MESLIER Nicolas Basile	Fr	Barbezieux, Charente	*1779*	Mobile, AL	1849		*1815*	Fr/StD	Am
312	MESTAYER Michel	Fr	Gironde (Fr)	*/1770*	LA		Bordeaux (Fr)	*1794*	Fr	Am
313	MÉTAIS Étienne Jean Baptiste	Fr	Loiret (Fr)	*1785*						
314	MÉTÉYÉ Jean-Pierre	Fr	Cap Français, St-Do	1784	Martinique	1828	Guadeloupe	1808	Cre	Fr
315	MEYNIÉ Jean-Ulysse	Fr	LaTeste, Gironde (Fr)	*1792*	Phila	1832	Phila	1828	Fr	Am
316	MIGNON Jean	Fr	Southwest France							
317	MILON Pierre Solidor	Fr	Italy	*1788*	Phila	1880/	Phila	/1849	Ir	Am
318	MIOT or MILLOT or MILHAUD	Fr	France		Am	/1821				
319	MOCQUARD Antoine-Marie	Fr	Nantes, Loire-Atl (Fr)	1792	France					
320	MONCRAVIÉ Jacques	Fr	Agen, Lot-Grne (Fr)	*1767*	France	/1806	France		Fr	Am
321	MONDIN	Fr	France?							

02	OCCUPATION 11 occ. ca. 1815	EMIGRATION 12 date	13 acc	14 departure	15 ship	16 arrival	SET 17 place	NAT 18 Am	RET 19 Fr
301	tradesman	1806	fam; col	Bordeaux (Fr)	*Charleston Packet (Am)*	Phila	AL; PA		Fr
302	merchant	1793	fam	St-Do		Phila	DE; PA		—
303	foundry worker or owner	1793	fam	St-Do		Phila	PA		—
304	tradesman	/1815		St-Do			NY		—
305	tradesman	1810		St-Do	via a French port	Noflk	AL; VA	1810/	—
306	physician	/1802		St-Do		Phila	PA	1805	—
307									
308	tradesman	—	—	—	—	—	LA		—
309	confectioner					Phila	AL; PA		—
310	non-commissioned officer	1815/					CdA		—
311	clockmaker jeweler	1816	fam	Bahamas	*Mary Ann (Am)*	Phila	AL	Am	—
312	tradesman	1808	—	Bordeaux (Fr)	*Charleston Packet (Am)*	Phila	PA	1808	—
313	tanner	1817	col	Bordeaux (Fr)	*Susquehanna (Am)*	Phila	AL; PA; LA	—	1820
314	merchant	1815	fam	LeHavre (Fr)		NY	CdA; LA	—	1818
315	tradesman	1806	—	Martinique		Phila	NY		—
316	music teacher	1817	col	Bordeaux (Fr)	*Hunter (Am)*	Phila	PA	1817	—
317		1817	col	Bordeaux (Fr)	*Hunter (Am)*	Phila	PA		—
318		1815/		Bordeaux (Fr)			AL	1841	—
319	physician/surgeon	1815/		Nantes (Fr)?	*Ariadne (Am)*	Phila	VA; PA / NY		Fr
320	bakery shop owner	1816		Marseille (Fr)		Phila	AL; PA; LA		1819
321							—		

01 NAME	BIRTH			DEATH		MARRIAGE			
1 Given name(s)	2 ori	3 place	4 date	5 place	6 date	7 place	8 date	9 spo	10 desc
322 **MONNOT** Charles	Fr	Vaufrey, Doubs (Fr)		LA	/1860				
323 **MONTALLEGRI** Hyacinthe	Ita	Faenza (Ita)		Italy					
324 **MONTELIUS** William	Pru	Reamstown, PA (Am)	1782	Phila	1864	Phila		Am	Am
325 **MONY** Dominique Victor de	Fr	Paris	*1795*	France					
326 **MOREL-GUIRAMAND** Jean Marie	Fr	Lyon, Rhône (Fr)	1767	LA?					
327 **MORIN** Charles	Fr	France?	1796						
328 **MOULIN** Sébastien	Fr	Donnery, Loiret (Fr)							
329 **MOYNIER** Théodore Joseph Aristide	Fr	Paris	1790/	LA	1824	—	—	—	—
330 **MURAT** Jean Baptiste	Fr	France							
331 **NARDEL** François	Fr	France							
332 **NARTIQUE** Jean Juste	Fr	Jérémie, St-Do (Fr)	1786	NoLA	1841				
333 **NEEL** Jean Baptiste	Fr	Rennes, Ille-Vil (Fr)	*1794*						
334 **NELSON**	En	Am?		Am					
335 **NIDELET** Étienne François	Fr	Port-de-Paix, St-Do	1789	StLouis, MO	1856	StLouis, MO	1826	Cre	Am
336 **NOËL** Thomas	Fr	Port-au-Prince, St-Do	1770	AL	1846	St-Do	1800	Fr	Am
337 **OLIVIERI** Joseph	Fr	Corsica (Fr)		France or Italy					
338 **ONFROY** Jean Baptiste	Fr	France							
339 **PAGAUD** Pierre	Fr	Bordeaux, Gironde (Fr)	1782	France		Bordeaux	1807	Fr	Fr
340 **PAGNERRE** Jeune André Michel	Fr	Paris	*1786*	LA	1821	France	/1815	Fr	Fr
341 **PAGNERRE** Alexandre Jacques	Fr	Paris	*1785*						
342 **PAGUENAUD** Édouard	Fr	Bordeaux, Gironde (Fr)		AL	/1834				
343 **PAPILLOT** Étienne	Fr	Givry, Saône-et-Loire (Fr)	1770	Bat Rge	1825	Givry	1797	Fr	Fr
344 **PARAT** François Romain	Fr	Hts-Pyr (Fr)		France	1866/				
345 **PARMANTIER** Nicolas Simon	Fr	Lorient, Morbihan (Fr)	1772	FL?	1831/	Lorient	1793	Fr	Fr
346 **PASCHAL** Paul	Fr	France							

| 02 OCCUPATION | EMIGRATION | | | 15 | 16 | SET | NAT | RET |
| | 12 | 13 | 14 | | | 17 | 18 | 19 |
occ. ca. 1815	date	acc	departure	ship	arrival	place	Am	Fr
322 physician/surgeon	1815/					CdA; PA; LA	Am	—
323 officer	1817					PA	—	/1830
324 tobacco merchant	—	—	—		Phila	PA	Am	—
325 cavalry lieutenant	1817	col	Antwerp (Bel)	Wm. P. Johnson (Am)	Phila	CdA; PA; LA	—	1819
326 lawyer for Parliament; notary	/1814		St-Do		Balt	AL; PA; LA	—	—
327						AL		—
328 carpenter	1817	col	Bordeaux (Fr)	Hunter (Am)	Phila	AL; PA	—	1826
329 veterinarian	1817					CdA; LA	—	—
330 tailor?	1815/					PA		
331	1815/				Phila	PA		
332 tradesman	/1803		St-Do		Phila	PA; LA	1806	—
333 soldier	1816		Amsterdam (Ne)			NY		
334 tradesman						PA?	Am	—
335 tradesman	1800	fam; col	St-Do		Phila	AL; MO	—	—
336 tradesman	/1804	fam	St-Do		NY	AL; NY; VA	1827	—
337 soldier?	1815/				NY	NY	—	/1830
338						—		—
339 milliner	1816	col	Bordeaux (Fr)	Bainbridge (Am)	Phila	PA	—	1817
340 pork butcher shop owner	1816	fam; col	LeHavre (Fr)		NY	CdA; LA	—	—
341 pork butcher shop owner	1816	fam; col	LeHavre (Fr)		NY	CdA; LA		
342 cartographer	1816		Bordeaux (Fr)			AL; PA		
343 cavalry captain	1816	col	LeHavre (Fr)	Eugene (Am)	NY	LA		
344 tradesman	1816	col	Bordeaux (Fr)	Susquehanna (Am)	Phila	AL; PA; LA	—	/1832
345 distiller	1808				Phila	AL; PA; FL	1809	—
346	1815/				Phila	PA; LA	—	—

01	1 Given name(s)	2 ori	3 place (BIRTH)	4 date (BIRTH)	5 place (DEATH)	6 date (DEATH)	7 place (MARRIAGE)	8 date (MARRIAGE)	9 spo	10 desc
347	PASTOL Pierrette Julienne "Julie" née Basire, Baroness (widow)	Fr	Dijon, Côte d'Or (Fr)	1784	France		France	/1802	Fr	Fr
348	PAYEN Claude (father)	Fr	Troyes, Aube (Fr)	*1767*	Mobile, AL?	1834/	Sens	1793	Fr	Am
349	PAYEN César (elder son)	Fr	Sens, Yonne (Fr)	1794	NoLA	1822	—	—	—	—
350	PAYEN Augustin (younger son))	Fr	Sens, Yonne (Fr)	1797	Mobile, AL?	1840/				
351	PELAGOT Antoine Zacharie	Fr	Paris	1780	France					
352	PENARD Jean	Fr	Southeastern France		AL	1819	—	—	—	Fr
353	PÉNIÈRES DELZORS Jean Augustin	Fr	Lacou, Corrèze (Fr)	1766	St Aug	1821	Paris	1793	Fr	Fr
354	PÉNIÈRES, Jean Baptiste Auguste (son), called "Émile"	Fr	Paris	1794	France?	/1850				
355	PENNAZZI Louis, Count	Ita	Duchy of Parma (Ita)	1790						
356	PERALDI Toussaint	Fr	Corsica (Fr)		France					
357	PERDREAUVILLE René Élisabeth DE DAVID DE	Fr	Versailles, Yvelines (Fr)	1776	Am		France	1801	Fr	Am
358	PETITVAL Jean Baptiste	Fr	France		Am	1837/				
359	PFISTER Armand, father	Fr	Alsace (Fr)	1752	Mobile, AL	1835	Phila	1795	Fr/StD	Am
360	PICHON Jean Claude Charles	?.	Brienne, Aube (Fr)		France	1834/				
361	PIERCE Jonathan	?.	Am?							
362	PIERCE (brother)	?.	Am?							
363	PILLERO Pierre	Fr	Bagnères-de-Bigorre, Hts-Pyr	1781						
364	PLAIDEAU François	Fr	Cte-Mtme (Fr)				Am	1815/	Fr/StD	
365	PLANTEVIGNE Jean Jacqu. Justin	Fr	Bordeaux, Gironde (Fr)	*1787*	NoLA?	1860/	NoLA	1817	Cre	Am
366	PLINVILLE Guillaume Victor de	Fr	France		LA?	*1821*				
367	POCHARD Augustin François	Fr	Paris	*1788*	France					
368	POCULOT Benoît Marguerite	Fr	Lyon, Rhône (Fr)	1778	AL	1818	Lyon (Fr)	1796	Fr	Fr
369	POTHIER Simon	Fr	St-Do	1758	NY	1830/	St-Do	1788	Cre	Am

02	OCCUPATION 11 occ. ca. 1815	EMIGRATION 12 date	13 acc	14 departure	15 ship	16 arrival	SET 17 place	NAT 18 Am	RET 19 Fr
347	none	1815/				NoLA	LA		1818
348	tanner	1816	fam; col	Rouen (Fr)	*Providence (Fr)*	NoLA	AL	—	—
349	tanner	1816	fam; col	Rouen (Fr)	*Providence (Fr)*	NoLA	AL; LA	—	—
350	tanner	1816	fam; col	Rouen (Fr)	*Providence (Fr)*	NoLA	AL; LA	—	—
351	building contractor	1817	—	Nantes (Fr)	*Confiance (Am)*	Charlstn	NY	—	1818
352		1818		France		NY	AL; NY; PA	—	—
353	Chamber member	1816	serv; col	Bordeaux (Fr)	*Harriet (Am)* wrecked	NoLA	AL; PA; FL	—	—
354	soldier	1817		Leghorn (Ita) France	*Hannah (Am)*	Phila	CdA; LA	—	/1820
355	infantry captain	1816				Phila	CdA; PA	—	
356	valet	1817	col	Antwerp (Bel)	*Saunders (Am)*	Phila	PA	—	1820
357	tutor of pages at imperial court	1801		Martinique		NY	LA	—	—
358	engineer	1815	col	Bordeaux (Fr)	*Hunter (Am)*	Phila	PA; SC	Am	—
359	tradesman	/1795		St-Do?		Phila	AL; PA	—	—
360	engineer	1815		LeHavre (Fr)		NY	NY	—	1827
361							AL	Am	
362								Am	
363	tradesman	1818		Havana		NY	NY		Cuba
364	tradesman	1815/					PA	—	—
365	tradesman	1816		Bordeaux (Fr)		NoLA	LA	—	—
366		1815/		France			CdA; LA	—	—
367	lawyer	1815		Guadeloupe		NY	PA; NY	—	1818
368		1815/	fam				AL	—	—
369	man of law	1796		St-Do		Phila	PA; NY	—	—

| 01 | 1 Given name(s) | BIRTH | | | DEATH | | MARRIAGE | | | |
		2 ori	3 place	4 date	5 place	6 date	7 place	8 date	9 spo	10 desc
370	PROMIS Guillaume	Fr	Bordeaux (Gironde)	1785	AL	/1824	Phila	1809	Fr/StD	Am
371	PROMPT Pierre	Fr	France							
372	PRUDHOMME Charles Barthélémy	Fr	Nantes, Loire-Atl (Fr)	1784			Nantes (Fr)	1809	Fr	Fr
373	PUEEK or PUECH Jean	Fr	Corrèze (Fr)?		AL	1819				
374	QUESSART Jean Sr.	Fr	Gours, Gironde (Fr)	1770	NoLA	1827	NoLA	1815	Fr/StD	none
375	RAGON Pierre	Fr	Belvès-de-Castillon, Gironde	*1800*	France					
376	RAOUL Nicolas Louis, known as Colonel and Count	Fr	Ronceux, Vosges (Fr)	1788	Paris	1850	Mobile, AL; Guatemala; France	1824; 1832; 1840	Ita; Gua; Fr	none; none; Fr
377	RAPIN Joseph	Fr	PA	/1765	Am		Phila	/1800		Fr
378	RAPIN Mathieu (nephew)	Fr	France							
379	RAVESIES Frédéric Guillaume Marie	Fr	Jean-Rabel, St-Do	1782	Mobile, AL	1857	Phila; Phila	1800/; 1851/	Fr/StD; Fr/StD	Am; Am
380	RAVESIES Jean Émile (nephew)	Fr	Bordeaux, Gironde (Fr)	1795	Bordeaux (Fr)	1877	Phila	1821	Fr/StD	Fr
381	RÉAL Pierre François	Fr	Chatou, Yvelines (Fr)	1757	Paris	1834	Paris	*1785*	Fr	Fr
382	REGNAUD DE SAINT-JEAN D'ANGÉLY Michel Louis Étienne	Fr	Saint-Fargeau, Yonne (Fr)	1760	Paris	1819	France		Fr	Fr
383	REGNAUD DE SAINT-JEAN D'ANGÉLY Louis Auguste Michel	Fr	Paris	1794	Nice (Fr)	1870	France; Sne-Mtme (Fr)	C; 1851	Fr; Fr	Fr; Fr
384	REINGEARD Mathurin	Fr	St-Do	/1780					Fr/StD	
385	REYNAUD DE SAINT-FÉLIX Jean Baptiste Gilles	Fr	Paroisse de Torbeck, St-Do	1790			Phila	1818	Am	
386	RICHARD Étienne	Fr	France					Y		

02	OCCUPATION	EMIGRATION					SET	NAT	RET
	11	**12**	**13**	**14**	**15**	**16**	**17**	**18**	**19**
	occ. ca. 1815	**date**	**acc**	**departure**	**ship**	**arrival**	**place**	**Am**	**Fr**
370	jeweler	1800	fam	St-Do	via Puerto Rico	Phila	PA		
371	gold/silversmith	1818	fam	Bordeaux (Fr)	Eagle (Am)	Phila	AL	—	—
372	tinsmith	1818	fam; col	Bordeaux (Fr)	Père-de-Famille (Fr)	NoLA	LA	—	
373		1815		Nantes (Fr)	Tennessee (Am)	Phila	AL; PA	—	—
374	chocolate dealer	1815/		Bordeaux (Fr)		NoLA	AL	—	
375	husbandman	1809	col	St-Do	via Cuba	NoLA	LA	—	1824
376	artillery captain	1817	fam	Bordeaux (Fr)	Marie-Thérèse (Fr)	NoLA	AL; LA	—	1833
377	merchant	—		Le Havre (Fr)	Hunter (Am)	—	PA	Am	—
378		1815	col	Bordeaux (Fr)	Comet (Am)	Phila	—	—	—
379	merchant	1817	fam	St-Do	via Bordeaux	Phila	AL; PA	—	—
380	merchant	1796	—	Bordeaux (Fr)	Susquehanna (Am)	NY	PA	1803	1821
381	prefect of police	1816	fam; col	Antwerp (Bel)	Swift (Am)	NY	NY	—	1826
382	member of Council of State	1815	serv; col	LaRochelle (Fr)	Alex. Paulowitsch (Fr)	Phila	NY	—	1819
383	second lieutenant	1815		LeHavre (Fr)		NY	NY		1816
384	tradesman	/1803	fam	St-Do			GA; MD	1803	—
385	merchant	1800		St-Do		Phila	PA	1812	—
386		1817	col	Bordeaux (Fr)	Susquehanna (Am)	Phila	PA; NY		

01	NAME	BIRTH			DEATH		MARRIAGE			
	1 Given name(s)	2 ori	3 place	4 date	5 place	6 date	7 place	8 date	9 spo	10 desc
387	RICHARD George	Fr	France					/1816	Fr/StD	
388	RICOVIDAL Juan, "General"	Sp	Monovar (Sp)	1773	Monovar (Sp)	1847	Alabama	1819	Fr	none
389	RIEGERT Gabriel Valentin Philippe	Fr	Jallieu, Isère (Fr)	1794	France					
390	RIGAU Antoine, General Baron (father)	Fr	Agen, Lot-Grne (Fr)	1758	NoLA	1820	Maastricht (Ne)	1788	Bel	Fr
391	RIGAU Narcisse Périclès (son)	Fr	Lille, Nord (Fr)	1794			Ht-Rhn (Fr)	1805	Fr	Fr
392	RIVET Pierre George	Fr	France	*1787*	NY	1820	LA	1824	Cre	
393	RIVIÈRE Amédée	Fr	Gironde (Fr)	*1786*	NoLA	1846		Y	Fr?	
394	ROBAGLIA Dominique Joseph	Fr	Corsica (Fr)	*1792*	France	1824			Fr?	
395	ROBARD Joseph	Fr	St-Do	/1790	Phila					
396	ROBIN Thomas	Fr	France		NY	1850				
397	ROLAND DE BUSSY Jean François	Fr	Lons-le-Saunier, Jura (Fr)	1767	France	/1835	Paris	/1794	Fr	Fr
398	ROSTER John	Fr								
399	ROUDEL Nicolas	Fr	France		France					
400	ROUDET Jean-Claude called Corneille Cadet	Fr	Corenc, Isère (Fr)	1789	Mobile, AL	1839	Grenoble, Isère	1808	LA; Cre	Am
401	ROUGIER Amédée	Fr	Southeastern France							
402	RUFFIER Alexandre Ferdinand	Fr	Paris	1795						
403	SAGNIER Henri Antoine	Fr	France	1790/			Phila	1817	Am	Am
404	SAINT-GUIRONS DES TRAVERSES Pierre Pascal (elder son)	Fr	Roquefort-de-Marsan, Landes	1784	Mobile, AL	1849	Mobile, AL	*1832*	Cre	
405	SAINT-GUIRONS Gabriel Alexandre Didier (younger brother)	Fr	Roquefort-de-Marsan, Landes	1789	France					
406	SALAIGNAC Julien Léon	Fr	Phila	1796	Phila	1836/	Phila	1836	Fr	
407	SALMON François	Fr	Coussey, Vosges (Fr)	1797						Am

02	OCCUPATION	EMIGRATION			15	16	SET	NAT	RET
	11	12	13	14	ship	arrival	17	18	19
	occ. ca.1815	date	acc	departure			place	Am	Fr
387	merchant					Phila	PA		
388	Capuchin & deputy Cortes	1816		Cadiz (Sp)	Unknown (Am)	NY	AL; PA	—	1821
389	infantry lieutenant	1817	fam; col	Antwerp (Bel)		NY	CdA; PA	—	1820/
390	brigadier	1817			Tybee (Am)		CdA; LA	—	—
391	adjutant-major	1817	fam; col	Antwerp (Bel)	Tybee (Am)	NY	CdA; LA	—	1831
392	Spanish teacher	1815/				NY	NY	—	—
393	teacher	1815/					AL; LA	—	—
394	artillery lieutenant	1816	col	LeHavre (Fr)	Deux-Frères (Fr)	NY	PA	—	1821
395	tradesman	/1808				Phila	PA	1808	—
396	jeweler						NY		—
397	police captain	1816	fam; col	Antwerp (Bel)	Swift (Am)	NY	NY	—	1821
398	plowman	1817		Amsterdam (Ne)	Francis (Am)	Phila	PA		
399		1815/					AL		
400	nurseryman	1818	fam; col	Bordeaux (Fr)	Magnet (Am)	Phila	AL; PA		—
401		1818	fam; col	Bordeaux (Fr)	Magnet (Am)	Phila	AL; PA		
402	soldier?	1815/		LeHavre (Fr)			NY		1818
403	riding master	1815/				Phila	PA		
404	soldier?/merchant	1816	col	Bordeaux (Fr)	William (Am)	Phila	AL; PA		—
405	soldier?/merchant	1815/		Bordeaux (Fr)?		Phila	AL; PA		1820
406	merchant	—		—		Phila	PA	Am	
407	soldier?/tradesman	1817	col	LeHavre (Fr)	Manchester (Am)	NY	PA	—	—

	NAME	BIRTH			DEATH		MARRIAGE			
01	1 Given name(s)	2 ori	3 place	4 date	5 place	6 date	7 place	8 date	9 spo	10 desc
408	**SARI** Jean Mathieu Alexandre	Fr	Ajaccio, Corsica (Fr)	1792	Paris	1862	Phila	1824	Fr/StD	Fr
409	**SAVARY** Pierre Joseph	Fr	St-Do?							
410	**SAVOURNIN** Joseph	Fr	Bchs-du-Rhô (Fr)	1792			Phila	1815/	Am	Am
411	**SCASSO** Vincencio (Vincent)	Ita	Italy	1790/						
412	**SCHOEN** Johann Sebastian	Pru	Bremen (Pru)							
413	**SCHUBART** Henri	?	France or Germany							
414	**SCHULTZ** Jean	Pol	Warsaw (Pol)	1768	Italy	1821				
415	**SÉVELINGE** Joseph de	Fr	Cap Français, St-Do (Fr)	1779	Phila	1836	Phila	/1815	Am	Am
416	**SIEBENTHAL** Jean François de	Sw	Montreux (Sw)	1785	OH	1857	KY	1806	Sw	Am
417	**SIEBENTHAL** Jean Louis de	Sw	Montreux (Sw)	1788	IN?					
418	**SIMON** [Mathurin ?]	Fr	Gironde (Fr)	/1785			Bordeaux (Fr)	/1806	Fr	Am
419	**SOULAS** Jean Baptiste	Fr	Nantes, Loire-Atl (Fr)					/1816	Fr	Am
420	**STEPHENS** Samuel James	Ir	Dublin (Ir)	1790	AL		LA	1819	Am	Am
421	**STEWART** George Noble	Ir	Burlington, NJ	1799	AL		AL	1826	Fr	Am
422	**STOLLENWERCK** Pierre Anne Chevalier	Ge; Fr	Cap Français, St-Do	1788	AL	1817		W		none
423	**STOLLENWERCK** Pierre François (brother)	Ge; Fr	Cap Français, St-Do	1772	AL	1832	NY	1809	Fr/StD	Am
424	**STOLLENWERCK** Louis Marie Auguste (brother)	Ge; Fr	Cap Français, St-Do	1784	AL	1834	NY	1808	Am	Am
425	**TABELE** William	En	NY?	1776			NY	1818	Fr/StD	Am
426	**TAVERLY** Philippe	Fr	France				NY?	/1799	Am	Am
427	**TAILLADE** Louis François	Fr	Lorient, Morbihan (Fr)	1775	Balt	1819	Elba	1809	Elb	Elb
428	**TASCA** Jean Baptiste	Ita	Italy		Mexico City	1824	France or Italy?	/1815	Elb	none
429	**TEISSEIRE** Antoine	Fr	St-Do	/1796	Paris	1834	Phila	1813	Am	Am
430	**TERRIER** Alexandre René	Fr	France		AL?	1827				

02 OCCUPATION	EMIGRATION					SET	NAT	RET
11	12	13	14	15	16	17	18	19
occ. ca. 1815	date	acc	departure	ship	arrival	place	Am	Fr
408 ship's lieutenant	1815		LeHavre (Fr)		NY	PA; NJ	—	1835
409 tradesman	1816				Phila	AL; PA; LA	1816	—
410			Marseille (Fr)?		Phila	PA		
411 soldier?	1817	col	Leghorn (Ita)	Gen. Jackson (Am)	Phila	PA		
412 soldier?	1816	—	Bremen (Ge)	Hannah (Am)	Phila	PA		
413 soldier?						PA		
414 captain	1817					AL; CdA; PA	—	—
415 distiller	/1813		St-Do		Phila	PA	1813	—
416 winegrower	1800	fam; col	LaRochelle (Fr)	Voodsop (Fr)	Noflk	KY; IN	—	—
417 winegrower	1817	fam	Bordeaux (Fr)			IN	—	—
418 shoemaker?	?					AL		
419 clockmaker	1815/				NoLA	AL; LA		
420 second lieutenant	1815/	fam; col				CdA; PA; LA		
421 lawyer	—		—	—	—	AL; NJ; PA	Am	—
422 gold/silversmith jeweler	1793	fam; col	St-Do	Charming Betsy (Am)	Phila	AL; NY	Am	—
423 gold/silversmith jeweler	1793	fam; col	St-Do	Charming Betsy (Am)	Phila	AL; NY		—
424 gold/silversmith jeweler	1793	fam; col	St-Do	Charming Betsy (Am)	Phila	AL; NY		—
456 broker; postal worker	—		—	—	—	AL; NY	Am	—
426	1815/					PA		
427 ship's lieutenant	1817	col	Leghorn (Ita)	Saunders (Am)	Phila	PA; MD		—
428 soldier?/confectioner	1817	fam; col	Gibraltar (Sp)	Savannah Packet (Am)	Phila	PA	1817	1833
429 merchant	1796	fam	St-Do		Phila	PA	1803	—
430	?		St-Do?		Phila	AL; PA	1818	—

01	NAME	BIRTH			DEATH		MARRIAGE			10
	1	2	3	4	5	6	7	8	9	desc
	Given name(s)	ori	place	date	place	date	place	date	spo	
431	TÊTE Jean Baptiste François	Fr	Port-au-Prince, St-Do	1798	Phila	/1860	Phila	/1830	Fr	Am
432	TETEREL François Hyacinthe	Fr	Le Havre, Sne-Mtme (Fr)	1757	Phila	1824	St-Do	/1800	Fr/StD	Am
433	TETEREL François son	Fr	St-Do							
434	TEXIER Jean	Fr	St-Do?	1771						
435	TEXIER DE LA POMMERAYE Arnaud	Fr	Poitiers, Vienne (Fr)	1768	Belleville, Seine (Fr)	1843	Paris	1797	Fr	Fr
436	THOURON Pierre (father)	Fr	LaRochelle, Cte-Mtme (Fr)	1757	Phila	1824	LaRochelle (Fr)	1782	Fr	Am
							Phila	/1804	Fr	Am
437	THOURON Nicolas Elisée (son)	Fr	LaRochelle, Cte-Mtme (Fr)	1788	Phila	1866	Phila	1826	Fr	none
							Phila	1837	Fr/StD	Am
438	TORTA Giovanni	Ita	Piedmont (Ita)	1795.						
439	TOURNELLE Jacques	Fr	said to be Spanish							
440	TRANSON Jean Marc	Fr	Nantes, Loire-Atl (Fr)	1783	AL	1821	Nantes (Fr)	1807	Fr	Fr
441	TROY Lin	Fr	France	1775	AL?	/1827		Y		none
442	TRUC Jean Joseph	Fr	Sahune, Drôme (Fr)	1782	NoLA	1818	NoLA	1818	Fr	Am
443	TULASNE Louis Etienne	Fr	Cherry Valley, NJ	1795	Am	1873	Am	/1825	Am	Am
444	TULASNE Victor	Fr	Cherry Valley, NJ	1795	NoLA	1837	Am	/1830	Am	Fr
445	VALLOT Joseph	Fr	Dijon, Côte d'Or (Fr)	1794	France	/1861	France		Fr	Fr
446	VANDAMME Dominique Joseph René, General Count	Fr	Cassel, Nord (Fr)	1770	Cassel (Fr)	1830	France	1800	Bel	Fr
447	VASQUEZ Juan Manuel, "Colonel"	Sp	Malaga (Sp)	/1775			Spain	1795	Sp	Fr
448	VAUGINE DE NUISEMENT Ch. Fr	Fr	Bayou Teche, LA	1768	AR	1831	LA	1790	Fr	Am
449	VERNHES Jean-Vincent	Fr	Villaudric, Hte-Grne (Fr)	1777	Phila	1839	Hte-Grne (Fr)	/1816	Fr	Fr
450	VERRIER François Jean	Fr	Nantes, Loire-Atl (Fr)	1782			Nantes (Fr)	1807	Fr	Am
							NoLA	/1825	Fr	Am
451	VIAL Antoine	Fr	Isère (Fr)		LA	1819	—	—	—	—
452	VILLAR Charles	Fr	Lyon, Rhône (Fr)	1764	Am	1840		—	—	Am

02	OCCUPATION 11 occ. ca.1815	EMIGRATION 12 date	13 acc	14 departure	15 ship	16 arrival	SET 17 place	NAT 18 Am	RET 19 Fr
431	merchant	1802		St-Do		Phila	PA	Am	—
432	tradesman			St-Do		Phila	PA	Am	—
433	tradesman			St-Do?		Phila	AL; PA	Am	—
434	merchant	/1808		St-Do			AL; NY		
435	infantry lieutenant colonel	1816		LeHavre (Fr)?		NY	PA	—	1828
436	merchant	1802	fam; col	LeHavre (Fr)	*Tryphena (Am)*	Phila	PA		—
437	merchant	1802	fam; col	LeHavre (Fr)	*Tryphena (Am)*	Phila	PA	1815	—
438	soldier?/merchant	1817	col	LeHavre (Fr)	*Elisabeth (Fr)*	NoLA	CdA; LA		—
439	soldier	1817		Bordeaux (Fr)		NoLA	CdA; LA		—
440	gold/silversmith	1815	fam; col	Nantes (Fr)	*Tennessee (Am)*	Phila	AL; PA		
441	office worker	1817		LeHavre (Fr)			AL; PA		
442	soldier?/merchant	1816	—	Marseille (Fr)		NoLA	LA	—	
443	tradesman	—	—	—	—	—	LA	Am	—
444	tradesman	—	—	—	—	—	LA	Am	—
445	second lieutenant	1817	—	StThomas	*Com. Barney (Am)*	Phila	CdA; PA; LA	—	1821
446	lieutenant general	1817	—	Amsterdam (Ne)	*John (Am)*	Phila	PA	—	1819
447	officer	1815/				—	CdA; PA; LA		
448	warrant officer	—				—	LA	—	
449	textile artisan	1816	col	Bordeaux (Fr)	*James Murdock (Am)*	Phila	PA	—	
450	merchant	1816	col	Nantes (Fr)	*Athalante (Fr)*	Phila	PA; LA		
451	husbandman	1817	col	Marseille (Fr)	*Sachem (Am)*	Phila	LA	—	
452	tradesman	/1798				Phila	AL; LA	1806	—

01 NAME	1	BIRTH			DEATH		MARRIAGE			
		2	3	4	5	6	7	8	9	10
	Given name(s)	ori	place	date	place	date	place	date	spo	desc
453	VILLEMONT Charles Melchior de	Fr	Burgundy (Fr)	1767	Am		AR	1802	Cre	Am
454	VIOLLE François, Dr.	Fr	Mauriac, Cantal (Fr)	1793	LA	*1845*	AL	1818	Fr	none
455	VITALBA Jean Baptiste	Ita	Italy		AR	1823				
456	VOESTER Émile	Ge	Schweln, duchy of Berg	1788	TX	1818				
457	VOGELSANG Daniel	Fr	Eastern France							
458	VORHEES Samuel	Du	LA		LA					
459	WEILL Jacques	Fr	France							
460	WELLS Edward B.	En	France							

02	OCCUPATION	EMIGRATION					SET	NAT	RET
	11	12	13	14	15	16	17	18	19
	occ. ca. 1815	date	acc	departure	ship	arrival	place	Am	Fr
453	soldier	/1794				NoLA	AR	Am	—
454	guard of honor	1817		France		NoLA	AL; CdA; LA	—	—
455	second lieutenant	1817	col	Leghorn (Ita)	*Saunders (Am)*	Phila	CdA; PA; LA	1825	—
456	infantry major	1817	col	Amsterdam (Ne)	*Amazon (Am)*	Phila	CdA; PA; LA	—	—
457	cabinetmaker	1815/				Phila	PA		—
458		—	—	—	—	—	AL; LA	Am	
459	soldier?	1817	col	Antwerp (Bel)	*Wm. P. Johnson (Am)*	Phila	PA		
460	soldier?	1815/		France		Phila	PA		

Notes

ABBREVIATIONS

ADAH   Alabama Department of Archives and History, Montgomery

ADG   Archives départementales de la Gironde, Bordeaux, France

ADLA   Archives départementales de la Loire-Atlantique

ADSM   Archives départementales de la Seine-Maritime, Rouen, France

AGI   Arquivo general de Indias, Seville

AMSF   Archives municipales de Sainte-Foy-la-Grande

AN   Archives nationales, Paris

APS   American Philosophical Society, Library Hall, Philadelphia

BMN   Bibliothèque municipal de Nantes

CADN   Centre des Archives diplomatiques de Nantes, France

FHS   Filson Historical Society, Louisville

LC   Library of Congress, Washington, D.C.

MAE   Ministère des Affaires étrangères, Paris

MD   France Mémoires et documents, France

NA   National Archives, London

NARA   National Archives and Records Administration, Washington, D.C.

NONA   New Orleans Notarial Archives

RG   Record Group

SHD/DAT   Service historique de la défense/Département de l'armée de terre, Vincennes, France

SHD/DM   Service historique de la défense/Département de la marine, Vincennes and Brest, Cherbourg, Lorient, Rochefort, Toulon, France

INTRODUCTION

1. Speech by General de Gaulle at Phnom-Penh, September 1, 1966, stating the French position on the Vietnam War.

2. French ambassador Charles Lucet (1965–72) was accompanied by the consuls of Montgomery, Birmingham, and Mobile, Alabama, and the consul general of New Orleans, Louisiana.

3. Ambassade de France, Service de presse et d'information, New York, per William Henry Britton, former president of the Marengo County Historical Society, "Address Delivered by His Excellency Charles Lucet, French Ambassador to the United States, in Demopolis, Alabama, on October 28, 1967."

4. The legend survived. Demopolis, which had sacrificed its sons to France, among them Adrian Samuel Pizer (born in Demopolis in 1895), killed at Saint Mihiel in the battle of the Meuse on September 12, 1918, and buried in Demopolis's Jewish cemetery, elected another as mayor from 1948 to 1952: Napoleon Bonaparte Fields (1894–1964), called Boney.

5. Saugera, "'She Is a Black Woman.'"

6. *Demopolis Times, Tuscaloosa News,* and *Birmingham News.*

7. Chappet, Martin, and Pigeard, *Le guide Napoléon,* 550.

8. The term "Bonapartist" was created in 1814 to designate men claiming to be followers of Napoleon and claiming his legacy. See Bluche, *Le bonapartisme.*

9. The name *Chickasaw Gallery* is strange, since this part of Alabama was occupied by the Choctaw nation.

10. The "villages of the nation of the Chaquetas" are indicated at the confluence of the Tombigbee and Black Warrior on Nicolas de Fer's 1709 Map of Louisiana and Mississippi (Carte de la Louisiane et du Mississippi), in the collection of the Musée du château des ducs de Bretagne. Guillet and Pothier, *France/Nouvelle-France,* 94.

11. For the terminology, see Groppo, "Exilés, réfugiés, émigrés, immigrés."

12. Chaudron, *Éloge funèbre de l'Empereur Bonaparte.*

13. The Demopolis Public Library has a local history collection in the Gwyndolyn Collins Turner Room.

14. Pickett, *History of Alabama,* 2:386–99.

15. "Aigleville," ca. 1818, 18 x 26 cm wash drawings, France, lithographed by Charles Aubry, professor at the École militaire de Saumur after 1822. Château de Blérancourt (Aisne), Musée national de la coopération Franco-Américaine.

16. As an example, see Lyon, "The Bonapartists in Alabama."

17. J. W. Beeson, "The Vine and Olive Company," a series of twelve articles in the *Demopolis Express,* 1895, reprinted in the *Demopolis Times* as "Early Demopolis," April–July 1950; Whitfield, "French Grant in Alabama"; Reeves, *Napoleonic Exiles in America.*

18. Judge S. G. Wolf and Colonel F. G. Jonah, "Demopolis Founded in 1818 by Exiled French Royalists," *Selma Times-Journal,* March 2, 1927; Franck Willis Barnett, "Demopolis, Site of Early French Settlement, among the Most Romantic Spots in Alabama: Venture in Growing Olives and Grapes Turned Out to Be Failed," *Birmingham News,* May 21, 1932.

19. *The Fighting Kentuckian,* directed by George Waggner, with John Wayne and Oliver Hardy (1949).

20. Emerson, "Bonapartist Exiles in Alabama."

21. W. Smith, *Days of Exile.*

22. Joseph Étienne Gardien, born near Rambouillet in 1788, died in Alabama in 1831; see Natalie B. Webb Cocke and Kent Gardien's article on the Thouron, Gardien, and Roudet families, "French Cemetery Is Part of Vine and Olive," *Greensboro Watchman*, January 21, 1971.

23. Gardien, "The Splendid Fools" and "The Domingan Kettle."

24. A microfilm of the document is at ADAH.

25. Blaufarb, *Bonapartists in the Borderlands*.

26. Ocampo, *Emperor's Last Campaign*.

27. Doher, *Proscrits et exilés;* Petiteau, *Lendemains d'Empire;* Bruyère-Ostells, "Les officiers de la Grande Armée."

28. Hartmann and Millard, *Le Texas*.

29. Philips, *Les réfugiés bonapartistes en Amérique;* Casenave, "Les émigrés bonapartistes de 1815"; Soulié, *Autour de l'aigle enchaîné;* Brice, *Les espoirs de Napoléon*.

30. See the dictionaries of Jourquin, Kuscinski, Pigeard, Quintin and Quintin, Six, and Tulard listed in the Bibliography.

31. Beaucour, "Qui était Roul?"; Beaucour, "Dans le sillage de la Révolution française"; Beaucour, *Un fidèle de l'Empereur*.

32. Guillot, *Le général Lefebvre-Desnoëttes;* Planchot-Mazel, *Simon Bernard*.

33. See, respectively, in the Bibliography: Dossios-Pralat (for Regnaud de Saint-Jean d'Angély), Bigard (for Réal), Florange (for Hentz), Faure (for Pénières), and Jouin (for Lakanal).

34. I regularly consulted three websites, one French, geneanet.fr, for public records on the émigrés; the others American: ancestry.com (Genealogy, Family Trees and Family History Records online) and familysearch.org (Family History and Genealogy Records) of the Church of Jesus Christ of Latter-Day Saints (Mormons), to follow the immigrants and their descendants in the United States. The Mormon website must be used very carefully, as it contains a great many errors and approximations.

35. For these letters see Saugera, "Renaître en Amérique?" vol. 4 (appendix 3), "Lettres d'exil: La correspondance de Jacques Lajonie (1816–1829)."

36. See Tulard, *Nouvelle bibliographie critique*.

37. Saugera, "Renaître en Amérique?" vols. 2 and 3 (appendixes 1 and 2), "Répertoire biographique des réfugiés et des exilés fondateurs des colonies du Tombigbee et du Champ d'Asile."

38. For studies on the Champ d'Asile see Penot, "Les relations entre la France et le Mexique de 1808 à 1840"; Gardien, "Take Pity on Our Glory"; and Klier, "Champ d'Asile, Texas." Blaufarb's *Bonapartists in the Borderlands* offers the best recent approach. For a fictionalized account, see Soublin, *Le Champ d'Asile*, republished as *La république des vaincus*.

CHAPTER 1

Epigraph: Chateaubriand, *Mémoires d'outre-tombe*, 2:201.

1. Stendhal attributes the phrase to a "royal duke" in *Vie de Napoléon*, 309. Vaulabelle calls it "30 years of banditry" in *Histoire des deux Restaurations*, 151.

2. A marshal is a general officer with the highest rank in the French army: Napoleon had twenty-six marshals.

3. Eymery et al., *Dictonnaire des girouettes*. Also see Serna, *La république des girouettes*, 229.

4. On April 13 he drew up the following certificate for General Bernard, his aide-de-camp: "You will uphold the good opinion I have of you by serving the new sovereign of France with the same loyalty and the same devotion that you have shown me." Quoted in Planchot-Mazel, *Simon Bernard*, 1:44.

5. Six, *Les généraux de la Révolution*, 166–67.

6. Chateaubriand, *Mémoires d'outre-tombe*, 2:494.

7. Louis XIV had instituted this royal military order in 1683.

8. Dupont, *Napoléon et la trahison des maréchaux*, 228–29.

9. SHD/DAT, 7 Yd 303 (2), Dominique Vandamme file.

10. According to Pasquier, head of the highways department, to the prime minister, in *Histoire de mon temps*, 3:46.

11. Tulard, "Les épurations de 1814 et 1815."

12. William Lee to James Monroe, Bordeaux, March 12, 1815, in Lee, *A Yankee Jeffersonian*, 166–67.

13. AN, F^7 3784, police report no. 15; see also Le Gallo, *Les Cent-Jours*, 26–29.

14. Gotteri, *Le maréchal Soult*, 581.

15. Chateaubriand, *Mémoires d'outretombe*, 1:196.

16. Besides Fouché, the triumvirate was composed of Maret, Duke of Bassano, an ex-minister and zealot for the Emperor; and Thibaudeau, regicide and erstwhile president of the National Convention and of the Five Hundred and prefect in Marseille during the Empire.

17. Napoleon's marriage to Marie-Louise, daughter of the emperor of Austria, had produced a son, born in Paris in 1811 and proclaimed king of Rome at his birth.

18. Bertier, *Souvenirs inédits d'un conspirateur*, 234.

19. AN, F^7 3784, police reports, octobre–décembre 1814.

20. Quoted by Lavalette, former postmaster general, in Lavalette, *Mémoires et souvenirs*, 2:140.

21. Germain, *J.-B. Drouët d'Erlon*.

22. Staël-Holstein, *Considérations sur la Révolution française*, 269.

23. Lavalette, *Mémoires et souvenirs*, 2:145.

24. Napoleon was on the brig *Inconstant* with Lieutenants Sari and Taillade, whom we will meet later in this narrative.

25. For the period to follow, see works offering two different views, the first highlighting the Emperor, the second, the king and his partisans: Villepin, *Les Cent-Jours;* Waresquiel, *Cent Jours*.

26. Marchand, *Mémoires*, 1:127.

27. As they were both in Paris with Lefebvre-Desnoëttes, this is the most plausible option, but it is not certain. Perhaps only one of them went.

28. Thiébault, *Mémoires*, 5:113, in Pigeard, *Les étoiles de Napoléon*, 431; Frisenberg, *Journal des Sciences Militaires*, 9th series, 65:445, in ibid.

29. Thibaudeau, *Mémoires*, 436.

30. AN, F⁷ 3784, police reports of November 17, 1814 (no. 6), and December 8, 1814 (no. 5).

31. Under Napoleon this was a military school, training artillery and engineering officers.

32. Jérôme Bonaparte, *Mémoires*, 2:28, in Pigeard, *Les étoiles de Napoléon*, 451.

33. See Proctor Patterson Jones, "The Story of No. 6, rue Chantereine (Hôtel de Beauharnais)," in P. P. Jones, *Napoleon*, 422–23.

34. CADN, consulat de France à Philadelphie, Légation 7, Lefebvre-Desnoëttes to Hyde de Neuville, for the king, Washington, December 11, 1818: "Note sur ma conduite et mes sentiments lors de la restauration et du retour de Napoléon."

35. Inspection report of Marshal Ney, colonel general of the royal cavalry corps of France, in Guillot, *Le général Lefebvre-Desnoëttes*, 106.

36. Lefebvre-Desnoëttes to Hyde de Neuville, Washington, December 11, 1818.

37. Dandré's police reports in Guillot, *Le général Lefebvre-Desnoëttes*, 108 n. 2.

38. Lefebvre-Desnoëttes to Hyde de Neuville, Washington, December 11, 1818.

39. Chevalier, *Souvenirs des guerres napoléoniennes*, 311.

40. During the Alençon stop, from February 7–9, 1815, in their card games the royal chasseurs substituted for the number 18 the expression "fat pig," cited in Guillot, *Le général Lefebvre-Desnoëttes*, 110.

41. SHD/DAT, C¹⁸ 34, war minister to Lieutenant General Count d'Erlon, February 1815.

42. Lefebvre-Desnoëttes to Hyde de Neuville, Washington, December 11, 1818. Desnoëttes here contradicts Savary, who claimed that Lallemand had told the Emperor that he had plotted for the Duke of Orleans.

43. Lefebvre-Desnoëttes to Hyde de Neuville, Washington, December 11, 1818.

44. Permanent court-martial of the first military division, session of August 20, 1816, trial of François-Antoine Lallemand, *Le Moniteur Universel*, August 22, 1816, 947. At this time five francs equaled one dollar.

45. Dumas, *Mes mémoires*, 2:63–65.

46. Pion des Loches, *Mes campagnes*, 451.

47. SHD/DAT, 8 Yd 1099, "État des recettes et dépenses faites par le général de brigade Rigau commandant le départment de la Marne depuis le 28 janvier 1815 jusqu'au 21 mars suivant."

48. Thibaudeau, *Mémoires*, 451.

49. François-René de Chateaubriand, *De la monarchie selon la charte*, chap. 34, reprinted in Chateaubriand, *Grands écrits politiques*, 2:361.

50. Thiers, *History of the Consulate and the Empire*, 11:157.

51. For the marshals' attitude see Chardigny, *Les maréchaux de Napoléon*, 321–30.

52. O'Meara, *Napoléon en exil*, in Murat, *Napoléon et le rêve américain*, 149.

53. Cornet, "Le maréchal Grouchy."

54. Godlewski, "Emmanuel de Grouchy," 13–15.

55. Drouët d'Erlon, *Vie militaire*, 101–2, quoted in Pigeard, *Les étoiles de Napoléon*, 70.

56. Claude-François de Méneval, *Mémoires pour servir à l'histoire de Napoléon depuis 1802 jusqu'à 1815*, 3 vols. (Paris: E. Dentu, 1893–94), 2:354, in Madelin, *Fouché*, 2:430.

CHAPTER 2

Epigraph: Chateaubriand, "Rapport sur l'état de la France au 12 mai 1815, fait au roi dans son conseil à Gand," in *Mélanges politiques*, 11:268.

1. The declarations of Cateau-Cambrésis of June 25 and Cambrai of June 28, 1815.

2. Vidalenc, *Les demi-solde*, 34.

3. Las Cases, *Mémorial de Sainte-Hélène*, in Vaulabelle, *Histoire des deux Restaurations*, 2:290.

4. *Bulletin des Lois du Royaume de France*, no. 9, 7th ser., vol. 1, no. 41, "Ordonnance du roi qui prescrit l'arrestation . . .", at the Château des Tuileries, July 24, 1815, 89–91.

5. On July 1, 1815, he had, with seventeen other generals, signed a petition to the Chambers in which he did not accept the Bourbons, whose return would be "the testament of the army and the dishonor of France." Quoted in Waresquiel and Yvert, *Histoire de la Restauration*, 141.

6. The Chamber of 1814 had been dissolved, and replaced, according to new voting methods applied to the elections of August 14 and 22, 1815, by an ultra-conservative Chamber.

7. For the ravages of the Terror and the opposition they aroused, see Schneider, "Les complots sous la Terreur Blanche."

8. Waresquiel, *Le duc de Richelieu*, 272. Under the ancien régime the Chambre Ardente (Burning Chamber) was an extraordinary commission of justice that could apply to a condemned prisoner death by burning.

9. Madelin, *Fouché*, 2:460.

10. *Bulletin des Lois*, no. 10:99. See Doher, *Proscrits et exilés*, 37–43, for an account of the trials.

11. The Chamber was "more counter-revolutionary than the Holy Alliance, more legitimist than Louis XVIII." Bonnal, *Manuel et son temps*, 74–75.

12. In 1815, bills were passed against seditious speeches and shouts (October 16), for the suppression of personal freedom (October 18), and for the temporary reestablishment of provost courts (November 17).

13. Discussion of November 11, 1815, in La Gorce, *La Restauration*, 53.

14. Condemned to death on November 21, 1815, for having on March 20 once more accepted the position of postmaster that he had held during the Empire, Lavalette escaped wearing his wife's clothes. See Lorédan, *Madame de Lavalette*. He then slipped into Belgium thanks to three British officers, among them General Robert Wilson. Wilson was charged with complicity and imprisoned, then sent to England, where he received a hero's welcome in July 1816. Sir Robert Thomas Wilson (1777–1849) subsequently campaigned actively in favor of Napoleon, exiled to St. Helena, and is said to have been involved in plans for his escape contrived notably with Bonapartists who had emigrated to North and South America. See Ocampo, *Emperor's Last Campaign*, 93, 96, 175.

15. Armand Emmanuel de Vignerot du Plessis, Duke of Richelieu (1766–1822), was the grandson of the Marshal of Richelieu and a lateral descendant of the cardinal.

16. *Le Nain Jaune réfugié* (Brussels), 1816, 24, in Bronne, *L'Amalgame*, 156; "Sur l'amnistie royale," *Aurora* (Philadelphia), April 23, 1816.

17. "Like they were in 93, so they were on March 20," fulminated the Count of Castelbajac. *Archives Parlementaires,* vol. 15, January 3, 1816.

18. *Archives Parlementaires,* vol. 16, session of January 9, 1816; *Bulletin des Lois,* no. 58 (no. 349), Paris, January 12, 1816, p. 18: "Loi qui accorde, sauf les exceptions y contenues, une Amnistie pleine et entière à tous ceux qui, directement ou indirectement, ont pris part à la rébellion et à l'usurpation de Napoléon Bonaparte" [Law which grants, except for the exceptions contained therein, a full and complete Amnesty to all those who, directly or indirectly, took part in the rebellion and the usurpation of Napoleon Bonaparte].

19. Bellet and Imbert, *Biographie des condamnés.* The Lallemand brothers were the first to enjoy the title "martyrs of loyalty" (103–4).

20. Macdonald to Davout and the field officers in Bourges, August 1, 1815, in Houssaye, *1815,* 435.

21. SHD/DAT, C^{18} 72, in Guillot, *Le général Lefebvre-Desnoëttes,* 143.

22. AN, F^7 6682, in ibid., 139.

23. Drouët d'Erlon, *Vie militaire,* 102.

24. Chevalier, *Souvenirs des guerres napoléoniennes,* 331.

25. SHD/DAT, C^{16} 42, Commission d'examen (1815–17): information sent by the prefects, no. 43; ADG, 1 M 91, prefect of Gironde to the war minister, Bordeaux, July 29, 1815.

26. Planat de la Faye, *Vie de Planat de la Faye,* 253. See "Journal de ma captivité" (1815–16), ibid., 249–96.

27. MAE, MD France, vol. 2138, letter from Constantinople, May 10, 1816; AN, F^7 6681, Lallemand brothers file; see also Doher, *Proscrits et exilés,* 90–94.

28. AN, F^7 6678, folder 3, the prefect of Ariège to the minister of police, Foix, October 15, 1815.

29. AN, F^7 6682, circular sent on . . . January 1816 by the minister of police, Count de Cazes.

30. AN, F^7 6678, folder 5, the prefect of Corsica to the minister of police, Ajaccio, March 4, 1816.

31. *Le Moniteur Universel,* May 12, 1816, trial of Lefebvre-Desnoëttes.

32. SHD/DAT, 7 Yd 475, Charles Lefebvre-Desnoëttes file.

33. For Rigau in flight see MAE, MD France, vol. 2138, fol. 116, 262, 272, 279.

34. SHD/DAT, 8 Yd 1099, Decazes to the Duke of Richelieu, Paris, May 4, 1816; MAE, MD France, vol. 2138, fol. 272, Duke of Feltre to the Duke of Richelieu, Paris, May 24, 1816.

35. Publication of the sentence in the *Le Moniteur Universel,* May 17, 1816, 571–72.

36. Lefebvre-Desnoëttes to Davout, Prince of Eckmül, April 14, 1816, in *Le Moniteur Universel,* May 16, 1816; also see SHD/DAT, 8 Yd 1099, Rigau file, letter from Lieutenant General Gérard to Davout, Chassons, April 3, 1815.

37. *Le Moniteur Universel,* August 11, 1816, vol. 2: 906; SHD/DAT Drouët d'Erlon file.

38. AN, F^7 6683, file 9, Vandamme to the minister, Limoges, August 7, 1815.

39. Ibid., *Exposé de la conduite du lieutenant général comte Vandamme* (Paris: Impr. Brasseur aîné, December 1815).

40. MAE, MD France, vol. 2138, fol. 24, Countess Vandamme to the king, n.d.; AN, F⁷ 6683, another petition to the king, Paris, January 6, 1816: "The general took no part in the events that drove His Majesty from his kingdom. He has always conducted himself and still conducts himself as a loyal subject."

41. His wife was from Ghent, but Vandamme himself was of Belgian descent. Jean Vandamme, his earliest known ancestor, was a notable and cloth manufacturer in Poperinge, canton of Ypres, in western Flanders, where he was born ca. 1550.

42. See Bronne, *L'Amalgame,* chapter 9; Marquiset, *Une Merveilleuse;* Duvivier, *Les anciens conventionnels;* Duvivier, *L'exil du comte Sieyès;* Duvivier, *L'exil du comte Merlin.*

43. MAE, MD France, vol. 2138, fol. 297, Golay and Vincent to Richelieu, Paris, June 19, 1816; ibid., fol. 298, Richelieu to the ministers, Paris, June 22, 1816.

44. MAE, MD France, vol. 2139, fol. 178–79, Decazes to the minister of foreign affairs, Paris, April 17, 1817.

45. AN, F⁷ 6680, Garnier de Saintes to the Duke of Otranto, minister of police, Paris, April 26, 1815.

46. Ibid., July 27, 1815.

47. Ibid., September 8, 1815.

48. Ibid., on February 1, 1816, the police chief of La Rochelle informed the minister of police of the younger Garnier's departure, judging him "suspicious and quite worthy of such a father," and on April 8, 1816, the prefect of Charente-Inférieure in Saintes described him to the minister as an "enterprising, exasperated young man of means."

49. According to Tournier, 136 former Convention members had already died. *Les conventionnels en exil,* 24.

50. AN, F⁷ 6683, file 4, Pierre-François Réal to the minister of police, Eunery, January 21, 1816.

51. The Tribunate, instituted by Napoleon's constitution of 1799, discussed bills and sent spokesmen to defend or attack them in the Legislative Body. Its one hundred members were appointed by the Senate. It was abolished in 1807 because it opposed too many of the Emperor's projects.

52. Jean-Augustin Pénières to his brother Luc-François Pénières, known as Dumoulin, Valette, January 8, 1816, quoted in Faure, *Pénières-Delzors,* 183.

53. For the complete file on this matter see AN, F⁷ 6714, plaq. 6, fol. 638–52.

54. Prefect of Corrèze to the minister of police, Tulle, January 29, 1816, in Kuscinski, *Dictionnaire des conventionnels,* 480; also see AN, F⁷ 6709, Corrèze, list of individuals included in the purview of article VII of the law of January 12: "[concerning the help that Pénières] gave to a great number of persons of all classes in all the most difficult times."

55. Jean-Augustin Pénières to his brother Raymond Pénières, known as Dubois, Bordeaux, March 7, 1816, in Faure, *Pénières-Delzors,* 184.

56. Sapori, *L'exil et la mort de Joseph Fouché,* 23.

57. Villèle, December 2, 1815, in Madelin, *Fouché,* 2:495.

58. The Duke of Otranto to Princess Élisa, August 6, 1810, in ibid., 2:217.

59. On this episode, see Pasquier, *Histoire de mon temps,* 3:419–20; and Arnaud, *Mémoires et relations politiques,* 3:198–200.

60. *Le Conservateur impartial,* February 3, 1816, in Madelin, *Fouché,* 2:505, mentioned the possibility of a refuge in London or New York.

CHAPTER 3

Epigraphs: B. Clauzel, *Exposé justificatif,* 21; Lajonie Papers (Gironde), Lajonie Lapeyre, *Mémoires justificatifs.*

1. Marzagalli, "Bordeaux et les Etats-Unis," 2:494–502 (appendixes 16–17).

2. Crouzet, "Bilan de faillite"; Butel, "Le temps des revers," 173–75; Butel, "Guerre et commerce."

3. Bertier de Sauvigny, *La Restauration,* 13.

4. Meaudre de Lapouyade, "Voyage d'un Allemand à Bordeaux en 1801," 171.

5. Report by the prefect of Landes, March 11, 1813, in Redon, "La révolution bordelaise," 197.

6. AN, F^7 3784, in a note on Bordeaux, given to the police and included in the "bulletin d'analyse de la direction générale de la police du 22 novembre 1814, no. 6."

7. J.-S. Rollac, *Exposé fidèle des faits authentiquement prouvés qui ont précédé et amené la journée de Bordeaux au 12 mars 1814* (Paris: A. Egron, 1816), in Redon, "La révolution bordelaise," 203.

8. Rollac, Louis de la Rochejaquelein, Alexandre de Lur-Saluces, the Chevalier de la Barthe and Taffard, known as de Saint-Germain.

9. Jean Baptiste Lynch (1749–1835), descended from an Irish Catholic family in exile in Bordeaux, councilor-general of Gironde during the Consulate, appointed mayor of Bordeaux by the Emperor from 1808 to 1815, count of the Empire in 1810, honorary mayor of Bordeaux and peer of France during the second Restoration.

10. Beauchamp, *La Duchesse d'Angoulême à Bordeaux.*

11. AN, F^7 7023, appeal of March 15, 1814, in Redon, "La révolution bordelaise," 496, appendix 16.

12. William Lee to James Monroe, La Rochelle, March 20, 1814, in Lee, *A Yankee Jeffersonian,* 163–64.

13. Redon stresses the uniqueness of this event: "it is the first, and till now the only, time that the signal for a change of regime has been given in France by a provincial city" ("La révolution bordelaise," 18).

14. This expression of the king's gratitude of March 30, 1814, is quoted in Redon, "La révolution bordelaise," 309–10.

15. This distinction, created on June 5, 1815, was granted to all those who participated in the day of March 12, that is, 1,247 people (Redon, "La révolution bordelaise," 347–48).

16. Martignac, *Bordeaux au mois de mars 1815,* 7.

17. B. Clauzel, *Exposé justificatif,* 21.

18. Count of Tournon, prefect of Gironde, to Prosper de Barante, Bordeaux, July 30, 1815, quoted in Barante, *Souvenirs,* 2:183–84.

19. See Blondy, "Les ultra-royalistes bordelais."

20. William Lee to Tournon, Bordeaux, August 14, 1815, in Lee, *A Yankee Jeffersonian,* 168.

21. "Down with the Americans! They are rogues who should be hanged! Long live the English!" William Lee to Henry Jackson, secretary of the American legation in Paris, Bordeaux, September 20, 1815, in ibid., 172.

22. The son of Abraham Sasportas, a refugee from Saint-Domingue, had settled in Charleston, South Carolina, where he had married an American and started a family. ADG, series 1, "Déclaration de personnes du culte hébraïque," in *Bordeaux, la Guyenne et les Etats-Unis*, 50, no. 133.

23. Lee, *A Yankee Jeffersonian*, 173.

24. Barante, *Souvenirs*, 2:183–84.

25. Lee, *A Yankee Jeffersonian*, 173.

26. AN, F^7 6618, file 2615, Bayonne, April 3, 1813: the consul's infatuation with the Empire was well known. In connection with issuing passports to children born to naturalized American citizens, Lee asked Savary for instructions "compatible with my duties toward my government . . . considering that I am never happier than when I can prove to His Majesty's officers my respect for his government and my affection for the nation." His visceral Anglophobia was an open secret, particularly after he wrote *Les États-Unis et l'Angleterre, ou Souvenirs et réflexions d'un citoyen américain . . .*, published in December 1814 in Bordeaux. Lee had been forced by the royalists to add a chapter favorable to the Bourbons. On December 20, 1814, he wrote to Thomas Jefferson: "I take the liberty to send you a copy of a work which I have published here. . . . I beg you will read it with indulgence, particularly that part relating to the Bourbons, which the authorities here insisted on my inserting before they would permit me to print it." Lee, *A Yankee Jeffersonian*, 165–66.

27. On October 20, 1815, complaining to James Monroe of the allies' behavior in Paris, William Lee expressed his true feelings: "The conduct of the Allies at Paris has retrieved in some measure the character of Napoleon. His generosity and magnanimity towards them as a conqueror is openly spoken of, while their disgraceful conduct towards their ally, Louis XVIII, is reprobated." Lee, *A Yankee Jeffersonian*, 175.

28. For the case of the Faucher brothers see Dulaure, *Histoire des Cent-Jours*, vols. 3 and 4; Lucas-Lebreton, "Les frères Faucher"; L. Jamet a special issue of the *Cahiers du Réolais*, September 1960; Bécamps, "Les frères Faucher"; and Schneider, "Les complots sous la Terreur Blanche," Vol. 1, 182–88.

29. Jacques Marie François Étienne de Faucher, known as César, and Pierre Jean Marie de Faucher, known as Constantin, sons of Jean-Étienne de Faucher de Ligerie and Marie Françoise Constance Faugeroux, were born in La Réole (Gironde) on September 12, 1760. L. Jamet, special issue of *Les Cahiers du Réolais*, September 1960, 3–4.

30. In 1793 there was a royalist and religious revolt against the Revolution in the western province of Vendée. After bitter fighting, it was viciously put down toward the end of the year by a Revolutionary army.

31. In the Revolutionary army, this was a rank between colonel and brigadier general.

32. For the trial, see "Procès des frères Faucher"; for the sentence, see *Le Moniteur Universel*, October 1, 1815, 1080–81; Faucher, *Procès des frères Faucher;* and Schneider, "Les complots sous la Terreur Blanche," Vol. 1, 182–88.

33. Dalbaret, *Un assassinat juridique.*

34. ADG, 2 U 591, cour d'Assises, Casimir Rougé, 33, of La Rochelle, December 5, 1815.

35. Sainte-Foy-en-Agenais or Sainte-Foy-sur-Dordogne became Sainte-Foy-la-Grande in the mid-nineteenth century. Corriger, *Sainte-Foy-la-Grande.*

36. AMSF, Jurades, Registre B.B. 2, election of Simon de la Johannye as consul, August 15, 1567 and 1578 and January 1, 1587.

37. The Edict of Nantes, proclaimed by King Henri IV in 1598, granted religious freedom to Protestants. It was revoked by Louis XIV in 1685, resulting in the expatriation of many Protestants.

38. The Lajonies lived in the heart of a Protestant stronghold of eight parishes on both sides of two tributaries of the Dordogne, the Durège and the Soulège: Flaujagues, Pessac, Juillac, Gensac, and Saint-Laurent were over 75 percent Protestant; Massugas, over 50 percent; Sainte-Radegonde and Pellegrue, 25 percent. Valette, "La situation religieuse."

39. Lajonie Papers, "Lycée de Bordeaux. Trimestre de Germinal an 13 (March–April 1805). Notes trimestrielles relatives à M. Lajonie." The Bordeaux lycée, which opened in July 1803, did not at first have many students due to the high fees for room and board. P. Courteault, "Les origines du lycée de Bordeaux."

40. Created by Napoleon in 1803 in Fontainebleau, the school was moved to Saint-Cyr near Versailles in 1808. Lajonie's regimental number was 1308. For his military career, see SHD/DAT, 2Ye, personnel file; 4 YB 29, École spéciale militaire, vol. 1, part 1, army file of 16 floréal, year XI to 25 December 25, 1806; 4 YB 32, École spéciale impériale militaire de Fontainebleau, register B containing the students' conduct and academic grades; 2 YB 252, 7e régiment de Dragons (1788–1811).

41. Quintin and Quintin, *La tragédie d'Eylau.*

42. SHD/DAT, 4M 134, Cap. Allenou, "Historique du 7e Dragons," 1880, 632 manuscript pages; Cossé-Brissac, *Historique du 7e régiment de Dragons.*

43. Lajonie Papers, manuscript Memoir on his campaigns in 1809.

44. Martinien, *Tableaux,* 545: Lajonie is listed as injured on July 11, 1809, at the battle of Znaim, but it appears to have been on the tenth.

45. AN, base Léonore, L1447063.

46. Lajonie Papers, Paul Martin Collection (Sainte-Foy-la-Grande), Lajonie to his mother, Crema, November 16, 1811.

47. Ibid.

48. Jean Pierre Taupier-Letage, born in Flaujagues (Gironde), ca. 1769, died in Juillac at the château of Le Soulat on July 25, 1849, married in Gensac on May 21, 1793, to Marie Lajonie, born in Sainte-Foy-la-Grande (Gironde) on February 21, 1777, died at the château of Le Soulat on July 26, 1861.

49. Cayre, *Sainte-Foy et ses environs,* 187–90.

50. Vidalenc, *Les demi-solde,* 49–62.

51. Ibid., 63.

52. Lajonie Papers, Me Pierre Martin aîné, imperial notary in Gensac, January 26, 1809. Division was into five parts "as equal as justice and their locality permit" among the five children of Jacques Lajonie, deceased in 1804. The inherited lands and house were at Caville, between Sainte-Foy and le Fleix.

53. The livre was worth about 19 cents when it was replaced by the franc in 1795.

54. ADG, 3 E 21100, Garrau notaire, Lajonie/Noguey-Marasin marriage contract, April 21, 1814.

55. Camille Marcellin Casimir, Count of Tournon-Simiane, auditor of the council of state from 1806 to 1810, then prefect of Rome, where he had a hand in kidnapping Pope Pius VII, appointed prefect of Gironde on July 12, 1815, owed his career to Napoleon. *Nouvelle biographie générale*, 45:546; Moulard, *Le comte Camille de Tournon*.

56. Tournon to Pasquier, September 2, 1815, in Tudesq, "La Restauration," 37.

57. AN, F1cIII/Gironde/6: On February 15, 1816, Prefect Tournon announced the dismissal of 69 mayors, 77 deputy mayors, 269 municipal councilors, and a great number of civil servants.

58. AMSF, Registre des délibérations municipales, 1815: address to "His Majesty Napoleon," April 9, p. 11, and address to "His Majesty Louis XVIII," July 15, p. 10 bis. On July 25 the mayor, Étienne Jauge, mentions a great festival in honor of the king at which people shouted "Long live the king, the Bourbon family" when Louis XVIII's bust was returned to the town hall and the white flag to the end of the building.

59. Address to the king from the city of Bordeaux, April 13, 1814, in Redon, "La révolution bordelaise," 497: "The scepter of Tyranny is finally broken. France can breathe again. The reign of happiness will once more begin."

60. ADG, 3 M 472, "État des fonctionnaires municipaux du département de la Gironde qui ont été provisoirement éloignés . . . "; ADG, 3 M 477, "Liste des candidates présentés par M. le sous-préfet de Libourne . . . le 5 avril 1816."

61. Petiteau, *Lendemains d'Empire,* writes that after returning to civilian life, many officers had difficulty conforming "to the norms of these times of the Restoration and reaction," and gives two examples of hunting misdemeanors committed by two captains (144–45).

62. Lajonie, *Mémoires justificatifs.*

63. Lajonie Papers, Paul Martin Collection, *Pétition pour Jacques Lajonie . . .* , by 120 signatories of the towns of Gensac, Juillac, Pessac, Coubeyrac, Sainte-Radegonde and Flaujagues.

64. ADG, 1 M 91, this was Alexandre de la Salle, appointed by the Count of Damas, civilian and military governor of the Eleventh Division, confirmed by the prefect of Gironde on July 29, 1815.

65. Lajonie, *Mémoires justificatifs.*

66. AMSF, état civil, death of Thomas Dagnac at the hospice of Sainte-Foy, August 31, 1816.

67. Lajonie, *Mémoires justificatifs.*

68. *La Minerve française,* November 1819, vol. 8: 333–36, letter from the chevalier de Bacheville, Lyon, November 27, 1819; Bacheville, *Pétition de M. Barthélémy Bacheville* [for the recall of his brother Antoine]; Bacheville, *Voyage des frères Bacheville;* Bellet and Imbert, *Biographie des condamnés,* 13; also see Doher, *Proscrits et exilés,* for an account of their odyssey; Schneider, "Les complots sous la Terreur Blanche," Vol. 1, 247–49.

69. When Decazes became minister of police, he had relaxed the enforcement of the law of general safety by limiting the right of arrest to investigating magistrates. Waresquiel and Yvert, *Histoire de la Restauration,* 177.

70. The *James Murdock*, 160 tons, Captain John E. Mathieu, owned by Ducoing and Lacombe of Philadelphia, left the Gironde estuary after November 4, 1816, and arrived in Philadelphia on January 6, 1817. Cargo: red wine, dry goods, skins, books, prints, notions, etc. to the order of some twenty-five consignees, including many Philadelphia merchants of French origin, particularly from the southwest: Nartigue, Dubarry, Pagaud, Bousquet, Beylard, Duval, Bujac, Ravesies, Fournier & George . . . Nine registered passengers: Mr. Allard, Mr. Gravier, Mr. [Jean Vincent] Vernhes, Mr. G[ermain] Combes, Mr. Combes, Mr. Nartigue, Mr. Fournier, Dr. Boujar, and a boy. Tepper and Bentley, *Passenger Arrivals at the Port of Philadelphia*, 740; NARA, *Passenger Lists of Vessels Arriving at Philadelphia, Pa., 1800–1882*, M 425, film 24, "Report and Manifest of the Cargo laden on board of the brig *James Murdock* . . .". Original lists of passengers disembarking in Philadelphia are available, ship by ship, at www.ancestry.com.

71. Lajonie Papers, Le Verdon (Gironde), November 4, 1816, first letter from Lajonie to Taupier at Le Soulat.

72. Although Lajonie addressed the letters to specific individuals—his brother-in-law, sister, wife, and friends—each letter was intended to be read by all. The Lajonie Papers include seventy-one other letters addressed to Lajonie and to the château of Le Soulat by nine correspondents.

73. Grouchy, *Mémoires*. Volume 5 (1874) includes the thirty-seven letters to his wife, all but two of which are from the United States.

74. His handwriting is often difficult to decipher. A former army comrade, A. Lerminier, wrote him from Paris on February 1, 1830, "You write like a cat." Lajonie Papers.

CHAPTER 4

Epigraphs: Las Cases, *Mémorial de Sainte-Hélène*, 1:648; MAE, MD France, vol. 2139, fol. 40, Savary to the French ambassador, Smyrna, August 8, 1816.

1. *Mercure-Surveillant*, February 28, 1816, passing on news of February 18 from Lausanne: "The Swiss have lately shown themselves so inhospitable toward the French, that the latter must not feel much like staying in this country."

2. Lajonie Papers, Lajonie to his family at Le Soulat, New Orleans, September 1, 1817.

3. See Poussou, *Bordeaux et le Sud-Ouest;* Cauna, *L'Eldorado des Aquitains;* for a case study, see Bourrachot and Poussou, "Bordeaux et l'émigration."

4. Fouché, Duke of Otranto, to Napoleon, April 23, 1814, printed in the *Moniteur* in August 1814 and quoted in J. Fouché, *Mémoires*, 2:171.

5. Hyde de Neuville, *Mémoires et souvenirs*, 2:128.

6. Georges Firmin Didot, *Royauté ou Empire: La France en 1814, d'après les rapports inédits du comte d'Anglès* (Paris, 1898), 53, in J. Bonaparte, *Le roi Joseph Bonaparte*, 54.

7. Conversation between Napoleon and Monge recounted by François Arago, *Biographie de Gaspar Monge* (Paris, 1853), 131–32, in Pairault, *Gaspar Monge*, 506–7. Arago told Pasquier that the Emperor urged him to accompany him to America. Pasquier, *Histoire de mon temps*, 3:273.

8. Général Becker, *Relation de la mission auprès de Napoléon*, 86–87, in Houssaye, *1815*, 359. For Baudin see Taillemite, *Dictionnaire des marins français*, 31–32.

9. Baudin's answer, received in Rochefort on July 5, 1815, in Bordonove, *La vie quotidienne de Napoléon*, 35.

10. Captain Charles Baudin to William Lee, American consul in Bordeaux, Blaye, July 16, 1815, in *Bordeaux, la Guyenne et les États-Unis*, 72.

11. Bainville, *Napoleon*, 473.

12. James Caret's account, written for Charles J. Ingersoll, *History of the Second War between the United States of America and Great Britain* (Philadelphia: Lippincott, Grambo, 1852), 2nd series, vol. 1, cited in Bertin, *1815–1832*, 14.

13. Article published on August 30, in Bertin, *1815–1832*, 3–4.

14. Guillaume Tell Lavallée-Poussin, born in Poissy in 1794, was a descendant of the painter Nicolas Poussin. See Bénézit, *Dictionnaire critique et documentaire*.

15. Poussin, *Les Etats-Unis d'Amérique*, 37.

16. Planchot-Mazel, *Simon Bernard*, 1:104.

17. SHD/DAT, 2 Ye, personnel file of Édouard René Pierre Charles Dubois de Montulé, born in 1792 near Le Mans (Sarthe), an officer's son,; in 1803 entered the Collège de Vendôme, then the cavalry school of Saint-Germain from 1810 to 1811; a second lieutenant with the Nineteenth Horse Chasseurs in the Russian campaign, sustained a serious injury to his left arm, promoted to lieutenant and made knight of the Legion of Honor in 1813, resigned in 1815 "due to injuries that barely permitted the use of one arm, and to engage in commercial speculation."

18. Montulé, *Travels in America*, 13.

19. ADLA, Inscription maritime, quartier de Nantes, Marine 120 J, rôle d'armement 2377: the *Virginie*, built in 1815, 143 tons, Captain Servanteau, eleven crewmen, thirteen passengers, sailed from Paimbœuf on September 19, arrived in New York on November 6, 1816, after a forty-eight-day crossing.

20. Lajonie Papers, Le Verdon, Lajonie to the château of Le Soulat, November 4, 1816.

21. ADG, 1 M 346, deputy customs inspector to the prefect of Gironde, Pauillac, March 15, 1816; AN, F^7 6709, prefect to the minister of police, Bordeaux, March 18, 1816.

22. *Le Nain Jaune ou Journal des Sciences, des Arts et de la Littérature* (Paris: Eymery-Dantu-Delaunay, 1815). The first editors and writers were Cauchois-Lemaire, the owner, Dirat, Étienne de Jouy, Bory-Saint-Vincent, Harel, Merle, and Lefebvre-Duruflé. See Thiessé, *M. Étienne*, 91–95; Serna, *La république des girouettes*, 194–223.

23. ADG, 8 M 199, *Relevé des navires sortis du port de Bordeaux*, the *Magnet*, Captain Garwood, shipowner Lafitte, 255 tons, 13 crewmen, cargo of wine and various merchandise, departure from Bordeaux for Philadelphia noted on March 1, 1816; according to NARA, *Passenger Lists of Vessels Arriving at Philadelphia, Pa. 1800–1882*, M 425, film 16, the passengers included Dirat and his wife, née Pougeois, and Colonel Pierre Roul.

24. *L'Abeille Américaine*, May 16, 1816, 74.

25. AN, F^7 6709, the prefect of Charente-Inférieure to the minister, La Rochelle, February 1, 1816.

26. AN, F^7 6710, Bernard de Saintes to the minister, January 1816, in Tournier, *Les conventionnels en exil*, 263.

27. ADG, 8 M 199, *Relevé des navires sortis du port de Bordeaux*, the *Harriot* [sic] of Bos-

ton, Captain Pillsbury, shipowner Lafitte, 300 tons, 9 crewmen, cargo of wine, brandy, and other merchandise, departure from Bordeaux for New Orleans on February 29, 1816.

28. *Aurora*, July 22, 1816, "Loss of the ship *Harriet* of Boston."

29. CADN, consulat de Lisbonne, B, 15, letter from the survivors, transcribed by Monsieur de la Tuellière, consular agent in Madeira, to Monsieur Lesseps, consul general in Lisbon, May 16, 1816.

30. AN, F⁷ 6710, Bernard de Saintes to the minister of the interior, Madera, May 16, 1816; Bernard de Saintes to the minister of foreign affairs, October 3 and November 9, 1816, in Tournier, *Les conventionnels en exil*, 263–65.

31. He died on October 19, 1818. See the French consular agent in Funchal, M. de la Tuellière, register 4, no. 1, fol. 44, in Kuscinski, *Dictionnaire des conventionnels*, 51.

32. NARA, *Passenger Lists of Vessels Arriving at Philadelphia, Pa., 1800–1882*, M 425, film 22, the American sloop *Hannah*, 61 tons, Captain John Scheer, arrived in Philadelphia on July 9, 1816.

33. *L'Abeille Américaine*, July 18, 1816, 224. A *questeur* is an administrative and financial officer elected to the French Parliament.

34. An analysis of passports to the United States issued in Bordeaux in the nineteenth century shows that 85 percent were for men and 92 percent were for people traveling alone. N. Fouché, "Les passeports délivrés à Bordeaux," 199. This is confirmed for the emigrant from Lorraine to America during the same period: he was a young man of about twenty-five, generally single, who left alone; if he was married, his wife and children joined him later. Maire, *L'émigration des Lorrains en Amérique*, 75.

35. A. Montholon, *Journal secret*, 67.

36. APS, Girard Papers, reel 15, no. 201. On July 19, 1817, Girard wrote to Joseph Bonaparte that the captain had informed him, in a letter from Le Texel dated May 16, 1817, that Madame J. Bonaparte had made arrangement for a passage on the *Montesquieu* but that he had been unable to consent to this because of his instructions. "Ever since the English have started threatening ships sailing under our flag," deplored Girard, "on the pretext that there were enemy passengers or chartered cargo aboard, I have decided to order all my captains to take on only American seamen and merchandise that would be charged to my account."

37. MAE, CP EU, vol. 74, fol. 49, Consul Pétry to the minister of foreign affairs, Philadelphia, May 24, 1817.

38. Merle d'Aubigné, *La vie américaine de Guillaume Merle d'Aubigné*, 45.

39. MAE, MD France, vol. 2139, fol. 95, *Observations* . . . .

40. AN, F⁷ 6679, file 7 (Clauzel), Count of Tournon to the minister, Bordeaux, February 19, 1816.

41. Ibid.

42. Pasquier, *Histoire de mon temps*, 3:272.

43. Grouchy, *Mémoires*, 4:5–7.

44. AN F⁷ 6681, Grouchy file.

45. On January 19, 1816, the Brussels *Surveillant* published a letter from Guernsey, dated December 27, 1815, stating that Marshal Grouchy had left "for North America about

three weeks previously on the *Two Brothers*, Captain Brown"; AN, F⁷ 6681, Grouchy file; CADN, consulat de Baltimore, série A, no. 47, letter to the Duke of Richelieu, February 15, 1816, announcing Grouchy's arrival in Annapolis.

46. Grouchy to his son Alphonse, November 5, 1815, in Grouchy, *Mémoires*, 4:10.

47. Grouchy to his wife, November 1815, in ibid., 4:11.

48. *L'Abeille Américaine*, December 5, 1816, 127.

49. SHD/DAT, 7 Yd 1163, Alphonse de Grouchy asks Marshal Gouvion Saint-Cyr for leave, Paris, April 3, 1817; SHD/DAT, 7 Yd 1284, Victor de Grouchy makes the same request, April 4, 1817.

50. SHD/DAT, 2 Ye, Philippe-Gustave Le Doulcet de Pontécoulant (1795–1874), graduate of the École polytechnique, son of the Convention member who had helped his brother-in-law Grouchy flee to the United States.

51. SHD/DAT, 2 Ye, on April 19, 1817, from Le Havre, Alphonse de Grouchy informed the war minister of his departure for the United States the next day; for his arrival, CADN, consulat de New York, B, no. 110, French consul in New York to Hyde de Neuville, May 17, 1817; MAE, CP EU, vol. 74, fol. 29v, Hyde de Neuville to Richelieu, Washington, May 20, 1817.

52. AN, F⁷ 6683, Regnaud de Saint-Jean d'Angély to Fouché, Duke of Otranto, Paris, August 20, 1815.

53. AN, F⁷ 6715, fol. 101, special police superintendent to the minister, Le Havre, June 10, 1816.

54. Lakanal to Marc-Antoine Jullien (Robespierre's agent and Saint-Just's friend during the Revolution), Le Havre, January 9, 1816, *Revue des autographes* 150 (March 1895), in Jouin, *Lakanal en Amérique*, 13.

55. AN, F⁷ 6713, plaq. 4, fol. 330–42, for Lakanal's attitude during the Hundred Days as seen by the second Restoration; see also Dawson, *Lakanal the Regicide*, 82–97.

56. Kuscinski, *Dictionnaire des conventionnels*, 329.

57. Florange, *Le conventionnel Hentz*, 161.

58. BMN, the *Feuille commerciale de Nantes*, January 29, 1816, announced the departure from the port of Le Havre on January 23, 1816, of the American brig *Eugene*, Captain P. Destebecho, bound for New York; Florange cites a letter from Hentz to his family, dated March 15, 1816, telling of his arrival in New York in six weeks (Florange, *Le conventionnel Hentz*, 165); CADN, consulat de New York, B, no. 110, the French consul announced the arrival of the *Eugene* in a dispatch of March 20, 1816.

59. SHD/DAT, 2 Ye, personnel file of Jean-Jérôme Cluis.

60. Various writers (following historian Albert James Pickett, ADAH, Pickett Papers "Notes Taken from the Lips of the Honorable George N. Stewart") claim that Cluis was for a time the jailer of the Spanish prince and future Ferdinand VII, held at the château of Valençay (Indre) from 1808 to 1813. This information could not be verified (particularly in AN, sous-série F⁷, "délivrance des passeports et listes des voyageurs du dép. de l'Indre"), but it is plausible: Cluis was from La Châtre near Valençay and was working under the direct orders of Savary, minister of police, involved in the kidnapping of the Spanish prince. Family tradition has it that Napoleon thanked him with the gift of a silver service that his

French widow later asked for. Communication from Daniel Cluis (Quebec) in reference to a letter of January 22, 2005, from Ann Kendall Ray.

61. AN, F⁷ 6678, file 1; BMN, *Feuille commerciale de Nantes*, March 1816, departure from Le Havre on March 17, 1816, of the American brig *Shakespeare*, 300 tons, Captain Curtis Holmes; *Feuille commerciale de Nantes*, January 26, 1816.

62. Charles-François Riffardeau, Marquis of Rivière (1763–1828), appointed ambassador to the Ottoman government at the first Restoration, arrived in Constantinople on June 4, 1816.

63. Marquis of Rivière to the minister of foreign affairs, in Melchior-Bonnet, *Un policier dans l'ombre de Napoléon*, 290.

64. MAE, MD France, vol. 2139, fol. 40, Rovigo to the Marquis of Rivière, Smyrna, August 8, 1816.

65. Savary, *Mémoires*, 285.

66. *L'Abeille Américaine*, December 26, 1816, 176, citing information of December 20 from Boston concerning the *Only Son* of Philadelphia.

67. Jerome Bonaparte (1784–1860), at the time a young midshipman, had stayed in Saint-Domingue only briefly in the winter of 1802 before being sent back to France to give the First Consul an early report on the results of the military expedition. The next autumn he returned to the West Indies on a corvette that he was soon to command, with the mission of touring the islands. When the Peace of Amiens collapsed, rather than risk capture by the English on his way back, he abandoned his ship and took a Danish ship to the United States, a country he dreamed of visiting. In Baltimore he met Elizabeth Patterson, known as Betsy, whom he married on December 24, 1803. For this episode see Saffell, *Bonaparte-Patterson Marriage*; Boudon, *Le roi Jérôme*, chapter 3, "L'aventure américaine."

68. SHD/DAT, 7 Yd 851, Charles François Antoine Lallemand file.

69. For Henri's arrival see the *Mercure-Surveillant*, July 11, 1816, repeating information from American papers published in London on June 6. For his port of departure, see the *Mercure-Surveillant*, July 2, 1816, repeating information printed in London papers on June 25, announcing that Henri Lallemand had arrived in Philadelphia from Hamburg the previous month.

70. *New York Advertiser*, April 30, 1817, in H. G. Warren, *The Sword Was Their Passport*, 190 (Warren says this was Henri Lallemand rather than his brother Charles); *New York Evening Post*, April 29, 1817, in Ocampo, *Emperor's Last Campaign*, 141.

71. Bibliothèque Royale Albert Ier, Brussels, J. B. 180, extract from the *Libéral*, June 10, 1817, citing a London paper of June 3, 1817.

72. CADN, consulat de France à New York, B, no. 110, French consul in New York to Hyde de Neuville, May 12 and 17, 1817.

73. AN, F⁷ 6683, file 4, extract from a report from police headquarters, November 30, 1815: "Master Réal is requesting a passport for America, where he wishes to settle, but as the season is already too far advanced, he wishes to retire provisionally to a farm he owns in Belgium."

74. Ibid.

75. Jacques Donatien Le Ray de Chaumont (1760–1840), known as James Le Ray in

the United States, son of Jacques-Donatien Le Ray de Chaumont (1725–1803), shipowner and merchant in Nantes, who amassed a huge fortune with which he supported the American cause during the War of Independence. Arriving in the United States for the first time in 1785, Le Ray de Chaumont junior married Grace Coxe of Burlington, New Jersey, in 1789. He then moved back and forth between France—where he was born and died—and the United States. After a stay in France from 1810 to 1816, he sold a tiny part of the 600,000 acres he owned on the shores of Lake Ontario through two companies—the Castorland Company and the Antwerpen Company. See Schaeper, *France and America in the Revolutionary Era;* Clarke, *Émigrés in the Wilderness.*

76. AN, F[7] 6683, file 4, extract from the visa registry; Stadsarchief Antwerpen, MA 2644/5, passport register 1816, no. 5506 bis and 5507 bis; arrival announced by the *New York Colombian* and reprinted by the *Aurora* on August 15, 1816.

77. Kuscinski, *Dictionnaire des conventionnels,* 280–81. Garnier, who had already published *Le retour de la vérité en France* (The Return of Truth in France) in Paris in 1815, published *La dette d'un exilé, ou plan nouveau d'éducation nationale* (The Debt of an Exile or New Plan for National Education) in Brussels the same year as *Adieux.*

78. Stadsarchief Antwerpen (Antwerp), MA 2644/6, passport register, no. 6456, no. 6469; NARA, *Passenger Lists of Vessels Arriving at Philadelphia, Pa., 1800–1882,* M 425, film 23.

79. NARA, *Passenger Lists of Vessels Arriving at Philadelphia, Pa., 1800–1882,* M 425, film 22.

80. CADN, consulat de New York, série C, no. 1, consul in New York to the minister of foreign affairs, New York, November 10, 1817; Williams Research Center, New Orleans, *L'Ami des Lois et Journal du Soir,* December 16, 1817, quoting the *Papier de New York.*

81. CADN, consulat de New York, série C, no. 1, consul in New York to the minister of foreign affairs, New York, July 19, 1817.

82. MAE, MD France, vol. 2139, fol. 252, Minister of police, Decazes, to the minister of foreign affairs, Paris, June 4, 1817, and ibid., fol. 256, minister of foreign affairs to Hyde de Neuville, Paris, June 14, 1817.

83. CADN, consulat de New-York, série C, no. 1, consul in New York to the minister of foreign affairs, July 31, 1817: "While General Vandamme embarked in Amsterdam under a pseudonym, on a vessel that arrived here a few days ago [the *John* of Baltimore], with a great many passengers, he is using his true name again here. It is said that he wants to purchase property and that he is expecting his wife" (see also his dispatch of August 4, 1817); *L'Ami des Lois* (New Orleans), September 6, 1817; NARA, *Passenger Lists of Vessels Arriving at Philadelphia, Pa., 1800–1882,* M 425, film 24.

84. *L'Abeille Américaine,* July 31, 1817, 49.

85. NARA, *Passenger Lists of Vessels Arriving at Philadelphia, Pa., 1800–1882,* M 425, film 24.

86. Ibid., film 25.

87. Bertrand, *Cahiers de Sainte-Hélène,* 1:93–94, July 29, 1816.

88. "France," *Niles' Weekly Register,* September 6, 1817, 7.

89. Vauthier, "Notes sur les Français," 47, in Rémond, *Les États-Unis devant l'opinion française,* 44.

90. N. Fouché, "Les passeports délivrés à Bordeaux," 190.

91. Tepper and Bentley, *Passenger Arrivals at the Port of Philadelphia, (1800–1819).*

92. Bentley, *Passenger Arrivals at the Port of New York;* Tepper and Bentley, *Passenger Arrivals at the Port of Baltimore.*

93. The passport registers in Antwerp and Bordeaux, the lists of passengers on French ships sailing from Nantes and Le Havre, or disembarking from American ships in Philadelphia.

94. The Archives départementales de la Loire-Atlantique, série Inscription Maritime, rôles d'armement et de désarmement de bord et de bureau, quartier de Nantes, have lists of passengers embarked, with their names, professions, places of origin, ages, and relationships.

95. Boucaud, "Les armaments nantais," 12–49. Eighty percent of the emigrants sailed to New Orleans, Savannah, and Charleston.

96. A perusal, in the Archives départementales d'Ille-et-Vilaine, of the "rôles d'armement et de désarmement du quartier de Saint-Malo" from 1815 to 1820 reveals only a very small number of French ships going to or coming from New Orleans, for the most part.

97. NARA, *Passenger Lists of Vessels Arriving at Philadelphia, Pa., 1815–1818.*

98. Basing his analysis on American statistics (*Reports of the Immigration Commission,* Washington, D.C., 1911), Henri Bunle estimates the number of French subjects who emigrated to the United States between 1820 and 1824 at 1,920 persons, that is, 384 per year; during the reign of Charles X, from 1825 to 1830, emigration rose considerably, with a total of 6,920 persons, that is, 1,153 per year. Bunle, "L'immigration française aux États-Unis," table on p. 203.

99. During the first Empire, passport legislation was very prescriptive, writes Maire (*L'émigration des Lorrains en Amérique,* 41–43), whereas in the United States "neither passports nor papers of any sort are asked for," according to the guidebooks consulted by Rémond (*Les Etats-Unis devant l'opinion française,* 32).

100. "In this country you land, sojourn, and travel about without a passport," writes Montulé (*Travels in America,* 19).

101. Creagh, *Nos cousins d'Amérique,* 216.

102. Chateaubriand, *Mémoires d'outre-tombe,* 1:69.

103. His work (1803–8) can be found in half a dozen American collections, and the name Clorivière regularly appears on lists of southern artists.

104. See Hulot, *Le général Moreau.*

105. Daudet, *L'exil et la mort,* 113, in Beaucour, "Moreau aux Etats-Unis," 286–87.

106. Daudet, *L'exil et la mort,* 113, ibid.

107. Chateaubriand, *Mémoires d'outre-tombe,* 2:179.

108. Quoted in Baeyens, *Sabre au clair,* 131.

109. Leclerc to the First Consul, October 7, 1802, ibid., 102.

110. Ibid., 128.

111. Philadelphia, July 29, 1813, Humbert to Maret, minister of foreign affairs, ibid., 131–32.

112. Ibid., 146–47.

113. For Hyde, see the outstanding work by Watel, *Hyde de Neuville.*

114. Anne Marguerite Henriette Rouillé de Marigny, born ca. 1749, was forty-five at the time of her marriage in 1794 to Jean Guillaume Hyde de Neuville, who was only eighteen.

115. Hyde de Neuville, *Mémoires et souvenirs,* 1:452.

116. Hyde de Neuville to Thomas Jefferson, December 22, 1807, quoted in Watel, *Hyde de Neuville,* 46.

117. Hyde de Neuville, *Mémoires et souvenirs,* 1:486.

118. Hyde de Neuville, *Éloge historique du general Moreau.*

119. Watel, *Hyde de Neuville,* 55–77.

120. Hyde de Neuville, *Mémoires et souvenirs,* 2:1.

121. Watel, *Hyde de Neuville,* 110.

122. Barante, *Souvenirs,* 2:220; Hyde admitted in his *Mémoires* that he had been one of the deputies carried away by their inexperience, the heat of their passions, their lack of discipline, and a spirit of vengeance and anger, motivated by the desire "to become peacemakers" (2:155).

CHAPTER 5

1. This is known as the French and Indian War in the United States.

2. Creagh, "La Révolution," 166.

3. Prospectus for the *Courier de Boston: L'utilité des deux mondes* (Boston: Guérard de Nancrède, 1789), 4–5, in Creagh, "La Révolution," 169.

4. See Fohlen, *Jefferson à Paris.*

5. Jefferson represented his country from May 7, 1784, to September 28, 1788, and Monroe, succeeding Gouverneur Morris, from August 1, 1794, to December 30, 1796. Monroe returned to France in 1803 to discuss the sale of Louisiana to the United States.

6. Lhéritier, "Bordeaux et les États-Unis," 179–80.

7. Bonnel, *La France,* 33–41.

8. When the Jacobins in Paris indicted Genêt, the United States granted him political asylum. Ammon, *The Genet Mission,* 76, in Lentz, "Relations américano-françaises," 11.

9. Peter P. Hill, "La suite imprévue de l'alliance: L'ingratitude américaine (1783–1798)," *La Révolution américaine et l'Europe* (Paris: Éditions du CNRS, 1979) 385–398, in Lentz, "Relations américano-françaises," 11.

10. Creagh, "La Révolution," 184.

11. See Bonnel, *La France;* also Egan, *Neither Peace Nor War.*

12. Mazel, "Joseph Bonaparte et Mortefontaine."

13. Kaspi, *Les américains,* 123.

14. Ibid., 125.

15. MAE, CP EU, vol. 72, fol. 263, January 26, 1816.

16. Hyde de Neuville, *Mémoires et souvenirs,* 2:181.

17. Ibid., 182 n. 1: the king informed him of his appointment to the United States on October 15, 1815.

18. Ibid., 183–84.

19. Ibid., 195.

20. The attachés were Bourqueney, recommended to Hyde by Chateaubriand, and Buchet-Martigny, whose family in Berry had distinguished itself "by its devotion to the royal cause during the insurrection in Sancerre." Ibid., 192. The novice consuls were Vol-

nais and Angelucci. The most experienced consul was Jean Baptiste Pétry, vice-consul in Wilmington, Delaware, in 1783. In 1816 he became consul general in Philadelphia. Mézin, *Les Consuls de France.*

21. MAE, CCC New York, vol. 4, fol. 319, January 6, 1816, in Watel, *Hyde de Neuville,* 122.

22. *Aurora,* October 8, 1818, commenting on a brief item in the *New York Evening Post* noting the Cazeaux family's departure for Bordeaux on the brig *Reindeer.*

23. MAE, CP EU, vol. 73, fol. 16, Roth to Richelieu, Washington, April 11, 1816.

24. In the *Boston Patriot* of June 4, 1814, we read: "Louis, bred up in England, amid all the prejudices of the Britons, and owing to that country his fortune, his prosperity, nay, his very throne, . . . cannot of course become ungrateful to her. England will sway every action of the court of France; she will rule her councils, govern her king, and steer the nation by her own chart and compass. Louis will be but the shadow of the French government, England the substance." Watel, *Hyde de Neuville,* 121 n. 3.

25. Adams to James Monroe, April 9, 1816, in J. Q. Adams, *Writings,* 6:17.

26. James Madison to Richard Rush, July 22, 1823, in H. M. Jones, *America and French Culture,* 560–61.

27. MAE, CP EU, vol. 73, fol. 154, Hyde to Richelieu, New Brunswick, October 9, 1816.

28. William Lee to Henry Jackson, secretary of the American legation in Paris, Bordeaux, September 20, 1815, in Lee, *A Yankee Jeffersonian,* 173.

29. *Aurora,* November 1816, in Philips, *Les réfugiés bonapartistes en Amérique,* 45.

30. In 1795, Madame Campan, a former lady's maid to Queen Marie Antoinette, had founded the Institution nationale de Saint-Germain, a boarding school for girls from excellent families. Aglaé Auguié, Ney's future wife, was Madame Campan's niece.

31. MAE, CP EU, vol. 72, fol. 269, Charles Roth to Richelieu, Washington, [late January 1816]. MAE, CP EU, vol. 73, fol. 143v, in the summer of 1816, Hyde de Neuville refers to "newspapers that, every day, contain coarse insults, cowardly slander against the king, the royal family, France . . . , for example, today an article in the *National Intelligencer* of Washington: 'The execution of Marshal Ney, and many other *great, good* and *brave* men, excites, in all *honest* hearts, feelings of *sorrow,* mingled with *indignation.*'"

32. Arriving in Paris as official representative in March 1789, Morris was appointed ambassador to France in February 1792 and remained in Paris until July 1794, when Robespierre was arrested and executed, putting an end to the Reign of Terror. See Morris, *Le journal de Gouverneur Morris.*

33. Morris, *An oration.*

34. MAE, CP EU, vol. 72, fol. 253–66, "Instructions de Richelieu, président du Conseil des ministres, à Hyde de Neuville, ministre de France aux États-Unis"; these are found in their entirety in Watel, *Hyde de Neuville,* 223–29.

35. CADN, consulat de France à New York, B, 1, Annexe 2, Richelieu to Mr. d'Espinville, French consul in New York, Paris, February 1816.

36. CADN, consulat de New York, B, 110, Annexe 3, circular from Hyde de Neuville to Mr. d'Espinville, consul in New York, Neuville's Farm, New Brunswick, New Jersey, July 18, 1816.

37. Hyde de Neuville, *Mémoires et souvenirs,* 2:197.

38. Ibid., 198.

39. MAE, CP EU, vol. 73, fol. 85, Hyde gave the minister of foreign affairs a report of his visit in a letter of July 25, 1816.

40. Ibid., fol. 66, 69, 70, and 76, for these successive exchanges.

41. Bertin, *1815–1832,* 7.

42. Rush wrote to Charles J. Ingersoll on January 4, 1817: "After all, Joseph was only half a king. In the first place, he was not of royal blood, which, for a king, is absolutely everything. Secondly, during the short period of time that his strings were being pulled to make him act, he appears to have succumbed to all the passionate appetites inherent in royal duties." Bertin, *1815–1832,* 11.

43. Dallas, *Life and Writings of Alexander James Dallas,* 445.

44. For the marriage, see chapter 4, note 67. Having other marriage plans for his brother, Napoleon denounced this "so-called marriage" that Jerome had entered into abroad, while still a minor. Arriving with his wife in Lisbon in early April 1805, Jerome was unable to win over the Emperor, who refused to let Elizabeth enter France and had the marriage annulled in 1806. On July 7, 1805, near London, Elizabeth gave birth to Jerome Napoleon Bonaparte. Handsomely compensated, she died in Baltimore at the age of ninety-four. Boudon, *Le roi Jérôme,* 87–98.

45. Bertin, *1815–1832,* 12, quotes the *United States Gazette*'s report of the emotions aroused on September 16, 1815, when Joseph Bonaparte passed through Lancaster, a small town near Philadelphia: "He went to the home of Mr. Slaymaker and soon the latter's house was filled by a crowd of people, incited by curiosity, whether distrustful or friendly, to come and study him close up. And they saw nothing but a man. Joseph is about five feet ten inches tall; he is rather thickset, well-proportioned and has a swarthy complexion; his physiognomy is neither attractive nor ugly; in short, nothing remarkable and strong features."

46. Hulot, *Le général Moreau,* 175.

47. MAE, CP EU, vol. 73, fol. 97, Hyde de Neuville to the minister of foreign affairs, New Brunswick, August 3, 1816.

48. MAE, CP EU, vol. 72, fol. 283v–284v, Roth to Richelieu, Washington, February 13, 1816.

49. Ibid., fol. 288–288v, Roth to Richelieu, Washington, February 29, 1816.

50. Ibid., fol. 295–297v, Roth to Richelieu, Washington, March 18, 1816.

51. CADN, consulat de Philadelphie, 45, Légation de France, exercice Pétry, p. 29, Pétry to Hyde de Neuville, Philadelphia, May 22 1817.

52. CADN, consulat de New York, B, 106, Hyde de Neuville to Pétry, Washington, May 24, 1817.

53. Ibid., Hyde to Pétry, Washington, June 5, 1817.

54. CADN, consulat de Philadelphie, 45, Légation de France, exercice Pétry, p. 4, Pétry to the minister of foreign affairs, Philadelphia, January 2, 1817.

55. CADN, consulat de New York, série C, 1, Espinville to the minister of foreign affairs, New York, November 20, 1817.

56. Vauthier, "Notes sur les Français," 45.

57. CADN, consulat de Philadelphie, 45, p. 27, Pétry to the minister of foreign affairs, Philadelphia, May 16, 1817.

58. CADN, consulat de New York, B, 114, register containing the correspondence with the Boston consulate, Pétry to Volnais, May 26, 1817; for another account of this dinner, see CADN, consulat de Philadelphie, Légation de France, exercice Pétry, 45, letter of May 16, 1817.

59. MAE, CP EU, vol. 73, fol. 192, April 18, 1817.

60. Ibid, fol. 91, Framery d'Ambreucq to Richelieu, Philadelphia, July 27, 1816.

61. John Stuart Skinner (Maryland, 1788–1851), lawyer, then writer and editor of journals, was postmaster in Baltimore from 1816 to 1849. In 1824 he welcomed to his Baltimore home General Lafayette, who was beginning his triumphal tour of the United States. See Wilson and Fiske, *Appleton's Cyclopaedia*, 5:545.

62. *Aurora*, November 9, 1816, quoting the *Baltimore Patriot* of November 6, 1816.

63. MAE, CP EU, vol. 73, fol. 134–35, Hyde de Neuville to Monroe, July 21, 1816.

64. *L'Abeille Américaine*, November 14, 1816. The November 9, 1816, issue of *Niles' Weekly Register* saw in Hyde's "insane" demand the desire to please his master even more.

65. MAE, CP EU, vol. 73, fol. 136, Monroe to Hyde de Neuville.

66. Ocampo, *Emperor's Last Campaign*, 70.

67. CADN, légation et consulat général de Philadelphie, register 6, fol. 213, A. Bourqueney to Hyde de Neuville, Washington, August 24, 1816.

68. Ibid., fol. 215, Bourqueney to Hyde, Montpelier, August 27, 1816.

69. James Madison to James Monroe, Montpelier, August 28, 1816, in Madison, *Writings*, 8:636.

70. Lee to Madison, Bordeaux, February 16, 1816, in Lee, *A Yankee Jeffersonian*, 178.

71. ADG, 8 M 199, *Le relevé des navires sortis du port de Bordeaux*, an American ship, Captain Norton, owner Lafitte, 317 tons, eleven crewmen, noted as bound for New York on June 2, 1816. The ship sailed on June 3 and reached New York on August 2, 1816.

72. William Lee to Thomas Jefferson, New York, October 25, 1816, in Lee, *A Yankee Jeffersonian*, 181: "chemists, mineralogists, naturalists, engineers, geographers, mechanicians, engravers, sculptors, dyers, opticians, weavers of cloth and stocking knit, gold beaters, hatters, tanners, gun and locksmiths, cutlers, distillers, gilders in wood and metal, founders, glove makers, fringe makers, glass makers, gardeners, vignerons cultivators, lampists, surgeons and dentists."

73. *L'Abeille Américaine*, August 8, 1816, 272.

74. SHD/DAT, 2 Ye, J. Roul file: Born in Villard-Saint-Pancrace (Hautes-Alpes) on June 9, 1775, enlisted in the Thirteenth Regiment of Hussars in 1793, promoted to second lieutenant in the chasseurs of the Imperial Guard, and awarded the Legion of Honor in 1805; fought in the Prussian, Polish, and Spanish campaigns, 1806–9, and in the French campaign in 1814; Quintin and Quintin, *Dictionnaire des colonels de Napoléon*, 757–58.

75. *Aurora*, August 2, 1816, reprinting the Baltimore announcement of June 14.

76. See Beaucour, "Qui était Roul?"

77. *L'Abeille Américaine*, August 8, 1816, 257–59: "Ode to the city of Bordeaux. Return,

return, go back to dust, / Enslaved city! . . . in irons rust! . . . / Today your well-earned just deserts / Are the scorn of the Universe."

78. MAE, CP EU, vol. 73, fol. 81, Hyde to Monroe, New Brunswick, July 21, 1816.

79. Ibid., fol. 82, Monroe to Hyde, Department of State, August 15, 1816.

80. AN, F^7 9994, Cour d'Assises de la Gironde, 4e trimestre 1815; *Annales politiques, morales et littéraires,* December 26, 1815, in Schneider, "Les complots sous la Terreur Blanche," Vol. 1, 188.

81. MAE, CP EU, vol. 73, fol. 132, certified copy of the original certificate by Bourqueney, attaché at the Philadelphia Legation.

82. Ibid.

83. Ibid., Hyde to Richelieu, New Brunswick, October 7, 1816.

84. Lee to Monroe, Bordeaux, October 20, 1815, in Lee, *A Yankee Jeffersonian,* 176.

85. On this subject see Weil, "L'état et l'émigration de France."

86. Article 417 of the 1810 Penal Code: "Whoever, with the intent of harming French industry, has sent to a foreign country directors, clerks or workers of an establishment, will be punished by six months to two years in prison and a fine of 50 to 300 francs."

87. AN, F^7 3517, passport issued at the Paris police headquarters on April 23, 1816, to Pierre Vauversin for travel abroad. The Belgian newspaper *Le Mercure-Surveillant,* July 4, 1816, states: "A certain Vauversin whom Louis XVIII has employed in his service in his secret police, has been, they say, attached to Monsieur Hyde de Neuville who will employ him under his orders, in the United States of America, where he will appear under the pseudonym of Mr. Pierre de la Marne." Bibliothèque Royale Albert Ier, Brussels, J. B. 286.

88. MAE, MD France, vol. 2139, fol. 94v, Viaud's record closes the "Etat nominatif et alphabétique des individus que [Vauversin] a été à même de fréquenter aux États-Unis."

89. Ibid., fol. 68, Vauversin to Hyde de Neuville, New York, August 7, 1816.

90. Ibid., fol. 69v–70, Vauversin to Hyde de Neuville, Philadelphia, August 17, 1816.

91. The gourde, a silver coin used in Saint-Domingue and Cuba, was worth about one dollar.

92. MAE, MD France, vol. 2139, fol. 69v–70, Vauversin to Hyde de Neuville, Philadelphia, August 17, 1816.

93. MAE, CP EU, vol. 73, fol. 130, Hyde to Richelieu, New Brunswick, October 7, 1816.

CHAPTER 6

Epigraph: General Bernard to Lieutenant Colonel (Engineers) Huart, assistant manager of the fortifications, Belfort (Haut-Rhin), in Bertin, *1815–1832,* 201–2.

1. MAE, CP EU, vol. 73, fol. 199, Hyde to Richelieu, Washington, D.C., January 10, 1817.

2. Lajonie Papers, Lajonie to the château of Le Soulat, Philadelphia, early January 1817. Born in Gironde, Jean Latour had married Grace Smith in Baltimore on October 23, 1795. They had two sons and two daughters; Lajonie was much attracted to Elisa the elder (March 1817): "Mademoiselle Elisa is charming. She plays the pianoforte and sings most pleasantly." John Latour is regularly listed in Baltimore business directories between 1799 and 1823, as well as in the *Baltimore City Directory* of 1817–18, p. 111: "Latour John, com-

mission merchant, 1 Commerce—dwelling Green near Mulberry." John Latour also appears in the census with two slaves. 1820 U.S. Federal Census, Maryland, Baltimore Ward 12. It is known that his brothers and sisters were living in Juillac, in particular at the des Faures estate, in "Fonvidal" (château of Font Vidal), and that on October 31, 1845, a Jean de Latour who had come back to live there, died a widower at eighty-three years of age (thus born ca. 1762). Was this John Latour, of whom Lajonie said, in March 1817 that, having made his fortune, he wished to return home?

3. Lajonie Papers, Lajonie to Pierre Audubert, Baltimore, April 13, 1817.

4. John Latour to Jacques Lajonie, Baltimore, July 21, 1818, Jean-Marie Léonard Collection (Isère).

5. Taillemite, *La Fayette*, 438.

6. Kramer, *Lafayette in Two Worlds*.

7. Mahlon Dickerson, governor of New Jersey, to Joseph Bonaparte, Trenton, January 28, 1817, in Bertin, *1815–1832*, 34.

8. Statesmen like Henry Clay, Daniel Webster, John Quincy Adams, Edward Livingston, and Senator Richard Stockton, who lived near Bordentown; Admiral Charles Stewart, another neighbor and hero of the War of 1812, and General Thomas Cadwalader were close friends; Joseph Hopkinson, a jurist and writer, his authorized representative; Nathaniel Chapman, a leading medical expert; William Short, a Francophile diplomat; and historian and jurist Charles Jared Ingersoll. See Bertin, *1815–1832*, 171–81.

9. Ann Savage (Philadelphia, 1800–New York, 1865), whom he met in 1818, gave him daughters Pauline Josephe Anne (1819–23) and Caroline Charlotte (1822–90). She was succeeded by Émilie Lacoste (the wife of merchant Félix Lacoste), who in 1825 bore him twins, one of whom survived, Félix Joseph Lacoste; and probably Emma Saint-Georges, a Creole, wife of Mathieu Sari, a Corsican sailor in Joseph's service. See Stroud, *The Man Who Had Been King*.

10. CADN, consulat de France à Baltimore, A, no. 47, consul to Richelieu, February 15, 1816.

11. Grouchy to Jefferson, Wilmington, Delaware, October 20, 1817, Jefferson Papers, Bureau of Rolls and Library, Department of State, in Reeves, *Napoleonic Exiles in America*, 24–25.

12. Jefferson to Grouchy, Monticello, Virginia, November 2, 1817, in ibid., 25.

13. Lakanal to Jefferson, Gallatin County, Kentucky, June 1, 1816, ibid., 31.

14. Jefferson to Lakanal, Monticello, July 30, 1816, ibid., 32.

15. Philips, *Les réfugiés bonapartistes en Amérique*, 28–29.

16. Issue of December 28, 1818, in Murat, *Napoléon et le rêve américain*, 84.

17. See chapter 4, note 75.

18. Bigard, *Le Comte Réal*, 173–74; Beaucour, *Un fidèle de l'Empereur*, vol. 1, part 1:597–98.

19. See Wildes, *Lonely Midas*; and McMaster, *Life and Times of Stephen Girard*.

20. The Bank of the United States, the chief American financial institution until 1812, had opened in Philadelphia in 1791 with a twenty-year charter. Its concession had not been renewed. D. R. Adams, *Finance and Enterprise*, 4.

21. Wildes, *Lonely Midas*, 192.

22. APS, Girard Papers, Girard to the Count of Survilliers [Joseph Bonaparte], Philadelphia, November 1, 1817.

23. Ibid., Girard's active correspondence with Joseph in the registers containing the letters of 1817 and 1818, microfilms 15 and 16.

24. Jacques Laffitte (1767–1844), associate and then successor of the banker Perrégaux, regent and governor of the Bank of France during the Empire; Jean-Baptiste Laffitte (1775–1843), stockbroker, married around 1800 to Antoinette-Louise-Euphrasie Lefebvre-Desnoëttes (1784–1824). See Douyrou, "Jacques Laffitte"; Brun, *Le banquier Laffitte.*

25. Jean (John) Girard (Bordeaux, 1751–St. Vincent, 1803), naturalized as an American citizen in 1792.

26. William Girard (1790–1807), Marie Antoinette Victoire (1796–1871), Caroline Eleanor (1798–1878), and Henriette Marie, called Harriet (1800–1880). Wildes, *Lonely Midas*, 306.

27. Supercargo (agent in charge of the cargo) on the *Rousseau* (1809), the *Voltaire* (1809–11), the *Montesquieu* (1815–20), and the *Superb* (1820–23). Ibid. 311–12.

28. "As I think the first of duties to confide in a benefactor and speak to him as a father, I own then candidly that the general's attentions were not received with indifference. You have given me reason to believe that you will never cross my inclinations, but still I wish to have your advice." Ibid. 248.

29. Ibid. 248. Lallemand first learned of Girard's consent from Harriet, then from Girard himself, who gave him his word. Information from two undated letters from Henri Lallemand to Harriet Girard, in the papers of Mrs. Plowden (Mayfield, Sussex), copies of which are in the collection of the American Philosophical Society in Philadelphia.

30. CADN, consulat de France à New York, C, register no. 1, consul to the minister of foreign affairs, September 25, 1817.

31. CADN, consulat de France à New York, C, 1, correspondence with the ministry of foreign affairs, letter of November 10, 1817.

32. CADN, Série Actes notariés, sous-série Philadelphie, register 2, fol. 8v–9.

33. Charles Frédéric Billon, watchmaker, born in Locle, canton of Bern, ca. 1766, died in St. Louis, Missouri, in 1822; in 1795 he immigrated to Philadelphia, where he married Jeanne-Charlotte "Sophie" Stollenwerck (Saint-Domingue, 1781–New York, 1880), who bore him thirteen children.

34. Chaudron's work can be seen in the collections of the following American museums and historical societies: New York Historical Society, Newark Museum, St. Louis Museum, Art Institute of Chicago, Henry Francis DuPont Winterthur Museum, Missouri Historical Society, Museum of Mobile, Alabama Department of Archives and History. See Britton, "French Refugee Chaudron's Silver Superb."

35. Anthony Rasch von Tauffkirchen, silversmith, born ca. 1780 near Passau (Bavaria), emigrated to Philadelphia ca. 1801–3, died in 1858 in New Orleans. For hallmarks, see Wyler, *Book of Old Silver*, 280–81 (Chaudron), 299–300 (Rasch).

36. Fred Billon quoted in Stollenwerck and Stollenwerck, *Stollenwerck, Chaudron and Billon Families*, 20.

37. Ibid., 20.

38. Dallett, "French Benevolent Society of Philadelphia."

39. MAE, MD France, vol. 2143, fol. 264, Hyde de Neuville to Minister of Foreign Affairs Dessole, Washington, July 25, 1819.

40. Chaudron, "To Bonaparte, on the Abandonment and the Massacres of Saint-Domingue," Ode IV, *Poésies choisies*, 15–20.

41. *L'Abeille Américaine: Journal Historique, Politique et Littéraire*, printed by A. Blocquerest, 130 South Fifth Street, Philadelphia, April 15, 1815, 15.

42. Ibid., September 9, 1815, 349.

43. Ibid., August 1815.

44. Ibid., October 7, 1815, 413.

45. Ibid., 419.

46. Ibid., September 2, 1815, 337, February 24, 1816, 315.

47. Ibid., June 6, 1816, 128.

48. Ibid., August 21, 1817, 98.

49. Ibid., September 16, 1815, 372–73.

50. Ibid., June 11, 1817, 7. Simon Chaudron sold his newspaper to the printer A. J. Blocquerest and to F. Maligot, agreeing to continue to edit it for one year from that day.

51. Fred Billon (Philadelphia, 1801–St. Louis, 1895), the son of Charles Billon and Sophie Stollenwerck, remembers Joseph Bonaparte's visit in person. He came with a morocco leather box containing six tumblers and twelve silver spoons and forks (later distributed as part of the estate). Unfinished manuscript of a "History of Philadelphia" at the Pennsylvania Historical Society, cited in Stollenwerck and Stollenwerck, *Stollenwerck, Chaudron and Billon Families*, 18.

52. Pierre Samuel Dupont, known as de Nemours (Paris, 1739–Wilmington, Delaware, 1817), in 1766 married Charlotte Marie Louise Le Dée de Raucourt (1743–84), who bore him two sons: Victor Marie (Paris, 1767–Philadelphia, 1827) and Éleuthère Irénée (Paris, 1771–Philadelphia, 1834).

53. Dupont to Benjamin Franklin, May 10, 1768, in Jolly, *Du Pont de Nemours*, 37.

54. Written on February 22, 1790, ibid., 6.

55. On December 17, 1800, the elder Dupont assured Jefferson that his son would send "cannonballs one fifth farther than English and Dutch cannonballs." Quoted in Krebs, *Dictionnaire de biographie française*, vol. 12, article 95: 471.

56. July 25, 1815, in Jolly, *Du Pont de Nemours*, 269.

57. Letter of July 1816 in Hyde de Neuville, *Mémoires et souvenirs*, 2:207–9. Hyde, who claimed to have a good relationship with Dupont despite their disagreements, seems to have censored from his letter passages that would contradict this assertion.

58. MAE, MD France, vol. 2139, fol. 71–85, from August to October 1816, the spy Pierre Vauversin reported to Hyde the regular visits an officer named Durand made to the Duponts.

59. Pierre Samuel Dupont de Nemours to Hyde de Neuville, Eleutherian near Wilmington, Delaware State, August 3, 1816, in Jolly, *Du Pont de Nemours*, 278.

60. CADN, commissariat des relations commerciales de la république française près des États de New York et New Jersey, État-civil, register 2, fol. 1–2, marriage on April 30, 1804.

61. Letter of August 3, 1816, in Jolly, *Du Pont de Nemours*, 278. Dupont's punning is untranslatable. During the French Revolution, the *sans-culottes* were the most ardent repub-

licans. The name comes from the fact that the common people wore trousers (*pantalons*), whereas the aristocrats wore knee-breeches (*culottes*). The female members of Dupont's family would naturally not wear knee-breeches, so they were *sans-culottes*—without *culottes*.

62. Victor Dupont de Nemours to Hyde de Neuville, Philadelphia, August 5, 1816, in Jolly, *Du Pont de Nemours*, 282.

63. Victor Dupont to Hyde de Neuville, Brandywine, October 23, 1816, in ibid., 284.

64. *National Intelligencer*, August 17, 1817; Pierre Samuel Dupont died on August 6, 1817.

65. Hyde de Neuville to the Duke of Richelieu, Washington D.C., August 19, 1817, in Jolly, *Du Pont de Nemours*, 290.

66. *Niles' Weekly Register*, April 4, 1818, 103–4. The explosion occurred on March 19, 1818.

67. In January 1788, degree of Master Mason in Blue Lodge no. 8, attached to the Grand Lodge of Ancient York Masons in Charleston. See Barratt and Sachse, *Freemasonry in Pennsylvania*, 3:246–47.

68. See Cauna, "La franc-maçonnerie burdigalo-aquitaine."

69. Chaudron, *Oraison funèbre du Frère George Washington*.

70. *L'Abeille Américaine*, December 15, 1815.

71. A member of the Égalité lodge in Saint-Jean d'Angély (Seine-Maritime), he had, in France, made several speeches at the order's most important ceremonies. Collaveri, *Napoléon franc-maçon?* 71.

72. François Collaveri is certain that Napoleon was a Mason, probably initiated in Egypt. Napoleon protected Freemasonry during the Empire and turned it into a particularly effective instrument for creating the Great Empire, thanks to relationships developed between French civil servants and military men on one hand and foreign "collaborators" on the other, writes J. Tulard in his introduction to Collaveri's work. The name Napoleon was given to dozens of lodges: "Brother Napoleon the Great" or "Napoleon of All Rites."

73. Based on the registers of the lodges of the Grand Orient of France, a list of 347 French and foreign generals and admirals in French service belonging to Freemasonry between 1792 and 1804 has been drawn up by Quoy-Bodin, "La franc-maçonnerie."

74. Grouchy, at the time a second lieutenant of the company with the rank of lieutenant colonel, was a member of the Héroisme lodge of the Orient of the Scots Company of the Royal Bodyguard garrisoned in Beauvais in 1787. Ibid., 76.

75. Quoy-Bodin lists him as honorary absent member of the Les amis de l'honneur et de la vérité lodge of the Orient of Madrid in January 1813, with the degree of grand inspector inquisitor. Ibid., 78.

76. Hagley Museum and Library, W3-2953. In Freemasonry, Victor Dupont was "Grand Marshal of the State of Delaware," the title he gives himself in a letter to Hyde de Neuville. Jolly, *Du Pont de Nemours*, 290.

77. Pierre Étienne Du Ponceau (Ile de Ré, 1760–Philadelphia, 1844), jurist and philologist, member (1791), secretary, and president of the American Philosophical Society (1828–44).

78. *L'Abeille Américaine*, May 30, 1816, 98–99.

79. "Ode recitée le 2 mai 1816, au diner donné au Maréchal de Grouchy, & aux généraux Lefebvre-Desnoëttes & Clausel [*sic*], par leurs compatriotes établis à Philadelphie," ibid., May 23, 1816, 81–84; "Ode, récitée par un Vieux Français, à la fête donnée à la Philadelphie, aux généraux proscrits de France," ibid., May 30, 1816, 98–99.

80. Du Ponceau, "English Phonology." See M. D. Smith, "Peter Stephen Du Ponceau."

81. Stephen Peter Du Ponceau, Philadelphia, May 12, 1836, in "Notes and Documents," 195–96.

82. Watel, *Hyde de Neuville*, 95.

83. Webb Papers, Chapron to his son, M. Dortenil's Boarding School in Germantown, Cap Français, December 30, 1800.

84. Webb Papers, John M. Chapron to Emilie Catherine and Jane Browder, daughters of Juliette Chapron and Dr. Browder, Hawthorn, December 24, 1867.

85. Jolly, *Du Pont de Nemours*, 209.

86. Hagley Museum and Library, Group 2, Du Pont de Nemours, Series A, Correspondence, Box 21, W2-4366, M. Regnaud de Saint-Jean d'Angély to Pierre Samuel Du Pont (de Nemours), New York, February 17, 1816.

87. *L'Abeille Américaine*, November 28, 1816.

88. SHD/DM (Vincennes), Taillade's personnel file, Taillade to Forestier, head of the first division of the offices of the ministry of the navy, Paris, 3 ventôse, year 12 [February 23, 1804].

89. For Lefebvre-Desnoëttes's stay in England, see Guillot, "Une évasion romanesque," 16–18.

90. "He therefore felt this loss keenly; he alone knew English perfectly and translated it to his satisfaction; the Emperor owed him the English lessons that enabled him to read the papers when they arrived." Marchand, *Mémoires*, 2:137.

91. See chapter 4, note 12. In 1827 James Caret, the interpreter, joined his brother in Cuba, where he had made his fortune. Bertin, *1815–1832*, 12.

92. According to John Fanning Watson, *Annals of Philadelphia and Pennsylvania in the Olden Time* . . . (Philadelphia, 1845), in Stroud, *The Man Who Had Been King*, 140.

93. Garnier de Saintes, *La dette d'un exilé*.

94. Philadelphia, September 10, 1817, in Grouchy, *Mémoires*, 5:63.

95. Bigard, *Le Comte Réal*, 175.

96. Pierre Étienne Chazotte (ca. 1770–ca. 1849), living at Cap Dame Marie et Jérémie, a refugee in Charleston in 1798, returned to Saint-Domingue in 1800, arrived in Philadelphia in 1803. In 1817 he published *An essay on the best method of teaching foreign languages as applied with extraordinary success to the French language . . . : to which is prefixed a discourse, on the formation and progress of languages* (Philadelphia: Edward Earle).

97. "Inquire, in the afternoon, at 10 South Sixth Street, and at all hours of the day, at 16 Lombard Street," *L'Abeille Américaine*, April 18, 1816.

98. Ibid., May 30, 1816.

99. Ibid., October 3, 1816.

100. Lajonie Papers, Lajonie to the château of Le Soulat, Burlington, New Jersey, January 20, 1817.

101. Lajonie Papers, Lajonie to the château of Le Soulat, plantation of Sainte-Hélène, Alabama, September 18, 1821.

102. Ibid., Lajonie to Pierre Audubert, Baltimore, April 13, 1817.

103. "Passengers Arrived in the U.S. from Foreign Countries, September 30, 1820," *Niles' Weekly Register,* March 24, 1821, 52–53.

104. *L'Abeille Américaine,* November 28, 1816.

105. In *Naissance et déclin,* P. M. Kennedy states that the American army was 16,000 men strong in 1816—excluding the 2,000 engineers, the army's elite; its size had been reduced by 75 percent. In 1914 the American army was composed of only 75,000 men (134).

106. Milbert, *Itinéraire pittoresque,* vii.

107. CADN, légation et consulat général de Philadelphie, register 6, fol. 23, attestation of Dewitt Clinton concerning Lieutenant Colonel Garin, New York, October 3, 1817.

108. Simon Bernard to General Lafayette, Paris, November 8, 1815, in Planchot-Mazel, *Simon Bernard,* 1:54.

109. Napoleon's opinion expressed in 1811 and reported by Molé in ibid., 27.

110. Lafayette to Madison and Crawford, La Grange-Bléneau, November 11, 1815, ibid., 59.

111. Rémond, *Histoire des États-Unis,* 39.

112. For the development of American national character and patriotic feeling before the War of 1812, see Rossignol, *Le ferment nationaliste.*

113. Christopher Vandeventer to "Friends," Engineers' Headquarters, New York, June 4, 1816, in Planchot-Mazel, *Simon Bernard,* 1:72–73.

114. *Aurora,* November 21, 1816.

115. The remarkable American careers of the French officers Pierre Benjamin Buisson (ca. 1793–1874) and Benoît Claudius Crozet (1789–1864), both graduates of the École polytechnique, should be noted, however.

116. Poussin, *Les États-Unis d'Amérique,* 51.

117. British-born Benjamin Henry Boneval Latrobe (1764–1820), America's first professional architect and engineer, was asked in 1815 to restore the interior of the Capitol, destroyed by the English attack in 1814. He descended from a Protestant family driven from France after Louis XIV revoked the Edict of Nantes.

118. Poussin to Girard, August 16, 1816, in Planchot-Mazel, *Simon Bernard,* 1:107.

119. Poussin, *Les États-Unis d'Amérique,* 73–74.

120. Planchot-Mazel, *Simon Bernard,* 1:108.

121. Poussin, *Les États-Unis d'Amérique,* 78–79.

122. Just a few titles in a vast bibliography: Chinard, *L'Amérique et le rêve exotique;* Durand Echeverria, *Mirage in the West: A History of the French Image of American Society to 1815* (Princeton: Princeton University Press, 1957); Gérard Defamie, "La mode des États-Unis et le voyage en Amérique pour les libéraux français à la veille de la Révolution" (diss., Lille III, 1973).

123. Benjamin Franklin, *Information to those who would remove to America* (London: printed for John Stockdale, 1784), appearing simultaneously in French as *Avis à ceux qui voudraient s'en aller en Amérique* (Passy: printed by Benjamin Franklin, 1784).

124. Moreau-Zanelli's *Gallipolis* is an invaluable resource on the subject.

125. Saint-John de Crèvecoeur, *Lettres d'un cultivateur américain.* This is a translation of the work published in London in English in 1782.

126. In 1787, Jacques Pierre Brissot de Warville (1754–1793), along with Clavière, Crèvecoeur, and Bergasse, founded the Société gallo-américaine to bring together France and the United States in their mutual interest. He wrote *Nouveau voyage dans les États-Unis de l'Amérique septentrionale fait en 1788,* 3 vols. (Paris: Buisson, 1791), published the following year in English as *New Travels in the United States of America performed in 1788* (London: J.S. Jordan, 1792).

127. Jacques Pierre Brissot de Warville, "Plan of a Society for promoting the emigration from Europa in the United-States," in *Correspondance et papiers précédés d'un avertissement et d'une notice sur sa vie,* ed. Claude Perroud (Paris: A. Picard & fils, 1912), 458–61, in Moreau-Zanelli, *Gallipolis,* 70; Guillotin to Benjamin Franklin, August 5, 1787, in John F. McDermott, "Guillotin Thinks of America," *Ohio Archaeological and Historical Society Quarterly* 47, no. 2 (1938): 138, in Moreau-Zanelli, *Gallipolis,* 73; Gustave Rudler, *La jeunesse de Constant (1767–1794)* (Paris: Colin, 1908), in ibid., 74.

128. Chateaubriand, *Atala;* Chinard, *L'exotisme américain.*

129. Hackensmith, *Biography of Joseph Neef;* Gutek, *Joseph Neef.*

130. Joseph Alphonse, born in Nancy ca. 1786, writer and Pestalozzian pedagogue; cf. two letters to Pierre Samuel Dupont de Nemours in Wilmington, Delaware, one from Louisville, dated August 20, 1816, the other from Philadelphia, dated October 17, 1816. Hagley Museum and Library, W2-4317 and 4327.

131. Among others: his sister Marie Perrine (Cirode) Thomazeau, twenty-eight, his brother-in-law, François Sébastien Thomazeau, a dealer in hardware, arriving in New York on November 16, 1816, on the *Virginie;* his brother Yves Cirode, twenty-nine, a lace merchant, his mother Perrine (Blanchet) Cirode, fifty-five, his niece Célina Thomazeau, three, arriving in New York on July 7, 1818, on the *Virginie.* ADG, Marine 120 J 2377 & 2387, quartier de Nantes, rôles d'armement 1817, 1818.

132. NARA, *Passenger Lists of Vessels Arriving at Philadelphia,* M 425, reel 10: Embarked in Bordeaux, with his wife, Henrietta Martin-Picquet, and five children, on the Philadelphia ship *Charleston-Packet,* arrived in Philadelphia on June 28, 1806.

133. François Désiré Dusouchet (Angoulême, 1778–Mt. Vernon, Ind., 1841) had emigrated from Saint-Domingue to the United States in 1803 or 1804. He married Catherine Sarchet in Cincinnati in 1816. For his projects in Louisville see the *Western Courier* (Louisville), August 1, 1814, and October 12, 1814.

134. NARA, *Passenger Lists of Vessels Arriving at Philadelphia, Pa.,* M 425, reel 21, line 3: embarked in Nantes with his wife, Michelle Thérèse Bonamy, and two children, Denise and Juste, in October 1815 on the Philadelphia ship *Tennessee,* 274 tons, Captain Peter Bell, arrived in Philadelphia November 17, 1815.

135. APS (Philadelphia), Girard Papers: reel 61, Lecoq Dumarselay to Girard, Louisville, January 6, 1817; reel 62, February 17, 1818; and reel 71 September 17, 1819.

136. In 1796 Lakanal had acquired the priory of the domain of Villarceaux, near Chaussy (Val d'Oise), a nationalized property, which he sold in 1812 for 30,000 francs in gold. Welvert, "Lakanal à Villarceaux."

137. Lakanal to Baron Bignon, February 26, 1838, in Dawson, *Lakanal the Regicide,* 99.

138. *L'Abeille Américaine,* April 6, 1816, 419.

139. Warden to General Mason in Virginia, Paris, December 15, 1815, in Dawson, *Lakanal the Regicide,* 96.

140. Michaux, *Voyage,* in Dawson, *Lakanal the Regicide,* 99. His father was the royal botanist and explorer André Michaux (1746–1802) with whom he had gone to the United States in 1785 to look for plants and trees for the Versailles palace park. André Michaux made an inventory of hundreds of species of eastern North America and founded Charleston's botanical gardens in 1786. His son continued his work by publishing, in 1805 and from 1810 to 1813, two works on the oaks and forest trees of North America.

141. "M. Lakanal, a distinguished French gentleman (member of the national institute of France, and of the legion of honor), remarkable for his republican principles, has lately arrived here with his family. He has purchased an estate on the bank of the Ohio, two miles above Vevay, on the Kentucky side. M. Vairin, a professor of mathematics, has also arrived from France, with a part of his family—he has purchased a farm on the river, one mile below Vevay. May happiness attend them in our land of liberty—their adopted country." Reprinted in *Niles' Weekly Register,* July 20, 1816, and in Dawson, *Lakanal the Regicide,* 101.

142. Hagley Museum and Library, Group 3, Du Pont de Nemours, Series A, Correspondence, Box 16, W3-2964, Lakanal to Victor Marie Dupont, Vevay, Indian Territory, July 5, 1816.

143. Joseph Lakanal to Thomas Jefferson, June 1, 1816, in Dawson, *Lakanal the Regicide,* 102.

144. In 1793, in Bergerac (Dordogne), where he had been sent on a mission by the Convention, he had created a *maison d'économie rurale* (rural economics center), a sort of model farm to teach the art of agriculture.

145. *L'Abeille Américaine,* April 6, 1816, 419.

146. Joseph Lakanal, "Rapport sur le Jardin national des plantes et le Cabinet d'histoire naturelle de Paris, 10 juin 1793," in *Procès-verbaux du Comité d'Instruction,* 1:481–83.

147. Jefferson to Lakanal, Monticello, July 30, 1816, in Dawson, *Lakanal the Regicide,* 102.

148. Lakanal to Victor Marie Dupont, Vevay, Indian Territory, July 5, 1816 (see note 142).

149. "États-Unis d'Amérique," *L'Abeille Américaine,* May 9, 1816, 60–62.

CHAPTER 7

Epigraphs: N. Bonaparte, *Oeuvres littéraires et écrits militaires,* 1:286; Franklin quoted in Lichine, *New Encyclopedia of Wines and Spirits,* 145.

1. Marchand, *Mémoires,* 2:69.

2. APS, Girard Papers, letters received, reel 60, Lefebvre-Desnoëttes, alias Charles Bernard, to Stephen Girard, received in Philadelphia on September 12, 1816.

3. "Le Chêne et le Laurier réconciliés par l'Olivier et la Vigne" [The Oak and the Laurel Reconciled by the Olive and the Vine], *L'Abeille Américaine,* September 18, 1817, 147–49.

4. L. D. Adams, *Wines of America.*

5. Ibid., 19.

6. Jean Louis Gibert, born in 1722 in Lunès, near Alès in the Cévennes region of south-central France. See Khaoua, "Conception, vie et mort d'un projet colonial."

7. Patrick Calhoun, the father of a vice-president of the United States, had chosen the site.

8. N. M. Davis, "French Settlement at New Bordeaux," 40–41.

9. Khaoua, "Conception, vie et mort d'un projet colonial," 101.

10. Louis de Saint Pierre, *The Art of Planting and Cultivating the Vine* (London: Wilkie, 1772), xxx, in N. M. Davis, "French Settlement at New Bordeaux," 48. Goldsmith's Library of Economic Literature at the University of London has a copy of this book, said to have belonged to the brother of George III.

11. Mills, *Atlas of the State of South Carolina.*

12. H. C. Martin, "Jefferson's Italian Vigneron," 23.

13. Mazzei, in Gabler, *Passions,* 5.

14. Jefferson to Albert Gallatin, 1793, in H. C. Martin, "Jefferson's Italian Vigneron," 29.

15. Latrobe to Mazzei (1805), in ibid., 29.

16. Stanton, "Reforming His Nation's Taste," 102.

17. Among the many works devoted to the subject, see Galtier, "La viticulture de l'Europe occidentale."

18. Jefferson, "Memoranda"; Galtier, "La viticulture de l'Europe occidentale," 54–76.

19. See De Treville Lawrence, "Wine Advisor to Five Presidents: Jefferson the Premier Connoisseur," in his *Jefferson and Wine,* 175–201.

20. Lichine, *New Encyclopedia of Wines and Spirits,* 15.

21. "Initially proposed by Peter Legaux at a meeting of the American Philosophical Society in 1793, the Vine Company of Pennsylvania was a stock company that encouraged the domestic production of grapes, wines, and brandy, and dissemination of knowledge about viticulture. After its incorporation in 1802, the Company operated vineyards on Legaux's farm at Spring Mill, 13 miles northwest of Philadelphia, until it failed in 1822." Website of the American Philosophical Society: www.amphilsoc.org.

22. Jefferson to Monroe, April 8, 1817, in *The Complete Jefferson,* 373, in Galtier, "La viticulture de l'Europe occidentale," 52.

23. Jefferson to Gallatin, June 23, 1807, in Jefferson, *The Jeffersonian Cyclopedia,* 2:no. 9152.

24. Jefferson to William Drayton, July 30, 1787, in ibid., no. 9149.

25. Jefferson to Lasteyrie, July 15, 1808, in Howland, "The Oenologist of Monticello," 7.

26. The hybrid vine called "Red Alexander" is probably the result of an accidental cross-breeding with the vine vinifera.

27. His estate, The Vineyard, located at the mouth of the Potomac, is now part of Rock Creek Park in Washington, D.C. Adlum was perhaps the first in the United States to succeed in growing a strong hybrid species from only indigenous vines, the Catawba. In the 1820s, near Cincinnati, this vine delighted another renowned American horticulturist and viticulturist, Nicholas Longworth.

28. Bowes, "Searching for the Best Wines," 220.

29. Jefferson to Adlum, October 7, 1809, in Jefferson, "The Garden Book," 44.

30. Jefferson to Dortie, October 1, 1811, in ibid., 46.

31. In 1816 French émigré Jean Jaques, a cobbler's son, was the first to plant vines in Orange County, New York, on the banks of the Hudson River.

32. Jefferson to William Johnson, Monticello, May 10, 1817, in Penney, "North Carolina Scuppernong," 239.

33. James Blount, "North Carolina Wine from Native Grapes," *Raleigh Star,* January 31, 1811, ibid., 243.

34. De Treville Lawrence, *Jefferson and Wine,* 247.

35. Jefferson's answer to John David, December 25, 1815, in Jefferson, "The Garden Book," 48.

36. De Treville Lawrence, *Jefferson and Wine,* 251.

37. On this subject, see J. J. Dufour, *American Vine-Dresser's Guide; History of Switzerland County;* P. Dufour, *Swiss Settlement of Switzerland County;* Knox, *The Dufour Saga;* Butler and Butler, *Indiana Wine.*

38. See "Brouillard de voyage pour Jean-Jacques Dufour de Sales à Montreux, au bailliage de Vevey (1796–1816)," in P. Dufour, *Swiss Settlement of Switzerland County,* 237–347.

39. A Virginian, soldier under Washington and Lafayette, friend of John Adams, supporter of Jefferson; brother of James, a senator from Louisiana, and Samuel Brown, a physician and planter in Alabama.

40. The Kentucky legislature passed a law incorporating the Vineyard Society on November 21, 1799: "An Act for incorporating the Vineyard Society." The company was listed under various names: Kentucky Vineyard Society, Kentucky Wine Industry Association, and Society for Promoting the Cultivation of the Vine. L. H. Bailey, *Sketch of the Evolution of Our Native Fruits* (London: McMillan, 1898), in J. J. Dufour, *American Vine-Dresser's Guide,* 20.

41. J. J. Dufour, *American Vine-Dresser's Guide,* 35.

42. Suzanne Marie Dubochet, Jean Jacques Rodolph Dufour's second wife, bore him, from 1781 to 1791, Antoinette, Jean François, Suzanne Marguerite, Jeanne Marie, Jean David, and Aimé Dufour, the youngest who left later.

43. François Louis Siebenthal, winegrower, born in Montreux, canton of Vaud (Switzerland), on September 30, 1763. His two sons were born in Montreux: Jean François (April 15, 1785) and Jean Louis (February 4, 1788). On June 7, 1817, the police lieutenant of Lyon sent the passport of Jean Louis Siebenthal, Swiss farmer, with six other Swiss, to the minister of the General Police; the minister had the passports sent to the mayor of Bordeaux. AN, F^7 2550b, no. 21,472.

44. Sakolski, *Great American Land Bubble,* 1.

45. Voltaire to the Count d'Argental, February 9, 1761, in J. Q. Adams, *Memoirs,* 4:4.

46. Dufour to Jefferson, First Vineyard, Kentucky, February 1, 1801, LC, Thomas Jefferson Papers, Series 1 (General Correspondence, 1651–1827).

47. Ibid., "To the Honorable Senate and House of Representatives of the United States of America, the petition of John James Dufour. . . ."

48. Ibid., Dufour to Jefferson, First Vineyard, Kentucky, January 15, 1802.

49. "A Covenant of Association for the settlement of the lands of Switzerland, on the Ohio River," in *History of Switzerland County,* 993–95.

50. Library of Congress, Washington, D.C., American State Papers, House of Repre-

sentatives, 9th Congress, 1st session, Public Lands, vol. 1, no. 118, "Cultivation of the Vine. Communicated to the House of Representatives, February 3, 1806."

51. Jean Jacques Dufour quotes an American newspaper that in 1816 reported the remarks of a Frenchman who judged the Vevay wine to be inferior to the great French wines like the Médoc, but superior to most of the wines that could be bought on the Paris market. J. J. Dufour, *American Vine-Dresser's Guide,* 49.

52. Dawson, *Lakanal the Regicide,* 101.

53. P. Dufour, *Swiss Settlement of Switzerland County,* 29.

54. Library of Congress, Washington D.C., Public Lands, 13th Congress, 1st session, no. 216, "Application of the cultivators of the Vine in Ohio for a remission of the debt for lands, or an extension of time for payment." The petition was presented by Jonathan Jennings, Representative of the Indiana Territory—Journal of the House of Representatives, Tuesday, January 12, 1813.

55. J. J. Dufour, *American Vine-Dresser's Guide,* 31.

CHAPTER 8

Epigraph: *L'Abeille Américaine,* August 22, 1816, 292, excerpt of a letter "Aux français habitans des États-Unis" [To the French living in the United States], signed, p. 294, "A Subscriber." APS.

1. Roger G. Kennedy has misinterpreted this, explaining that Thebaid was, for the honest citizens of Philadelphia, an allusion to the Thebes described by Alexander Pope in his translation of the *Iliad* (1715–20), that is, an implicit allusion to conquest and war. For Kennedy, this reading underpins the idea that the émigrés' sole aim was the conquest of Mexico as a launching point for the liberation of Napoleon from St Helena. R. G. Kennedy, *Orders from France,* 359.

2. Ibid., 358.

3. *L'Abeille Américaine,* September 12, 1816, 351. The Garniers and Taillefer had come on the *Prince of Orange,* commanded by Captain Smith, leaving Antwerp in July 1816. NARA, *Passenger Lists of Vessels Arriving at Philadelphia, Pa., 1800–1882,* M 425, reel 23.

4. Chaudron to Lakanal, Philadelphia, January 16, 1817, in Jouin, *Lakanal en Amérique,* 14.

5. McMaster, *Life and Times of Stephen Girard,* 2:324–25.

6. Their shortened forms are Colonial Society, Tombigbee Association, and Tombigbee Company.

7. *L'Abeille Américaine,* January 9, 1817, "Procès-verbaux de la Société coloniale des Émigrés français, séance du [ca 22] octobre 1816," 203–4.

8. It is not clear who presided at this meeting. While the heading of the minutes indicates Garnier de Saintes as presiding, the document is signed by the treasurer, Joseph Martin, "president *ad hoc.*" If it was Martin who chaired, this does not mean, as Kent Gardien states ("The Splendid Fools," 495), that Garnier had already left for the American West, since his presence was required for meetings of the executive committee.

9. CADN, consulat de France de Philadelphie, register 45, Légation de France, exercice Pétry, letter of October 10, 1816, p. 18.

10. Writing from New York on October 25, 1816, Lee complained to Jefferson that he had been forced to give up, in Bordeaux, "a situation which was in every point of view agreeable" to him, because of the viciousness of a party, in Lee, *A Yankee Jeffersonian*, 180.

11. Ibid., 181.

12. On March 28, 1811, Lee wrote to his wife, Susan Palfrey Lee, in Bordeaux: "He is probably the next president, so that my standing will be kept us." Ibid., 136.

13. William Lee to Susan (Palfrey) Lee, Paris, November 22, 1811, ibid., 144–45.

14. William Lee to Susan (Palfrey) Lee, Paris, February 10, 1812, ibid., 153.

15. William Lee to Thomas Jefferson, New York, October 25, 1816, ibid., 182.

16. William Lee to James Madison, New York, November 8, 1816, ibid., 183.

17. William Lee to James Madison, New York, November 17, 1816, ibid., 185.

18. *L'Abeille Américaine*, January 9, 1817, "Séance du 2 janvier 1817: Présidence de M. W. Lee, vice-président, en l'absence du président."

19. On March 4, 1817, his mission completed, Lee was appointed auditor of the Treasury Department. See *Le Moniteur Universel*, April 21, 1817, citing the *National Intelligencer* of March 10, 1817, for the list of appointments of March 4, 1817.

20. CADN, consulat de France à La Nouvelle-Orléans, Série D, no. 742, Parmentier: Paris, 3 octobre 1825, Mabille, général inspecteur of the forests and domains of the King, to the consul.

21. Nicolas-Simon Parmantier, born March 17, 1772, in Lorient, where, on March 10, 1793, established as a tradesman, he married Françoise-Hélène Le Tallec, a twenty-three-year-old native of Lorient, with whom he had at least one son and two daughters.

22. He appears in the census of 1810 in the Philadelphia suburb "Northern Liberties" along with an unidentified person. 1810 U.S. Federal Census, Pennsylvania, Philadelphia.

23. *Index to Records of Aliens' Declarations of Intention and / or Oaths of Allegiance, 1780–1880 . . .* , vol. 9, Letter P, p. 12, no. 9305, *Court of Common Pleas*, Phila., April 26, 1809. Filby, *Philadelphia Naturalization Records*.

24. A few traces of this can be found in Stephen Girard's commercial correspondence: APS, Girard Papers, reel 8: Parmantier, N., Endorsements: L.B. 131, 158.

25. Rouillé, *Les notables ou la "seconde noblesse,"* 2:2, 520.

26. An inhabitant of Port-Margot, married to Charlotte Fillon, he owned a coffee plantation in Boucan Richard, in the parish of Gros-Morne. In 1830 he was compensated 24,900 francs for its loss. *État détaillé des liquidations*, no. 2744.

27. "The Assembly having learned with what courage you have devoted yourself to the defense of your country, your glorious actions have raised its admiration and you are worthy of its gratitude. The General Assembly of the French Part of St. Domingo. To Mons. Martin, Citizen of Port Margot. Le Cap, October 4th., 1791. Caduche, president." Ward, "The Germantown Road and its Associations," 121–25.

28. A census lists him as "gentleman." 1810 U.S. Census, County of Philadelphia—outside city of Philadelphia—Northern Liberties.

29. On June 5, 1793, the English had captured the warship *Gracieuse*, on which he was stationed, and had imprisoned him on the island of New Providence in the Bahamas. CADN, consulat général de Philadelphie, carton 73 (secours aux réfugiés . . .), Villar to Genêt (the French ambassador to the United States), Providence, July 25, 1793.

30. "Petition of Charles Villar, to become a citizen of the U.S., Sept. 19, 1806," *Petitions*

to the U.S. Circuit & District Courts for the Eastern District of Pennsylvania (1795–1951), Charles Villar, no. 395, 1806, September 19, NARA, M 248, reel 8; "Declaration and/or Oath of Allegiance of Charles Villar, 1806, Sept. 19," in Filby, *Philadelphia Naturalization Records,* 667.

31. NARA, *Passenger Lists of Vessels Arriving at Philadelphia, 1800–1882,* M 425, reel 3: on June 22, 1802, Charles Villar arrived in Philadelphia on the ship *Hope,* sailing from Havana.

32. Girard Papers, letter 347, reel 8, Girard to Hourquebie frères, Philadelphia, September 1, 1802–APS, Philadelphia. For Villar and Girard, also see, in these papers, L.B. 8: 347; L.B. 9: 150, 315, 494; L.B. 10: 293; 1801: 379; 1802: 468; 1804: 409, 484, 618.

33. After Bonnaffé in 1792–93 and Fenwick Mason & Co. until 1796, Hourquebie were the cosignatories of Girard's shipments to Bordeaux from 1796 to 1802. According to Silvia Marzagalli of the University of Nice, Girard seems to have fallen out with his cosignatories after a few years.

34. *The Philadelphia Directory for 1810 containing the names, trades and residence of the inhabitants of the City, Southward, Northern Liberties and Kensington.*

35. See Jean-Charles Assali, "Napoléon et l'antiquité," *Souvenir Napoléonien* 333 (January 1984): 2–23.

36. See Roger G. Kennedy, *Greek Revival America* (New York: Stewart, Tabori, & Chang, 1989).

37. *L'Abeille Américaine,* January 9, 1817, 203.

38. "Espionnage économique: Dénonciation no. 4," *L'Abeille Américaine,* October 10, 1816, 407.

39. Lajonie Papers, Lajonie to the château of Le Soulat, Philadelphia, January 31, 1817.

40. CADN, consulat de France à Philadelphie, exercice Pétry, register no. 45, October 10, 1816.

41. *Louisiana Gazette* (New Orleans), December 4 and 23, 1816, in H. G. Warren, *The Sword Was Their Passport,* 191.

42. *Louisiana Gazette,* December 4, 1816, quoting an item from the *Kentucky Reporter* (Lexington) in ibid., 191.

43. *Pittsburgh Mercury,* December 7, 1816, picked up by *Niles' Weekly Register,* December 28, 1816, in W. Smith, *Days of Exile,* 27. Was this the group that was expected in New Orleans in January 1817, according to Masot to Captain Général, Pensacola, January 11, 1817 (no. 31), AGI PC, leg. 1874, in H. G. Warren, *The Sword Was Their Passport,* 191? Among the exiles mentioned were Clauzel, Grouchy, and Lallemand.

44. Also see Seeber, "A Napoleonic Exile in New Albany," 175–77.

45. Montulé, *Travels in America,* 121, letter XVII, Louisville, July 2, 1817.

46. Ibid., 121–25, pages where he discusses meeting the family of "Monsieur L. C."

47. Ibid., 122.

48. Journals, or Socratic Evenings. This book was to follow *La dette d'un exilé, ou Plan nouveau d'éducation nationale, basé sur les principes de Socrate . . .* (Brussels, 1816).

49. They sailed on the brig *Nymphe,* registered in Nantes, 186 tons, shipper Trottier, Captain Redureau, leaving Mindin on May 17, 1816, and arriving in New York on July 1. ADLA, 120 J 2382.

50. General Clauzel wrote to Lakanal: "You have seen Generals Lallemand, Lefebvre-

Desnouettes [*sic*], Pénières, Garnier de Saintes, etc.; you are the only one who has acted well; it is up to us to imitate you." Dawson, *Lakanal the Regicide,* 105.

51. Chaudron to Lakanal, Philadelphia, January 16, 1817, in Jouin, *Lakanal en Amérique,* 14–15.

52. Parmantier to Lakanal, Washington, February 25, 1817, in Dawson, *Lakanal the Regicide,* 107.

53. Jean Augustin Pénières to Raymond Pénières-Dubois, Philadelphia, October 22, 1816, in Faure, *Pénières-Delzors,* 188 n. 2.

54. Hagley Museum and Library, Group 2, Du Pont de Nemours, Series A, Correspondence, Box 21, W2-4328, Pénières to Pierre Samuel Dupont (de Nemours), Philadelphia, October 24, 1816.

55. Jean Augustin Pénières to Raymond Pénières-Dubois, Pittsburgh, November 16, 1816, in Faure, *Pénières-Delzors,* 189.

56. *L'Abeille Américaine,* November 28, 1816, 108.

57. Mignet, *Notice historique,* in Dawson, *Lakanal the Regicide,* 103–4.

58. Lakanal in Dawson, *Lakanal the Regicide,* 104.

59. Ibid., 104–5.

60. Faure, *Pénières-Delzors,* 189.

61. *L'Abeille Américaine,* January 9, 1817, 204.

62. Ibid.

63. Ibid., 206.

64. W. Smith, *Days of Exile,* 29.

65. The treaty was not signed at Fort St. Stephens but rather at the trading post of Fort Confederation, on the Tombigbee, in the heart of the Choctaw nation, from which it takes its name: Treaty of Choctaw Trading House.

66. Kappler, *Indian Affairs,* vol. 2.

67. Joseph Lakanal, "Rapport sur la translation au Panthéon des cendres de J. J. Rousseau," September 15, 1794, in *Procès-verbaux du Comité d'Instruction,* 43.

68. Rousseau, *Social Contract,* 163.

69. Ibid., 178 (book 2, chapter 6, "The Law").

70. Ibid., 180 (book 2, chapter 7, "The Legislator").

71. Ibid., 181 (book 2, chapter 7, "The Legislator").

72. Jean Baptiste Auguste, called Émile Pénières, was born in Paris on September 23, 1794; the following October 11 Rousseau's ashes were transferred to the Panthéon in Paris.

73. His novel *Paul et Virginie,* an idyllic story of the innocent love of two young people in the edenic nature of Mauritius in the Indian Ocean (1787) was tremendously successful.

74. Rousseau, *Social Contract,* 182 (book 2, chapter 7, "The Legislator").

75. *L'Abeille Américaine,* January 9, 1817, 207.

76. Jefferson to Lee (Washington, D.C.), Monticello, January 16, 1817, MS, Jefferson Papers, in Reeves, *Napoleonic Exiles in America,* 37–38.

77. Ibid., 37–38. See Rousseau, *Considérations sur le gouvernement de Pologne;* Gabriel Bonnot de Mably, *Du Gouvernement et des Lois de Pologne* (rpt. of the edition of 1794–95; Darmstadt: Sciencia Vlg. Aalen, 1977).

78. Blaufarb, *Bonapartists in the Borderlands,* 47.

CHAPTER 9

1. Dangerfield, *Era of Good Feelings.*

2. For a fine account and synthesis of this subject, see Howe, *What Hath God Wrought.*

3. See Beaucour, "Les projets de deliverance"; Brice, *Les espoirs de Napoléon;* Ocampo, *Emperor's Last Campaign.* For Napoleon's statements on the United States (1816–21) and the presence there of his brother, officers, and the possibility of his own, see Murat, *Napoléon et le rêve américain,* 35–63, quoting Las Cases.

4. MAE, CP Angleterre, vol. 607, fol. 126 bis, Marquis d'Osmond, French ambassador in London, to Richelieu, prime minister, March 17, 1816.

5. Bagot to Castlereagh, Washington, April 25, 1817, NA FO 5/122, fol. 57, in Ocampo, *Emperor's Last Campaign,* 144.

6. *Niles' Weekly Register,* March 29, 1817.

7. MAE, CP EU, vol. 72, fol. 262v, Richelieu to Hyde de Neuville, Paris, January 26, 1817.

8. MAE, CP EU, vol. 73, fol. 27, Hyde de Neuville to Richelieu, Brest, May 7, 1816. In this letter, Hyde expanded the proposal to the usefulness of having at his constant disposal a light, well-armed ship to prevent the formation of a hostile plan, such as the kidnapping of Bonaparte: "such a maneuver would dampen the zeal of certain merchants who, while wanting to serve the Emperor and his accomplices" would not want to see one of their ships sent to the bottom.

9. Ibid., fol. 20, excerpt of a letter from the French ambassador in Madrid of April 22, 1816, concerning a letter from M. de Onís to his government dated March 6, 1816; ibid., fol. 29, Richelieu to Roth, consul general, Paris, May 8, 1816.

10. A piaster was a Spanish silver coin equal in worth to the dollar.

11. MAE, CP EU, vol. 73, fol. 29, Richelieu to Roth, Paris, May 8, 1816.

12. Ibid., fol. 55–57, Hyde de Neuville to Richelieu, New York, June 22, 1816.

13. Ibid., fol. 62, Hyde to Richelieu, New York, July 12, 1816.

14. Ibid., Hyde de Neuville to Richelieu, Washington, July 12, 1816.

15. According to John Wilson Croker, secretary of the British admiralty, to Robert Peel, August 8, 1816, in Ocampo, *Emperor's Last Campaign,* 94.

16. According to Las Cases to Lucien Bonaparte in a letter intercepted on St. Helena and given to Hudson Lowe, November 1816, in ibid., 108.

17. Gourgaud, *Journal de Sainte-Hélène,* 1:169.

18. C. de Montholon, *Histoire de Napoléon,* 2:471–472.

19. MAE, CP EU, vol. 73, fol. 204, Hyde de Neuville to Richelieu, Washington, January 10, 1817.

20. Ibid.

21. In 1817, due to very strict conditions, there were few men receiving a pension in comparison with the total number of soldiers who had served during the Republic and the Empire, and the annual amount of their pensions came to 333 francs, which was very little given the thirty years of service required for a right to retirement income. Petiteau, *Lendemains d'Empire,* 94.

22. MAE, CP EU, vol. 73, fol. 292, Richelieu to Hyde de Neuville, Paris, April 18, 1817.

23. William Lee himself to Timothy Pickering (congressman from Massachusetts, former secretary of state in the Washington and Adams administrations), Washington, February 25, 1817, reel 31, Pickering Papers, Massachusetts Historical Society, in Blaufarb, *Bonapartists in the Borderlands,* 47 n. 82.

24. Lajonie Papers, Jacques Lajonie to Le Soulat, Baltimore, March 1817.

25. *Journal of the Senate of the United States of America, being the Second Session of the Fourteenth Congress, begun and held in the city of Washington, Dec. 2d, 1816* (Washington, 1816); *Journal of the House of Representatives of the United States of America, being the Second Session of the Fourteenth Congress, begun and held in the city of Washington, Dec. 2d, 1816* (Washington, 1816).

26. Chaired by Jeremiah Morrow, a Democratic-Republican senator from Ohio.

27. Lafayette to Henry Clay, La Grange, December 26, 1815, in Clay, *Papers,* 2:115.

28. Parmantier to Lakanal, Washington, February 25, 1817, in Dawson, *Lakanal the Regicide,* 106–7.

29. Writing from New York on November 8, 1816, William Lee spoke glowingly to James Madison of the project of the company of French immigrants that he had formed. Lee, *A Yankee Jeffersonian,* 184.

30. Chaudron to Lakanal, Philadelphia, January 6, 1817, in Nigoul, *Lakanal,* 149.

31. Condict (1772–1862), "congressman from New Jersey," elected as an "Anti-Federalist" to the Twelfth Congress beginning March 4, 1811, and reelected to the Thirteenth and Fourteenth Congresses. *Biographical Directory of the American Congress.* All the information on congressmen in this section is from this work.

32. See also *Debates and Proceedings in the Congress of the United States, 14th Congress, 2nd Session* (Washington, D.C.: Gales and Seaton, 1854), 108, 114, 136–37, 139, 1019.

33. Cyrus King (1782–1817): congressman from Massachusetts, elected as a Federalist to the Thirteenth and Fourteenth Congresses. He died three weeks after the beginning of the Fifteenth, on April 25, 1817. Bolling Hall (1767–1836): congressman from Georgia, elected as a "War Democrat" to the Twelfth Congress, reelected to the Thirteenth and Fourteenth.

34. Quoted in *Niles' Weekly Register,* January 31, 1818.

35. John C. Calhoun (1782–1850): congressman from South Carolina (1811–17), Monroe's secretary of war (1817–25), John Quincy Adams's and Andrew Jackson's vice-president (1825–32), John Tyler's secretary of state (1844). John Forsyth (1780–1841): Democratic congressman from Georgia (1813–18), secretary of state in the Jackson and Van Buren administrations (1834–41). William Henry Harrison (1773–1841): congressman from Virginia (1816–19), ninth president of the United States (March 4–April 4, 1841). John Robertson (1787–1873): congressman from Virginia (1816–19), a descendant of Pocahontas. Thomas Bolling Robertson (1779–1828): a Virginian, John Robertson's older brother, third governor of Louisiana, first congressman from Louisiana (1812–18), visited France while in Congress and gave his impressions in letters published in the *Richmond Enquirer* (September 30–December 23, 1815) and in a book, *Events in Paris* (Philadelphia, 1816). Samuel Smith (1752–1839), senator (1803–15) and then congressman from Maryland (1816–22).

36. "Congress. House of Representatives—Thursday, February 27. Cultivation of the vine," *Aurora,* March 4, 1817.

37. Only details of the Senate vote are known. In favor of the bill were Senators Ashmun, Barbour, Brown, Campbell, Chace, Condit, Dana, Fromentin, Gaillard, Goldsborough, Horsey, Howell, Hunter, King, Lacock, Macon, Morrow, Noble, Roberts, Sanford, Stokes, Tait, Talbot, Taylor, Tichenor, Troup, Varnum, and Wells. Voting against it were Senators Daggett, Hardin, Mason, Ruggles, and Smith.

38. Lajonie Papers, Jacques Lajonie to the château of Le Soulat, Baltimore, March 9, 1817.

39. *Huntsville Republican,* October 28, 1817, in Blaufarb, *Bonapartists in the Borderlands,* 50.

40. *Alabama Republican* (Huntsville), April 4, 1818, in ibid., 50.

41. "French Emigrants," *Niles' Weekly Register,* August 8, 1818.

42. Montulé, *Travels in America,* 123, letter from Louisville, July 2, 1817.

43. John Stuart Skinner to José Miguel Carrera, November 27, 1817, in Carrera, *Archivo,* 19:187–91, in Ocampo, *Emperor's Last Campaign,* 190.

44. W. Smith, *Days of Exile,* 30; Blaufarb, *Bonapartists in the Borderlands,* 49.

45. Onís to Cevallos, Philadelphia, November 16, 1816, AHN Estado 5641, in Ocampo, *Emperor's Last Campaign,* 91.

46. See chapter 5, note 74, for a biographical note on Colonel Roul. Leaving Philadelphia, he arrived in Buenos Aires on October 24, 1816, having established contact with the insurgents led by Juan Martín de Pueyrredón, who had him imprisoned because he supposedly refused to serve. He was freed by General Brayer two days after he arrived (February 24, 1817) with the Carrera expedition, and returned to the United States, where he landed in Annapolis in late April 1817. He returned to Europe via Amsterdam on November 20, 1817. See Beaucour, "Qui était Roul?"

47. Paul Albert Marie Raymond de Latapie (1786–1849), generally called "colonel" but actually an infantry major, should not be confused with cavalry second lieutenant Antoine Latapie (1787–1849), a member of the Colonial Society who settled permanently in Alabama, where he died. Along with other officers, Raymond de Latapie and Lieutenant Philippe Gustave de Pontécoulant, a nephew of Grouchy, left New York on June 4, 1817, aboard the *Paragon* bound for Pernamboue in northeastern Brazil. The final destination of the expedition, supposedly instigated by Joseph Bonaparte, was said to be St. Helena. But the project came to a sudden end: Latapie and Pontécoulant were arrested in Brazil and taken to Lisbon, where they were imprisoned. Latapie escaped and, after all sorts of incidents, was back in Philadelphia in December 1818.

48. Proclamation of José Álvarez de Toledo, Philadelphia, December 1, 1816, in Ocampo, *Emperor's Last Campaign,* 92.

49. This did not seem at all fantastic to Onís, who had learned that Lucien had proposed to an Italian living in Philadelphia that they exchange his, Lucien's, properties in Rome for land in the United States. Onís to Cevallos, Philadelphia, December 18, 1816, AHN Estado 5641, in ibid., 92.

50. This new neutrality law, passed by Congress on March 3, 1817, imposed harsh penalties on any American citizen who engaged in an armed enterprise against nations not at war with the United States.

51. Samuel Engle Burr, *Napoleon's Dossier on Aaron Burr* (San Antonio: Naylor, 1969) 37, in Ocampo, *Emperor's Last Campaign,* 137.

52. Manuel Ortuño Martinez, *Xavier Mina: Fronteras de Libertad* (Mexico City: Editorial Porrúa, 2003) 205, in ibid., 138.

53. Details of a conversation with McGregor, Bagot to Castlereagh, April 25, 1817, NA FO 5/122, fol. 57, in ibid., 139. See also Hurtado, *Les soldats de Napoléon.*

54. Ocampo, 142.

55. Memorandum from Álvarez de Toledo, Madrid, April 8, 1817, in Carlos M. Trelles y Govin, "Un Precursor de la Independencia de Cuba, Don Álvarez de Toledo," *Academia de la Historia,* Havana, June 1926, 114–19, in ibid., 141.

56. Onís to Spanish Secretary of State José García de León y Pizarro, March 3, 1817, AHN Estado 5642, I, in ibid., 141.

57. Ocampo, *Emperor's Last Campaign,* 91, 109.

58. Ibid., 216, 307, 309. Ocampo overestimates the elder General Lallemand's reputation given the way he failed in the northern conspiracy and lost face with the gendarmes who arrested him and his brother.

59. According the *Anti-Gallican* of London, November 10, 1816, cited in ibid., 98.

60. Henry, "Henry Clay et la South American Question."

61. Beauchef, *Mémoires.*

62. Carrera to Grouchy, August 28, 1816, in Carrera, *Archivo,* 17:82–83, 85, and Grouchy to Carrera, Bordentown, September 6, 1816, in Carrera, *Archivo,* 16:103–4, both in Ocampo, *Emperor's Last Campaign,* 86.

63. Carrera, *Archivo,* 17:96–98, memorandum from Grouchy to the Supreme Director of Buenos Aires, Philadelphia, September 1, 1816, in Ocampo, *Emperor's Last Campaign,* 86–87. A duro was worth about a dollar.

64. Stroud, *The Man Who Had Been King.*

65. Bertrand, *Cahiers de Sainte-Hélène,* 2:42.

66. MAE, CP EU, vol. 74, fol. 82, Hyde to Richelieu, New Brunswick's Farm, July 13, 1817.

67. See chapter 10 for a biographical sketch.

68. Onís to Pizarro, no. 175, Philadelphia, October 9, 1817, AHN Estado 5642/1, in Ocampo, *Emperor's Last Campaign,* 189.

69. CADN, consulat de France à Philadelphie, register 6, fol. 326–60, which includes D, no. 7, "Rapport adressé à S. M. le roi des Espagnes et des Indes" [Report addressed to H. M. the King of Spain and the Indies]; one of four envelopes is marked: "To the Count of Survilliers. For his eyes only."

70. R. G. Kennedy, *Orders from France,* 362–63.

71. In addition to *Bonapartists in the Borderlands,* see his article "The Western Question."

72. The following paragraphs summarize Blaufarb's argument. See *Bonapartists in the Borderlands,* 51–57.

73. Dodd and Dodd, *Historical Statistics,* 2.

74. In 1820 there were 85,451 whites; in 1830, 190,406. M. B. Owen, *Alabama Census Returns,* 10–14.

75. Andrew Jackson to James Monroe, Nashville, November 12, 1816, reel 21, Jackson Papers, in Blaufarb, *Bonapartists in the Borderlands,* 54 n. 102.

76. ADAH, SG 24709, Message of Governor Bibb to the Gentlemen of the Legisla-

tive Council and the House of Representatives, St. Stephens, November 3, 1818, in ibid., 54 n. 99.

77. MAE, CP Espagne, vol. 701, fol. 196, Duke of Fernan-Nunez and Montellano to the Duke of Richelieu, Paris, March 20, 1818.

78. Henry Clay to Lakanal, Washington, D.C., March 20, 1817, in Clay, *Papers*, 2:328. On March 29, Clay informed Lafayette that Congress had granted to the French émigrés 92,000 acres of land on the Tombigbee and that Lakanal, as he had just written him, was about to leave for this river and its region, pleased with its excellent soil and climate, which were said to resemble those of southern France. Ibid., 331.

79. Lajonie Papers, Jacques Lajonie to the château of Le Soulat, Aigleville, "Nauzelbine," October 1, 1818. The original document of his commission, in the Lajonie Papers, reads: "William W. Bibb, Governor of the Alabama Territory and Commander in Chief of the Militia thereof. To All who shall see these presents, greeting: Konw [*sic*] ye, That reposing special trust and confidence in the patriotism, valor, fidelity and abilities, of Lajoinie [*sic*], I do appoint him Lieutenant of B. in the first Battalion of the 9th. Regiment of the Militia of the Alabama Territory . . . Town of St. Stephens, the 19th. day of August 1818 . . . , William W. Bibb." See also July 31, 1818, White Bluff, Lajonie named lieutenant of the Third Company of the militia, in Carter, *Territorial Papers*, 18:389; April 16, 1819, James Lajonie appointed lieutenant of the Third Company of the militia, ibid.

80. William Lee to Josiah Meigs, Washington, April 25, 1817, in Carter, *Territorial Papers*, 18:90. Meigs's answer is on page 91.

CHAPTER 10

Epigraph: *L'Abeille Américaine*, November 14, 1816, 80.

1. *L'Abeille Américaine*, January 9, 1817, 203.

2. Ibid., October 10, 1816, 407, and November 14, 1816, 80.

3. Lajonie Papers, Lajonie to the château of Le Soulat, Burlington (Del.), January 20, 1817.

4. Ibid., Lajonie to Le Soulat, Burlington (Del.), January 31, 1817.

5. *L'Abeille Américaine*, January 23, 1817.

6. *Le Moniteur Universel*, September 28, 1817, reprinted in the *Affiches, Annonces et Avis divers de Nantes* on October 5, 1817, carried an excerpt of a letter written on August 1, 1817, by a Frenchman in Charleston to one of his friends, in which he mentions the Alabama Territory, created from fifty million arpents of land ceded by the "Cheroekes" and given this name "because this river is the largest of the region and after it the Tombighee [*sic*]"; the lands suitable for the cultivation of cotton, tobacco, and maize are attracting émigrés from Virginia and the two Carolinas, but "it is to the French that this territory has been granted; they hope to naturalize the vine and the olive there."

7. Nicolas-Félix Desportes (1763–1849), a diplomat during the Revolution, baron of the Empire, prefect of the department of Haut-Rhin (1802–13), deputy in the Chamber of the Hundred Days, banished by article 2 of the edict of July 24, 1815.

8. MAE, MD France, vol. 2140, fol. 7–7v.

9. Lajonie Papers, Pierre Fougnet to Jacques Lajonie in Mobile, July 21, 1819.

10. "Rapport [by Durand] fait au nom du Comité de la Société Française Agricole & Manufacturière du Tombigbee, aux Membres de cette Société, assemblés," *L'Abeille Américaine,* November 13, 1817, 277–84.

11. "List of the Shares of the Tombeckbee Company" in letter from the secretary of the treasury transmitting information of the progress that had been made under the act of Congress of March 3, 1817, entitled "An Act to set apart and dispose of certain public lands for the encouragement of the cultivation of the vine and olive," December 14, 1818, Washington, printed by E. de Krafft, 1818, pp. 7–17.

12. The transcription errors occurred when names were copied from the handwritten papers entitled "Tombeckbee Allotments."

13. Whitfield, "French Grant in Alabama," 4:321–55, and Barefield, *Old Demopolis Land Office Records,* 72–80, both reproduce the list from Lowrie et al., *American State Papers: Public Lands,* 3:396–99; Dawson (*Lakanal the Regicide,* 193–202) reproduces the list published in *Documents, Legislative and Executive, of the Congress of the United States, in Relation to the Public Lands* (1834); W. Smith, *Days of Exile,* 83–96, reproduces the two preceding lists and blends them with the list appended to a map drawn by a French colonist, Édouard Paguenaud, long in the possession of the Cobbs family of Greensboro, Alabama; Blaufarb, *Bonapartists in the Borderlands,* uses all of these lists (176–82), and most particularly (188–227) the biographical notes of Kent Gardien (see next note).

14. Gardien, who was the first to study in detail the American, Spanish, and French archives on this subject, did not publish the biographical notes resulting from his research, but had them microfilmed by ADAH.

15. In public archives, the study of passport registries, certificates of identity, and the lists for fitting and laying up of ships were the most helpful: in France, these were found chiefly in the Archives nationales, the Centre des Archives diplomatiques de Nantes, and the Archives départementales of Gironde, Loire-Atlantique, and Seine-Maritime; in Belgium, the archives of the port of Antwerp. In private archives, the study of hundreds of letters written by émigrés was very profitable. I also found helpful the sites www.geneanet.fr and www.culture.gouv.fr/documentation/leonore/pres.htm (records of the Legion of Honor at the Centre historique des archives nationales); www.ancestry.com and www.familysearch.org (Church of Jesus Christ of Latter-Day Saints).

16. They were George Strother Gaines, John Roster, Samuel Jackson, William Tabele, George Billington, Jonathan Pierce and his brother, Samuel Voorhees, William Montelius, and Kimbal. Rush and Adams, *Report from the Secretary of Treasury,* 36.

17. The departments of Savoie, Haute-Savoie, and Alpes-Maritimes have not been included, since in 1789 they were part of the Duchy of Savoy and the Earldom of Nice.

18. An analysis of the passports issued by the prefecture of Gironde during the nineteenth century shows that 55 percent of the passengers emigrating to the United States from Bordeaux were from Aquitaine and 41 percent from Gironde itself. N. Fouché, "Les passeports délivrés à Bordeaux," 196.

19. Baron of the Empire Yves-Marie Pastol de Keramelin, born in Guingamp (Côtes d'Armor) on March 5, 1770, brigadier general in 1809, married to Julie Basire on August 26, 1800, in Dijon, who bore him a son. Died on May 31, 1813, in Lützen in the battle of

Neukirch, near Breslau in Silesia. Révérend, *Armorial du Premier Empire*, 4:13; Six, *Dictionnaire biographique*, vol. 2.

20. Chevalier of the Empire Henri David, born in Toulon on March 28, 1762, promoted to *adjudant general chef de brigade* (a rank between colonel and brigadier general in administrative and staff positions in divisions of the Revolutionary army), *adjudant commandant* (the same rank, renamed) in the Army of Italy on February 10, 1800, transferred to the Army of Saint-Domingue from March to November 1803, married in Philadelphia on March 29, 1804, to Adèle de Sevré, who bore him two daughters. Died in Bordeaux on July 29, 1816. Quintin and Quintin, *Dictionnaire des colonels de Napoléon*, 250.

21. Socio-professional classification is a term I have borrowed from statisticians and sociologists for practical reasons, and adapted to the sources and social structure of the eighteenth and nineteenth centuries. In this I follow Daumard, "Une référence pour l'étude des sociétés urbaines." For a comparative model, see the study on French émigrés in Canada by Choquette, *Frenchmen into Peasants*.

22. In French, *marchand* can mean tradesman, shopkeeper, stall-holder, or merchant. In this sentence it is translated as tradesman, the most neutral of these terms. The word *merchant* in this sentence is *négociant* in the French original. In the sentence following, *milliner* is *marchand de modes*, and dealer in porcelain is *marchand de porcelaine*.

23. Me. Chavet (Paris), C. M. ET/LII/593, May 9, 1785, website of Bertrand Cor via geneanet.fr.

24. He was the first to have borne this name, that of the estate of Noëttes, near Laigle (Orne), which Charles Lefebvre had acquired in 1743.

25. ADG, 3 E 24087, Gabriel Séjourné, notary in Bordeaux, marriage contract of September 4, 1788, between George Philippe Batré, born in Stettin on September 12, 1757, and Élisabeth Félicité Lafitau-Gimon, born in Grande Terre (Guadeloupe), daughter of Pierre Lafitau-Gimon, bourgeois of Bordeaux, and Marie Claire Lubet, domiciled on Grande Rue, suburb and parish of Saint-Seurin.

26. Public secondary schools created by law in 1795.

27. Boudon, *Napoléon et les lycées*.

28. Stadsarchief Antwerpen, MA 2644/5, passports no. 5506 bis, 5507 bis, Jean François Roland.

29. AN, F⁷ 3523, list of passports for travel abroad issued by the Paris Police Prefecture, March 8, 1817, Charles-François Jénin [*sic*].

30. *L'Abeille Américaine*, November 14, 1816.

31. Ibid., January 23, 1817, 239.

32. Ibid., October 23, 1817.

33. Blaufarb, *Bonapartists in the Borderlands*, 117–18.

34. SHD/DAT, Douarche file, General Tilly's inspection (1811), note for the minister, September 9, 1815.

35. CADN, postes, New York, sous-série C, no. 2*, Pétry to the Duke of Richelieu, New York, April 28, 1818.

36. Brandywine, October 23, 1816, quoted in Jolly, *Du Pont de Nemours*, 283.

37. Members of the elite Guard were paid at the rank above them in the regular army.

38. AN, F⁷ 2550b, correspondence received by the minister of the interior, no. 22,150, June 26, 1817, authorization of the minister of police to issue him a passport.

39. For Combe, see Barou, *Le colonel Michel Combe;* Quintin and Quintin, *Dictionnaire des colonels de Napoléon,* 222; SHD/DAT, 3 Yf 53 221, 1st series of pension funds, Michel Combe's personnel file: the extract of the register of the Infantry and Cavalry Committee's deliberations, session of April 7, 1834, describes Colonel Combe as "a fiery man, incapable of mastering his temper," "with a violent and irascible character, a despot toward his subordinates and independent with regard to his superiors."

40. SHD/DAT, personnel file GB 3040, Nicolas Raoul.

41. MAE, MD France, vol. 2143, fol. 106, French ambassador to Marquis Dessolle, Minister of Foreign Affairs, Rome, March 4, 1819, referring to his dispatch to the Duke de Richelieu of December 31, 1818.

42. SHD/DM, personnel file. In 1809 Taillade married a Mademoiselle Fortini, daughter of the mayor of Longonne in Elba, who bore him a daughter, Olimpia. After he was exiled, his wife and child lived in Florence.

43. *Niles' Weekly Register* of September 27, 1817, announced his arrival: "Among the emigrants who have lately reached the United States from France, is capt. Bailliard [*sic*], who conveyed Bonaparte from Elba to France."

44. MAE, MD France, vol. 2143, fol. 194, Gruchet to the French consul, Philadelphia, April 28, 1819.

45. Galabert to the Duke d'Angoulême, June 10, 1814, quoted by Jean-François-Joseph Massié, "Louis Galabert, 1773–1841," 1–24. This very detailed article on Louis Jacques Galabert is part of the Kent Gardien archive. There is no indication as to the journal in which it appeared.

46. For the nineteenth century as a whole, this underrepresentation of the primary sector in emigrants from southwestern France has been observed: only 10 percent of those requesting passports in Bordeaux for the United States were farmers, and winegrowers were a very small minority of this group. N. Fouché, "Les passeports délivrés à Bordeaux," 203, table 9.

47. Lee to Madison, New York, November 8, 1816, in Lee, *A Yankee Jeffersonian,* 184.

48. For his life until 1813, see Bazin, "Bouteilles à la mer."

49. *Mémoires de M. de Bourrienne, minister d'État, sur Napoléon* . . . (Paris: Ladvocat, 1829), in Bazin, "Bouteilles à la mer," 169; *Mémoires du Roi Joseph,* in ibid., 172.

50. On October 9, 1810, in Paris, he had married Anne Geneviève Caroline Boscary de Villeplaine, a banker's daughter, placed after her father's death in 1797 under the guardianship of Brillat-Savarin. The marriage contract was signed by the Queen of Spain (Julie Clary, Joseph's wife), the Infantas Zenaide and Charlotte, the Princess Royal of Sweden (Désirée Clary, Bernadotte's wife), Prince Oscar (the future Oscar I of Sweden), Archchancellor Cambacérès, Minister of Finance Gaudin, etc. Bazin, "Bouteilles à la mer."

51. Stadsarchief Antwerpen, MA 2644/4, passport no. 3239 issued in Antwerp on August 5, 1815, for Liège (Belgium), James Debellièvre, landowner, born in Chagny (Switzerland) [*sic*], domiciled in Geneva.

52. See Pierre Pinon, *Louis-Pierre et Victor Baltard* (Paris: Éditions du Patrimoine, 2005).

CHAPTER 11

Epigraph: Bernard quoted in Planchot-Mazel, *Simon Bernard*, 1:121.

1. Israel Pickens to General William Lenoir at Fort Defiance (N.C.), St. Stephens (Ala.), January 18, 1818, Pickens Family Papers, LPR 46, ADAH.

2. Carter, *Territorial Papers*, 18:240, in Southerland and Brown, *Federal Road*, 66.

3. Peter A. Brannon, "The Federal Road—Alabama's First Improved Highway," *Alabama Highways*, April 1927, 7, in Southerland and Brown, *Federal Road*, 2.

4. Montulé, *Travels in America*, 77–78.

5. The *Jeune-Henriette*, Captain Pronck, left Antwerp for Charleston on October 15, 1816. See Archives générales du Royaume de Belgique (Bruxelles), commissariat général de la Justice, carton 2, dossier Politie 16, Bruxelles, October 21, 1816, the governor of southern Brabant to the Count de Thiennes, fol. 15–17; ibid., Antwerp, October 19, 1816, the maritime bailiff to the Count de Thiennes; Stadsarchief Antwerpen, Politie-Bannelingen, MA 464/3 (2), Outlawed men in Antwerp, the police commissioner to the mayor of Antwerp, October 5 and 13, 1816.

6. This first appeared in the *National Intelligence*, followed by the *Niles' Weekly Register* (September 6, 1817) and the *Ami des Lois* (October 9, 1817).

7. Michaux, *Voyage*, quoted in Dawson, *Lakanal the Regicide*, 100.

8. Dawson, *Lakanal the Regicide*, 99.

9. CADN, consulat de France à New York, série B, no. 122, letter of May 24, 1817. The consul mentions Garnier and Dularauday [Delaroderie from Charente, a cavalry captain in Naples].

10. Jean-Marie Léonard Collection (Isère), John Latour to Jacques Lajonie, Baltimore, July 21, 1817: "We had here with us for a few days M. Garnier de Saintes, a charming young man. His father is to go to New O. He is one of your society members for your Tomgueby [*sic*] lands. The son almost fell in love with, you can guess I think, but he has no P[ecuniary] means. He was in Philadelphia."

11. Blaufarb, *Bonapartists in the Borderlands*, 160, writes that Garnier de Saintes died in a steamboat explosion on the Ohio in 1818.

12. MAE, CP EU, fol. 39, Pétry to Richelieu, Philadelphia, August 29, 1818 (copy of a letter to Hyde de Neuville); AN F[7] 6680, Garnier de Saintes file, the minister of foreign affairs to Messrs. Bernard jeune and Cie, merchants in Paris, Paris, December 3, 1818 ; CADN, consulat de France à La Nouvelle-Orléans, série D, carton 202, file no. 1397, Saintes, May 10, 1832, André-François Godet fils, captain ret., member of the Legion of Honor, to the French consul in New Orleans, Saintes, May 10, 1832, concerning the disappearance of the Garniers, father and son, citing a letter from Lecoq Dumarselay to Basile Meslier, Louisville, Kentucky, Jan. 19, 1818.

13. AN, F[7] 6680, Garnier de Saintes file, Paris, December 3, 1818, the police minister to Messrs Bernard jeune & Cie, merchants in Paris.

14. MAE, MD France, folios 39–39v, Pétry to Richelieu, Philadelphia, August 29, 1818 (copy of a letter to Hyde de Neuville).

15. Lajonie Papers, Jacques Lajonie to the château of Le Soulat, Baltimore, April 19, 1817.

16. Poussin, *De la puissance américaine,* 2:84.

17. Ibid., 84–86.

18. Lajonie Papers, Jacques Lajonie to the château of Le Soulat, New Orleans, ca. June 25, 1817.

19. Montulé, *Travels in America,* 75.

20. For the relationship between the river and the city, see Kelma, *A River and Its City.* Also see Langlois, *Des villes pour la Louisiane française.*

21. Lajonie Papers, Jacques Lajonie to the château of Le Soulat, New Orleans, ca. June 25, 1817.

22. Ibid., Jacques Lajonie to the château of Le Soulat, New Orleans, July 20, 1817.

23. U.S. Department of Commerce, Bureau of the Census, *Negro Population, 1790–1915,* 51, 57.

24. The 1820 census indicated a population of 27,000, that is, 10,000 more than in 1810, but this did not take into account the many people who had left the sweltering summer city and the risk of yellow fever at the time of the census. According to the census there were 13,584 whites and 13,592 blacks, and of the blacks, 7,355 were slaves and 6,237 free.

25. Lajonie Papers, Lajonie to Pierre Audebert, Muset, Juillac, June 1817, Lajonie to the château of Le Soulat, August 19, 1817, and Lajonie to the château of Le Soulat, September 1, 1817.

26. Simon Bernard to General Swift, New Orleans, October 4, 1817, in Planchot-Mazel, *Simon Bernard,* 1:122.

27. Duffy, "Nineteenth Century Public Health," 332.

28. Pierre-Louis Berquin-Duvallon, *Vue de la colonie espagnole du Mississipi, ou des provinces de la Louisiane et Floride occidentale, en l'année 1802, par un observateur resident sur les lieux* (Paris: Duvallon, 1803), in Carrigan, *Saffron Scourge,* 19. Carrigan dates the first epidemic to 1739, without excluding the possibility of earlier cases of fever.

29. Carrigan, *Saffron Scourge,* 38.

30. Latrobe, *Impressions Respecting New Orleans,* 141–42.

31. The last epidemic, in 1905, claimed almost one thousand lives. See Moe, "Yellow Fever in New Orleans," 8.

32. Born in 1769 in the department of Aude, Lafon died in New Orleans on September 29, 1820, of yellow fever. See Bos, "Barthélémy Lafon."

33. Pierre François Du Bourg, chevalier de Sainte-Colombe (Bordeaux, 1762–New Orleans, 1830), a colonist in Saint-Domingue, married in 1797 to Élisabeth Étiennette Charest de Lauzon; refugee in the United States in 1798, settled in New Orleans in 1800, merchant, planter, broker to wealthy Louisiana planters, in particular the Bringiers: in 1812 his daughter Louise Élisabeth Aglaé married Michel Doradou-Bringier, brother of Louis Bringier, a topographical engineer and member of the Tombigbee Society in 1819. In the 1820 census, Du Bourg de Sainte-Colombe was listed as living on Dumaine Street in the French Quarter of New Orleans. Sainte-Marie, "Descendance aux États-Unis."

34. Pierre Augustin Charles Bourguignon Derbigny (Laon, 1769–Jefferson Parish, La., 1829), emigrated to Saint-Domingue, then to the United States during the Revolution, married in 1791 in Pittsburgh to Jeanne Odile Félicité de Hault de Lassus de Luzières (born in 1773), seven children; in 1797 arrived in New Orleans, where he had a brilliant ca-

reer: first as interpreter, then a well-known jurist and politician, secretary of state of Louisiana from 1821 to 1828, and then its sixth governor from December 15, 1828, to October 6, 1829. Derbigny was in contact with the Tombigbee Society members, since on October 13, 1817, he was a witness at the New Orleans marriage of Nicolas Alphonse Desmare, a former guard of honor from Caudebec-en-Caux (Seine-Maritime), to Odile Marie Pitot. On May 17, 1820, his wife became the godmother of their first child, Marie Odile Desmare, born in New Orleans on October 11, 1819.

35. He built a house for merchant Vincent Rillieux, painter Edgar Degas' great-grandfather, on the corner of Royal and Conti Streets, known today as the Rillieux-Waldorn House.

36. *Map of the Mississippi River, from North to the sea, with all passes and entrances of said river; and a plan of the Bay of Spiritu Sancto, surveyed with accuracy in the years 1804 & 1805. The whole forming an atlas containing seventeen fol.s, Philadelphia Commercial Advertiser,* December 2, 1805; Lafon, *Carte générale du Territoire d'Orléans.*

37. Like another French architect, Arsène Lacarrière Latour, who was born in Aurillac in 1778, naturalized in 1812, and is often associated with Lafon. Garrigoux, *Un aventurier visionnaire.*

38. For Lafon's relationship with the Baratarians and Lafitte, see ibid.

39. Address found in the *Courrier de la Louisiane,* October 20, 1815.

40. The *Ami des Lois* of July 14, 1817, carried the following notice: "The undersigned [Lafon, engineer] wishes to take on 12 Negroes, good woodcutters, only to clear land in the town of Chetimata. They will be well fed and cared for."

41. Jean-Marie Léonard Collection (Isère), Baltimore, July 21, 1817, John Latour to M. Lajonie Lapeyre, New Orleans, c/o M. Derbigni [*sic*], judge.

42. The house, designed in 1811 by Arsène Lacarrière Latour and Hyacinthe Laclotte, belonged to Marie Louise de la Ronde, the wealthy widow of Don Andres Almonester y Roxas, who had married Jean-Baptiste Castillon, the French consul. In 1828, shortly after the building passed to their daughter, the Baroness of Pontalba, it was partly destroyed in a fire and reconstructed. See Samuel Wilson in Latrobe, *Impressions Respecting New Orleans,* 23–24 nn. 4 and 5.

43. Jean-Bernard Trémoulet and Marie-Victoire Soubie, born in Mirande (Gers) in 1756 and 1772, respectively, married in New Orleans in 1785 (Trémoulet-Soubie marriage contract, Davidson-Tremoulet Papers). Of eight children born between 1788 and 1814, three were living in 1817, among them Antoine Cyprien Trémoulet (1809–82), from whom are descended Marie-Louise Tremoulet-Davidson, born in New Orleans in 1927, whom I thank for her genealogical assistance, and her sister Doris Marie Tremoulet Cost (1931–2002), mother of Celeste Marie Cost Cook (born in 1965), the last descendant of the Trémoulets.

44. Latrobe, *Impressions Respecting New Orleans,* 23–26, 53–54.

45. Castellanos and Reinecke, *New Orleans as It Was,* 312–13.

46. Attributed to José de Salazar; see Bruns, *Louisiana's Portraits,* 248.

47. Montulé, *Travels in America,* 76.

48. Montlezun, *Voyage,* 1:342.

49. Lajonie Papers, Lajonie to the château of Le Soulat, New Orleans, August 19, 1817.

50. According to a witness cited by La Souchère-Deléry, "Le thème napoléonien dans la poésie louisianaise," 34.

51. CADN, consulat de La Nouvelle-Orléans, 2, Consul Guillemin to the political administration, May 7, 1817. For a slightly different version see MAE, CP EU, v. 74, fol. 7: "Précis de ce qui s'est passé à la Nouvelle-Orléans le 4 mai 1817."

52. *L'Ami des Lois,* August 14, 1817, announcement by Pierre Caillou, faubourg Sainte Marie.

53. Ibid., August 9, 1817, announcement by Mr. Renault.

54. Ibid., August 15, 1817.

55. Ibid., October 14, 1817; see the song "Adieux des missionnaires aux Bordelais" (Missionaries' farewell to the people of Bordeaux).

56. Ibid., December 23, 1817, for the Christmas performance, at the St. Philippe Theater, of *Édouard en Écosse ou la nuit d'un proscrit* (Edward in Scotland or The Night of an Outlawed Man), in which "the perfect resemblance of his destiny with that [of Napoleon] could not help but sharply arouse the spectator's curiosity."

57. Lajonie Papers, Lajonie to the château of Le Soulat, New Orleans, June 27, 1817.

58. Montulé, *Travels in America,* 79.

59. Ibid.

60. "General Bernard's Report on the Defense of the Gulf of Mexico Frontier," National Archives, Washington, Department of War, Record of the Office of the Chief of Engineers, RG 77, Entry 221, p. 75, in Planchot-Mazel, *Simon Bernard,* 1:130. The study, including field explorations and writing the cartographic report, took more than one year of work.

61. Lajonie Papers, Lajonie to the château of Le Soulat, New Orleans, September 1817.

62. Two men and two women between the ages of fourteen and twenty-six, and two men and two women between twenty-six and forty-five. U.S. Census Office, Fourth Census, 1820, *Population Schedules of the Fourth Census of the United States 1820,* Louisiana, vol. 3.

63. John Adams Paxton, *The New-Orleans directory and register; containing the names, professions, & residences, of all the heads of families, and persons in business, of the city and suburbs; notes on New-Orleans; with other useful information* (New-Orleans: Benj. Levy & Co., 1822), 24.

64. The *Rebecca* left Bordeaux commanded by Captain Turner, 210 tons, ten crewmen; on board were eleven persons whose passage had been paid by Madame Taupier-Letage: 800 francs to lodge a young lady in the ship's cabin, 2,500 francs to lodge ten men in steerage. ADG, 8 M 200: *Relevé bi-mensuel des navires sortis du port de Bordeaux en 1817.* The *Rebecca*'s arrival was announced in the *Courrier de la Louisiane* and the *Ami des Lois* of August 29, 1817.

65. Lajonie Papers, Lajonie to the château of Le Soulat, New Orleans, September 1817.

66. Cauna, *L'Eldorado des Aquitains,* 131.

67. Lajonie Papers, Lajonie to the château of Le Soulat, New Orleans, September 1817.

68. The *McDonough*: 74 tons, two masts, 17 meters long, 6 meters wide, and 2.3 meters high. *Ship Registers and Enrollments of New Orleans, Louisiana,* 6 vols. (University: Hill Memorial Library, Louisiana State University, 1941–42), 1:82.

69. *L'Abeille Américaine,* August 28, 1817, 114.

70. Benoît Poculot (1778–1818) had married Louise Bichat, born in 1778 in Sainte-Foy-lès-Lyon (Rhône), on February 8, 1796. They had at least one daughter, Marguerite

Poculot, born in Lyon in 1799. At this time he appears on the census as a merchant (trading in yarn) in rue Lainerie, where his father Pierre Poculot, who died shortly after Benoît's birth, had already been in business. As Charles Villar was the brother of Anne (Villar) Bichat, Louise Bichat Poculot's mother, he was thus Benoît Poculot's uncle by marriage. It is not known whether Louise Bichat Poculot and her daughter Marguerite, who were in Lyon in 1821 at the very latest, went to the Tombigbee settlement.

71. "Notes from the Lips of the Honorable George Noble Stewart," Pickett Papers, ADAH.

72. Michel Le Bouteillier (ca. 1766–Aigleville, Ala., 1819), a refugee from Saint-Domingue (?), living in Philadelphia in 1793, captain of American and French ships, sailing in 1816. In Philadelphia in 1806 he married Helena Counsell (Cádiz, 1778–Mobile, Ala., 1856), widow of Noble Caldwell Stewart, who died in Philadelphia in 1804. For the Counsell-Stewart-Le Bouteilliers, see *Demopolis Times*, July 29, 1975.

73. Gardien, "Who Were the McDonough Passengers?" *Demopolis Times*, Sept. 21, 1983.

74. Pickett, *History of Alabama*, 2:387–88.

75. Parmantier to a friend in Philadelphia, Mobile Bay, May 26, 1817, in Whitfield, "French Grant in Alabama," 327–28.

76. The article "The Sea-ports in the Gulf of Mexico" was reprinted in the *Mobile Argus* on December 30, 1822, 2.

77. *Niles' Weekly Register*, December 12, 1818: "The population of the town of Mobile by a late census is as certain to consist of 604 white persons, 149 free people of color, and 374 slaves." But the *Mobile Commercial Register*, February 1822, 3, gave it only 800 in 1819. In 1822 the population had risen to 2,800 inhabitants.

78. "The Tombigby Country," *National Intelligencer*, August 30, 1817 (Extracts, by translation, of a letter from Colonel Parmantier, one of the commissioners of the French Tombigby Vine Company, to his friend in this city, dated White Bluff, July 14).

79. Lajonie Papers, Lajonie to the château of Le Soulat, New Orleans, August 1, 1817.

80. *Gazette de la Mobile*, August 23, 1817; *L'Ami des Lois*, September 3, 1817.

81. *L'Ami des Lois*, October 9, 1817.

82. *Mobile Argus*, December 30, 1822, 2.

83. Parmantier's long letter published by the *National Intelligencer*, August 30, 1817, begins at this point.

84. Holmes, "Fort Stoddert in 1799."

85. For Harry Toulmin (1766–1824) see Ball, *Glance into the Great South-East*, 439–40.

86. Gaines, *Reminiscences*, 44.

87. Edmund Pendleton Gaines (1777–1849) married Frances Toulmin, one of the judge's seven daughters. He gained fame in 1807 when he captured Aaron Burr, Jefferson's former vice-president. *Dictionary of American Biography*, 11 vols. (New York: Scribner's, 1964), 4:92–93.

88. Southerland and Brown, *Federal Road*, 27–28, 33.

89. McDonald, "Jackson: General History," 199.

90. Ball, *Glance into the Great South-East*, 459.

91. Young Gaines (1760–1829) emigrated from South Carolina in about 1790 with his wife, Esther Lawrence Gaines, and their children, to the lower Tombigbee. He prospered in trade with the Choctaws, whose language he learned, and acquired twenty-four hundred

acres of land on the Tombigbee and twelve hundred acres in present-day Perry County, Mississippi. His daughter, Anne Gaines (1795–1868), married George Strother Gaines in 1812. See Gaines, *Reminiscences*, 161, 180 n. 13.

92. Jackson, formerly Republicville, then Pine Level, which gave its name to the Pine Level Land Company, an organization established on lands in part purchased from Carney that was responsible for the birth of the town. "Records of the General Office, Journal and Report of James Leander Cathcart and James Hutton, agents appointed by the Secretary of the Navy to survey timber resources between the Mermentau and Mobile Rivers, November 1818–May 1819," National Archives, Washington, D.C., in Strickland and Edwards, *Residents of the Southeastern Mississippi Territory*, book 4, *The Journals*.

93. Gaines Plantation, no. 60, *Map of Tensaw region with list of houses on east side of Tensaw and West side of Tombigbee Rivers*, National Archives, Label-Map Alabama 27, RG 49, in Matte, *History of Washington County*, 17, 401–2.

94. *Territorial Papers*, 5:363, quoted by Pate in Gaines, *Reminiscences*, 179 n. 6.

95. Matte, Brown, and Wadell, *Old St. Stephens*.

96. *Alabama Republican* (Huntsville), September 30, 1817, in Griffith, *Alabama*, 178.

97. In 1809 Magoffin came from Philadelphia to St. Stephens, where he represented Washington County in the Mississippi territorial assembly, and then the government as register of the United States Land Office. T. Owen and Owen, *History of Alabama*, 4:1147.

98. Their names can be found in the transactions and founding documents of various establishments in St. Stephens and the region. See Matte, Brown and Wadell, *Old St. Stephens*.

99. Samuel Dale (1772–1841), famous for the episode known as the "Canoe Fight" on the Alabama River (1813), when he and two lieutenants killed nine Creeks, and for his December 1814 ride from Milledgeville, Georgia, to New Orleans, to inform General Jackson of the signing of the Treaty of Ghent, ending the war against England.

100. Robert Lee Hudson (Demopolis, 1885–1973), son of James Albert Hudson (1850–1927) and grandson of Richard Henry Hudson (1815–63), who came from Virginia to settle in Uniontown, Perry County, Alabama, where he was a prosperous merchant in the years preceding the Civil War. Information on Hudson comes from James and Emogene Armistead's article in the *Demopolis Times* of August 30, 1979; Hudson's will in Linden, Alabama (Will Record 118 E, fol. 250–51); family documents kindly made available by Mrs. Martha Griffith of Demopolis; and interviews with several people who knew Robert Hudson well, such as his cousin Henry Clay Graves, and Reginald McKey, who worked with him for many years. I am grateful to them all for their assistance.

101. "Captain" François Bierne (Lille, 1788–Demopolis, 1824). He is said to have been wounded and made prisoner during the Russian Campaign. He was in Mobile in 1820, from where he sent Stephen Girard in Philadelphia 480 five-franc coins. No reliable source indicates that he was actually an officer. Bierne was a friend of Lajonie's, perhaps from his army days, as suggested by the letter he sent from Demopolis on March 13, 1826, humorously addressed to Mr. Lapeyra Lageone [*sic*]—whom he calls "my old dragoon," in Tombigby, Alabama State. See, successively, James Armistead and Emogene Armistead, "Napoleon Officer Man of Wealth," *Demopolis Times*, April 27, 1978; *Mobile Commercial Register*, August 15, 1826, Bierne obituary; APS, Girard Papers, 1829, 1205; Lajonie Papers.

CHAPTER 12

Epigraph: Chateaubriand, *Atala. René,* 54–55.

1. ADAH, Alabama Territory, SG 3672, folder 5: "The Members of the French Tombigbee Company in the County of Marengo, to his Excellency the Governor of Alabama Ty." This is an undated petition, drawn up by Pénières and signed by twenty-five other society members, for the appointment of a justice of the peace. It was submitted between the creation of Marengo County on February 6, 1818, and Parmantier's letter of April 14, 1818, to Governor Bibb refusing his appointment as justice of the peace.

2. Colonel Parmantier to Major Thomas Freeman, Surveyor General of the Mississippi Territory, St. Stephens, Demopolis on the White Bluff, August 16, 1817. Letter transcribed, commented on, and published by M. Clinton McGee in *The Dixie Philatelist,* Summer 1978, 17–20. Mr. Van Koppersmith of Mobile has graciously furnished a copy of the original letter.

3. J. A. Pénières to his brother Raymond Pénières, Aigleville, June 18, 1818, in Faure, *Pénières-Delzors,* 192.

4. Lajonie Papers, Jacques Lajonie to Jean Quessart, Mobile, January 26, 1819.

5. *L'Ami des Lois et Journal du Commerce,* April 25, 1820, letter to the editors from L** T*** [Lin Troy].

6. *L'Abeille Américaine,* December 18, 1817, 368.

7. Lajonie Papers, Lajonie to the château of Le Soulat, Demopolis, January 19, 1818.

8. Parmentier to Freeman, August 16, 1817.

9. Lowrie et al., *American State Papers: Public Lands,* 3:388.

10. Parmantier's letter to Freeman demonstrates what a rough estimate the French had of the land they wanted. He erroneously places the confluence of the Tombigbee and the Black Warrior very near the thirty-third degree of north latitude.

11. Secretary of the Treasury to Josiah Meigs, June 7, 1817, in Carter, *Territorial Papers,* 6:794–95.

12. Ibid., 812.

13. The Alabama Land Office maps showing the progression of the surveying and settling of Alabama in the nineteenth century are reproduced in Barefield, *Old Demopolis Land Office Records,* ix–xiv.

14. Ibid., 190.

15. Carter, *Territorial Papers,* 18:202.

16. CADN, consulat de New York, série C, register no. 1, correspondence with the minister of foreign affairs, letter of September 25, 1817. Quotations in the next six paragraphs are from this document.

17. Ibid., letter of October 19, 1817. The quotations in this paragraph are all from this document.

18. J. Q. Adams, *Memoirs,* 4:11: "W. Lee came and read me an account, drawn up by him, of all the information obtained by him from the French exiles and their projects here" (September 28, 1817).

19. Lee to Monroe, Monroe Papers, Library of Congress, in Reeves, *Napoleonic Exiles in America,* 64–65.

20. J. Q. Adams, *Memoirs,* 4:11: "Crawford seemed dissatisfied that the Commissioners were not yet gone" (September 29, 1817).

21. *L'Abeille Américaine,* November 6, 1817, 269.

22. CADN, consulat de New York, série C, register 1, correspondence with the minister of foreign affairs, November 24, 1817.

23. *L'Abeille Américaine,* October 23, 1817, 241.

24. "Discours de Monsieur le G^{ral} Ch^{les} Lallemand, à l'assemblée générale des Sociétaires de la compagnie agricole & manufacturière du Tombigbee," ibid., October 30, 1817, 245–48; "Discours de Monsieur Villar . . ." ibid., November 6, 1817, 267–70; "Rapport fait au nom du Comité de la Société Française, Agricole & Manufacturière du Tombigbee, aux Membres de cette Société, assemblés," ibid., November 13, 1817, 277–84.

25. CADN, Consulat de New York, série C, register no. 1, correspondence with the minister of foreign affairs, November 24, 1817.

26. Lyon, "The Bonapartists in Alabama."

27. For a detailed account of the Champ d'Asile episode, see Blaufarb, *Bonapartists in the Borderlands.*

28. For the list of subscribers and the amounts of their donations published in the *Minerve française,* November 1818, go the the website "Gallica" of the Bibliothèque nationale de France.

29. Carter, *Territorial Papers,* 18:169.

30. Ibid., 254–55.

31. Lajonie Papers, Lajonie to the château of Le Soulat, Demopolis, November 1817.

32. Carter, *Territorial Papers,* 18:255.

33. Lajonie Papers, Lajonie to the château of Le Soulat, Demopolis, December 17, 1817.

34. Ibid., Lajonie to the château of Le Soulat, Demopolis, January 10, 1818.

35. Carter, *Territorial Papers,* 18:259–60.

36. "Message from the President of the United States transmitting a Statement of the Proceedings which have been had under the Act of Congress passed On the 3d of March 1817 . . . ," Lowrie et al., *American State Papers: Public Lands* 3:387.

37. Carter, *Territorial Papers,* 18:390–92.

38. The dividing line between Ranges 2 and 3 is Front Street in present-day Demopolis.

39. Carter, *Territorial Papers,* 18:511.

40. Lajonie Papers, Jacques Lajonie to the château of Le Soulat, Mobile, August 16, 1819.

41. For the shape and boundaries of the colony, see H. Cobbs, "Geography of the Vine and Olive Colony."

42. George E. Phillips (Paleoecology Group, Department of Marine, Earth and Atmospheric Sciences, North Carolina State University) has graciously made available information from his unpublished study "The Black Prairie of Mississippi and Alabama" (June 2003).

43. According to an analysis by chemist Matt Taylor of the Cemex plant in Demopolis, it is 45.27 percent CaO, along with silica, aluminum, iron, sulfur, sodium, potassium, magnesium, etc.

44. The plant on Arcola Road has changed ownership several times, but throughout

the twentieth century it has played a major role in the economic life of the city thanks to the large number of direct and indirect jobs it has created. The first plant, founded by three Englishmen—Messrs. Spoor, Carey, and Cans—gave birth to a little town, Spocari, entirely dependent on cement making. When the plant, jointly owned by Lone Star Industries and Canada Cement Lafarge, was purchased by the Cemex group, in 2000, it employed ninety-six persons and produced 950,000 tons annually.

45. This soil is also called rendzina. See the commentary, map, and legend of the soil distribution in Lineback, *Atlas of Alabama*, 8–9.

46. Rostlund, "The Myth of a Natural Prairie Belt"; Mayers, "Geography of the Mississippi Black Prairie."

47. Goodell, "Mission among the Choctaws," 223–24.

48. Lajonie Papers, Lajonie to the château of Le Soulat, successively, beginning with the first quotation of the paragraph, November 1817, April 24, 1819, August 16, 1819, and March 2, 1820, in a letter begun on and dated February 2, 1820.

49. A list of species categorized as "bearing trees" in Township 18, Range 3 East at the time of the French colonization: ash, lynn, dogwood, sweet gum, red oak, white oak, post oak, chestnut oak, blackjack, and cedar.

50. ADAH, Montgomery, SG 4352, Alabama Secretary of State Lands Division, US St. Stephens Land Office, Surveyor's boundary line five, notes 1817–1821: "Township 18 Range 3 East in the land district of Alabama: quality of the land on the boundaries of the township."

51. In a personal communication, George Phillips writes that "the relationship between heavy, limed soils and open grassland prairies along high stream divides is obvious from the maps created by Myers in 1948."

52. Claiborne, *Mississippi*, 484.

53. Southeast Regional Climate Center: "Historical Climate Summaries and Normal for the Southeast," for Demopolis the observations cover the period 1951–2003; for Greensboro, 1890–2003.

54. Carter, *Territorial Papers*, 18:525.

55. *American State Papers: Public Lands*, vol. 5, in Whitfield, "French Grant in Alabama," 331–32, or W. Smith, *Days of Exile*, 50–51.

56. Names deleted: Martin Pignet Joseph, Wiles and Leclerc, V. M. Garesché, Jacques Brand and John Roster, Jean Thomas Carré, Laurent Faures, Englebert, Samuel Jackson, Joseph Robard, Pierce brothers, Jean Baptiste Neel, William Tablee, Bellington, George Gaines, S. Voorhees, Guillaume Montelius, Kimbal; names added: Jacques Moncravie, R. A. Terrier, Charles Brugière, Joseph Ducommun, Madame George, Pierre Garesché, J. Bonno, Pierre Drouet, Émely and Condé.

57. *Aurora*, March 30, 1818.

58. Carter, *Territorial Papers*, 18:161.

59. Ibid., 257.

60. "A Map of Alabama constructed from the Surveys in the General Land Office and other Documents." A copy is at ADAH.

61. "Lime" is a component of many Alabama place names and there is a Chalkville in Jefferson County. See Foscue, *Place Names in Alabama*, 30, 84.

62. Abner Smith Lipscomb (1789–1856), a judge on the Supreme Court of Alabama (1820–24), then chief justice of the court until 1832. See T. Owen and Owen, *History of Alabama*, 4:1052.

63. Helmer, *Lipscomb*, 252–53.

64. Screamersville, whose name came from the nocturnal cries of wild animals, was located south of Chickasaw Bogue, in the northeast corner of Section 32 of Township 16, Range 3 East, three hundred feet from the house of a certain Mrs. Irby, which was chosen to house the first court of Marengo County.

65. For the creation and transformation of Alabama counties (1800–1903), see the maps in Barefield, *Old Demopolis Land Office Records*, xxiv–xl, and Foscue, *Place Names in Alabama*, 169–74.

66. Foscue, *Place Names in Alabama*, 65. Greene County was created on December 13, 1819. The Alabamians were honoring General Nathaniel Greene, who had defeated the English at Eutaw Springs, South Carolina, in 1781.

67. James T. May, quoted in the Greensboro *Alabama Beacon* 17 (1849), in Harris, *Dead Towns of Alabama*, 78.

68. "Erie's 'glory has departed . . . and now a few dilapidated tenements, the ruins of the old courthouse and jail, one old homestead and its hospitable occupants . . . are all that are left to bear witness that Erie rejoiced in a prosperous existence.'" Snedecor, *Directory of Greene County For 1855–56* (Mobile, 1856), in Harris, *Dead Towns in Alabama*, 78.

69. The Russell brothers, whose land near Greensboro bore their name: "Russell's Ridge." See Yerby and Lawson, *History of Greensboro*, 1.

70. Record of Deeds for Hale County, Book I: 141, in ibid., 8.

CHAPTER 13

Epigraphs: Jean Quessart's letters are part of the Lajonie Papers. To avoid a multiplicity of notes, each quotation from the letters of Jacques Lajonie and his associates will simply include the date. All come from the privately owned Lajonie Papers.

1. On this topic see, for Louisiana, Aubert, "'Français, nègres et sauvages.'"

2. Moundville received its name in homage to the Mound Builders who there erected dozens of mounds, twenty of which are still visible, the largest sixty feet in height. See Walthall, *Moundville*, 7.

3. The name *Choctaw*, in its English version, was first used to designate the tribe, or in any case a part of it, around 1700. For the history of the Choctaws see Debo, *Rise and Fall*.

4. Cyrus Byington translates Chahta as Choctaw, without further explanation, in *A Dictionary of the Choctaw Languages*, 96–97.

5. "It is considered beautiful among these Peoples to have a flat head," wrote French voyageur Jean-Bernard Bossu in *Nouveaux voyages aux Indes Occidentales, contenant une relation des différens peuples qui habitent les environs du grand fleuve Saint-Louis appelé vulgairement le Mississipi* (1768; Paris: Aubier, 1980), in Swanton, *Source Material*, 263.

6. Swanton, *Source Material*, 261. Swanton also cites (251–52) a similar account in a "Relation de la Louisiane" by an anonymous French officer, which he dates to 1755, and an-

other (265–66) by Jean-Antoine Leclerc, aka Milfort, aka Tastanéguy, *Coup d'oeil rapide sur mes voyages parmi les peuplades sauvages de l'Amérique septentrionale* (Paris, 1802), republished by Christian Buchet as *Chefs de guerre chez les Creek* (Paris: France-Empire, 1994).

7. Byington, *Dictionary of the Choctaw Languages,* 216; Read, *Indian Place Names in Alabama,* 69.

8. While estimations of the Choctaw population generally run from 20,000 to 25,000, Villebeuvre's census of November 1795 lists 7,870, including 2,200 warriors. See Pate, "The Fort of the Confederation," 183.

9. Sharing the same language and traditions, occupying the same banks of the Tombigbee and the Yazoo, the two tribes had separated before the whites arrived, but this arrival exacerbated their reciprocal hostility as the Chickasaw warriors joined the camp of the English whom they met toward the end of the seventeenth century.

10. Jean Baptiste Le Moyne, sieur de Bienville, founded the Louisiana colony early in the eighteenth century with his brother d'Iberville, then experienced six years of disfavor in France before returning in 1733. See Wilkins, *Colonial Wars of North America,* 372–76.

11. Wilkins, "Outpost of Empire," 133.

12. Bernard Romans, *A Concise Natural History of East and West Florida* (New York, 1775) 325, in Rea, "Trouble at Tombeckby," 39.

13. See Pate, "The Fort of the Confederation," 171–86.

14. Guice, "Face to Face in Mississippi Territory."

15. Wallace, *Jefferson and the Indians.*

16. See Plaisance, "Choctaw Trading House."

17. Remini, *Andrew Jackson,* 232.

18. Serme, "Le traité de Fort Jackson."

19. The first, in 1786, Hopewell; the next four, between 1801 and 1805, under Jefferson: Fort Adams, Fort Confederation, Hoe Buckintoopa, and Mount Dexter (ratified in 1808). For their territorial implications see Ferguson, "Treaties between U.S. and Choctaw Nation," 214–30.

20. Remini, *Andrew Jackson,* 227.

21. Wells, "Federal Indian Policy," 194.

22. See Kidwell, *Choctaw and Missionaries in Mississippi.*

23. Plaisance, "Choctaw Trading House," 415.

24. "Plat of Choctaw Trading House, surveyed by Thomas Malone in June 1816," in Gaines, *Reminiscences,* 72.

25. Kappler, *Indian Affairs,* vol. 2.

26. The treaty forcibly took from the Choctaws almost 10.5 million acres. Totally, chiefly between the Hopewell Treaty in 1786 and the Treaty of Dancing Rabbit Creek in 1830, the Choctaws ceded to the United States more than 25 million acres in the states of Alabama and Mississippi, and received approximately 5 million in the Indian Territory of Oklahoma, to which they were exiled after 1830.

27. "Relation de la Louisiane," in Swanton, *Source Material,* 247.

28. J. Lakanal to M. Geoffroy Saint-Hilaire, president of the Académie des Sciences in Paris, Mobile, August 1, 1834, in Jouin, *Lakanal en Amérique,* 45–47.

29. The sol or sou was a French coin worth about one cent; one-twentieth of a franc.

30. Swanton, *Source Material,* 265–66.

31. CADN, consulat de France, Phila., register 6, "Vocabulaire des Indiens nomades, dans le voisinage du Mexique" fol. 339–44, and "Tableau des nations indiennes qui demeurent dans la Louisiane Septentrionale, depuis le cours du Missouri et à l'ouest de ce fleuve, jusqu'aux montagnes qui bornent à l'est, le nouveau Mexique," fol. 345–47.

32. Lakanal to Saint-Hilaire, August 1, 1834.

33. French travel accounts frequently mention fraternal feelings, such as that of Paul Du Ru, a missionary in Louisiana in 1700, who wrote: "We live with them like brothers, and I would far sooner be alone in their lands and in their midst at nine in the evening than in rue Saint-Jacques in Paris." Paul Du Ru, *Journal de voyage,* f. 57, in Havard and Vidal, *Histoire de l'Amérique française,* 221.

34. Governor Claiborne, Town of Washington, October 13, 1802, in Rowland, *Territorial Archives* 520, quoted in Guice, "Face to Face in Mississippi Territory," 180.

35. Brown and Owens, *World of the Southern Indians,* 49.

36. Serme, "Le traité de Fort Jackson," 12.

37. Dinsmoor to Armstrong, August 4, 1813, in Carter, *Territorial Papers,* 6:391.

38. Number of whites: 1810: 6,422; 1820: 85,451; 1830: 190,406. Dodd and Dodd, *Historical Statistics,* 2; M. B. Owen, *Alabama Census Returns,* 10–14.

39. Abernethy, *Formative Period in Alabama,* 26.

40. The portion of Greene County east of the Black Warrior was detached to form a large portion of the newly created Hale County in 1867.

41. In 1856 Greene County numbered 357 persons native to Alabama, 438 from South Carolina, 348 from North Carolina, 92 from Georgia, 45 from Tennessee, 24 from Kentucky, 12 from Connecticut, 37 from Ireland, and 10 from Germany. Abernethy, *Formative Period in Alabama,* 32.

42. The 1830 census places more than 12,000 whites in Marengo and Greene Counties, that is, about 6.4 percent of the 190,000 whites in the state, and places 41,000 whites in the four counties along the Tennessee, that is, about 22 percent.

43. Remington and Kallsen, *Historical Atlas of Alabama,* Greene, Hale and Marengo Counties.

44. Colonel Alexander McAlpine (South Carolina, 1780–Alabama, 1858), see Russell, *McAlpin(e) Genealogies, 1730–1990.*

45. Lilian T. Baskin, "Helpless without Slaves," *Birmingham News,* October 9, 1967.

46. Saugera, "Histoire de la traite négrière française"; Bénot and Dorigny, *1802.*

47. Daget, *La répression de la traite des noirs.*

48. Slaves that the northern slave trading ports, Providence and Newport, willingly supplied: the lure of gain long linked northern Puritans and southern planters. See Farrow, Lang, and Frank, *Complicity.*

49. U.S. Department of Commerce, Bureau of the Census, *Negro Population, 1790–1915,* 25, 53.

50. Wood, *Radicalism of the American Revolution,* 7.

51. Washington, Jefferson, Madison, and Monroe, all Virginians elected to two terms; Jackson (two terms) and Polk from Tennessee; and Taylor from Louisiana. Fohlen, *Histoire de l'esclavage aux États-Unis,* 222.

52. As early as 1688, Quakers in Germantown, near Philadelphia, drew up a petition denouncing the slave trade and slavery. In 1758, at their headquarters in Philadelphia, they decided to exclude slave traders from their membership. Supported in their position by the Methodists and other evangelical groups, they formed the first antislavery association, which, in 1784, became the first American abolitionist organization, the Pennsylvania Abolition Society.

53. See Rossignol's "Le contexte nord-américain," which analyzes in detail the ideas of a great many American historians specializing in these questions, such as D. B. David Brion Davis, *The Problem of Slavery in Western Culture* (1966; New York: Oxford University Press, 2008) and *The Problem of Slavery in the Age of Revolution 1770–1823* (1975; New York: Oxford University Press, 1999); and Christopher Leslie Brown, *Moral Capital: Foundations of British Abolitionism* (Chapel Hill: Published for the Omohundro Institute of Early American History and Culture, Williamsburg, Virginia, by the University of North Carolina Press, 2006).

54. See Sala-Molins, *Les misères des Lumières.*

55. Jefferson to Edward Coles, Monticello, 1814, in Jefferson, *Works,* 11:418.

56. Ibid., 417.

57. MAE, CP EU, vol. 78, fol. 152–53; Daget, *Le répertoire des expéditions négrières,* 190–91, 166, 197–98, 169–70; Daget, *La répression de la traite des noirs,* 207–10.

58. MAE, CP EU, vol. 79, fol. 196, Hyde de Neuville to Pasquier, Washington, October 11, 1821.

59. Ibid., fol. 220, Hyde to Pasquier, October 20, 1821.

60. CADN, consulat de France à Philadelphie, register 6, fol. 328, Lakanal to Monsieur le Comte de Survilliers [alias Joseph Bonaparte], [1817].

61. Mobile, May 1, 1830, Lakanal to Étienne Geoffroy Saint-Hilaire, in Jouin, *Lakanal en Amérique,* 28–30.

62. Gravatt, *L'église et l'esclavage,* 8–9.

63. Arriving by sea in Mobile, where she spent Christmas of 1819, she went by "wagon" to her husband Édouard George's allotment, Range 3 East, Section 8, no. 22, bordering the Black Warrior River.

64. See chapter 11, note 34.

65. APS, Girard Papers, Victoire George to Stephen Girard, Mobile, December 24, 1819. All of our subsequent references for Madame George come from letters she wrote to Stephen Girard and occasionally John Henry Roberjot and can be found in APS. To reduce the number of notes, they will simply be identified by date.

66. ADSM, 6 P 6/28 (6 Mi 231), rôle de désarmement in Le Havre of the *Marie-Thérèse,* désarmement no. 168.

67. Dodd and Dodd, *Historical Statistics,* 2.

68. Gehman, *Free People of Color in New Orleans,* 49.

69. The Bourse Maspéro, or Maspero's Exchange (for slaves), located at 440 Charles Street in the Vieux Carré, belonged to Jean Paillet in 1788. The family owned it until 1878. Toledano, *National Trust Guide to New Orleans,* 21.

70. *L'Ami des Lois et Journal du Soir,* October 4, 1817: "Ventes à l'encan."

71. See Cunz, *Maryland Germans;* Furer, *Germans in America.*

72. MAE, CP EU, vol. 74, fol. 144, consulat de France à Philadelphie, Pétry report, August 19, 1817.

73. *L'Abeille Américaine,* September 1817, 130: "there is a course of 40 lessons to perfect the pupil's hand . . ."

74. Township 18, 3rd range East, Section 5, allotment 17, 120 acres, acquired from the government, and Section 15, allotment 42, 120 acres, paid in cash to a private individual (Dieudonné Menou).

75. *L'Ami des Lois et Journal du Commerce* (New Orleans) 25 (April 1820), letter to the editors from Lin Troy.

76. CADN, consulat de France à New York, série B, no. 122, Angelucci, consul in Baltimore, to the consul general in Philadelphia, Baltimore, October 10, 1818; see also the letter of October 30, 1818.

77. APS, Girard Family Papers (Mrs. Plowden), Lefebvre-Desnoëttes to Henri Lallemand, Demopolis, June 22, 1821.

78. The United States counted much less than the other slaveholding countries of the American continent on the slave trade to renew its supply of slaves. Curtin, in *The Atlantic Slave Trade,* establishes the fact that 6 percent of the Atlantic slave trade (i.e., 600,000 to 650,000 captives) was shipped to the United States. With an equal number of men and women, childbearing was encouraged and thus the country became self-sufficient.

79. Fogel and Engerman, *Time on the Cross,* 78–86 ("The Myth of Slave-Breeding").

80. Ibid., 48, 79.

81. Douglass, *Narrative,* 73–74.

82. APS, Girard Papers, General Henri Lallemand to Stephen Girard, Tombigbee, September 20, 1821.

83. APS, Girard Papers, Lallemand to Girard, Cap Vincent, along the Saint Lawrence River, November 18, 1821.

CHAPTER 14

1. Lajonie Papers, Pierre Fougnet Sr. to J. Lajonie, La Fourche, Black Warrior, June 6, 1818. To avoid a multiplicity of notes, quotations from the letters of Jacques Lajonie and his associates will simply be identified parenthetically by date wherever possible. All come from the privately owned Lajonie Papers.

2. Ibid., on February 7, 1818, Jean Quessart wrote to Marie "Nauzille" Taupier, telling her that her letter of November 19, 1817, that is, written almost three months earlier, had arrived in New Orleans. He now had to forward it to Demopolis via Mobile.

3. "The log cabin is only a temporary shelter for the American, a concession circumstances have forced on him for the moment," pointed out Alexis de Tocqueville in the early 1830s. Tocqueville, *Journey to America,* 334.

4. Adams to the secretary of the treasury, in Whitfield, "French Grant in Alabama," 342.

5. Lajonie was punning. In French, scabies is *gale;* the country of Wales is *Galles,* pronounced identically. Lajonie wrote that in two weeks he would no longer be "le prince de gale."

6. Louis Leroy, *La médicine curative du Dr Leroy* (Paris, 1817), *Le charlatanisme dé-*

masqué ou la médecine appréciée à sa juste valeur (Orléans, 1819), and *La médecine curative prouvée et justifiée par les faits,* 4 vols. (Paris, 1823–24). Leroy was a surgeon, with an office at 49 rue de Seine, Saint-Germain quarter, in Paris.

7. Lajonie Papers, Jean Quessart to Pierre Taupier, New Orleans, April 20, 1824. According to Quessart, Laclaverie practiced medicine in Saint-Domingue for thirty years, and after that in New Orleans, leaving for Bordeaux when he became ill. There Dr. Leroy's remedy cured him.

8. The dried tuberous root of any of several plants of the morning glory family, or the light yellowish powder derived from it, used in medicine chiefly as a purgative.

9. Gensac Municipal Archives, January 6, 1834, at the Gensac town hall, the Lajonie couple registered "a female child, born of their marriage on July 31, 1820, in Demopolis, Marengo County, State of Alabama (North America) where they were living subsequent to the deplorable reaction of 1815."

10. Lajonie Papers, birth certificate, May 15, 1833, certified by the official seal of the French consulate of New Orleans, Mobile, May 16, 1833, A. Batré French, consular agent in Mobile.

11. MAE, CP EU, vol. 74, fol. 144, Pétry (informed by Colonel Roul) to Hyde de Neuville, August 19, 1817. Is this François Desportes, fifty, clerk, residing in Paris, emigrated to the United States via Le Havre, passport issued August 26, 1816, by the Paris Police Prefecture? AN, F⁷ 3519, statement of passports issued by the department of police.

12. ADC, Clément-Simon Collection, subseries 6 F, Paris, August 5, 1807, Pénières to Mme Agathe Pénières-Stach, in Faure, *Pénières-Delzors,* 176.

13. ADAH, Gardien microfilm, Philadelphia, November 20, 1818, General Clauzel to Mme Démérest, Mobile.

14. Southern Historical Collection at the University of North Carolina Library, Chapel Hill, Charles Lefebvre-Desnoëttes Papers, Lefebvre-Desnoëttes to Clauzel, Demopolis, August 25, 1819.

15. The French word *source* means both source and spring, as in water flowing naturally from the earth.

16. Yerby and Lawson, *History of Greensboro,* 2.

17. Carter, *Territorial Papers,* 18:389.

18. The size of his holding had increased slightly due to acquisitions.

19. The reference is to Napoleon Bonaparte, who died on May 5, 1821.

20. Lajonie Papers, Partnership contract signed in Demopolis on November 1, 1819, between Fougnet and Lajonie, with Mathieu Labrousse as witness. The draft of the contract is on the back of a letter written to Lajonie by Charles Batré on January 7, 1820.

21. *Journaux* is the plural of *journal,* the amount of land a man could till in a day.

22. Lajonie Papers: this consisted of six hundred pounds of pork in foodstuff, a salt cow, a barrel of flour, an ample provision of maize, a third measure of rum, seven sacks of fine salt, and fodder for the animals, Lajonie specified to his family in Le Soulat on February 2, 1820.

23. Eutaw Courthouse, Greene County, Alabama, Deeds, A 220-1, Aigleville, October 26, 1819.

24. NARA, RG 49, TAL, PC file 66, Demopolis, Bazile Meslier (justice of the peace) to Charles & Adolphe Batré, Demopolis, March 9, 1822 (per Kent Gardien Papers).

25. Located in the commune of Pineuilh, south of the Dordogne River, on hillsides overlooking Sainte-Foy-la-Grande, "Château Les Mangons" was in 2007 a 79-acre enterprise with a square courtyard of eighteenth-century buildings and a 44.5-acre vineyard. In the fall of 2000 its owners, Michel and Brigitte Comps, converted their vineyard to biodynamic agriculture.

26. A French vine bearing red grapes, better known as *béquignol,* grown in the Bordeaux region and southwest France, vigorous, fairly resistant to mildew and winter frosts, but not very productive.

27. ADSM, 6P6/28 (6 Mi 231), rôle de désarmement, in Le Havre of the *Marie-Thérèse* of Bordeaux, Captain Bazin: Antoine Fougnet, age fourteen, of Coubeyrac, and Pierre Ragon, age twenty-one, of Belvès-de-Castillon, both farmers.

CHAPTER 15

Epigraph: Pierre Mangon Sr. to Monsieur Lajonie de Lapeyre, per steamship, in Gensac, via Castillon, Gironde (France), Lajonie Papers. The letter was stamped in New Orleans on November 10, in Toulouse on December 8, and in Castillon on December 11.

1. See chapter 12, note 1.

2. Lowrie et al., *American State Papers: Public Lands,* 3:396–99 (December 14, 1818, Treasury Department, "Land allotted to encourage the Cultivation of the Vine and Olive").

3. Lefebvre-Desnoëttes to William H. Crawford, Secretary of the Treasury, Demopolis, December 8, 1821, in Desnouettes and Crawford, *Report of the Secretary,* 8.

4. For a biographical sketch of François Bierne, see chapter 11, note 101.

5. APS, Girard Papers, letters received, 1829, 1205: on December 10, 1829, François Bierne's wife wrote to Stephen Girard from Dunkirk, referring to her husband's letter of June 24, 1826, from Mobile.

6. Marengo County, Miscellaneous Records, Book I, fol. 130: August 21, 1826, estate of Francis Bierne, of Marengo County, deceased. Administration bond: Achille George, of Mobile, administrator, C. C. Stone & Nathan Lipscomb, of Marengo County.

7. APS, Girard Family Papers/Mrs. Plowden Papers: Bierne wrote to Lefebvre-Desnoëttes in Washington (about to return to Europe) on March 14, 1822, from Demopolis, asking him to support his request for a 240-acre allotment. On April 13, 1822, from Demopolis, Basile Meslier wrote to Desnoëttes on the same subject.

8. Linden Courthouse, Marengo County, Probate Office, Miscellaneous Records, Marengo County (1823–40), vol. 1: 128–74.

9. These are thirty-three of the sixty-five military men on the list, or 51 percent. Blaufarb, *Bonapartists in the Borderlands,* 118.

10. Lajonie Papers, Aigleville, February 12, 1818: Jacques Lajonie mentions "the arrival of Generals Clauzel and Lefebvre-Desnouettes" with men they indentured for next to nothing.

11. Southern Historical Collection at the University of North Carolina Library, Chapel Hill, Charles Lefebvre-Desnoëttes Papers, Lefebvre-Desnoëttes to Clauzel, Aigleville, March 10, 1819.

12. SHD/DAT, 20 Yc 172, fol. 64, no. 380: François Violle, born July 12, 1793, in Mauriac (Cantal), arrived June 15, 1813, in the fourth regiment of the honor guard, Imperial Guard. AN, F[7] 2550b, no. 16,489: On February 10, 1817, the minister of general police authorized the prefect of Cantal to issue François Violle a passport for travel abroad.

13. In 1825 he had bought a small lot in Demopolis, see NARA, RG 49, TAL, Small Lots, miscellaneous papers: May 20, 1825, Demopolis, Johannes Bütche (a Swiss) sells to Francis Martin, both of Demopolis, a small lot, $52.

14. Frédéric Guillaume Marie Ravesies, born in Jean-Rabel (Saint-Domingue) on March 11, 1782, son of Jean Ravesies, known as "the American" (Bordeaux, ca. 1750–Philadelphia, 1809). First marriage to Marie Roan, niece and adopted daughter of John Soullier, a Domingan refugee, and Marie (Roan) Soullier, who died in Philadelphia on May 27, 1815. They had four children. ADG, 3 L 184: on April 22, 1796, in Bordeaux, the central administration of the department of Gironde issued Jean Ravesies a passport "to go to Louisiana, taking with him his son Frédéric Ravesies, age 13, passing through New England."

15. Dorothy Garesché Holland, *The Garesché, Bauduy and des Chapelles Families* (St. Louis: privately printed by the Schneider Print Co., 1963), 72.

16. For Henri David, see chapter 10, note 20. The David-Sevré couple had two daughters, born in France: Cécile Agnès David (ca. 1810–Mobile, ca. 1890) and Marie Pauline David, married in 1826 to George Noble Stewart, for a time the secretary of the Vine and Olive Colony. They had three children.

17. SHD/DAT, 2 Ye, carton 1024, Descourt's personnel file: this was the French department of Roër.

18. AN, F[7] 3514: On October 6, 1815, Alexandre Léonard Descourt, thirty-four, merchant, residing in Paris at 12 rue du Coq Héron, asked the police prefecture for a passport to the United States, going there on business; the passport was issued on October 17, for departure via Le Havre.

19. MAE, CP EU, vol. 74, fol. 155: on August 25, 1817, the French consul in New York informed Hyde de Neuville that these were Generals Grouchy, Lallemand, Clauzel, and Lefebvre-Desnoëttes and Colonels Galabert, Douarche, and Forni.

20. Ibid., fol. 182: on September 1, 1817, the New York consul informed the Duke of Richelieu that Aimée Éléonore Chesne, residing in New York for the past year, left on August 31, 1817, for Le Havre on the American ship *Favorite.*

21. It will be recalled that Cluis had emigrated with his mistress, Émilie Louise Mézières, and their son, Frédéric Victor Cluis, born in France ca. 1815; a daughter, Émilie Éléonore Cluis, was born on September 30, 1820, in New York City.

22. Jean Claude Roudet, known as Corneille Jr., born in La Tronche (Isère) April 18, 1789, died in Mobile September 13, 1839, son of Pierre Roudet, caterer, known as Corneille Roudet Sr., grandson of Corneille Roudet, winegrower in Saint-Ferjus (Isère).

23. Blaufarb, *Bonapartists in the Borderlands,* 132.

24. Article by Corneille Roudet for the *Tuscaloosa Intelligencer,* reprinted by the *Southern Advocate* (Huntsville, Ala.), September 10, 1831.

25. NARA, *Passenger Lists of Vessels Arriving at Philadelphia, 1800–1882,* M425, film 27: Amédée Rougier, a relative of Marie Thérèse (Melizet) Roudet, Mr. Hurtel, and his son,

Mr. David, other passengers of the *Magnet* of Philadelphia, 255 tons, Captain Richard Garwood, owners William Leedom and R. Garwood, arrived at Philadelphia November 2, 1818.

26. NARA, RG 49, TAL, PC file 136.

27. Ibid., Deeds, P/Attys, Affidavits, file 104, document of November 21, 1819.

28. This is substantiated by Jacques Lajonie's letter of January 2, 1821, stating that in 1820 he bought one hundred young vines from Pierre Hurtel for twenty-five sols (cents) apiece, most probably coming from the Bordeaux region.

29. In 1826, Jean Jacques Dufour wrote that only the vine called "Cape of Good Hope" could succeed: "I believe that only the Cape grape can do well for the French at Demopolis in Alabama, but they will face the same obstacles as others have." J. J. Dufour, *American Vine-Dresser's Guide*, 44.

30. The date of 1821 makes the reference to Galveston impossible, since Violle returned to Demopolis from Texas in May 1818 at the very latest. Violle may have gone to New Orleans in 1821.

31. On January 3, 1833, Pierre Corneille Roudet married Marie Anne Ursule Uranie Thouron, the widow of Joseph Étienne Gardien, with whom she had had four children. Roudet is buried in a small French cemetery north of the French colony, then located in Greene County, and now in Hale. The gravestone, redone at the urging of Kent Gardien, bears this inscription in French: "Devoted to the memory of Pierre Corneille Roudet, born in France, died on September 16, 1837, at the age of 28 years."

32. Lucien Ensfelder, born in Strasburg in 1796, married to Marie Adélaïde Rivière (living in New Orleans in April 1819). Probably a military man. Settled in Aigleville before April 1818. Lowrie et al., *American State Papers: Public Lands*, 3:399. On July 3, 1818, appointed ensign, Third Company of the militia. Carter, *Territorial Papers*, 18:389. On December 12, 1817 [1818], in Mobile, Lucien Ensfelder, Alabama Territory, sold to Lefebvre-Desnoëttes 160 acres for one hundred dollars. NARA, RG 49, TAL, PC file 76.

33. Washington, April 18, 1820, Secretary of the Treasury Crawford to the president of the Senate; letter read in the Senate on April 21, 1820. Cf. 16th Congress, 1st session, no. 320: "Lands allotted to the cultivation of the Vine and Olive."

34. Aigleville, December 12, 1821, Ch. Villers [*sic*], Agent of the Colony, to William H. Crawford, Secretary of the Treasury, in Desnouettes and Crawford, *Report of the Secretary*, 5–6.

35. "Extract from the contract entered into with the French Emigrant Association, on the 8th. day of January, 1819," ibid., 6–7.

36. Department of the Treasury, William H. Crawford to the president of the Senate, March 18, 1822, ibid., 4.

37. Spring Grove, Tuscaloosa, February 1827, William L. Adams to Richard Rush, secretary of the treasury, introducing the *Report of the special agent appointed by the Secretary of the Treasury respecting the condition of the grants of land made to French emigrants, for the encouragement of the cultivation of the vine and olive*, submitted to the Senate December 24, 1827, 20th Congress, 1st session, no. 592.

38. Ibid.

39. Treasury Department, May 17, 1827, Richard Rush, secretary of the treasury, to

William L. Adams, in "Grants made to French Emigrants for the Cultivation of the Vine and the Olive," 20th Congress, 1st session, no. 649.

40. Aigleville, January 16, 1828, F. Ravesies, Agent of the Tombeckbee Association, to Richard Rush, in "Grants made to French Emigrants for the Cultivation of the Vine and the Olive," 20th Congress, 1st session, no. 649.

41. Aigleville, February 27, 1827, report by F. Ravesies, Agent of the Tombeckbee Association, to Richard Rush, in "Grants made to French Emigrants for the Cultivation of the Vine and the Olive," 20th Congress, 1st session, no. 592.

42. Aigleville, January 16, 1828, report by Frédéric Ravesies "in addition and explanatory to the one made by the late Wm. L. Adams, esq., relative to the Tombeckbee Association, made in obedience to the request of the Secretary of the Treasury in his letter to Mr. Adams bearing date May 17, 1827," in "Grants made to French Emigrants for the Cultivation of the Vine and the Olive," 20th Congress, 1st session, no. 649.

43. Creek, "Bombyx Mori and Americans."

44. ADAH, John M. Chapron, Letter Book, Philadelphia (1838–41), Chapron to John McRae (Capt.) in Arcola, Philadelphia, May 25, 1839. John McRae (Petersburg, Va., 1795–"Athol," Gallion, Ala., 1854) married Marie Joséphine Ravesies, born in Pennsylvania, ca. 1813, daughter of Frédéric Ravesies and Mary Roan on June 9, 1833, in Marengo County. They had five children. Gandrud, *Marriage Records of Marengo County,* 77.

45. Chapron's plantation, on which Martin was living must have taken up most of Section 36 and half of Section 30, Township 18, Range 3 East, according to a June 19, 1981, letter from Kent Gardien to Winston Smith in Reynolds, *Heritage of Marengo County,* n.p.

46. ADAH, John M. Chapron, Letter Book, Chapron to James Martin, Macon (now Prairieville), Philadelphia, June 5, 1840.

47. *The American Silk Grower and Farmer's Manual* . . . (Philadelphia: Charles Alexander, 1838–39).

48. ADAH, John M. Chapron, Letter Book, Chapron to James Martin, Macon, Philadelphia, August 7, 1841.

49. CADN, consulat de France à Philadelphie, Dossiers d'Affaires particulières, 116, Benjamin Dubuisson to the French consul in New Orleans, December 14, 1844: Dubuisson states that Villar had died in Texas four years earlier, thus around 1840.

50. Archives municipales Lyon, Lyon mairie unique, registre mariages 1821, fol. 77v–78: in connection with a "document signed Lefebvre-Desnoëttes, judge of Marengo County, Alabama Territory in America, June 25, 1818, certifying" the death of Benoît Poculot; Consul Pétry to Richelieu, Philadelphia, August 29, 1818, MAE, CP EU, vol. 76, fol. 39–39v: Poculot drowned in the Tombigbee "on his way back from taking M. Grouchy, Jr. downstream. Grouchy was going to France via New Orleans."

51. W. Smith, *The People's City,* 23.

52. Lajonie Papers, Jacques Lajonie to the château of Le Soulat, Demopolis, November 3, 1820.

53. Ibid., J. Mangon to James Lajonie in Gensac, McAlpin Bluff, Alabama, November 7, 1833.

54. Ibid., James Mangon to James Lajonie in Juillac, on the *Petit Bibi* at McAlpin Bluff,

June 10, 1832. Jean Mangon at times called himself James. The *Petit Bibi* is his boat, probably a flatboat for hauling freight.

55. Ibid., J. Mangon to James Lajonie in Gensac, McAlpin Bluff, Alabama, November 7, 1833.

56. Eutaw Courthouse, Greene County, Alabama, Marriages 1836–1848, pp. 58, 89: marriage of Mathew Labrousse, on November 16, 1841, Greene County, to Mrs. Matilda Latham, née Martin, widow of Melville Latham, whom she married on December 27, 1836.

57. Lajonie Papers, James Mangon to James Lajonie in Juillac, on the *Petit Bibi* at McAlpin Bluff, June 10, 1832.

58. Ibid., J. Mangon to James Lajonie in Gensac, McAlpin Bluff, Alabama, November 7, 1833.

59. FHS, A/B 229, folder 2, document 22, Alexander Fournier to Joseph Martin-Picquet in Louisville, Demopolis, August 6, 1842.

60. Ibid., Alexander Fournier to Joseph Martin-Picquet, Demopolis, November 7, 1842.

61. Ibid., Alexander Fournier to Joseph Martin-Picquet, Demopolis, July 2, 1844.

62. Hewett, *Roster of Confederate Soldiers,* 9:383.

63. Lajonie Papers, Francis L. Constantine to Jacques Lajonie, Erie, March 18, 1829: "Je vous anvois $71 dollars pour vos chevo que jais vandu ou pour mieu dier done á un vande publique; car il nont pas raporte autant que jaurais desiré, mai j'ai feai pour le mieu . . ." Spelled correctly, this would read: "Je vous envoie $71 dollars pour vos chevaux que j'ai vendus ou pour mieux dire donnés à une vente publique; car ils n'ont pas rapporté autant que j'aurais désiré, mais j'ai fait pour le mieux . . ." [I am sending you $71 for your horses that I sold or rather gave at public auction; for they did not bring in as much as I would have wished, but I did my best . . .]

64. Noxubee County, 1860 Slave Census Schedule.

65. Lancaster, *Eutaw,* lot 142. A photograph of the house exists, taken by Robert T. Cargo in 1972.

66. Hewett, *Roster of Confederate Soldiers,* 4:87.

67. "French-born Mangone Was Well-known in This Area," *Demopolis Times,* October 7, 1976.

68. Hewett, *Roster of Confederate Soldiers,* 10:148.

69. Tharin, *Directory of Marengo County for 1860–61,* 23: Peter Mangone, wheelwright . . .

70. McGuire, Spear, and Fay, *Mobile Directory.*

71. Nelson and Nelson, *Church Street Graveyard,* 20.

72. For a list of those buried there and the location of their graves, see ibid.

73. See his obituary in the *Mobile Daily Register,* October 29, 1846, available in the Mobile Public Library, Local History and Genealogy.

74. "The French Connection: Jean Simon Chaudron Returns to Mobile," *Mobile Press Register,* October 11, 1996.

75. Victoire George's pastel self-portrait is reproduced on the cover of Kent Gardien's *The Châtelaine of George Villa.* See also B. S. Wilson, "Catherine-Victoire George."

76. Lajonie Papers, A. McAlpine to M. James Lajonie, Juillac, care of M. Étienne La-

roque, rue Chay des Farines, Bordeaux, via New York and Le Havre, Orange Grove, Greene County (Ala.), November 5, 1830.

77. CADN, consulat de France à La Nouvelle-Orléans, série D, carton 199, no. 705, letters from 1836 to 1839.

78. Robichaux, *Civil Registration of Orleans Parish,* 688.

79. Jean Quessart Jr. died on July 27, 1828, and Jeanne Élisabeth Quessart née Berquin died on September 3, 1828. Nolan and Woods, *Sacramental Records,* vol. 18.

80. *Diocese of Baton Rouge,* 251.

81. East Baton Rouge Parish, Conveyances, Judges Book I, 251; Judges Book G, 179, 184.

82. NONA, Orleans Parish, Louisiana, Notarial Archives, Michel de Armas, 17, act 227.

83. On February 1, 1820, a prospectus was printed, entitled *College of Baton Rouge, on the Plan of Those of Europe, Under the Plan of The Rev. Mr. Martial* (Baton Rouge: College of Baton Rouge, 1820).

84. NARA, RG 49, TAL, PC file 33: Bellièvre's authorization to Descourt, East Baton Rouge Parish, September 26, 1825.

85. Phillips, *American Negro Slavery,* chap. 20, "Town Slaves."

86. CADN, consulat de France à La Nouvelle-Orléans, série D, carton 202, no. 1360, Messrs. Courajod, rue des Enfants Rouges, no. 11, au Marais, to the French consul in New Orleans, Paris, November 30, 1831.

87. According to an article signed J.S.P. in *La revue du Lyonnais,* 1836, vol. 4, p. 231.

88. Thérèse Aimée Audibert, born in Saint Georges de Montaigu (Vendée) June 29, 1799.

89. Of these, Alfred Alphonse Delaroderie (1818–83) was successively an engineer, a manufacturer, and a lumber merchant; Rodolphe Gabriel (or Gustave) Delaroderie (1822–91) was successively a jeweler, a farmer, and a retail grocer.

90. New Madrid County, Missouri, Deeds 10, 285: New Madrid, sale on May 30, 1818, $30 by Isaac Philpot and his wife Anne. Witnesses: Godfrey Lesieur, François Lesieur. The property became known as the DeLaroderie Wood Yard.

91. François Lesieur, founder of New Madrid in 1789 and Point Pleasant in 1815; born in Quebec in 1762, married in Little Prairie in 1791, died in Little Prairie, New Madrid County, in 1826.

92. Eutaw Courthouse, Greene County, Deeds A, 182, June 6, 1820, New Madrid, Missouri Territory, Alphonse Delaroderie sells to Joseph Barbaroux of Philadelphia, allotment 341, 240 acres, plus a town lot and a 10-acre lot near Demopolis, $310. Witness: François Thomazeau. Notary: Stephen Mitchell, Justice of the Peace, New Madrid, Missouri.

93. New Madrid County, Missouri, Deeds 12, 24.

94. His obituary in the Baton Rouge *Weekly Gazette and Comet* on October 29, 1864, 2, states: "Mr. Delaroderie, a native of France, was born in the Department of Perigord. . . . Descended from the 'ancien noblisse' [ancienne noblesse] of France, he was, in every sense of the term, 'one of Nature's noblemen.'" There was indeed, in Dordogne (present-day Périgord) in the sixteenth and seventeenth centuries, a family "du Faure de la Roderie."

95. For the French presence see http://www.rootsweb.ancestry.com/~monewmad/nm-baptisms/nm-baptisms.htm for the baptismal registry (1821–35) of the Church of the Immaculate Conception (formerly St. Johns), New Madrid, Missouri.

96. East Baton Rouge Parish, Louisiana, Conveyances D, 289; ibid. Q, 441.

97. http://www.delaroderie.com/index.html.

98. CADN, consulat de France à La Nouvelle-Orléans, série D, carton 20, file 149, May 6, 1856, from Paris, Mme Vve Rosalie de Chappotin, to the French consul in New Orleans; Orleans Parish Probate Court, Index of All Successions 1805–1846, inventory after decease, 1831, A. Chapotin.

99. Louis Charles Chapotin (1822–79), Pierre Henry Chapotin (1826–91), Joseph Alfred Chapotin (b. 1828).

100. "The Southern Historical Society: Its Origin and History," Southern Historical Society Papers, Vol. 18, January–December 1890.

101. CADN, consulat de France à La Nouvelle-Orléans, série D, carton 45, file 33, Louis Coquillon to the French consul in New Orleans, Mandeville, February 29, 1870. The reference is probably to the battle of Toulouse, April, 10, 1814, lost by the French imperial army to the forces of the Duke of Wellington. It was fought after the allied armies had already entered Paris.

102. Jean Charles Coquillon (ca. 1750–1826), his father, and his uncle François Barthélémy Coquillon (1748–Augusta, Georgia, 1827), grantees of the 240-acre allotment 64, unfarmed in 1827. François Barthélémy Coquillon had emigrated to Georgia from Saint-Domingue. Dumont, "French Emigrants to Richmond County," 75.

103. Eléonore C. Coquillon (Les Cayes, Saint-Domingue ca. 1786–Mandeville, Louisiana, January 25, 1866).

104. CADN, consulat de France à La Nouvelle-Orléans série D, Carton 45, file 33, Louis Coquillon to the French consul in New Orleans, February 29, 1870.

105. In 1822, in New Orleans, a twenty-eight-year-old black woman who said she had acquired her freedom in the free state of Ohio, where she had come from, filed a petition against Louis Coquillon, accusing him of "unjustly and illegally" keeping her in slavery. Online at http://library.uncg.edu/slavery_petitions/details.aspx?pid=16260.

106. From the Scrapbook of Mrs. Z. Taylor Davis, "Louis Coquillion [sic], Sr., d 12 Dec, 1882, a 87 y," online at http://files.usgwarchives.net/la/sttammany/cemeteries/mandevil.txt.

107. CADN, commissariat des relations commerciales de la république française près des États de New York et New Jersey, register 2, fol. 22v–23: François Cavaroc, an inhabitant of Saint-Domingue, is in New York on January 2, 1805, to declare under oath the murder of French colonists in Saint-Domingue.

108. Philadelphia County, Miscellaneous Deeds MR1, 42.

109. Cavaroc, Charles M., fancy store, 71 Chartres corner Toulouse; Baulos & Cavaroc, fancy store, 71 Chartres, idem. 1822 New Orleans City Directory. Cavaroc, Charles, 92 Levee, dry goods. 1832 New Orleans City Directory.

110. Pierre Charles Cavaroc (New Orleans, 1828–95) married in 1848 in New Orleans, Marie Heloise Lamothe (1830–81). They had at least two children, Charles Cavaroc (New Orleans, 1849–99) and Gabrielle Cavaroc (New Orleans, ca. 1860–1928), who married Louis Dessommes (1859–83). The Dessommes had descendants.

111. Toledano, National Trust Guide to New Orleans, 32, shows the facade of the 1866 building.

112. For this complex subject see U.S. Supreme Court, Casey v. Cavaroc 96 U.S. 467 (1877), online at http://supreme.justia.com/us/96/467/.

113. CADN, consulat de La Nouvelle-Orléans, série D, carton 107, files 186/1891 and 408/1900: concerning a firm in Cognac asking for information on the legal and financial situation of Mr. P. A. Cavaroc, successor of C. Cavaroc, importer of wines and liquors in New Orleans, with a view to establishing commercial relations.

114. NARA, RG 49, TAL, PC file 154.

115. Jean François Canonge (1785–1848), married in New Orleans, August 17, 1816, to Jeanne Amélie Mercier, sister of Marie Justine Mercier, wife of Justin Plantevigne. See note 117 below.

116. APS, Mrs. Plowden Papers, Francis Bierne to Lefebvre-Desnoëttes, Demopolis, March 14, 1822, concerning his departure: "I wrote to Messrs. Batré to send them [the general's trunks] immediately to M. Plantevigne."

117. Justin Plantevigne married Marie Justine Mercier, and they had at least three daughters (Marie Zoé, 1818, Jeanne Amélie, 1822, and Jeanne Éliza, 1826) and two sons (Jean Charles Eugène, 1820, and Louis Henri, 1824); Antoine Plantevigne married Adeline Ledoux (1801–41) in Pointe Coupée in 1819, and their children were Catherine Anaise (1821–66) and Pierre Edouard (1825–70). They have descendants.

118. 1822 New Orleans City Directory: Plantevignes [sic] John, merchant, 21 Levee; Plantevignes [sic], Antoine, Jr., merchant, 5 Conde below St. Ann. 1823 City Directory: Plantevignes [sic], Antoine, Jr., merchant; Plantevignes [sic], Justin, merchant. 1824 City Directory: Justin Plantevigne, 55 St. Peter.

119. Justin Plantegivne, listed in 1820 in New Orleans, Levee Street with four slaves. 1820 U.S. Federal Census, Louisiana, Orleans Parish. They were three members of one family, purchased from Baptiste Wiltz on May 7, 1817, for $1,500 (the mother, Marianne, forty, born in Louisiana, and her twins, a boy and a girl, age seven and a half), plus a fifteen-year-old slave named Polly, for $800, June 27, 1818. NONA, Michel de Armas, act 250; NONA, Mᶜ Lynd, act 55.

120. Plantevigne, James, private, Pointe Coupee Artillery, enlisted August 15, 1861, St. Joseph (La.), taken captive near the Mississippi River February 13, 1864; paroled in New Orleans and exchanged July 22, 1864, at Red River Landing.

121. CADN, consulat de France à La Nouvelle-Orléans, série D, carton 197, letter from Jean Justin Richard, a Frenchman from Saintonge, residing in Veracruz.

122. Pierre Edouard Plantevigne (1825–70) had in 1852 married Pauline Tassin, a Creole.

123. John Joseph Plantevigne (1871–1913), born on a small farm near Chenal, Pointe Coupee Parish.

124. Alexandre Fournier (Tours, 1755–Philadelphia, 1819), emigrated to Port-au-Prince in 1785, married Marthe Hérils in 1788, probably both arriving in Charleston in early 1794. In Philadelphia Fournier had a silversmith and jewelry shop at 102 North Second Street, according to the *Courrier de la France et des Colonies* of October 28, 1795, in Childs, *French Refugee Life*, 91.

125. Lorensky Alexander Fournier (Alabama, 1832–New Orleans, 1900) married Nathalie Fremineaux, daughter of Marie Louise Fremineaux, a native of Belgium. They had seven children.

126. Charles Antoine Mangon, born in Gironde on July 20, 1845, died in Lincoln, Lancaster County, Nebraska, on April 16, 1924, buried in Clifton Cemetery, Nemaha County,

Nebraska. Cf. "Burials in Nebraska of Civil War Veterans" at http://www.civilwarmuseumnc
.org/burial-index.html.

127. For Bertha Evelyn Mangon Thomson and her husband, Clinton Harris Thomson
(1886–1927), see http://www.usgennet.org/usa/ne/topic/resources/OLLibrary/Nebraskana/
pages/nbka0266.htm, website of the Nebraskana Society, and www.mun.ca/rels/restmov/
texts/world/WCIT.HTM, the "World Call Index 1919–1973."

CHAPTER 16

1. Alphonse de Grouchy to his father, Paris, May 25 and 30, 1818, in Grouchy, *Mémoires*, 5:69–70, 71–72.

2. Alphonse de Grouchy to his father, Paris, June 20, July 13, August 1, 1818, in ibid.,
73–78.

3. *La Minerve française*, by Messrs. Étienne Aignan, Benjamin Constant, Évariste
Dumoulin, Charles-Guillaume Étienne, Antoine Jay, Étienne de Jouy, Lacretelle senior,
Paris, 113 issues in 9 volumes, from February 1818 to March 1820.

4. Vapereau, *Dictionnaire universel des littératures*, 402. See also Yvert, *Politique libérale*.

5. *Biographie des journalistes*.

6. *La Minerve française*, February 1819, 94.

7. Ibid., February 1818, 353–55; "L'exilé" (written in January 1817); August 1818,
577–79, "Le champ d'asile," by Antony Béraud (1792–1860), captain; also published in
Metz by Mme Verronnais, November 14, 1818.

8. Ibid., August 1818, 257, editorial by Évariste Dumoulin.

9. Ibid., August 1818, 259–61, "Champ-d'Asile, May 11, 1818"; also see 525–26, an
excerpt from the letter of a French refugee at the Champ-d'Asile, to his mother in Paris,
June 11, 1818.

10. Ibid., August 1818, 350, subscription opened at Gros, Davillier and Company,
Boulevard Poissonnière 15, the chief European banks, and the *Minerve* office.

11. The *Minerve* published twenty statements of the sums received in Paris and the
provinces for the subscription for the French refugees at the Champ d'Asile in August 1818,
November 1818, February 1819, and May 1819.

12. Ibid., November 1819, 318.

13. For the distribution of funds, which caused problems, particularly since the refugees were widely dispersed, see ibid., November 1819, 316–21.

14. According to the minister of justice (Pierre-François-Hercule de Serre) in the
Chamber of Deputies, session of May 18, 1819, in *Annuaire Historique Universel 1819*, Session législative, Pétitions, 231.

15. Ibid., 224.

16. Ibid.: "The right of petition is established by the Charter; but should it therefore
be unlimited and without rules, and can it include and dare everything with impunity?"

17. *La Minerve française* 6 (May 1819): 442–44. Brigadier G. de Vaudoucourt, who
signed one of the petitions, cites these words of Caumartin defending the exiles in the con-

clusion of his letter of May 27, 1819, to the *Minerve,* concerning the Chamber's and government's reaction to the petitions.

18. *Annuaire Historique Universel 1819,* Session législative, Pétitions, 223. Royer-Collard's speech closed the session of May 18, 1819.

19. Ibid., 226, session of May 17, 1819.

20. AN, F⁷ 6683, file 5, Regnaud to a friend, New York, March 30, 1817: "My friend, I have been so ill these past days that I have been unable to take pen in hand . . . first I tell you in my letter that you can write me in Harlem." According to Arnault, Jay, and Jouy, *Biographie nouvelle des contemporains,* 17:311, Regnaud "was suffering from an inflammatory illness" and doctors felt that a cure was possible only in Europe.

21. CADN, consulat de France à New York, série B, no. 126, New York consul, d'Espinville, to Pétry, French consul general in Philadelphia, June 4, 1817.

22. Ibid., no. 110, French consul in New York to Hyde de Neuville, letters of June 13, July 1, 12, and 20, 1817, regarding Regnaud's insanity.

23. APS, Girard Papers, letters received, reel 63, Philadelphia, July 14, 1817.

24. He joined his wife in Antwerp on August 18, 1817, according to Dugenne, *Dictionnaire,* 4:1298, cited in Dossios-Pralat, "Regnaud de Saint-Jean d'Angély," 3:662.

25. CADN, consulat de France à Anvers, 1, liasse III, letters of the French legation in the Netherlands (1816–32), Count Roger de Caux (the king's chargé d'affaires in the Netherlands) to the consul in Antwerp, Mr. Despallières, The Hague, October 31, 1817; ibid., consulat de France à La Haye, 32, the consul general in Amsterdam, Mr. Desjobert, to the French chargé d'affaires in The Hague, Mr. de Caux, November 5, 1817.

26. See successively the petition in Dugenne, *Dictionnaire,* 4:1298, in Dossios-Pralat, "Regnaud de Saint-Jean d'Angély," 670–71; AN, F⁷ 6683, file 4; *Le Moniteur Universel,* February 16, 1819; Dugenne, *Dictionnaire,* 4:1298, in Dossios-Pralat, "Regnaud de Saint-Jean d'Angély," 672; MAE, MD France, vol. 2143, fol. 102; *Le Moniteur Universel,* March 10, 11, and 13, 1819.

27. AN, F⁷ 6683, file 4: he had been named police prefect during the Hundred Days "only because he had held a similar position in the past."

28. MAE, MD France, vol. 2143, fol. 222 and 227.

29. Ibid., fol. 264–65; on August 25, 1819, Hyde wrote to Dessolle that "Mr. Réal does not plan on leaving the United States, at least for the present." MAE, CP EU, vol. 76, fol. 299v.

30. APS, Girard (and Plowden) Family Papers, Lallemand file. Called Stone House, the house had been built by Jacques Donatien Le Ray, who had also sold him the land before Réal left for America. For Cape Vincent, see Chappet, Martin, and Pigeard, *Le guide Napoléon.*

31. New York, April 4, 1826, Réal to Eustache-Marie Courtin, author of the *Encyclopédie moderne,* in Bigard, *Le Comte Réal,* 175.

32. Ibid., 177.

33. AN, F⁷ 6683, file 4, report of June 15, 1827.

34. Ibid., letter from Chaumont to the minister of the interior asking his aid in recovering a sum of 3,829 francs owed by Réal for work done on his Paris house before his exile. Paris, September 18, 1827.

35. Ibid., report from the prefect of police to the minister of the interior, July 11, 1827.

36. Michaud, *Biographie universelle ancienne et moderne,* 35:284.

37. Stendhal mentions Réal in July 1830 in his *Souvenirs d'égotisme,* 36.

38. Bigard, *Le Comte Réal,* 208. His death was reported on May 7, 1834, by Claude-Charles Pichon, proprietor, age fifty-seven, 12 rue du Battoir St-André.

39. Burial at Père Lachaise on May 28, 1834; his tomb, a truncated column surrounded by a railing, is in the cemetery's fourteenth division.

40. *Aurora,* November 9, 1816: "Embroidery. Mrs. Dirat has the honor of informing the Ladies, that having prepared accommodious appartment [*sic*] for the reception of young ladies, in her house, no. 96, Chestnut Street, She will commence giving lessons in Embroidery on Monday next, the 11th. inst. from 10 o'clock"; Philadelphia City Directory 1819: "Mrs. Dirat, Milliner and mantuamaker from Paris, 69 S 2d."

41. AN, F^7 6680, Louis-Marie Dirat to His Excellency Mr. Hyde de Neuville, ambassador plenipotentiary of His most Christian Majesty to the United States, Philadelphia, July 25, 1819.

42. Ibid., Marquis Dessolle, minister of foreign affairs, to Decazes, minister of the interior, Paris, September 30, 1819.

43. The son returning with the Dirats was their youngest, Gustave. Their older son, Claude-Adolphe, born in Paris on November 29, 1803, a merchant's clerk, was still in Philadelphia on March 8, 1822, when he was issued a passport and went to New York to sail for France. CADN, consulat de France à Philadelphia (exercice Delaforest), 109, passport register . . . , fol. 31–no. 2.

44. CADN, consulate de France à New York, série B, 208, notebook no. 9, on the back of passport no. 405: New York, November 30, 1819, the consul issues a passport to Mr. Louis-Marie Dirat, former French officer; ADSM 6P6/33, arrival Le Havre 1820 of the *Bonne-Henriette,* 172 tons, owner Le Seigneur, Cap. Jappie, cargo of tea, cotton, potash, wax, etc., seventeen passengers including Marie-Victorine Volozan, née Reynaud de Saint-Félix, sister of Jean-Baptiste-Gilles Reynaud de Saint-Felix, Tombigbee subscriber, her husband, Denis-Auguste Volozan, and their three children.

45. Andrieu, *Bibliographie générale,* 1:238.

46. Grouchy to his wife, Philadelphia, February 15, 1817, in Grouchy, *Mémoires,* 5:57–58. The retreat in question is his château of La Ferrière, located a few miles north of Bayeux, Calvados.

47. Grouchy to his son, Colonel Alphonse de Grouchy, [Normandy], November 5, 1815, in ibid., 9.

48. Grouchy to his wife, Washington, February 20, 1816, in ibid., 41.

49. Grouchy to his wife, Baltimore, May 22, 1816, in ibid., 43.

50. Ibid., 43–44.

51. Grouchy had actually never been sentenced: the second court-martial of the First Military Division, in charge of judging Grouchy, twice declared itself incompetent.

52. Grouchy to his wife, Brandy-Wine (Delaware), May 26, 1817, in Grouchy, *Mémoires,* 5:61.

53. Grouchy to his wife, Baltimore, May 22, 1816, in ibid., 44.

54. CADN, consulat de France à New York, série C, register 1, December 31, 1817: writing to the minister of foreign affairs about the departure of the French officers for the

Tombigbee (actually for Texas via New Orleans), the consul said that "Generals Grouchy and Vandamme do not know about any of this."

55. CADN, consulat de France à La Nouvelle-Orléans, 2mi1961 (vol. 2, p.7), French consul to the minister of foreign affairs, April 10, 1817.

56. Grouchy to his son, Colonel Alphonse de Grouchy, Philadelphia, September 17, 1819, in Grouchy, *Mémoires,* 5:208.

57. Grouchy to his son, Colonel Alphonse de Grouchy, Clermont, on the North River, August 21, 1819, in ibid., 205.

58. Grouchy to his son, Colonel Alphonse de Grouchy, Boston, October 2, 1818, in ibid., 66–67. The dinner had been hosted by the former consul, Cazeaux.

59. Grouchy to his son, Colonel Alphonse de Grouchy, Philadelphia, March 25, 1820, in ibid., 226.

60. CADN, consulat de France à Philadelphie, register 46, p. 9, Philadelphia consul to Hyde de Neuville [addition to his letter of April 13], April 15, 1820. According to the consul, three of Joseph Bonaparte's men came to the church of Sainte Trinité to provoke Grouchy, among them "Colonel" Latapie, with a violet in his mouth—a gesture of loyalty to the Emperor who, when he returned from Elba in March 1815, was called "Father Violet" because the violet is reborn in the spring.

61. MAE, MD France, vol. 2143, fol. 295, Dessolle, minister of foreign affairs to Hyde de Neuville in Washington, Paris, September 22, 1819.

62. Grouchy to his wife, Philadelphia, October 25, 1819, concerning "an overture already made" for this position, in Grouchy, *Mémoires,* 5:209–10.

63. Grouchy to Colonel A. de Grouchy, Philadelphia, December 3 and December 28, 1819, in ibid., 212–13 and 214–15.

64. Hyde de Neuville, French ambassador to the United States, to the Count of Grouchy, Washington, February 14, 1820, in ibid., 224.

65. Grouchy to Hyde de Neuville, Philadelphia, February 18, 1820, in ibid., 225.

66. February 14, 1820, in ibid., 224.

67. MAE, CP EU, vol. 74, fol. 331v, Consul Pétry to Richelieu, Philadelphia, December 31, 1817; also see CADN, consulat de France à Philadelphie, no. 45, Légation de France, exercice Pétry, p. 59.

68. AN, F⁷ 6683, file 9, Countess Sophie Vandamme, to Decazes, minister of police, Cassel, August 20, 1818.

69. MAE, MD France, vol. 2143, fol. 130, minister of police to Dessolle, minister of foreign affairs, Paris, March 19, 1819.

70. Ibid., fol. 192, consul to Hyde de Neuville, Philadelphia, April 24, 1819.

71. Ibid., fol. 195, minister of foreign affairs to Hyde de Neuville, Paris, September 22, 1819.

72. CADN, consulat de France à New York, série C, no. 2, May 5, 1819; for his destination of Ghent, see CADN, consulat de France à Anvers, 1, liasse III, letters of the French legation in the Netherlands (1816–32), M. de la Tour du Pin, French ambassador, to Despallières, French consul in Antwerp, Brussels, May 22, 1819.

73. Upon her arrival she asked for a passport to Saint-Omer. Was she related to Auguste Rieffel, born in Liège, listed as an officer at the Champ d'Asile before getting into busi-

ness as a watchmaker and marrying in New Orleans in 1822? New Orleans City Directory 1822: "Rieffel, A., watchmaker, 46 Chartres below Conti."

74. *Le Moniteur Universel,* June 11, 1819: the authorities gave him "the city as a prison."

75. MAE, MD France, vol. 2144, January 1820; *Le Moniteur Universel,* December 22, 1819, reprinting information published in the Brussels *Le Vrai Libéral;* CADN, consulat de France à Anvers, 1, liasse III, letters of the French legation in the Netherlands, letters of December 14 and 19, 1819, The Hague, Count Roger de Caux (the king's chargé d'affaires in the Netherlands) to the consul in Antwerp, M. Despallières.

76. For the abandoned château of Cassel see Beaucour's article in *Bulletin historique de la Société de sauvegarde du château imperial de Pont-de-Briques* 2 (1969): 121. In 2008 the château, which is in private hands, was still in the same state of disrepair.

77. Du Casse, *Le général Vandamme,* 2:581.

78. In his *Exposé justificatif* (1816), Clauzel states that he had lived through the first Restoration "a stranger to politics," had been unaware of the existence "of a conspiracy to bring Bonaparte back to France," and had learned of his landing "only because people were talking about it." He even adds that he "vainly begged" of the king "the favor" of letting him fight Napoleon when he returned from Elba, and "again on the night of March 19" had offered his services "to protect his departure and see to the safety of his person" (8–9).

79. MAE, CP EU, vol. 74, fol. 28, Hyde de Neuville to Richelieu, Washington City, May 24, 1817.

80. CADN, consulat de France à New York, série B, 112, October 30, 1818, consul in Baltimore, M. Angelucci, to the consul general in Philadelphia, M. Pétry, referring to their arrival in Baltimore before October 10.

81. MAE, MD France, vol. 2142, fol. 288–89, Hyde de Neuville to Richelieu, Washington, October 28, 1818.

82. Ibid., fol. 249, or AN, F⁷ 6679, file 7, Clauzel to the king via the French consul in Baltimore and Ambassador Hyde de Neuville, Baltimore, October 23, 1818.

83. MAE, MD France, vol. 2143, fol. 295.

84. CADN, consulat de France à New York, série B, 209, notebook no. 11, no. 440.

85. Ibid., notebook 124, *minutier* 39, French consul in Antwerp, M. Despallière, to the minister of foreign affairs, June 22, 1820.

86. CADN, consulat de France à Anvers, 1, liasse III, letters of the French legation in the Netherlands (1816–32), Count Roger de Caux (the king's chargé d'affaires in the Netherlands) to the French consul in Antwerp, The Hague.

87. This contradicts Blaufarb's assertion that Clauzel's brilliant career shows the Bourbons' desire to rehabilitate their former enemies. See *Bonapartists in the Borderlands,* 160.

88. C. Clauzel, "Bertrand Clauzel, maréchal de France," 27. Also see B. Clauzel, *Correspondance du maréchal Clauzel.*

89. Napoleon left him 100,000 francs.

90. MAE, MD France, vol. 2142, fol. 292v, Hyde de Neuville to the Duke of Richelieu, Washington, October 28, 1818.

91. Ibid., fol. 360, Hyde de Neuville to Richelieu, Washington, December 12, 1818.

92. Ibid., Consul General Pétry to Richelieu, December 28, 1818.

93. MAE, MD France, vol. 2143, fol. 300–301, or MAE, CP EU, vol. 2147, fol. 179–179v, petition from Lefebvre-Desnoëttes to Hyde de Neuville, Demopolis, November 1, 1821.

94. Ibid., fol. 346, Hyde de Neuville to Lefebvre-Desnoëttes, Washington, December 1, 1821.

95. MAE, MD France, vol. 2148, fol. 32, Lefebvre-Desnoëttes to Hyde de Neuville, New Orleans, February 1, 1822.

96. Ibid., fol. 86, Lefebvre-Desnoëttes to Hyde de Neuville, Washington City, March 23, 1822.

97. Ibid., fol. 89, Lefebvre-Desnoëttes to Hyde de Neuville, Philadelphia, March 30, 1822.

98. The liner *Albion* of New York, Captain Williams, owners (Isaac and William) Wright Brothers.

99. MAE, CP EU, vol. 79, fol. 143–143v, Hyde de Neuville to Viscount de Montmorency, French foreign minister, Washington, June 11, 1822.

100. CADN, consulat de France à New York, exercice d'Espinville, État-civil, register EEE, begun August 6, 1818, fol. 33–34, [2 Mi 1469], declaration to the French consulate in New York by Hiram M. Raymond and Francis Bloom, American sailors residing in New York, survivors of the shipwreck of the *Albion*, July 19, 1822; also see Bourke, *Shipwrecks of the Irish Coast,* 1:116, 2:119.

101. See note 99.

102. See Chappet, Martin, and Pigeard, *Le guide Napoléon.* Today (2010) the monument is located on rue Charles-Alexandre Lesueur, named for the Le Havre traveler and artist who was also a member of the Tombigbee Society.

103. SHD/DAT, 7 Yd 475.

104. CADN, consulat de France à La Nouvelle-Orléans, série D, non classé, carton 205, no. 1717, minister of foreign affairs to the French consul in New Orleans, Paris, May 25, 1833.

105. Ibid., Countess Lefebvre-Desnoëttes to the French consul in New Orleans, Paris, February 11, 1834.

106. Ibid., French consul to Countess Desnoëttes, New Orleans, May 26, 1835.

107. On St. Helena Napoleon put Lefebvre-Desnoëttes into his will for the sum of 180,000 francs. His heirs received 62,143 francs from the funds deposited with Jacques Laffitte and 74,771 francs from those decreed by the Emperor Napoleon III. *Nouvelle biographie générale,* vol. 29.

108. The paddle steamer is mentioned in a letter from Lecoq Dumarcelay, dated Louisville, Kentucky, January 19, 1818, to Basile Meslier, and quoted by André-François Godet Jr., Garnier de Saintes' son-in-law, a retired captain and member of the Legion of Honor, in a letter addressed from Saintes on May 10, 1832, to the French consul in New Orleans regarding the deaths of Garnier and his son. Blaufarb writes that Garnier perished in the explosion of a steamboat on the Ohio in 1818. *Bonapartists in the Borderlands,* 160.

109. Marengo County, Deeds A, 1–2, sale of July 27, 1820, lot 44; NARA, RG 49, TAL, Small lot, Misc. Paper, deposition, July 15, 1837.

110. Paul Martin Collection (Sainte-Foy-la-Grande), Jean-Augustin Pénières-Delzors to Mr. Lajonie, Demopolis, Alabama State, Strasbourg (Virginia), September [*sic*] 27, 1820.

111. Pénières to his brother Raymond, known as Dubois, Washington, March 21, 1821, in Faure, *Pénières-Delzors,* 203.

112. Pénières to his brother Raymond, known as Dubois, November 18, 1820, in ibid., 203.

113. The United States had purchased Florida from Spain two years earlier. Pénières was under the authority of Andrew Jackson, the new governor of Florida.

114. Carter, *Territorial Papers,* 22:434: on September 14, 1821, Jean Augustin Peniers [*sic*] swears an oath of naturalization before James Grant Forbes, Esq., mayor of St. Augustine (Florida).

115. Carter, *Territorial Papers,* 22:240: letter from W. G. D. Worthington, secretary and governor of eastern Florida, dated October 6, 1821, in St. Augustine, announcing Pénières' death to Governor Jackson in Pensacola. Communal archives of Saint-Julien-aux-Bois, death register for 1826, no. 3, via Isabelle Desmaison, secretary of the organization "Généalogie en pays de Tulle": on January 27, 1826, the registry office of the commune of Saint-Julien-aux-Bois, misled by poor translations, gives Pénières' death date as August 6, 1823.

116. Marengo County, Orphans Court Minutes, Book A & B, 36, 72: January 1825, Basile Meslier, administrator of Pénière's will in Alabama.

117. Robichaux, *Civil Registration of Orleans Parish,* 703–4: declaration of death on September 5, 1820, by Jean Chabaud, proprietor, age fifty. Died on September 4 according to the religious certificate in Woods, Nolan, and Dupont, *Sacramental Records,* 14:343.

118. French consul Guillemin to the minister of foreign affairs, New Orleans, September 14, 1820, in *Le General Antoine Rigau,* a sixteen-page anonymous booklet.

119. MAE, MD France, vol. 2143, fol. 39, Guillemin to Ambassador Hyde de Neuville in Washington, sending him Rigau's petition to the king, New Orleans, January 13, 1819.

120. Ibid., fol. 265, Hyde to Dessolle, minister of foreign affairs, Washington, July 25, 1819.

121. SHD/DAT, 8 Yd 1099, Narcisse Rigau to Baroness Rigau, his stepmother, New Orleans, January 29, 1820.

122. Concerning Antoine Rigau: Marengo County, Alabama, Deeds A, 124–27, December 10, 1817, and Deeds A, 127–29, December 13, 1817. Concerning Narcisse Rigau: NARA, RG 49, TAL, PC file 74, December 11, 1817.

123. Could this be Sophie Rigau, mulatto, born in New Orleans ca. 1822, daughter of a white man and a black woman born in New Orleans? She appears in the 1880 census with, in her home, a twenty-five-year-old daughter, Laure Rigau. 1880 U.S. Federal Census, New Orleans District 50.

124. SHD/DAT, GB 1099, Rigau, Antoine: Paris, March 4, 1854, Madame Antonia Dubois, née Rigau, asks the war minister for a copy of the service record of her father, who "died in exile in New Orleans in 1820 as a result of the injuries received during the wars of the Empire."

125. Document shared by Olivier Dornès, a descendant of General Rigau through Eugénie Rigau Kestner.

126. Buried in the Bordentown cemetery on September 16, 1823, according to the *Na-*

tional Gazette, September 16, 1823, cited in *York County South Carolina Marriage and Death Notices, 1823, 1865,* 2, quoting *The Pioneer* of October 4, 1823. For Henri Lallemand's last days, see the letters from Roberjot to Girard of September 10, 11, and 12, 1823 (APS, Girard Papers, 1823: 732, 735, 738, 740, 741) and the announcement of his death by Harriet Lallemand, September 17 (APS, Girard Papers, 1823: 748).

127. At the age of twenty-one, Caroline Lallemand had, on February 24, 1840, married Hubert de Saint-Marsault, born July 14, 1812, in Eyburie (Corrèze). She died on September 20, 1863. See Beauchet-Filleau, Beauchet-Filleau, and Beauchet-Filleau, *Dictionnaire historique et généalogique,* 4:392; Révérend, *Armorial du Premier Empire,* 3:28.

128. Claude Joseph Brandelis Green de Saint-Marsault (Uzerche, 1807–Paris, 1866), subprefect and prefect from 1836 to 1865; at the time prefect of Seine-et-Oise since 1852, he was appointed a senator in 1865. Robert, *Dictionnaire des parlementaires français,* 5:242; Lamoussière, *Personnel de l'administration préfectorale,* 364.

129. Greene Co., Deeds B, fol. 29, October 7, 1824; Philadelphia Co., Pa, Administrations 1824, 307, Lallemand, Henry, General.

130. In the "District Court of the U.S.," according to the New Orleans *Ami des Lois,* December 21, 1818.

131. NONA, Philippe Pedesclaux VII (January 2–May 1, 1819), no. 180, March 5, 1819, Zenon Cavalier to Lallemand, a property in Métairie; NONA Michel de Armas, 18, act 57, October 1, 1819, Louis de Fériet, proprietor, to Lallemand, a farm at Bayou Saint Jean, $12,000. Deeds show the purchase of at least seven slaves from February 23 to June 18, 1819, whose number and names are repeated on February 15, 1820, NONA, Michel de Armas, 19, act 73.

132. La Souchère-Deléry, *Napoleon's Soldiers in America,* 71–72, citing a letter from Louis de Fériet to his sister.

133. Louis Lauret to Stephen Girard, New Orleans, March 10, 1824, reel 87, Girard Papers, APS. After reminding Girard that he had been the best man at Henri Lallemand's marriage, Lauret complains of having been talked into the Champ d'Asile venture by Charles Lallemand, then of wanting to do him a favor by endorsing some 2,327 piasters for him, despite his meager resources.

134. *La Minerve française,* November 1819, 319–20.

135. See the *Louisiana Courier* of May 8, 9, and 10 (published list) and June 3 and 9, 1820 (Lallemand's relinquishment), and the *Ami des Lois* of May 28, 1820.

136. NONA, Michel de Armas, 19, act 32: on June 28, 1820, Lallemand sold to Charles Amédée Hurtel allotment 69, 240 acres, which he had acquired from Jean Juste Nartigue, and his own allotment, no. 166, 480 acres, near Aigleville, plus 12 acres and a town lot in Aigleville, for $2,980. For this sale, see also ibid., 20, act 4, January 6, 1821.

137. AN, F[7] 6681, the consulate to the ministry of foreign affairs in Paris, New Orleans, June 1, 1821.

138. APS, Plowden Papers via Girard Family Papers, Nicolas Raoul to Lefebvre-Desnoëttes, Demopolis, January 24, 1822, informing him of the news of General Charles Lallemand's arrival in Philadelphia.

139. APS, Plowden Papers, Pierre François Réal to Henri Lallemand, New York, June 21, 1822.

140. CADN, consulat de France à New York, série B, no. 110, consul to Hyde de Neuville, June 15, 1822.

141. "Not more than 20 men," according to Colonel Fabvier, writing from London in June 1823 to Lallemand, quoted in Bruyère-Ostells, "Les officiers de la Grande Armée," 80; "a dozen French and Italian scoundrels," according to Sergeant Roses, whom he tried to recruit with a promise of money and an officer's rank, SHD/DAT, 7 Yd 851.

142. For this affair, see AN, F^7 6681, the minister of foreign affairs to the minister of the interior, Paris, October 31, 1826.

143. J. Lakanal to Geoffroy Saint-Hilaire, Mobile, May 1, 1830, in Jouin, *Lakanal en Amérique,* 30.

144. Lucien Charvet, President of Orleans College and mathematics professor, Bourbon Street, no. 183: 1824 New Orleans City Directory.

145. J. Lakanal to Geoffroy Saint-Hilaire, Mobile, November 25, 1830, in Jouin, *Lakanal en Amérique,* 34.

146. J. Lakanal to Geoffroy Saint-Hilaire, Mobile, August 1, 1834, in ibid., 47.

147. Marie Françoise Joséphine Lakanal, twenty-one, had married Jules Frédéric Guinand (1788–1845), a Swiss émigré, on February 6, 1817, in Port William, Kentucky. On October 30, 1821, in Vevay, Indiana, Alexandrine Lakanal had married Hyppolite Lucien Charvet (1795–1827), with whom she had two sons; widowed, she married Henri Germain (1791–1860), a member of the Tombigbee Society, on January 10, 1829, in Baton Rouge; they had two children.

148. Mobile, probably 1832, J. Lakanal to Geoffroy Saint-Hilaire, Paris, in Jouin, *Lakanal en Amérique,* 39–42.

149. Mobile, August 1, 1834, J. Lakanal to Geoffroy de Saint-Hilaire, in ibid., 45.

150. On December 21, 1842, the mayor of the 8th arrondissement of Paris united Joseph Lakanal and Rosalie Céleste Bienaimée Lepelletier (1806–80), his mistress since 1838, in marriage. See Dawson, *Lakanal the Regicide,* 167.

151. He was buried at Père Lachaise, eleventh division; his grave is next to that of Frédéric Chopin, which attracts many visitors.

152. Marengo County, Deeds, A, 41–43 and 43–46.

153. His two stepchildren, Alexandre and Zoé Sinibaldi, remained in Guatemala City, where they were married, the first to Dolores Cladera Ferrer, the second to Juan Mathew Batellini, a native of Cádiz. In 1977 three generations of women directly descended from Zoé made a pilgrimage to Demopolis. Elma Bell, "Guatemalan Visits Her Ancestor's Home in State," *Birmingham News,* December 2, 1977; Ellen Byrd Rish, "Descendants of the Countess Come Home," *Demopolis Times,* December 8, 1977.

154. SHD/DAT, GB 3040, Nicolas Raoul's personnel file.

155. Ibid.

156. Joséphine Françoise Octavie Pavans de Ceccaty, married on March 18, 1840, gave birth to Caroline Louise Raoul on January 15, 1841.

157. See his *Correspondance.*

158. Barou, "Le colonel Michel Combe," 9.

159. CADN, consulat de France à Philadelphie, 109, fol. 64v, no. 34: on July 29, 1831, the consul issued a passport to M. Barthélémy Colonna d'Ornano, former colonel, land-

owner, [born in Cognocoli-Monticchi, Southern Corsica, ca. 1778] with his wife [Maria Anna Cometta Nicolina Colombani, born in Leghorn, ca. 1798], and their son Émile [born in Leghorn April 27, 1820], "going to France, either via England, or via New York to Le Havre."

160. NARA, RG 49, TAL, PC file 178: on May 11, 1818, in Philadelphia, B. Colonna d'Ornano sold to Peter Lewis Pittori, of Philadelphia, 320 acres, 1600 dollars. Witnesses: L. M. Dirat; A. Caziano.

161. For diplomatic dispatches concerning Colonna see MAE, MD France, vol. 2144, fol. 109, 130–32, 160.

162. Archives nationales d'outre-mer, online, public records of Mustapha Supérieur (near Algiers), September 20, 1855, death of Barthélémy Colonna d'Ornano; public records of Mers-El-Kébir, death of Émile Colonna d'Ornano, age forty-seven, April 20, 1867.

163. CADN, consulat de France à New York, série B, 209, notebook no. 13, no. 493, passport issued to Jean François Roland de Bussy; no. 498, to Camille Arnaud; no. 502, to Alexandre Ambroise Germond.

164. AN, Léonore database, no. L2371031, "Service record of Mr. Jean François Roland de Bussy . . . , certified in Algiers, December 8, 1835."

165. Tomb Eighth Division, European Cemetery of Bab-El-Oued (Saint-Eugène), carré 4, CAP no. 501.

166. Lamoussière, *Personnel de l'administration préfectorale*, 623.

167. University of Notre Dame Archives, Calendar 1850, online: Paris, May 15, 1850, Father Louis Dufour to Antoine Blanc, archbishop of New Orleans, www.archives.nd.edu/calendar/c185005.htm. Centre des archives d'outre-mer Algeria Registry via Jacques Youx: Father Anduze died in the rectory of the town of Blida on January 23, 1847.

168. Lajonie Papers, Lajonie to Le Soulat, December 17, 1826.

169. Lajonie Papers. Thirteen letters relate to Lajonie's reinstatement into the Order of the Legion of Honor: the first from Lajonie, dated August 5, 1829, Paris, Hôtel des Sept frères, rue Grenelle St.-Honoré, to the secretary-general of the grand chancellery; the last from Vicount de Saint-Marc, dated June 8, 1830, grand chancellery of the Royal Order of the Legion of Honor, Third Division, First Bureau.

170. Ibid., A. Lerminier to Jacques Lajonie in Juillac via Castillon-sur-Dordogne, Grosbois, August 22, 1829,. Grosbois was the château belonging to the widow of Marshal Berthier, the Princess of Wagram, at the head of the order's grand chancellery.

171. Ibid., P. M. Mannoury, rue Neuve St-Eustache 36, to Monsieur Lajonie, Juillac, via Castillon-sur-Dordogne, Paris, November 10, 1834. This is the last in a series of nine letters from Mannoury to Lajonie, beginning on November 20, 1830.

172. ADSM, 6P6/42, arrival in Le Havre on June 25, 1824, of the Le Havre–registered ship *Roman*, with as passenger Pierre Mathieu Mannoury, merchant and planter, whose passport was issued by the New Orleans consulate on May 13, 1824.

173. NONA, Michel de Armas, 18, act 332: April 28, 1824, Marengo County, Pierre M. Manoury, of New Orleans, by his representative Jacques Lajonie, Marengo County, sells to Édouard Chaudron of Marengo County, allotment 69, 240 acres. Witnesses: John Hurtel; F. Stollenwerck.

174. ADG, 3 Q 13,607, notification of transfer by death, no. 14, January 19, 1850.

In the Lajonie Papers, Léopanno Lajonie's estate is described as a property of 106 acres in vineyards, fields, meadows, and timber at Le Soulat, Robert, Muset, Ponset, and Boutounelle with an estimated value of 266,562.26 francs. At that time the franc was worth about 20 cents.

175. Jean-Jacques Lajonie Lapeyre (1823–1911), son of Pierre Lajonie (nephew of Jacques Lajonie) and Jeanne-Élisabeth Dubreuilh, married in 1820. Marriage contract, February 19, 1851. Daughters: Marguerite Lajonie died single and childless; Thérèse (1853–1942), married to Dr. Louis Coustou (1850–1936), has descendants that continue into the twenty-first century.

176. Lajonie Papers, "Respected friend, I presume you are elected col. in your county, if so permit me so to title you," wrote Alexander McAlpin of Orange Grove, Greene County, Alabama, to Lajonie on November 25, 1831.

177. The "two strengths" he mentions were the material and moral strength of the law. Lajonie Papers.

178. Lajonie Papers.

179. Ibid., *Inventaire des biens assurés contre l'incendie* [Inventory of goods insured against fire], May 4, 1868, with La Paternelle (agency in Sainte-Foy-la-Grande): 6,000 francs on the house and outbuildings at the place called Lapeyre; 5,000 francs on furnishings consisting of furniture, clothing, linens, beds household utensils and supplies, clocks, watches, paintings, mantels, mirrors and ornaments, arms, silverware, and books.

180. Ibid. On September 29, 1841, at a meeting for settlement of the estate in the office of Saint-Jean Lestage, attorney-at-law in Gensac, Jacques Lajonie, his wife, and their daughter Ninon, residing at Lapeyre, Gensac, Léopanno Lajonie, their son, unemployed, residing at Le Soulat, Juillac, and Eugène Gentillot affirmed that Elbine Lajonie Gentillot died after drawing up a holograph will on September 15, 1840, filed with Attorney Brisson in Libourne.

181. Ibid. Holograph will of Dorothée Noguey-Maransin Lajonie, dated Lapeyre, March 5, 1854: "I give to Jeanne Elizabeth Ninon, my daughter, a quarter of all I shall possess at my death and I give half of the life interest to Jacques Lajonie, my husband."

182. Ibid.

CONCLUSION

1. W. Smith, *The People's City*, "Antebellum Good Times," 105–27.

2. The inscription "To the Beloved Memory of George Strother Lyon, 1800–1882" lines the bottom of a window of Trinity Episcopal Church in Demopolis. The church is located on Lyon Street.

3. Hellman, *The Little Foxes*, act 1, p. 13. First staged at the National Theater of New York in 1939. Hellman was the daughter of Sophia Newhouse Marx and the granddaughter of Isaac Marx, an 1844 Demopolis precursor of Jews emigrating from Germany to the southern United States. W. Smith, *The People's City*, 12–14. The play was made into a motion picture in 1941, starring Bette Davis and directed by William Wyler, whose wife was born in Demopolis.

4. Carmer, *Stars Fell on Alabama*, 113.

5. Arch Street follows the crest of the cliffs, from the present entrance of the Botanical Gardens or Lower Landing to the Upper Landing, that is, from the southern to the northern boundary of the town. On June 18, 1819, the commissioners of the Demopolis Town Company put Arch Street at the free disposal of the public in perpetuity. The street, which was meant to provide river access to the town's eventual inhabitants, was never more than a line and a name on maps, but its space fulfills its role of an economic (wharves for ships, cotton, wood, passengers), religious (baptisms), or simply recreational opening to the river.

6. They belonged to Harry Britton, a former president of the Marengo County Historical Society, who died in Demopolis in January 2008.

7. Probably also because the Swiss settled permanently and built houses, some of which still survive, Vevay has made better use of the fact that it has a twin, Vevey, on the shores of Lake Geneva. References to this relationship, as well as the Swiss flag, can be found everywhere, and ties between the Swiss nation and Switzerland County have endured via a museum, an association, and an annual August wine festival.

8. Local news item in Uniontown, May 5, 2004: a pregnant African American mother and her four children, aged one to nine, burned to death in a mobile home. They had been using candles, as their electric and water service had been shut off. *Demopolis Times,* May 6, 2004; *Democrat-Reporter,* May 13, 2004.

9. Two historical markers call attention to their contribution. In 1832, Allen Glover's slaves built the Greek Revival home of Bluff Hall, today one of Demopolis's historical gems, for his daughter Sarah and his son-in-law Francis S. Lyon. In 1863, after years of effort due to the rocky subsoil, Confederate general Whitfield's slaves completed construction of a drainage canal. One mile long, in places thirty feet deep and equally wide, the canal took the water that would have flooded his plantation of Marlmont, due to its impermeable limestone subsoil, to the Tombigbee River. *Demopolis Times,* October 3, 1974. Whitfield, who had come from North Carolina with his slaves, owned 249 of them in 1860. W. Smith, *The People's City,* appendix E, part 1: "Slave Owners in Demopolis and Number of Slaves Owned in 1850 and 1860," 348–52.

10. The Confederate War Memorial was erected in 1910 on the initiative of the Marengo Rifles Chapter, United Daughters of the Confederacy.

11. No one criticizes the slaveowning builders of Demopolis, because they populated, improved, and beautified it. They are the emblems of the town, and their descendants return in pilgrimage; the streets are named after them, along with the nation's founding fathers, Virginians who also owned slaves. Between a democratic republic and slavery followed by segregation, we are here at the center of the American contradiction.

Bibliography

LIBRARIES, MUSEUMS, AND COLLECTIONS OF ARCHIVES

Alabama Department of Archives and History, Montgomery
American Philosophical Society, Library Hall, Philadelphia, Pennsylvania
Archives départementales d'Ille-et-Vilaine, Rennes, France
Archives départementales de la Gironde, Bordeaux, France
Archives départementales de la Charente-Maritime, La Rochelle, France
Archives départementales de la Loire-Atlantique, Nantes, France
Archives départementales de la Seine-Maritime, Rouen, France
Archives générales du Royaume de Belgique, Brussels and Anderlecht, Belgium
Archives municipales de Bordeaux, Lorient, Nantes, Sainte-Foy-la-Grande, etc., France
Archives nationales, Paris, France
Archives nationales d'outre-mer, Aix-en-Provence, France
Arquivo general de Indias, Seville, Spain
Bibliothèque municipale de Bordeaux, de Nantes, France
Bibliothèque nationale de France, Paris, France
Bibliothèque Royale Albert Ier, Brussels, Belgium
Birmingham Public Library, Birmingham, Alabama
Centre des Archives diplomatiques de Nantes, France
Demopolis Public Library, Alabama
Escambia County Courthouse, Pensacola, Florida
Filson Historical Society, Louisville, Kentucky
Greene County Courthouse, Eutaw, Alabama
Hagley Museum and Library (until 1985 Eleutherian Mills Historical Library), Wilmington, Delaware
Hale County Courthouse, Greensboro, Alabama
Herman B. Wells Library, Bloomington, Indiana
Historic New Orleans Collection, Williams Research Center, Archives and Manuscripts, New Orleans
Historical Society of Pennsylvania, Philadelphia
Howard-Tilton Memorial Library, Manuscript Department, Tulane University, New Orleans, Louisiana

Library of Congress, Washington, D.C.
Marengo County Courthouse, Linden, Alabama
Marengo County Historical Society, Demopolis, Alabama
Ministère des Affaires étrangères, Paris
Missouri Historical Society, Saint Louis
Mobile Genealogical Society, Mobile, Alabama
Mobile Municipal Archives, Mobile, Alabama
Mobile Museum of Art, Mobile, Alabama
Mobile Public Library Local History and Genealogy, Mobile, Alabama
National Archives (formerly Public Record Office), London
National Archives and Records Administration, Washington, D.C.
Noel-Ramsey House (Museum) or Old French House, Greensboro, Alabama
Service historique de la défense/Département de l'armée de terre (SHD/DAT), Vincennes,
    France
Service historique de la défense/Département de la marine (SHD/DM), Vincennes and
    Brest, Cherbourg, Lorient, Rochefort, Toulon, France
Southern Historical Collection, University of North Carolina, Chapel Hill
Stadsarchief Antwerpen, Antwerp, Belgium
Switzerland County Historical Museum, Vevay, Indiana
University of Notre Dame Archives, Notre Dame, Indiana
Virginia Historical Society, Richmond

PRIVATELY HELD ARCHIVES IN FRANCE AND THE UNITED STATES

Martin (Paul) Collection, Sainte-Foy-la-Grande
Papiers Lajonie, Gensac and Sainte-Foy-la-Grande (Lajonie Papers)

Chapron Papers, Demopolis, Alabama
Constantine Papers, Birmingham, Alabama
Davidson-Tremoulet Papers, New Orleans, Louisiana
Gardien Papers, Cuernavaca, Mexico
George Papers, Mobile, Alabama
Hurtel-Noël Papers, Greensboro, Alabama
Jean-Marie Léonard Collection, Isère
Webb Papers, Demopolis, Alabama

PRINT SOURCES: NEWSPAPERS IN FRANCE
AND THE UNITED STATES

Affiches, Annonces et Avis divers de la ville de Bordeaux
Affiches, Annonces et Avis divers or *Feuille commerciale de Nantes*
Bulletin des lois
Gazette officielle (July 8, 1814–February 1, 1815)
Journal des arts, des sciences et de la littérature, later *Le Nain Jaune,* then *Le Nain Jaune réfugié*

Journal des Débats politiques et littéraires
Mémorial Bordelais. Feuille Politique, Littéraire, Administrative et Commerciale
Mercure-Surveillant, Liège, Belgium
Minerve française (1818 editions consulted for subscriptions in favor of the Champ d'Asile)
Le Moniteur Universel
Nain Jaune réfugié, third and fourth volumes, 1816
Quotidienne

L'Abeille Américaine (Philadelphia)
Alabama Republican
L'Ami des Lois (New Orleans)
Argus (Mobile), replaced *L'Ami des Lois* in April 1824
Attakapas Gazette (St. Martinville, La.)
Aurora (Philadelphia)
Birmingham News
Columbia Sentinel
Courrier de la Louisiane (New Orleans)
Democrat-Reporter (Linden, AL)
Demopolis Times
Louisiana Gazette (New Orleans)
Mobile Commercial Register
Mobile Daily Commercial Register and Patriot
Mobile Register
National Intelligencer (Washington D.C.)
Niles' Weekly Register (Baltimore)
Poulson's American Daily Advertiser (Phildelphia)
Selma Times Journal
Western Courier (Louisville)

BOOKS, MONOGRAPHS, AND ARTICLES

Abernethy, Thomas Perkins. *The Formative Period in Alabama, 1815–1818*. Publications of the Alabama State Department of Archives and History: Historical and Patriotic Series no. 6. Montgomery: Brown Printing Company, 1922.

Adams, Donald R., Jr. *Finance and Enterprise in Early America: A Study of Stephen Girard's Bank, 1812–1831*. Philadelphia: University of Pennsylvania Press, 1978.

Adams, John Quincy. *Memoirs of John Quincy Adams, Comprising Portions of His Diary from 1795 to 1848*. Ed. Charles Francis Adams. Vol. 4. Philadelphia: Lippincott, 1874–77.

———. *The Writings of John Quincy Adams*. Ed. Worthington Chauncy Ford. Vol. 6. New York: Macmillan, 1913–17.

Adams, Leon D. *The Wines of America*. 4th ed. New York: McGraw-Hill, 1990.

Agee, Rucker. *Maps of Alabama: The Evolution of the State Exhibited in Printed Maps from the Age of Discovery: A Preliminary Catalogue*. Birmingham, Ala.: Public Library, 1955.

Almanach de commerce pour La Nouvelle-Orléans . . . , 1822.

Ammon, Harry. *The Genet Mission*. New York: Norton, 1973.

Andrews, Johnnie. *Fort Tombeche Colonials: A Compendium of the Colonial Families of the Tombigbee River Valley, 1735–1825.* Prichard, Ala.: Bienville Historical Society, 1987.

Andrieu, Jules. *Bibliographie générale de l'Agenais et des parties du Condomois et du Bazadais incorporées dans le département de Lot-et-Garonne.* Vol. 1. Paris: A. Picard, 1886–91.

Annuaire Historique Universel 1819.

Antonetti, Guy. *Louis-Philippe.* Paris: Fayard, 1994.

Aragon, Louis. *La semaine sainte.* 1958. Paris: Folio-Gallimard, 1998.

Archives Parlementaires de 1787 à 1860, recueil complet des débats législatifs et politiques des Chambres Françaises. 49 vols. Paris, 1867–96.

Armistead, James, and Emogene Armistead. "Pioneers of Marengo County." *Demopolis Times,* 1974–76.

Arnaud, Eugène-François-Auguste, baron de Vitrolles. *Mémoires et relations politiques du baron de Vitrolles.* Vol. 3. Paris: G. Charpentier, 1884.

Arnault, V.-A., Antoine Jay, and Étienne de Jouy, *Biographie nouvelle des comtemporains.* Vol. 17. Paris: Librairie historique, 1820–25.

Artaud, Denise, and André Kaspi. *Histoire des Etats-Unis.* 6th ed. Paris: A. Colin, 1985.

Atkins, Leah Rawls. *The Warrior and the Tombigbee: Two Rivers Flowing through History.* Mobile, Ala: Warrior-Tombigbee Waterway Association, 2000.

Aubert, Guillaume. "'Français, nègres et sauvages': Constructing Race in Colonial Louisiana, 1699–1763." Diss. Tulane University, New Orleans, 2002.

Bacheville, Barthélémy. *Pétition de M. Barthélémy Bacheville à la Chambre des Députés.* Paris: Corréard, 1820.

———. *Voyage des frères Bacheville . . . en Europe et en Asie après leur condamnation par la cour prévôtale du Rhône en 1816.* Paris: Béchet aîné, 1822.

Baeyens, Jacques. *Sabre au clair: Amable Humbert, des Vosges à la Louisiane, 1789–1823.* Paris: Éditions Albatros, 1981.

Bainville, Jacques. *Napoleon.* Paris: Fayard/Le livre de Poche, 1955.

Ball, Timothy Horton. *A Glance into the Great South-East, or, Clarke County, Alabama and Its Surroundings from 1540 to 1877.* 1882. N.p.: Clarke County Historical Society, 1994.

Baltimore City Directory for 1817–18. Baltimore: J. Kennedy, 1817.

Balzac, Honoré de. *La rabouilleuse.* 1842. Paris: Folio Classique-Gallimard, 1972.

Barante, Prosper Brugière baron de. *Souvenirs du baron de Barante.* Vol. 2. Paris: Calmann-Lévy, 1890–97.

Barefield, Marilyn Davis. *Old Demopolis Land Office Records and Military Warrants 1818–1860 and Records of the Vine and Olive Colony.* Easley, S.C.: Southern Historical Press, 1988.

Barou, Joseph. *Le colonel Michel Combe.* Special issue of Village de Forez, Montbrison, 1995.

———. "Le colonel Michel Combe, 1787–1837." *Bulletin de la Diana* 53 (1993). www .forezhistoire.free.fr/images/colonelmichelcombe.pdf.

Barratt, Norris S., and Julius F. Sachse. *Freemasonry in Pennsylvania.* Vol. 3. Philadelphia, 1919.

Bazin, Jean-François. "Bouteilles à la mer: Le vin de Bourgogne durant l'expédition d'Égypte ou le singulier destin de Jean-Baptiste Jame." Association bourguignonne des sociétés savantes. *Les Bourguignons et le Levant.* 69ᵉ congrès de l'Association Bourguignonne des

Sociétés Savantes Dijon-Auxerre, October 23–25, 1998 (Dijon: Association bourguignonne des sociétés savantes; Auxerre: Société des sciences historiques et naturelles de l'Yonne, 2000), 169–79.

Beauchamp, Alphonse de. *La Duchesse d'Angoulême à Bordeaux, ou relation circonstanciée des événemens politiques dont cette ville a été le théâtre en mars 1815, suivie du rapport inédit de M. le comte de Lynch, maire de Bordeaux, sur ces mêmes événemens.* Versailles: J.-A. Lebel, 1816.

Beauchef, Georges. *Mémoires pour server à l'indépendance du Chili.* Paris: La Vouivre, 2001.

Beauchet-Filleau, Paul, Henri Beauchet-Filleau, and Joseph Beauchet-Filleau. *Dictionnaire historique et généalogique des familles du Poitou.* 2nd ed. Vol. 4. Poitiers: impr. P. et O. Lussaud frères, 1963.

Beaucour, Fernand. "Dans le sillage de la Révolution française à l'ombre de Napoléon: Le general Nicolas Raoul (1788–1850)." Levallois: Centre d'études napoléoniennes, 1989.

———. *Un fidèle de l'Empereur en son époque: Le général Jean-Mathieu-Alexandre Sari (1792–1862).* Vol. 1, part 1. Paris: Société de sauvegarde du Château impérial de Pont de Briques, 1972. 3 parts in 5 vols.

———. "Joseph Bonaparte, ex-roi d'Espagne, s'installe aux États-Unis en 1815." Levallois: Centre d'études napoléoniennes, 1995.

———. "Moreau aux Etats-Unis." Centre d'études Napoléoniennes (1990–91): 286–87.

———. "Les projets de délivrance de Napoléon de Sainte-Hélène." 31 typed folios. Levallois: Centre d'études napoléoniennes, 1985.

———. "Qui était Roul? Premier Officier d'Ordonnance de l'Empereur à l'île d'Elbe (1775–1840)." Nice: impr. A. Colle, 1966.

———. [Vandamme's abandoned château]. *Bulletin historique de la Société de sauvegarde du château imperial de Pont-de-Briques* 2 (1969): 121.

Beauharnais, Hortense de. *Mémoires de la reine Hortense.* Paris: Plon, 1927.

Bécamps, P. "Les frères Faucher: La commemoration du bi-centenaire de leur naissance à La Réole." *Revue historique de Bordeaux* 9 (1960): 271–75.

Bellet, B-L., and Auguste Imbert. *Biographie des condamnés pour délits politiques depuis la restauration des Bourbons en France jusqu'en 1827.* Brussels: Aug. Imbert, 1827.

Bénézit, Emmanuel. *Dictionnaire critique et documentaire des peintres, sculpteurs, dessinateurs et graveurs de tous les temps et de tous les pays.* Paris: Gründ, 1999.

Bénot, Yves. *La démence coloniale sous Napoléon.* Paris: La découverte, 1991.

Bénot, Yves, and Marcel Dorigny, eds. *1802: Le rétablissement de l'esclavage dans les colonies françaises.* Colloquium U Paris VIII, June 20–22. Paris: Maisonneuve et Larose, 2003.

Bentley, Elizabeth Petty, ed. *Passenger Arrivals at the Port of New York (1820–1829).* Baltimore: Genealogical Publishing Company, 1999.

Bertaud, Jean-Paul. *Le duc d'Enghien.* Paris: Fayard, 2001.

Bertier, Ferdinand de. *Souvenirs inédits d'un conspirateur.* Paris: Tallandier, 1990.

Bertier de Sauvigny, Guillaume de. *Histoire de la Restauration.* Paris: Flammarion, 1999.

Bertin, Georges. *1815–1832: Joseph Bonaparte en Amérique.* Paris: Librairie de la "Nouvelle Revue," 1893.

Bertrand, Henri-Gatien. *Cahiers de Sainte-Hélène.* Vols. 1 and 2. Paris: Sulliver; A. Michel, 1949–59.

Bertrand, Jean-Claude, Jean Heffer, and André Kaspi. *La civilisation américaine*. 4th ed. Paris: Presses universitaires de France, 1993.

Bigard, Louis. *Le Comte Réal ancien Jacobin (De la Commune révolutionnaire de Paris à la Police générale de l'Empire*. Paris: Firmin-Didot; Versailles, 1937.

Biographie des journalistes, avec la nomenclature de tous les journaux, et les mots d'argot de ces messieurs par une société d'écrivains qui ont fait tous les métiers, et qui se sont plies à toutes les circonstances. Paris: Imprimerie d'Auguste Barthélémy, 10 rue des Grands-Augustins, 1826.

Blaufarb, Rafe. "Alabama's Vine and Olive Colony: Myth and Fact." *Alabama Heritage* 81 (Summer 2006): 26–34.

——. *Bonapartists in the Borderlands: French Exiles and Refugees on the Gulf Coast (1815–1835)*. Tuscaloosa: University of Alabama Press, 2005.

——. "French Consular Reports on the Association of French Emigrants: The Organization of the Vine and Olive Colony." *Alabama Review* 56, no. 2 (2003): 104–24.

——. "The Western Question: The Geopolitics of Latin American Independence." *American Historical Review* 112, no. 3 (2007): 742–63. http://www.historycooperative.org/journals/ahr/112.3/blaufarb.html.

Blondy, Alain. "Les ultra-royalistes bordelais." Diss. Bordeaux III, 1978.

Bluche, Frédéric. *Le bonapartisme: Aux origines de la droite française (1800–1850)*. Paris: Nouvelles Éditions Latines, 1980.

Bonaparte, Joseph. *Le roi Joseph Bonaparte: Lettres d'exil inédites*. Ed. Hector Fleischmann. Paris: Charpentier et Fasquelle, 1912.

Bonaparte, Napoléon. *Napoléon Bonaparte: Oeuvres littéraires et écrits militaires*. Ed. Jean Tulard. Vol. 1. Paris: C. Tchou, 2001.

Bonnal, Edmond. *Manuel et son temps: Étude sur l'opposition parlementaire sous la Restauration*. Paris: E. Dentu, 1877.

Bonnel, Ulane. *La France, les États-Unis et la guerre de course (1797–1815)*. Paris: Nouvelles Éditions Latines, 1961.

Bordeaux, la Guyenne et les États-Unis (1750–1820). Bordeaux: Archives départementales de la Gironde, 1987.

Bordonove, Georges. *La vie quotidienne de Napoléon en route vers Sainte-Hélène*. Paris: Hachette, 1977.

Borel, Raymond C. *Death at French Creek*. New York: McGraw-Hill, 1975.

Bos, Harriet P. "Barthélémy Lafon." Master's thesis. Tulane University, 1977.

Boucaud, Serge. "Les armements nantais: Afrique-Amérique, 1815–1822." Master's thesis, University of Nantes, 1987–88.

Boudon, Jacques-Olivier, ed. *Napoléon et les lycées*. Actes du colloque tenu à la Bibliothèque Marmottan, November 15–16, 2002. Paris: Nouveau Monde éditions, 2004.

——. *Le roi Jérôme, frère prodigue de Napoléon*. Paris: Fayard, 2008.

Bourke, Edward J. *Shipwrecks of the Irish Coast, 1105–1993*. Vols. 1 and 2. Dublin: E. J. Bourke, 1994, 1998.

Bourrachot, L., and J.-P. Poussou. "Bordeaux et l'émigration foyenne au XVIIIe siècle." *Actes du XIXe Congrès, Fédération Hist. Sud-Ouest*. May 1966, 92–112.

Bowes, Mary M ."Searching for the Best Wines." In *Jefferson and Wine: Model of Moderation,* ed. R. De Treville Lawrence III, 220–38. The Plains, Va.: Vinifera Wine Growers Association, 1989.

Brice, Médecin général Raoul. *Les espoirs de Napoléon à Sainte-Hélène.* Paris: Payot, 1938.

Britton, William H. "French Refugee Chaudron's Silver Superb." *Antique Monthly,* June 1974, 19A.

Bronne, Carlo. *L'Amalgame: La Belgique de 1814 à 1830.* Brussels: A. Goemaere, 1948.

Brown, Virginia Pounds, and Laurella Owens. *The World of the Southern Indians.* Birmingham: Beechwoods Books, 1983.

Brun, Maurice. *Le banquier Laffitte.* Abbeville: F. Paillart, 1997.

Bruns, Mrs. Thomas Nelson Carter. *Louisiana's Portraits.* New Orleans: National Society of the Colonial Dames of America in the State of Louisiana, 1975.

Bruyère-Ostells, Walter. *La grande armée de la liberté.* Paris: Tallandier, 2009.

———. "Les officiers de la Grande Armée dans les mouvements nationaux et libéraux (1815–1833)." Diss. U. Paris IV, 2005.

———. "L'opposition bonapartiste sous la 1ère Restauration: Le complot du Nord." Master's thesis. U. Paris-IV, 1966.

Bunle, Henri. "L'immigration française aux Etats-Unis." *Bulletin de la statistique générale de la France et du service d'observation des prix* 14 (January 1925): 199–222.

Butel, Paul. "Guerre et commerce: L'activité du port de Bordeaux sous le régime des licences, 1808–1815." *Revue historique de Bordeaux* 19 (1972): 128–49.

———. "Le temps des revers: La Révolution et l'Empire." In *Histoire de Bordeaux,* ed. Robert Étienne, 173–75. Toulouse: Privat, 1990.

Butler, James L., and John L. Butler. *Indiana Wine: A History.* Bloomington: Indiana University Press, 2001.

Byington, Cyrus. *A Dictionary of the Choctaw Languages.* Ed. John R. Swanton and Henry S. Halbert. 1915. Washington, D.C.: G.P.O., 1976.

Carmer, Carl. *Stars Fell on Alabama.* New York: Farrar & Rinehart, 1934; Tuscaloosa: University of Alabama Press, 2000.

Carrigan, Jo Ann. *The Saffron Scourge: A History of Yellow Fever in Louisiana, 1796–1905.* Lafayette: Center for Louisiana Studies, University of Southwestern Louisiana, 1994.

Carter, Clarence Edwin. *The Territorial Papers of the United States.* Vol. 6, *The Territory of Mississippi, 1809–1817.* Washington, D.C.: G.P.O., 1938. Vol. 18, *The Territory of Alabama, 1817–1819.* Washington, D.C.: G.P.O., 1952. Vol. 22, *The Territory of Florida, 1821–1824.* Washington, D.C.: G.P.O., 1956.

Cartron, Michel Bernard. *Le roi inattendu: Louis XVIII en 1814.* Paris: Sicre, 2001.

Casenave, Georges. "Les émigrés bonapartistes de 1815 aux États-Unis." *Revue d'Histoire Diplomatique* 43 (1929): 20–32.

Castellanos, Henry C., and George F. Reinecke. *New Orleans as It Was: Episodes of Louisiana Life.* 1895. Baton Rouge: Published for the Louisiana American Revolution Bicentennial Commission by the Louisiana State University Press, 1978.

Cauna, Jacques de. *L'Eldorado des Aquitains: Gascons, Basques et Béarnais aux Iles d'Amérique (XVIIe–XVIIIe siècles).* Biarritz: Atlantica, 1998.

———. "La franc-maçonnerie burdigalo-aquitaine et ses réseaux à Saint-Domingue (XVIIIe siècle)." In *Le monde créole: Peuplement, sociétés, et condition humain: Mélanges Hubert Gerbeau,* ed. Jacques Weber, 125–37. Paris: Les Indes Savantes, 2005.

Cayre, André. *Sainte-Foy et ses environs dans le passé.* Sainte-Foy: Laulan, 1967.

Chappet, Alain, Roger Martin, Alain Pigeard. *Le guide Napoléon: 4 000 lieux pour revivre l'épopée.* Paris: Tallandier, 2005.

Chardigny, Louis. *Les maréchaux de Napoléon.* Paris: Tallandier, 2003.

Chateaubriand, François-René de. *Atala, ou les amours de deux sauvages dans le désert.* Paris, 1801.

———. *Atala. René.* Trans. Irving Putter. Berkeley: University of California Press, 1952.

———. *De Buonaparte et des Bourbons, et de la nécessité de se rallier à nos princes légitimes pour le bonheur de la France et celui de l'Europe.* Paris: Le Normand, 1814.

———. *Grands écrits politiques.* 2 vols. Paris: Imprimerie Nationale, 1993.

———. *Mélanges politiques.* In *Œuvres complètes de Chateaubriand.* Vol. 11. Paris: P.-H. Krabbe, 1851.

———. *Mémoires d'outre-tombe.* Vol. 2. Pléiade ed. Paris: Gallimard, 1946–48.

———. *Politique, opinions et discours.* In *Œuvres complètes de Chateaubriand.* Vol. 14. Paris: P.-H. Krabbe, 1851.

Chaudron, Jean-Simon. *Éloge funèbre de l'Empereur Bonaparte, Prononcé en Présence des Membres d'Une Association Française sur les Confins de l'Amérique Civilisée.* Philadelphia: J.F. Hurtel, Printer, at the Corner of Second and Dock Streets, 1823.

———. *Oraison funèbre du Frère George Washington prononcée le premier Janvier 1800, dans la Loge Française l'Aménité.* Philadelphia: John Ormond, 41 Chestnut Street, 1800.

———. *Poésies choisies de Jean Simon Chaudron, suivies de l'Oraison funèbre de Washington.* Paris: impr. De E.-B. Delanchy, 1841.

Chevalier, Jean-Michel. *Souvenirs des guerres napoléoniennes.* Ed. Jean Mistler and Hélène Michaud. Paris: Hachette, 1970.

Childs, Frances Sergeant. *French Refugee Life in the United States: An American Chapter of the French Revolution.* Baltimore: Johns Hopkins Press, 1940.

Chinard, Gilbert. *L'Amérique et le rêve exotique dans la littérature française au XVIIe et au XVIIIe siècle.* Paris: Hachette, 1913.

———. *L'exotisme américain dans l'oeuvre de Chateaubriand.* Paris: Hachette, 1918.

Chopin, Thierry. *La république une et indivisible: Les fondements de la fédération américaine.* Paris: Plon, 2002.

Choquette, Leslie. *Frenchmen into Peasants: Modernity and Tradition in the Peopling of French Canada.* Cambridge: Harvard University Press, 1997.

Claiborne, John Francis Hamtramck. *Mississippi as a Province, Territory, and State, with Biographical Notes of Eminent Citizens.* 1880. Baton Rouge: Published for the Mississippi Historical Society by Louisiana State University Press, 1964.

Clarke, T. Wood. *Émigrés in the Wilderness.* New York: Macmillan, 1941.

Clauzel, Bertrand (Maréchal Comte). *Correspondance du maréchal Clauzel, gouverneur général des possessions françaises dans le Nord de l'Afrique, 1835–1837.* 2 vols. Paris: Larose, 1948.

———. *Exposé justificatif de la conduite politique de M. le lieutenant-général Cte Clausel depuis le rétablissement des Bourbons en France jusqu'au 24 juillet 1815* . . . Paris: Pillet, 1816.

Clauzel, Comte. "Bertrand Clauzel, maréchal de France." *Le Souvenir Napoléonien* 292 (March 1977): 27.

Clay, Henry. *Papers.* Ed. James F. Hopkins and Mary W. M. Hargreaves. Vol. 2. Lexington: University of Kentucky Press, [1959]–88.

Cobbs, Hammer. "Geography of the Vine and Olive Colony." *Alabama Review* 14, no. 2 (1961): 87–89.

Cobbs, Nicholas H., Jr. "Alabama's 'Wonder of the Earth.'" *Alabama Review* 49, no. 3 (1996): 163–80.

Coleman, John S., Jr. "Banished Bonapartists in Alabama." Unpublished essay, 1948.

Collaveri, François. *Napoléon franc-maçon?* 2nd ed. Paris: Tallandier, 2003.

Collomp, Catherine, and Menéndez, Mario, eds. *Exilés et réfugiés politiques aux États-Unis, 1789–2000.* Paris: Éditions du CNRS, 2003.

Connelly, Owen. *The Gentle Bonapart: A Biography of Joseph, Napoleon's Elder Brother.* New York: Macmillan, 1968.

Constant, Benjamin. *Mémoires sur les Cent-Jours.* Ed. Olivier Pozzo de Borgo. Paris: Pauvert, 1961.

Cornet, Marc. "Le maréchal Grouchy (1766–1847): Des guerres de Vendée à Waterloo." *Napoléon Ier.* Special edition 2 (April 2005).

Corriger, Jean. *Sainte-Foy-la-Grande: Son histoire.* Sainte-Foy-la-Grande, 1976.

Cossé-Brissac, René de. *Historique du 7e régiment de Dragons.* Paris: Éd. Leroy, 1909.

Cotteril, R. S. *The Southern Indians: The Story of the Civilized Tribes before Removal.* Norman: University of Oklahoma Press, 1954.

Courteault, P. "Les origines du lycée de Bordeaux." *Le centenaire du Lycée de Bordeaux.* Bordeaux, 1905.

Craighead, Erwin. *From Mobiles's Past: Sketches of Memorable People and Events.* Mobile: Powers Printing Company, 1925.

Creagh, Ronald. *Nos cousins d'Amérique: Histoire des Français aux Etats-Unis.* Paris: Payot, 1988.

———. "La Révolution et les Français des Etats-Unis." In *Les Français aux Etats-Unis, d'hier à aujourd'hui,* ed. Ronald Creagh, 165–88. Montpellier: Éditions Espaces 34, U Montpellier III, 1994.

Creek, Alma Burner. "Bombyx Mori and Americans: Or, Here We Go Round the Mulberry Bush." *University of Rochester Library Bulletin* 39 (1986). http://www.lib.rochester.edu/index.cfm?PAGE=2407.

Crouzet, François. "Bilan de faillite: La ruine du grand commerce." In *Histoire de Bordeaux,* vol. 5, *Bordeaux au XVIIIe siècle,* 485–510. Bordeaux: Fédération historique du Sud-Ouest, 1968.

Cunz, Dieter. *The Maryland Germans: A History.* Princeton: Princeton University Press, 1948.

Curtin, Philip. *The Atlantic Slave Trade: A Census.* Madison: University of Wisconsin Press, 1969.

Cushman, Horatio Bardwell. *History of the Choctaw, Chickasaw and Natchez Indians.* 1899. Norman: University of Oklahoma Press, 1999.

Daget, Serge. *Le répertoire des expeditions négrières françaises à la traite illégale.* Nantes: Centre de recherche sur l'histoire du monde atlantique: Comité nantais d'études en sciences humaines, 1988.

———. *La répression de la traite des noirs au XIXe siècle: L'action des croisières françaises sur les côtes occidentales de l'Afrique, 1817–1850.* Paris: Karthala, 1997.

Dalbaret, Charles. *Un assassinat juridique (1815): Les généraux Faucher ou les jumeaux de La Réole fusillés à Bordeaux sous la terreur blanche.* Paris: A. Bellier, 1894.

Dallas, George Mifflin. *Life and Writings of Alexander James Dallas.* Philadelphia: Lippincott, 1871.

Dallett, Francis James. "The French Benevolent Society of Philadelphia and the Bicentennial." *Records of the American Catholic Historical Society* 90 (March–December 1979): 61–68.

Dangerfield, George. *The Awakening of American Nationalism (1815–1827).* New York: Harper Torchbooks, 1965.

———. *The Era of Good Feelings.* 1952. Chicago: Elephant Paperback, 1989.

Daudet, Ernest. *L'exil et la mort du general Moreau.* Paris: Librairie Hachette, 1909.

Daumard, Adeline. "Une référence pour l'étude des sociétés urbaines en France au XVIIIe et XIXe siècles: Projet de code socio-professionnel." *Revue d'Histoire moderne et contemporaine* 10 (July–September 1963): 185–210.

Davis, Nora Marshall. "The French Settlement at New Bordeaux." *Transactions of the Huguenot Society of South Carolina* 56 (1951): 28–57.

Davis, William C. *Way through the Wilderness: The Natchez Trace and the Civilization of the Southern Frontier.* New York: HarperCollins, 1995; Baton Rouge: Louisiana State University Press, 1996.

Dawson, John Charles. "A French Exile in Alabama [Col. Raoul]." *Alabama Historical Quarterly* 8, no. 1 (1946): 41–43.

———. *A French Regicide in Alabama (1824–1837).* Tuscaloosa: University of Alabama Press, 1939.

———. *Lakanal the Regicide: A Biographical and Historical Study of the Career of Joseph Lakanal.* Tuscaloosa: University of Alabama Press, 1948.

Debo, Angie. *The Rise and Fall of the Choctaw Republic.* 2nd ed. Norman: University of Oklahoma Press, 1967.

Degeorge, Frédéric. "Les proscrits de la Restauration." In *Paris Révolutionnaire,* ed. Godefroy Cavaignac 181–96. Paris: Pagnerre, 1848.

De Meulenaere, Philippe. *Bibliographie analytique des témoignages oculaires imprimés de la campagne de Waterloo.* Paris: Éditions historiques Teissèdre, 2004.

DeRosier, Arthur H., Jr. *The Removal of the Choctaw Indians.* Knoxville: University of Tennessee Press, 1970.

Desnouettes, Charles Lefebvre, and William Harris Crawford. *Report of the Secretary of the Treasury on the Petition of Lefebvre Desnoettes, and Others, French Emigrants in Alabama, engaged in the Cultivation of the Vine and Olive, praying a modification of the condition of their grant.* Washington, D.C.: Gales and Seaton, 1822.

De Treville Lawrence R. III, ed. *Jefferson and Wine: Model of Moderation.* The Plains, Va.: Vinifera Wine Growers Association, 1989.

Diocese of Baton Rouge: Catholic Church Records. Vol. 3 (1804–19). Baton Rouge: The Diocese, 1978.

Dodd, Donald B., and Wynette S. Dodd. *Historical Statistics of the South, 1790–1970.* University: University of Alabama Press, 1973.

Doher, Marcel. *Proscrits et exilés après Waterloo.* Paris: J. Peyronnet, 1965.

Dossios-Pralat, Odette. "Michel Regnaud de Saint-Jean d'Angély, éminence grise sous le Consulat et l'Empire." Vol. 3. Diss. U Paris-IV Sorbonne, 2002.

———. *Michel Regnaud de Saint-Jean d'Angély, serviteur fidèle de Napoléon,* Paris: Éditions historiques Teissèdre, 2007.

Douglass, Frederick. *Narrative of the Life of Frederick Douglass, An American Slave, Written by Himself.* Ed. David W. Blight. Boston: Bedford Books of St. Martin's Press, 1993.

Douyrou, Marcel. "Jacques Laffitte, le roi des banquiers, le banquier des rois." *Cercle Généalogique du Pays Basque et Bas-Adour* (Bayonne) 10, no. 2 (1991): 2–17.

Drouët d'Erlon, Jean-Baptiste. *Vie militaire écrite par lui-même et dédiée à ses amis.* Paris: Barba, 1844.

DuBose, John Witherspoon. "Chronicles of the Canebrake." *Alabama Historical Quarterly* 9, no. 4 (1947): 474–91.

Du Casse, Albert. *Le général Vandamme et sa correspondence.* Vol. 2. Paris: Didier, 1870.

Duffy, John. "Nineteenth Century Public Health in New York and New Orleans: A Comparison." *Louisiana History* 15 (Fall 1974): 325–37.

Dufour, John James. *The American Vine-Dresser's Guide being a treatise of the Cultivation of the Vine and the Process of Wine Making adapted to the soil and climate of the United States.* 1826. Ed. Carol Louise Hartman and annotated by Pierre Galet. Vevey, Switzerland: La Valsainte and West Lafayette, Ind.: Purdue University Press, 2003.

Dufour, Perret. *The Swiss Settlement of Switzerland County, Indiana.* Indianapolis: Indiana Historical Commission, 1925.

Dugenne, Paul-Camille. *Dictionnaire biographique, généalogique et historique du département de l'Yonne.* Vol. 4. Auxerre: Société généalogique de l'Yonne, 1996–2003.

Duhamel, Jean. *Les cinquante jours de Waterloo à Plymouth.* Paris: Plon, 1963.

Dulaure, Jacques-Antoine. *Histoire des Cent-Jours, de la Restauration et de la Révolution de 1830.* Vols. 3 and 4. Paris: Poirée, 1845.

Dumas, Alexandre. *Mes mémoires.* Vol. 2. Paris: Michel Lévy frères, 1863.

Dumont, William H. "French Emigrants to Richmond County, Georgia." *National Genealogical Society Quarterly* 52, no. 2 (1964): 74–78.

Du Ponceau, Peter Stephen. "English Phonology; or an Essay towards an Analysis and Description of the Component Sounds of the English Language." *Transactions of the American Philosophical Society* ns 1 (1818): 228–64.

Dupont, Marcel. *Napoléon et la trahison des maréchaux (1814).* Paris: Hachette, 1939.

Dusmukes, Camillus J. "The French Colony in Marengo County, Alabama." *Alabama Historical Quarterly* 22, nos. 1–2 (1970): 81–113.

Duvivier, Paul. *Les anciens conventionnels sous la Restauration: L'exil de Cambacérès à Bruxelles (1816–1818).* Paris: Picard, 1923.

———. *L'exil du comte Merlin dans les Pays-Bas.* Malines: Impr. L. et A. Godenne, 1911.

———. *L'exil du comte Sieyès à Bruxelles (1816–1830).* Malines: L. et A. Godenne, 1910.

Ébeyer, Pierre Paul. *Revelations concerning Napoleon's Escape from St. Helena.* New Orleans: Windmill, 1947.

Egan, Clifford L. *Neither Peace Nor War.* Baton Rouge: Louisiana State University Press, 1983.

Emerson, O. B. "The Bonapartist Exiles in Alabama." *Alabama Review* 11 (April 1958): 135–43.

État détaillé des liquidations opérées par la commission chargée de répartir l'indemnité attribuée aux anciens colons de Saint-Domingue . . . 1829.

Eymery, A., P.-J. Charrin, A. S. C. Tastu, R. Périn, C. de Proisy d'Eppe. *Dictonnnaire des girouettes, ou nos Contemporains peints d'après eux-mêmes . . . par une Société de girouettes.* 3rd ed. Paris: A. Eymery, 1815.

Farrow, Anne, Joel Lang, and Jenifer Frank. *Complicity: How the North Promoted, Prolonged, and Profited from Slavery.* New York: Ballantine Books, 2005.

Faucher, Casimir. *Procès des frères Faucher de La Réole, morts en 1815, victimes de la fureur des partis.* Bordeaux: Les principaux libraires, 1830.

Faure, Victor. *De la Corrèze à la Floride: Jean Augustin Pénières-Delzors, conventionnel et député d'Ussel (1766–1821).* Ussel: Musée de Pays d'Ussel, 1989.

Ferenczi, Imre. *International Migrations.* Vol. 1, *Statistics.* New York: National Bureau of Economical Research, 1929.

Ferguson, Robert B. "Treaties between the United States and the Choctaw Nation." In *The Choctaw before Removal,* ed. Carolyn Keller Reeves, 214–60. Jackson: University Press of Mississippi, 1985.

Filby, P. William, ed. *Philadelphia Naturalization Records.* Detroit: Gale, 1982.

Florange, Jules. *Le Conventionnel Hentz, député de la Moselle.* Metz: Librairie Sidot-Vannière, 1911.

Fogel, Robert William, and Stanley L. Engerman. *Time on the Cross: The Economics of American Negro Slavery.* 2nd ed. London: Wilwood House, 1976.

Fohlen, Claude. *L'Amérique anglo-saxonne de 1815 à nos jours.* Paris: Presses universitaires de France, 1965.

———. *Histoire de l'esclavage aux États-Unis.* Paris: Perrin, 1998.

———. *Thomas Jefferson.* Nancy: Presses Universitaires de Nancy, 1992.

———. *Thomas Jefferson à Paris: 1784–1789.* Paris: Perrin, 1995.

Fortier, Alcée. *A History of Louisiana.* Vol. 3. New York: Goupil and Company of Paris, Manzi, Joyant and Company, successors, 1904.

Foscue, Virginia O. *Place Names in Alabama.* Tuscaloosa: University of Alabama Press, 1989.

Fouché, Joseph. *Mémoires de Joseph Fouché, duc d'Otrante, ministre de la police générale.* 2nd ed. Vol. 2. Paris: Lerouge, 1824.

Fouché, Nicole. *Émigration alsacienne aux États-Unis, 1815–1870.* Paris: Publications de la Sorbonne, 1992.

———. "Les passeports délivrés à Bordeaux pour les États-Unis de 1816 à 1889." In *L'émigration française: Études de cas: Algérie, Canada, Etats-Unis,* 189–210. Paris: Publications de la Sorbonne, 1985.

Foucrier, Annick. *The French and the Pacific World, 17th–19th Centuries: Explorations, Migrations and Cultural Exchanges.* Aldershot, Hampshire, England; Burlington, VT: Ashgate, 2005.

———. *Le rêve californien: Migrants français sur la côte Pacifique (XVIIIe–XIXe siècles).* Paris: Belin, 1999.

Fournier, Marcel. *Les Bretons en Amérique française.* Rennes: Éditions Les Portes du Large, 2005.

Furer, Howard B. *The Germans in America, 1607–1970.* Dobbs Ferry: Oceana Publications, 1973.

Gabler, James M. *Passions: The Wines and Travels of Thomas Jefferson.* Baltimore: Bacchus Press, 1995.

Gaines, George Strother. *The Reminiscences of George Strother Gaines, Pioneer and Statesman of Early Alabama and Mississippi, 1805–1843.* Ed. James P. Pate. Tuscaloosa: University of Alabama Press, 1998.

Galtier, G. "La viticulture de l'Europe occidentale à la veille de la Révolution française, d'après les notes de voyage de Thomas Jefferson." *Bulletin de la Société Languedocienne de Géographie* 3 (1968): 43–86.

Gandrud, Paula Myra Jones. *Marriage Records of Marengo County, Alabama, 1818–1860.* Memphis: Milestone Press, 1970.

Gardien, Kent. *The Châtelaine of George Villa.* Demopolis, Ala.: Marengo County Historical Society, 1979.

———. "The Domingan Kettle: Philadelphian-Émigré Planters in Alabama." *National Genealogical Society Quarterly* 76, no. 3 (1988): 173–88.

———. "The Splendid Fools: Philadelphia Origins of Alabama's Vine and Olive Colony." *Pennsylvania Magazine of History and Biography* 104 (October 1980): 491–507.

———. "Take Pity on Our Glory: Men of Champ d'Asile." *Southwestern Historical Quarterly* 87, no. 3 (1984): 241–68.

Garnier de Saintes, Jacques. *La dette d'un exilé, ou plan nouveau d'éducation nationale. . . .* Brussels: A. Wahlen, printer-bookseller, 1816.

Garrigoux, Jean. *Un aventurier visionnaire: Arsène Lacarrière Latour (1778–1837).* Aurillac: Société "La Haute-Auvergne," 1997.

Gayarré, Charles. *Aubert Dubayet; or, the Two Sister Republics.* 1882. Whitefish, MT: Kessinger Publishing, 2007.

Geggus, David P., ed. *The Impact of the Haitian Revolution in the Atlantic World.* Columbia: University of South Carolina Press, 2001.

Gehman, Mary. *The Free People of Color in New Orleans.* New Orleans: Margaret Media, 1994.

Le General Antoine Rigau, 1758–1820. Paris: Quantin, 1886.

Gérard, Alain. "Un 'dur' de la grande Armée: Vandamme." *Le Souvenir Napoléonien* 318 (July 1981): 2–29.

Gerber, David A. *Authors of Their Lives: The Personal Correspondence of British Immigrants to North America in the Nineteenth Century.* New York: New York University Press, 2006.

Germain, Pierre. *J.-B. Drouët d'Erlon: Maréchal de France, général comte d'Empire, premier gouverneur de l'Algérie.* Paris: Fernand Lanore, 1985.

Godechot, Jacques. *L'Europe et l'Amérique à l'époque napoléonienne (1800–1815).* Paris: Nouvelle Clio–Presses universitaires de France, 1967.

Godlewski, Guy. "Emmanuel de Grouchy." *Le Souvenir Napoléonien* 329 (May 1983): 13–15.

———. *Napoléon à l'île d'Elbe: 300 jours d'exil.* Paris: Nouveau Monde éditions. Fondation Napoléon, 2003.

Goodell, William. "Mission among the Choctaws." *Missionary Herald* 18 (1922): 223–24.

Gotteri, Nicole. *Le maréchal Soult.* Paris: Bernard Giovanangeli, 2000.

Gourgaud, Gaspard. *Journal de Sainte-Hélène (1815–1818).* Vol. 1. Paris: Flammarion, 1947.

Gravatt, Patricia. *L'église et l'esclavage.* Paris: L'Harmattan, 2003.

Griffith, Lucille. *Alabama: A Documentary History to 1900.* University: University of Alabama Press, 1968.

Groppo, Bruno. "Exilés, réfugiés, émigrés, immigrés: Problèmes de définition." In *Exilés et réfugiés politiques aux Etats-Unis, 1789–2000,* ed. Catherine Collomp and Mario Menéndez, 19–30. Paris: CNRS Éditions, 2003.

Gros and Delettrez catalog for Drouot Richelieu auction, October 11, 2004.

Grouchy, Emmanuel-Henri de. *Mémoires du maréchal de Grouchy.* Vols. 4 and 5. Paris: E. Dentu, 1873–74.

Guice, John D. W. "Face to Face in Mississippi Territory, 1798–1817." In *The Choctaw before Removal,* ed. Carolyn Keller Reeves, 157–80. Jackson: University Press of Mississippi, 1985.

Guillet, Bertrand, and Louise Pothier, eds. *France, Nouvelle-France: Naissance d'un peuple français en Amérique.* Nantes: Musée du château des ducs de Bretagne, 2005.

Guillon, Edouard, *Les complots militaires sous la Restauration, d'après les documents des Archives.* Paris: Librairie Plon, Nourrit et Cie, 1895.

Guillot, Lucien. "Épisodes de la vie du général Lefebvre-Desnoëttes." *Le Souvenir Napoléonien* 377 (June 1991): 2–18.

———. "Une évasion romanesque." *Souvenir Napoléonien* 377 (June 1991): 16–18.

———. *Le général Lefebvre-Desnoëttes, 1773–1821.* Paris: n.p., 1961.

Guizot, François. *Memoirs to Illustrate the History of My Time.* Vol. 1. London: Richard Bentley, 1858.

Gutek, Gerald Lee. *Joseph Neef: The Americanization of Pestalozzianism.* University: University of Alabama Press, 1978.

Hackensmith, Charles William. *Biography of Joseph Neef, Educator in the Ohio Valley, 1809–1854.* New York: Carlton Press, 1973.

Hamilton, Peter J. "The Beginnings of French Settlement of the Mississippi Valley." *Gulf States Historical Magazine* 1, no. 1 (1902): 1–12.

———. *Mobile of the Five Flags: The Story of the River Basin and Coast about Mobile from the Earliest Times to the Present.* Mobile: Gill Printing Company, 1913.

Harris, W. Stuart. *Dead Towns of Alabama.* University: University of Alabama Press, 1973.

Hartmann, L., and Millard. *Le Texas, ou Notice historique sur le Champ d'Asile, comprenant tout ce qui s'est passé depuis la formation jusqu'à la dissolution de cette colonie.* Paris: Béguin, 1819.

Havard, Gilles, and Cécile Vidal. *Histoire de l'Amérique française.* Paris: Flammarion, 2003.

Hellman, Lillian. *The Little Foxes*. New York: Viking Press, 1961.

Helmer, Dorothy Garr. *Lipscomb: 300 Years in America, 1679–1979. English Background and Some Descendants of Ambrose II, William and John. The Three Sons of Our Immigrant Ancestor Ambrose Lipscomb I, Whose First Record in Virginia Was in 1679.* Indianapolis, 1979.

Henry, Monica. "Henry Clay et la South American Question." *Transatlantica* 1 (2002). http://transatlantica.revues.org/index404.html.

Hewett, Janet. *The Roster of Confederate Soldiers, 1861–1865.* Vols. 4, 9, and 10. Wilmington, N.C.: Broadfoot, 1995–96.

History of Switzerland County, Indiana. Chicago: Weakley, Harraman, 1885.

Holland, Dorothy Garache. *The Gareshe, Bauduy and des Chapelles Families* (St. Louis: Schneider Print Co., 1963).

Holmes, Jack J. D., ed. "Fort Stoddart in 1799: Seven Letters of Captain Bartholomew." *Alabama Historical Quarterly* 26 (Fall–Winter 1964): 231–52.

———. "French and Spanish Cartography in Alabama." *Alabama Historical Quarterly* 27 (Spring–Summer 1965): 7–21.

Horton, Patty McCoy. "Jean Simon Chaudron." *Alabama's Writers Conclave*, 1989.

Houssaye, Henri. *1815.* 7th ed. Paris: Perrin, 1905.

Howe, Daniel Walker. *What Hath God Wrought: The Transformation of America, 1815–1848.* New York: Oxford University Press, 2007.

Howland, William S. "The Oenologist of Monticello." *Jefferson and Wine: Model of Moderation,* ed. R. De Treville Lawrence III, 1–12. The Plains, Va.: Vinifera Wine Growers Association, 1989.

Hulot, Frédéric. *Le général Moreau.* Paris: Pygmalion, 2001.

Hurtado, Fernando Berguño. *Les soldats de Napoléon dans l'indépendance du Chili (1817–1830).* Paris: L'Harmattan, 2010.

Hyde de Neuville, Jean-Guillaume. *Éloge historique du general Moreau.* New York: 1814.

———. *Mémoires et souvenirs du baron Hyde de Neuville.* 3rd ed. 3 vols. Paris: Plon, 1893–1912.

Jamet, L. Special issue of *Les Cahiers du Réolais,* September 1960, 3–4.

Jefferson, Thomas. *The Complete Jefferson: Containing His Major Writings, Published and Unpublished, Except His Letters.* New York: Tudor, 1943.

———. "The Garden Book: Notes on Wines and Vine." In *Jefferson and Wine: Model of Moderation,* ed. R. De Treville Lawrence III, 32–51. The Plains, Va.: Vinifera Wine Growers Association, 1989.

———. *The Jeffersonian Cyclopedia: A Comprehensive Collection of the Views of Thomas Jefferson . . .* Ed. John P. Foley. Vol. 2. New York: Russell & Russell, 1967.

———. "Memoranda Taken on a Journey from Paris into the Southern Parts of France, and Northern of Italy, in the Year 1787." In *The Complete Jefferson: Containing His Major Writings, Published and Unpublished, Except His Letters,* 759–95. New York: Tudor, 1943.

———. *The Works of Thomas Jefferson.* Ed. Paul Leicester Ford. Vol. 11. New York: Putnam, 1904–5.

Jolly, Pierre. *Du Pont de Nemours, soldat de la liberté.* Paris: Presses universitaires de France, 1956.

Jones, Howard Mumford. *America and French Culture: 1750–1848.* Chapel Hill: University of North Carolina Press, 1927.

Jones, Proctor Patterson, ed. *Napoleon: An Intimate Account of the Years of Supremacy*. New York: Distributed by Random House, 1992.

Jouin, Henry. *Lakanal en Amérique, d'après sa correspondance inédite (1815–1837)*. Besançon: impr. Dodivert, 1904.

Jourquin, Jacques. *Dictionnaire des maréchaux du Premier Empire*. 5th ed. Paris: Christian/Jas, 2001.

Jubé, Charles. *Quelques observations sur la situation actuelle des militaires en retraite, par un maréchal-de-camp en retraite*. Paris: Imprimerie de Sétier, 1830.

Kappler, Charles J., ed. *Indian Affairs: Laws and Treaties*. Vol. 2, *Treaties*. Washington, D.C.: G.P.O., 1904.

Kaspi, André. *Les américains*. Vol. 1. *Naissance et essor des Etats-Unis (1607–1945)*. Paris: Points Seuil, 1986.

Kelma, Ari. *A River and Its City: The Nature of Landscape in New Orleans*. Berkeley: University of California Press, 2003.

Kennedy, Paul Michael. *Naissance et déclin des grandes puissances*. Paris: Payot, 1991.

Kennedy, Roger G. *Mr. Jefferson's Lost Cause: Land, Farmers, Slavery, and the Louisiana Purchase*. New York: Oxford University Press, 2003

———. *Orders from France: The American and the French in a Revolutionnary World, 1780–1820*. New York: Knopf, 1989.

Kennet, Lee. "Le culte de Napoléon aux États-Unis jusqu'à la guerre de Sécession." *Revue de l'Institut Napoléon* (January 1972): 145–56.

Khaoua, Myriam. "Conception, vie et mort d'un projet colonial à la fin the XVIIIe siècle: Jean-Louis Gibert et le township Huguenot de New Bordeaux en Caroline du Sud." Master's thesis, University of La Rochelle, 1999.

Kidwell, Clara Sue. *Choctaw and Missionaries in Mississippi, 1818–1918*. Norman: University of Oklahoma Press, 1995.

Kitchens, Bill. "Exiles in Alabama: The French Vine & Olive Colony." *Early American Homes* (Magazine), June, 1999; reprint *Tuskaloosa Magazine* 65, (2000): 12–15; 24–25; 37–39.

Klier, Betje Black. "Champ d'Asile, Texas." In *The French in Texas: History, Migration, Culture*, ed. François Lagarde, 79–97. Austin: University of Texas Press, 2003.

Knox, Julie LeClerc. *The Dufour Saga (1796–1942): The Story of the Eight Dufours Who Came from Switzerland and Founded Vevay, Switzerland County, Indiana*. Crawfordville, Ind.: Howell-Goodwin, 1942.

Kramer, Lloyd S. *Lafayette in Two Worlds: Public Cultures and Personal Identities in an Age of Revolutions*. Chapel Hill: University of North Carolina Press, 1996.

Krebs, A. Article 95. *Dictionnaire de biographie française*. Vol. 12. Paris: Letouzey et Ané, 1970.

Krettly, Élie. *Souvenirs historiques*. Paris: Nouveau Monde/Fondation Napoléon, 2003.

Kuscinski, Auguste. *Dictionnaire des conventionnels*. Brueil-en-Vexin [Yvelines]: Éditions du Vexin français, 1973.

Lafon, Barthélémy. *Carte générale du Territoire d'Orléans comprenant aussi la Floride occidentale et une Portion du Territoire du Mississipi*. Paris: Ch. Picquet, 1806.

Lagarde, François, ed. *The French in Texas: History, Migration, Culture.* Austin: University of Texas Press, 2003.

La Gorce, Pierre de. *La Restauration: Louis XVIII.* Paris: Plon-Nourrit et Cie., 1926.

Lajonie Lapeyre, Jacques. *Mémoires justificatifs pour Jacques Lajonie Lapeyre, membre de la Légion d'honneur, habitant de la commune de Juillac, arrondissement de Libourne.* Paris: J.-B. Imbert, 1816.

Lamoussière, Christiane. *Le personnel de l'administration préfectorale, 1800–1880.* Paris: Centre historique des archives nationales, 1998.

Lancaster, Clay. *Eutaw: The Builders and Architecture of an Ante-Bellum Southern Town.* Eutaw, Ala.: Greene County Historical Society, 1979.

Langlois, Gilles-Antoine. *Des villes pour la Louisiane française.* Paris: L'Harmattan, 2003.

Las Cases, Emmanuel, comte de. *Le mémorial de Sainte-Hélène.* Pléiade ed. 2 vols. Paris: Gallimard, 1956.

La Souchère-Deléry, Simone de. *Napoleon's Soldiers in America.* Paris, 1950. Reprint, Gretna, La.: Pelican, 1998.

———. "Le thème napoléonien dans la poésie louisianaise." *French American Review* 2 (April–June 1949).

La Tour du Pin Gouvernet, Henriette Lucie Dillon, marquise de. *Mémoires de la marquise de La Tour du Pin.* Paris: Mercure de France, 2002.

Latrobe, Benjamin Henry Boneval. *Impressions Respecting New Orleans, 1818–1820.* Ed. Samuel Wilson Jr. New York: Columbia University Press, 1951.

Lavalette, comte de. *Mémoires et souvenirs du comte de Lavalette.* 2 vols. Paris: H. Fournier jeune, 1831.

Lee, William. *A Yankee Jeffersonian: Selections from the Diary and Letters of William Lee of Massachusetts Written from 1796 to 1840.* Ed. Mary Lee Mann. Cambridge: Belknap-Harvard University Press, 1958.

Le Gallo, Émile. *Les Cent-Jours: Essai sur l'histoire intérieure de la France, depuis le retour de l'île d'Elbe jusqu'à la nouvelle de Waterloo.* Paris: Librairie Félix Alcan, 1924.

Lentz, Thierry. "Les relations américano-françaises de la Révolution à la chute de l'Empire (1783–1798)." *Souvenir Napoléonien* 405 (January–February 1996): 7–23.

———. *Savary: Le séide de Napoléon.* Paris: Fayard, 2001.

Lever, Évelyne. *Louis XVIII.* Paris: Fayard, 1988.

Lhéritier, Michel. "Bordeaux et les États-Unis d'Amérique, une fête de l'an II." *Revue Philomathique de Bordeaux et du Sud-Ouest* (1917): 179–80.

Lichine, Alexis. *Alexis Lichine's New Encyclopedia of Wines and Spirits.* 4th ed. New York: Knopf, 1985.

Lineback, Neal, ed. *Atlas of Alabama.* University: University Press of Alabama, 1973.

Long, Campbell. "Tales of Early Demopolis." *Greensboro Watchman,* 1969; reprint *White Bluff Chronicles,* Demopolis, 1978.

Lorédan, Jean. *Madame de Lavalette née Beauharnais.* Paris: Perrin, 1929.

Louis, Jérôme. "La dissolution de l'armée impériale, 1814–1824." *Revue Historique des Armées* 4 (1997): 29–34.

Lowrie, Walter, et al., eds. *American State Papers: Documents Legislative and Executive, of the*

Congress of the United States... Series 8, *Public Lands.* Vol. 3. Washington, D.C.: Gales and Seaton, 1832–61.

Lucas-Lebreton, J. *Le culte de Napoléon (1815–1848).* Paris: Albin Michel, 1960.

———. "Les frères Faucher. . . ." *Revue des Deux Mondes,* August 15, 1956, 589–603.

Lyon, Anne Bozeman. "The Bonapartists in Alabama." *Southern Home Journal,* Memphis, Tennessee, March 1900, reprinted in *Gulf States Historical Magazine,* Montgomery, Alabama, March 1903, and in *Alabama Historical Quarterly* 25 (1963): 227–41.

Madelin, Louis. *Fouché (1759–1820).* Vol. 2. Paris: Plon-Nourrit et Cie, 1923.

Madison, James. *The Writings of James Madison, Comprising His Public Papers and His Private Correspondence.* Vol. 8. New York: Putnam, 1900–10.

Maire, Camille. *L'émigration des Lorrains en Amérique.* Metz: Centre de recherches Relations internationales de l'Université de Metz, 1980.

———. *En route pour l'Amérique: L'odyssée des émigrants en France au XIXe siècle.* Nancy: Presses Universitaires de Nancy, 1993.

———. *Lettres d'Amérique: Des émigrants d'Alsace et de Lorraine écrivent au pays, 1802–1892.* Metz: Serpenoise, 1992.

Marchand, Louis-Joseph. *Mémoires de Marchand: Premier valet de chambre de l'Empereur et exécuteur testamentaire de Napoléon.* 2 vols. Paris: Plon, 1955.

Marino, Samuel J. "The French Refugees Newspapers and Periodicals in the United States, 1789–1825." Diss. University of Michigan, 1962.

Marquiset, Alfred. *Une Merveilleuse (Mme Hamelin).* Paris: H. Champion, 1909.

Martignac, Jean-Baptiste Sylvère de Gaye, vicomte de. *Bordeaux au mois de mars 1815, ou Notice, sur les événemens qui ont précédé le départ de S.A.R. Madame, Duchesse d'Angoulême.* Bordeaux: Lawalle Jeune, n.d.

Martin, H. Christopher. "Jefferson's Italian Vigneron: Philip Mazzei, Revolutionary Patriot." In *Jefferson and Wine: Model of Moderation,* ed. R. De Treville Lawrence III, 18–31. The Plains, Va.: Vinifera Wine Growers Association, 1989.

Martin, Thomas Wesley. *French Military Adventures in Alabama (1818–1828).* Birmingham: Birmingham Publishing Company, 1937.

Martinien, Aristide. *État nominatif des officiers tués et blessés de 1816 à 1911.* Paris: L. Fournier, 1915.

———. *Tableaux, par corps et par batailles, des officiers tués et blesses pendant les guerres de l'Empire (1805–1815).* Paris: Editions Militaires Européennes, n.d.

Marzagalli, Silvia. "Bordeaux et les États-Unis, 1776–1815: Politique et stratégies négociantes dans la genèse d'un réseau commercial." 2 vols. Diss. U. of Paris I–Panthéon Sorbonne, 2004.

Matte, Jacqueline Anderson. *The History of Washington County, First County in Alabama.* Chatom, Ala.: Washington County Historical Society, 1982.

Matte, Jacqueline Anderson, Doris Brown, and Barbara Wadell. *Old St. Stephens: Historical Records Survey.* Rev. ed. St. Stephens, Ala.: The Commission, 1999.

Mayers, M. W. "Geography of the Mississippi Black Prairie." Diss. Clark University, Worcester, Mass., 1948.

Mazel, Geneviève. "Joseph Bonaparte et Mortefontaine." *Revue de l'Institut Napoléon* 4 (1994): 23–71.

McCartney, Clarence, Edward Noble, and John Gordon Dorrance. *The Bonapartes in America*. Philadelphia: Dorrance, 1939.

McCorvey, Thomas Chalmers. "The Vine and Olive: A Sketch of the Colony of French Imperialists That Settled in Marengo County, Alabama." *Times-Democrat*, New Orleans, February, 22, 1885. Reprinted in *Alabama Historical Sketches*, 77–89. Charlottesville: University of Virginia Press, 1960.

———. "The Vine and Olive Colony." *Alabama Historical Reporter*, Tuscaloosa, April 1885.

McDonald, Wynona Bolen. "Jackson: General History." In *Historical Sketches of Clarke County, Alabama*, 198–224. Huntsville, Ala.: Strode, 1977.

McGuire, H. M., Dorothea N. Spear, and T. C. Fay. *Mobile Directory, Embracing Names of the Heads of Families and Persons in Business for 1837*. New Haven: Research Publications, 1969.

McMaster, John Bach. *Life and Times of Stephen Girard, Mariner and Merchant*. 2 vols. Philadelphia: Lippincott, 1918.

Meadows, R. Darell. "Engineering Exile/Social Networks and the French Atlantic Community, 1789–1809." *French Historical Studies* 23, no. 1 (2000): 67–102.

———. "The Planters of St-Domingue, 1750–1804: Migration and Exile in the French Revolutionary Atlantic." Diss. Carnegie Mellon U, 2004.

Meaudre de Lapouyade, Maurice. "Voyage d'un Allemand à Bordeaux en 1801." *Revue Historique de Bordeaux* 5 (1912): 164–81, 229–55.

Melchior-Bonnet, Bernardine. *Un policier dans l'ombre de Napoléon, Savary, duc de Rovigo*. Paris: Perrin, 1962.

Merle d'Aubigné, Guillaume. *La vie américaine de Guillaume Merle d'Aubigné. Extraits de son journal de voyage et de sa correspondance inédite, 1809–1817*. Paris: Droz, 1935.

Mézin, Anne. *Les Consuls de France au siècle des Lumières: 1715–1792*. Paris: Ministère des Affaires Étrangères, 1997.

Michaud, Louis-Gabriel. *Biographie universelle ancienne et moderne*. Vol. 35. Paris, 1843.

Michaux, François-André. *Voyage à l'ouest des monts Alleghany: Dans les états de l'Ohio, du Kentucky, et du Tennessee . . . entrepris pendant l'an X–1802*. Paris: Levrault, 1804. Translated as *Travels to the Westward of the Allegany Mountains, in the States of Ohio, Kentucky, and Tennessee, in the year of 1802*. London, 1805.

Mignet, François-Auguste. *Notice historique sur la vie et les travaux de M. Lakanal: Lue à la séance publique annuelle du 2 mai 1857*. Paris: Institut impérial de France, 1857.

Milbert, Jacques-Gérard. *Itinéraire pittoresque du fleuve Hudson et des parties latérales de l'Amérique du Nord*. Paris: Gaugain, 1828–29.

Mills, Robert. *Atlas of the State of South Carolina, 1825*. Easley, S.C.: Southern Historical Press, 1980.

Moe, Christine. "Yellow Fever in New Orleans." *Louisiana Historical Quarterly* 2nd ser. 7 (1974).

Montholon, Albine de. *Journal secret d'Albine de Montholon, maîtresse de Napoléon à Sainte-Hélène*. Paris: Albin Michel, 2002.

Montholon, Charles de. *Histoire de Napoléon, d'après les mémoires écrits à Sainte-Hélène, sous la dictée de ce prince, par les généraux Montholon et Gourgaud, le Cte Las Cases, le docteur O'Méara. . . .* Vol. 2. Paris: Constant-Chantpie, 1829.

Montlezun, Baron de. *Voyage fait dans les années 1816 et 1817, de New York à la Nouvelle-Orléans et de l'Orénoque au Mississipi par les petites et les grandes Antilles.* Vol. 1. Paris: Gide fils, 1818.

Montrol, François Mongin de. *Histoire de l'émigration (1789–1825).* 2nd ed. Paris: Ponthieu, 1825.

Montulé, Édouard de. *Recueil des cartes et des vues du voyage en Amérique, en Italie, en Sicile et en Egypte, fait pendant les années 1816, 1817, 1818, et 1819.* N.p., n.d.

———. *Travels in America (1816–1817).* Trans. Edward D. Seeber. Bloomington: Indiana University Press, 1950.

Moreau-Zanelli, Jocelyne. *Gallipolis: Histoire d'un mirage américain au XVIIIe siècle.* Paris: L'Harmattan, 2000.

Morris, Gouverneur. *Le journal de Gouverneur Morris, 1789–1792, ministre plénipotentiaire des États-Unis en France.* Paris: Mercure de France, 2002.

———. *An oration, delivered on Wednesday, June 29, 1814, at the request of a number of citizens of New-York in celebration of the recent deliverance of Europe from the yoke of military despotism.* Salem, Mass.: Cushing & Appleton, 1814.

Moulard, Jacques (abbé). *Le comte Camille de Tournon, préfet de la Gironde (1815–1822).* Paris: E. Champion, 1914.

Murat, Inès. "Les exilés bonapartistes aux États-Unis." *Le Souvenir Napoléonien* 294 (July 1977): 14–20.

———. *Napoléon et le rêve américain.* Paris: Fayard, 1976.

Nelson, Soren, and Lucy Green Nelson. *A History of Church Street Graveyard, Mobile, Alabama.* [Mobile?]: n.p., 1963.

Nigoul, Toussaint. *Lakanal.* 1879. Nimes: C. Lacour, 2003.

Nolan, Charles E., and Earl C. Woods. *Sacramental Records of the Roman Catholic Church of the Archdiocese of New Orleans.* New Orleans: Archdiocese of New Orleans, 1986–2008. 19 vols. (1718–1831).

"Notes and Documents." *Pennsylvania Magazine of History and Biography* 63, no. 2 (1939): 195–96.

Nouvelle biographie générale. Ed. Dr. Hoefer. Vols. 29 and 45. Paris: Firmin Didot, 1859, 1866.

O'Brien, Greg. *Choctaws in a Revolutionary Age, 1750–1830.* Lincoln: University of Nebraska Press, 2002.

Ocampo, Emilio. *The Emperor's Last Campaign.* Tuscaloosa: University of Alabama Press, 2009. Published in Spanish as *La última campaña del emperador: Napoleón y la independencia de América.* Buenos Aires: Editorial Claridad, 2007.

O'Meara, Barry E. *Napoléon en exil, ou l'écho de Sainte-Hélène.* Brussels, 1824.

Orléans, Louis-Philippe de. *Mon journal—Événements de 1815.* Paris: Michel Lévy frères, 1849.

Overton, Grant. "Madame Bonaparte of Baltimore." *The Mentor* (October 1927): 15–22.

Owen, Marie Bankhead. *Alabama Census Returns 1820 and An Abstract of Federal Census of Alabama 1830.* Baltimore: Genealogical Publishing Company, 1971.

Owen, Marie Bankhead, and Thomas McAdory Owen. *The Story of Alabama: A History of a State.* Vol. 3. New York: Lewis Historical Publishing Company, 1949.

Owen, Thomas, and Marie Bankhead Owen. *History of Alabama and Dictionary of Alabama Biography.* Vol. 4. Chicago: S. J. Clarke, 1921.

P., J. S. "Jean Marie-Gabriel Thibière. *Revue du Lyonnais* vol. 4 (1836): 217–31.

Pagé, Sylvain. *L'Amérique du Nord et Napoléon.* Paris: Nouveau Monde Éditions/Fondation Napoléon, 2003.

Pairault, François. *Gaspar Monge: Le fondateur de Polytechnique.* Paris: Tallandier, 2000.

Parent, Jacques. *Charles-Louis Huguet de Sémonville: De Mirabeau à Louis-Philippe, haute politique et basses intrigues.* Paris: Éditions S.P.M., Kronos, 2002.

Pasquier, Étienne-Denis. *Histoire de mon temps: Mémoires du chancelier Pasquier.* Vol. 3. Paris: E. Plon, Nourrit, 1893–95.

Pate, James P. "The Fort of the Confederation: Spain on the Upper Tombigbee." *Alabama Historical Quarterly* 44 (Fall–Winter 1982): 171–86.

Paxton, John Adams, *The New-Orleans directory and register; containing the names, professions, & residences, of all the heads of families, and persons in business, of the city and suburbs; notes on New-Orleans; with other useful information.* New Orleans: Benj. Levy & Co., 1822–1830s.

Penney, Annette A. "North Carolina Scuppernong: Jefferson's 'Exquisite Wine.'" In *Jefferson and Wine: Model of Moderation,* ed. R. De Treville Lawrence III, 239–46. The Plains, Va.: Vinifera Wine Growers Association, 1989.

Penot, Jacques. "Les relations entre la France et le Mexique de 1808 à 1840. Un chapitre d'histoire écrit par les marins et diplomates français." Diss., Paris X, 1976.

Petiteau, Natalie. *Élites et mobilités: La noblesse d'Empire au XIXe siècle (1809–1914),* Paris: La boutique de l'histoire, 1997.

———. *Lendemains d'Empire: Les soldats de Napoléon dans la France du XIXe siècle.* Paris: La boutique de l'histoire, 2003.

Philips, Edith. *Les réfugiés bonapartistes en Amérique (1815–1830).* Paris: Éditions de la "vie universitaire," 1923.

Phillips, Ulrich Bonnell. *American Negro Slavery: A Survey of the Supply, Employment and Control of Negro Labor as Determined by the Plantation Regime.* Baton Rouge: Louisiana State University Press, 1969.

Pickett, Albert James. *History of Alabama, and Incidentally of Georgia and Mississippi from the Earliest Period.* 3rd ed. Vol. 2. New York: Arno Press and the New York Times, 1971.

Pigeard, Alain. *Les étoiles de Napoléon: Maréchaux, amiraux, généraux.* Paris: Éd. Quatuor, 1996.

Pinney, Thomas. *A History of Wine in America: From the Beginnings to Prohibition.* 2nd ed. Berkeley: University of California Press, 2007.

Pion des Loches, Antoine Augustin. *Mes campagnes (1792–1815).* Paris: Firmin Didot, 1889.

Plaisance, Fr. Aloysius. "The Choctaw Trading House—1803–1822." *Alabama Historical Quarterly* 16, no. 1 (1954): 393–423.

Planat de la Faye, Nicolas-Louis. *Vie de Planat de la Faye: Souvenirs, lettres et dictées recueillis et annotés par sa veuve.* Paris: Ollendorff, 1895.

Planchot-Mazel, Françoise. *Un général français aux États-Unis de 1816 à 1831: Simon Bernard.* 2 microfiches. Lille 3: ANRT, 1989.

Poussin, Guillaume-Tell. *De la puissance américaine: Origine, institution, esprit politique, ressources militaires, agricoles, commerciales et industrielles des États-Unis.* Vol. 2. Paris, 1843.

———. *Les États-Unis d'Amérique.* Paris, 1874.

Poussou, Jean-Pierre. *Bordeaux et le Sud-Ouest au XVIIIe siècle: Croissance économique et attraction urbaine.* Paris: EHESS, 1983.

Prime, Alfred Coxe. *The Arts and Crafts in Philadelphia, Maryland and South Carolina 1786–1800.* Topsfield, Mass.: The Walpole Society, 1932.

"Procès des frères Faucher." *Bibliothèque historique ou recueil de matériaux pour servir à l'histoire du temps.* Paris, 1818–20, 1819: 173–98.

Procès ou assassinat juridique de Louis XVI, roi de France et de Navarre. Avignon: P. Chaillot Jeune, 1814.

Procès-verbaux du Comité d'Instruction publique de la Convention nationale. Vol. 1. Paris: M. J. Guillaume, 1891–1907.

Quintin, Danielle, and Bernard Quintin. *Dictionnaire des colonels de Napoléon.* Paris: SPM, 1996.

———. *La Tragédie d'Eylau, 7–8 février 1807: Dictionnaire biographique des officiers, sous-officiers et soldats tués ou blesses mortellement au combat.* Paris: Archives et Culture, 2006.

Quoy-Bodin, Jean-Luc. "La Franc-maçonnerie dans les armies de la Révolution et de l'Empire: Le cas des généraux." *Revue de l'Institut Napoléon* (1981): 68–89.

Ragsdale, Bruce A., and Kathryn Allamong Jacob. *Biographical Directory of the American Congress, 1774–1989.* Washington, D.C.: U.S. G. P. O., 1989.

Rea, Robert R. "The Trouble at Tombeckby." *Alabama Review* 21, no. 1 (1968): 21–39.

Read, William A. *Indian Place Names in Alabama.* 1937. Rev. ed., with a foreword, appendix, and index by James B. McMillan. University: University of Alabama Press, 1984.

Réal, Pierre-François, comte de. *Les indiscrétions d'un préfet de police de Napoléon* Paris: Tallandier, 1986.

Redon, Stephan. "La révolution bordelaise du 12 mars 1814: Ses racines, ses acteurs, son déroulement et sa portée." Diss. U. of Paris I–Sorbonne, 1993.

Reeves, Jesse S. *The Napoleonic Exiles in America: A Study in American Diplomatic History, 1815–1819.* Baltimore: Johns Hopkins Press, 1905.

Remington, W. Craig, and Thomas L. Kallsen, eds. *Historical Atlas of Alabama.* Vol. 1, *Historical Locations by County.* Tuscaloosa: Department of Geography, College of Arts and Sciences, University of Alabama, 1997.

Remini, Robert V. *Andrew Jackson and the Course of American Empire, 1767–1821.* New York: Harper and Row, 1977.

"Reminiscences of the Vine and Olive Colony and Early Demopolis on Its 150th Anniversary." Demopolis Chamber of Commerce, August 1967.

Rémond, René. *Les États-Unis devant l'opinion française: 1815–1852.* Paris: Armand Colin, 1962.

———. *Histoire des États-Unis.* 18th ed. Paris: Presses universitaires de France, 1999.

Révérend, Albert. *Armorial du Premier Empire.* Vols. 3 and 4. Paris: Au bureau de l'annuaire de la noblesse, 1894–97.

Reynolds, Louise Webb. *The Heritage of Marengo County, Alabama.* Demopolis, Ala.: privately printed, 2000.

Rigau, Dieudonné, Col. *Souvenirs des guerres de l'Empire*. 1846. Paris: Librairie des Deux Empires, 2000.

Risnick, Daniel. *The White Terror and the Political Reaction after Waterloo*. Cambridge: Harvard University Press, 1966.

Robert, Adolphe, ed. *Dictionnaire des parlementaires français*. Vol. 5. Paris: Bourloton, 1891.

Robichaux, Albert, Jr. *Civil Registration of Orleans Parish: Births, Marriages, and Deaths, 1730–1833*, Rayne, La.: Hebert, 2000.

Robinson, James. *The Philadelphia Directory for 1810 containing the names, trades and residence of the inhabitants of the City, Southward, Northern Liberties and Kensington*. Philadelphia: Wm. Woodhouse, 1810.

Rogers, William Warren. *Alabama. The History of a Deep South State*. Tuscaloosa: University of Alabama Press, 1994.

Rosengarten, J. G. *French Colonists and Exiles in the United States*. 1907. Bowie, Md.: Heritage Books, 1989.

Ross, Michael. *The Reluctant King: Joseph Bonaparte, King of the Two Sicilies and Spain*. New York: Mason/Charter, 1976.

Rossignol, Marie-Jeanne. "Le contexte nord-américain de l'antiesclavagisme brittanique: le débat atlantique sur l'esclavage et l'abolition 1688–1787." In *Le débat sur l'abolition de l'esclavage en Grande-Bretagne (1787–1840)*, ed. Françoise Le Jeune and Michel Prum, 63–88. Paris: Ellipses, 2008.

———. *Le ferment nationaliste. Aux origines de la politique extérieure des États-Unis: 1789–1812*. Paris: Belin, 1994.

Rostlund, E. "The Myth of a Natural Prairie Belt in Alabama: An Interpretation of Historical Records." *Annals of the Association of American Geographers* 47 (1957): 392–411.

Rouillé, Michel Dugast. *Les notables ou la "seconde noblesse."* Vol. 2. Nantes: M. Dugast Rouillé, 1979.

Rousseau, Jean-Jacques. *Considérations sur le gouvernement de Pologne et sur sa Réformation projetée, et Lettres sur la legislation de la Corse dans les quelles tous les souverains trouveront des choses utiles*. La Hay and Lausanne, 1783.

———. *The Social Contract and the First and Second Discourses*. Ed. Susan Dunn. New Haven: Yale University Press, 2002.

Rush, Richard, and William L. Adams. *Report from the Secretary of the Treasury (in compliance with a resolution of the Senate of 20th May, 1826) in relation to grants of land made to French emigrants, to encourage the cultivation of the vine and olive*. Washington, D.C.: Duff Green, 1827.

Russell, Doris McAlpin. *McAlpin(e) Genealogies, 1730–1990*. Baltimore: Gateway Press, 1990.

Saffell, William Thomas Roberts. *The Bonaparte-Patterson Marriage in 1803 and the secret correspondence on the subject never before made public*. Philadelphia: The Proprietor, 1873.

Sainte-Marie, Georges de. "Descendance aux États-Unis des Du Bourg de Sainte-Colombe." *G.H.C. Bulletin* 84 (July–August 1996): 1684–1688.

Saint-John de Crèvecoeur, J. Hector. *Lettres d'un cultivateur américain à W[illiam] S[etton] écuyer, depuis l'année 1770 jusqu'à 1781*. 2 vols. Paris: Cuchet, 1784.

Sakolski, Aaron M. *The Great American Land Bubble. The Amazing History of Land-Grabbing,*

Speculations and Booms from Colonial Days to the Present Time. New York: Harper & Brothers, 1932.

Sala-Molins, Louis. *Les misères des Lumières: Sous la raison, l'outrage.* 1992. Paris: Homnisphères, 2008.

Sanford, Thaddeus, Putnam P. Rea, and John W. Townsend. *Mobile Commercial Register.* Mobile, Ala.: J. Batelle & J. W. Townsend, 1821–32.

Sapori, Julien. *L'exil et la mort de Joseph Fouché.* Parçay-sur-Vienne: Editions Vitae-Anovi, 2007.

Saugera, Eric. "Pour une histoire de la traite négrière française sous le Consulat et l'Empire." *Revue française d'Histoire d'Outre Mer* 76 (1989): 203–29.

———. "Renaître en Amérique? Réfugiés et exilés français aux États-Unis: L'aventure de la vigne et de l'olivier (1815–1865)." Diss. U Nantes, 2007.

———. "'She Is a Black Woman': Les affrontements raciaux du côté de l'U.S. Route 80 en Alabama (mars–septembre 1965)." In *Le monde créole: Peuplement, sociétés, et condition humaine: Mélanges Hubert Gerbeau,* ed. Jacques Weber, 477–88. Paris: Les Indes Savantes, 2005.

Savary, Anne-Jean-Marie-René, duc de Rovigo. *Mémoires du duc de Rovigo.* 2nd ed. Vol. 8. Paris: A. Bossange, 1829.

Schaeper, Thomas J. *France and America in the Revolutionary Era: The Life of Jacques-Donatien Leray de Chaumont (1725–1803).* Providence: Berghahn Books, 1995.

Schneider, Christian. "Les complots sous la Terreur Blanche (1815–1818)." 2 vols. Diss. U. of Paris I–Panthéon Sorbonne, 2006.

Seeber, Edward D. "A Napoleonic Exile in New Albany." *Indiana Magazine of History* 44 (1948): 175–77.

Sellers, James Benson. *Slavery in Alabama.* Tuscaloosa: University of Alabama Press, 2002.

Serme, Jean-Marc. "Le traité de Fort Jackson, 9 août 1814: Andrew Jackson et la création d'un nouveau Sud." *2002 Jeune République. Transatlantica,* American studies journal: www.transatlantica.org.

Serna, Pierre. *La république des girouettes, 1789–1815 . . . et au-delà.* Seyssel: Champ Vallon, 2005.

Ship Registers and Enrollments of New Orleans, Louisiana. Vol. 1. University, La.: Hill Memorial Library, Louisiana State University, 1941–42.

Six, Georges. *Dictionnaire biographique des généraux et amiraux français de la Révolution et de l'Empire: 1792–1814.* 2 vols. Paris: Georges Saffroy, 1934.

———. *Les généraux de la Révolution et de l'Empire.* Paris: Bordas, 1947.

Smith, Murphy D. "Peter Stephen Du Ponceau and His Study of Languages: A Historical Account." *Proceedings of the American Philosophical Society* 127, no. 3 (1983): 143–80.

Smith, Winston. *Days of Exile: The Story of the Vine and Olive Colony in Alabama.* Tuscaloosa, Ala.: W. B. Drake, 1967.

———. *The People's City: The Glory and the Grief of an Alabama Town, 1850–1874.* Demopolis, Ala.: Marengo County Historical Society: 2003.

Snedecor, V. Gayle. *A Directory of Greene County for 1855–6: Embracing the Names of the Voters in the County (. . .).* Mobile: Strickland, 1856.

Soublin, Jean. *Le Champ d'Asile*. Paris: Seuil, 1985. Republished as *La république des vaincus*. Paris: Phébus, 2004.

Soulié, Maurice. *Autour de l'aigle enchaîné: Le complot du Champ d'Asile*. Paris: Marpon, 1929.

Southeast Regional Climate Center. "Historical Climate Summaries for Alabama." 2003. http://www.sercc.com/climateinfo/historical/historical_al.html.

Southerland, Henry deLeon, Jr., and Jerry Elijah Brown, *The Federal Road through Georgia, the Creek Nation, and Alabama, 1806–1836*. Tuscaloosa: University of Alabama Press, 1989.

Staël-Holstein, Germaine de. *Considérations sur la Révolution française: œuvres posthumes de Madame la baronne de Staël-Holstein*. Genève: Slatkine Reprints, 1967.

Stanton, Lucia C. "Reforming His Nation's Taste." In *Jefferson and Wine: Model of Moderation*, ed. R. De Treville Lawrence III, 102–7. The Plains, Va.: Vinifera Wine Growers Association, 1989.

Stendhal. *Souvenirs d'égotisme*. Paris: Folio classique-Gallimard, 1983.

——. *Vie de Napoléon*. In *Œuvres completes*. Ed. Victor del Litto and Ernest Abravanel. Vol. 39. N.p.: [Levallois-Perret] distribué par le Cercle du bibliophile, 1970.

Sterne, Emma Gelders. *Some Plant Olive Trees*. New York: Dodd, Mead, 1937.

Stollenwerck, Frank, and Dixie Orum Stollenwerck. *The Stollenwerck, Chaudron and Billon Families in America*. [Baltimore]: n.p., 1948.

Strickland, Jean, and Patricia N. Edwards. *Residents of the Southeastern Mississippi Territory*. Moss Point, Miss.: J. Strickland, 1995.

Strode, Hudson. "Bonapartists in the Alabama Canebrakes." *Travel*, April 1938.

Stroud, Patricia Tyson. *The Man Who Had Been King: The American Exile of Napoleon's Brother Joseph*. Philadelphia: University of Pennsylvania Press, 2005.

Summersell, Charles Grayson. *Alabama History for Schools*. 4th ed. Montgomery, Ala.: Viewpoint Publications, 1970.

Survey of Federal Archives in Louisiana. *Ship Registers and Enrollments of New Orleans, Louisiana*. Vol. 1. University: Hill Memorial Library, Louisiana State University, 1941–42.

Swanton, John R. *Source Material for the Social and Ceremonial Life of the Choctaw Indians*. Washington, D.C.: G.P.O., 1931.

Taillemite, Étienne. *Dictionnaire des marins français*. Paris: Tallandier, 2002.

——. *La Fayette*. Paris: Fayard, 1989.

Tepper, Michael H., and Elizabeth Petty Bentley, eds. *Passenger Arrivals at the Port of Baltimore (1820–1834), from Customs Passenger Lists*. Baltimore: Genealogical Publishing Company, 1982.

——, eds. *Passenger Arrivals at the Port of Philadelphia, 1800–1819: The Philadelphia Baggage Lists*. Baltimore: Genealogical Publishing Company, 1986.

Tharin, W. C. *A Directory of Marengo County, for 1860–61: Embracing the Names of the Voters in the County . . .* Mobile: Farrow & Dennet, 1861.

Thibaudeau, Antoine-Claire. *Mémoires de A.-C. Thibaudeau, 1799–1815*. Paris: Plon, 1913.

Thiers, Adolphe. *History of the Consulate and the Empire of France under Napoleon*. Vol. 11. Philadelphia: Lippincott, 1894.

Thiessé, Léon. *M. Étienne: Essai biographique et littéraire*. Paris: Firmin Didot, 1853.

Tocqueville, Alexis de. *Journey to America.* Trans. George Lawrence. Ed. J. P. Mayer. New Haven: Yale University Press, 1960.

Toledano, Roulhac. *The National Trust Guide to New Orleans.* New York: Wiley, 1996.

Tournier, Albert. *Les conventionnels en exil.* Paris: Flammarion, 1910.

Tudesq, André. "La Restauration: Renaissance et déceptions." *Histoire de Bordeaux.* Vol. 6. *Bordeaux au XIXe siècle.* Bordeaux: Fédération historique du Sud-Ouest, 1969.

Tulard, Jean. "Les épurations de 1814 et 1815." *Le Souvenir Napoléonien* 396 (July–August 1994): 5–21.

———. *Joseph Fouché.* Paris, Fayard: 1997.

———. *Napoléon ou le mythe du sauveur.* Paris: Fayard, 1987.

———. *Nouvelle bibliographie critique des mémoires sur l'époque napoléonienne.* Geneva: Droz, 1991.

———. *Les vingt jours: Napoléon ou Louis XVIII?* Paris: Fayard, 2001.

United States. Census Office. Fourth Census, 1820. *Population Schedules of the Fourth Census of the United States 1820.* Louisiana. Reel 3. Washington, D.C.: National Archives, GSA, 1958. 3 microfilm reels.

United States. Department of Commerce. Bureau of the Census, *Negro Population, 1790–1915.* Washington, D.C., 1918.

U.S. Census, 1810. County of Philadelphia—outside city of Philadelphia—Northern Liberties.

Valette, Jean. "La situation religieuse de l'archiprêtré de Pellegrue en 1765." *Revue Historique et Archéologique du Libournais* 169, no. 3 (1978): 90–93.

Vapereau, Gustave. *Dictionnaire universel des littératures.* Paris: Hachette, 1884.

Vaulabelle, Achille de. *Histoire des deux Restaurations.* Vol. 2. Paris: Perrotin, 1845–54.

Vauthier, Gabriel. "Notes sur les Français retirés aux États-Unis à l'issue des Cent-Jours de 1815, par M. de Maud'huy." *Revue des Études Napoléoniennes* 32 (1931): 45.

Vidalenc, Jean. *Les demi-solde.* Paris: Librairie Marcel Rivière et cie, 1955.

Villepin, Dominique de. *Les Cent-Jours ou l'esprit de sacrifice.* Paris: Perrin, 2002.

Wallace, Anthony F. C. *Jefferson and the Indians: The Tragic Fate of the First Americans.* Cambridge, Mass.: Belknap Press of Harvard University, 1999.

Walthall, John A. *Moundville: An Introduction to the Archaeology of a Mississippian Chiefdom.* Special Publication no. 1 of the Alabama Museum of Natural History. 2nd ed. Tuscaloosa: University of Alabama Press, 1994.

Ward, Townsend. "The Germantown Road and Its Associations." *Pennsylvania Magazine of History and Biography* 5, no. 2 (1881): 121–25.

Waresquiel, Emmanuel de. *Cent Jours, la tentation de l'impossible, mars–juillet 1815.* Paris: Fayard, 2008.

———. *Le duc de Richelieu: Un sentimental en politique.* Paris: Perrin, 1990.

———. *Talleyrand, le prince immobile.* Paris: Fayard, 2003.

Waresquiel, Emmanuel de, and Benoît Yvert. *Histoire de la Restauration.* Paris: Perrin, 2002.

Warren, Alfred P. "Arcola Cemetery" [Ravesies Cemetery]. Greensboro. Unpublished paper. April 2000.

———. *Something of Pride: An Historical Novel of the Hatch House.* Greensboro, Ala., 1999.

Warren, Harris G. *The Sword Was Their Passport.* Port Washington, N.Y.: Kennikat Press, 1972.

Watel, Françoise. *Jean-Guillaume Hyde de Neuville (1776–1857): Conspirateur et diplomate.* Paris: Ministère des affaires étrangères, Direction des archives et de la documentation, 1997.

Weil, François. "L'état et l'émigration de France." *Citoyenneté et émigration: Les politiques du depart,* ed. Nancy L. Green and François Weil, 119–35. Paris: Éditions de l'EHESS, 2006.

———. "Migrations francophones en Amérique du Nord: enjeux et questionnements." In *Les émigrants préférés: Les Belges,* ed. Serge Jaumain, 27–33. Ottawa: Presses universitaires d'Ottawa, 1999.

———. "Migrations, migrants, et ethnicité." In *Chantiers d'histoire américaine,* ed. Jean Heffer and François Weil, 407–32. Paris: Belin, 2000.

Wellington, Duke of (Arthur Wellesley). *Supplementary Despatches, Correspondence and Memoranda of Field Marshal Arthur Duke of Wellington.* 15 vols. London: J. Murray, 1858–72.

Wells, Samuel J. "Federal Indian Policy: From Accommodation to Removal." In *The Choctaw before Removal,* ed. Carolyn Keller Reeves, 181–213. Jackson: University Press of Mississippi, 1985.

Welvert, Eugène. "Lakanal à Villarceaux." *Mémoires de la Société Historique et Archéologique de Pontoise* 32 (1913): 86–119.

Whitfield, Gaius, Jr. "The French Grant in Alabama: A History of the Founding of Demopolis." *Transactions of the Alabama Historical Society, 1899–1903* 4 (1904): 321–55. Reprinted in Marie Bankhead Owen and Thomas McAdory Owen, *The Story of Alabama: A History of a State,* vol. 3 (New York: Lewis Historical Publishing Company, 1949), 10–29.

Wilburn, Bessie Paterson. "A History of the Old French Gun of Demopolis." *Alabama Historical Quarterly* 6, no. 1 (1944): 71–75.

Wildes, Harry Emerson. *Lonely Midas: The Story of Stephen Girard.* New York: Farrar & Rinehart, 1963.

Wilkins, Joe Bassette. *Colonial Wars of North America, 1512–1763. An Encyclopedia.* New York: Garland, 1996.

———. "Outpost of Empire: the Founding of Fort Tombecbé and de Bienville's Chickasaw Expedition of 1736." In *Proceedings of the 12th Colloquium of the French Colonial Historical Society, Ste-Genevieve, May 1986,* 133–53. Lanham, Md.: University Press of America, 1988.

Wilson, Allen Gay, and Roger Asselineau. *Saint John Crèvecœur: The Life of an American Farmer.* New York: Viking, 1987.

Wilson, Brenda S. "Catherine-Victoire George: A French Alabama Settler, a Southern Lady." March 1979. Paper written under the direction of Dr. Elizabeth Meese, Adjunct Professor of Women's Studies at the University of Alabama.

Wilson, James Grant, and John Fiske, eds. *Appleton's Cyclopaedia of American Biography.* Vol. 5. New York: D. Appleton, 1887–1900.

Wolfe, Suzanne. "Exiles in Alabama: Napoleonic Exile's Letters Tell Tale of Attempt to

Establish a Vine and Olive Colony in Demopolis." *Tuscaloosa News,* Sunday, February 23, 2003, section E.

———. "Vine and Olive Colony: The Unfolding Legacy." *Alabama Heritage* 81 (Summer 2006): 35.

Wood, Gordon. *The Radicalism of the American Revolution.* New York: Knopf, 1992.

Woods, Earl C., Charles E. Nolan, and Dorenda Dupont. *Sacramental Records of the Roman Catholic Church of the Archdiocese of New Orleans.* Vol. 14. New Orleans: Archdiocese of New Orleans, 1999.

Wyler, Seymour B. *The Book of Old Silver, English, American, Foreign.* New York: Crown, 1937.

Yans-McLaughlin, Virginia, ed. *Immigration Reconsidered: History, Sociology, and Politics.* New York: Oxford University Press, 1990.

Yerby, William Edward Wadsworth, and Mabel Yerby Lawson. *History of Greensboro, Alabama from Its Earliest Settlement.* 1908. Northport, Ala.: Colonial Press, 1963.

Yvert, Benoît. *Politique libérale: Bibliographie sélective du libéralisme français (1814–1875).* Paris: Librairie Historique le Conservateur, 1994.

———. "Les terreurs blanches: Essai sur une typologie des terreurs (1795–1815)." In *Les révolutions françaises,* ed. Frédéric Bluche and Stéphane Rials, 273–89. Paris: Fayard, 1989.

Zieseniss, Charles-Otto. "Fouché et Napoléon." *Le Souvenir Napoléonien* 382 (April 1992): 2–16.

Zins, Ronald. *Revue du Souvenir Napoléonien* 429 (June–July 2000): 3–33.

Zweig, Stephan. *Joseph Fouché.* 1929. Paris: Livre de Poche-Grasset, 1960.

Index

Parmantier group, 323
Pasquier, Étienne-Denis, 41, 439n7
Pastol de Keramelin, Pierrette Julienne Basire ("Julie"): as member of Colonial Society of French Emigrants, 187; wife of General Pastol de Keramelin, 185, 187, 470n19
Pastol de Keramelin, Yves-Marie (General; Baron): marriage, 185, 187, 470n19; as member of Colonial Society of French Emigrants, 185, 187
Paternelle, La (insurance company), 371, 506n179
Patterson, Elizabeth, 443n67; marriage between Bonaparte, Jerome, and, 68, 88
Patton, Reverend (Constantine's eulogizer), 335
Pavans de Ceccaty, Joséphine Françoise Octavie, 504n156
Payen, Auguste, as former colonist, 295
Payen, César, death of, 338
Payen, Claude, 321; as former colonist, 295; as member of Colonial Society, 186
Payen, Louise, 321
Payen family, burial of, 337
Pease, M., Jr., signing of contract and, 246
Pegaux, Pierre, 132
Pénières, Émile, 464n72
Pénières, Luc-François, 434n52
Pénières, Raymond, 434n55
Pénières Delzors, Jean Augustin, 6, 8, 141, 142, 143, 144, 148, 223, 434n52, 434n55, 502n114; on choice of town name, 221; correspondence of, 283–84; death of, 332, 359, 502n115; exile of, 39–40, 295–96; journey to Philadelphia, 63; land grant for, 220; land owned by, 236; as member of Colonial Society of French Emigrants, 142, 149, 150, 177; as regicide, 63, 352, 359; reservations plan for community system, 228; as subagent of Indian affairs in Florida, 331–32
Penn, John, 132
Pennsylvania, 3, 53, 99, 119, 121, 180, 453n51, 491n44; abolishment of slavery in, 271, 273–74; 278, 485n52; attorney general of, 111; as free state, 274; Germantown, 112; Grand Lodge of, 109–10; hospital for insanity of, 185; in the Revolutionary War, 145; winegrowing and making in, 130, 135,

139, 141, 313, 459n21. *See also* Philadelphia; Pittsburgh
Pennsylvania Abolition Society, 485n52
Pensacola, Florida, 166, 167
Peraldi, Toussaint: Colonial Society of French Emigrants membership and, 184; exile of, 70
Percier (architect), 62
Perdido Vineyards, 376
Perdreauville, René Élisabeth de David de, as member of Colonial Society of French Emigrants, 188
Pérignon, Marguerite Agnès, death of, 349
Pérignon, Marshal, 145
Périgord, cost of labor in, 310
Perpignan, 122
Perrégaux, 452n24
Perret, Elisabeth, 343
Perry County, Alabama, 477–78n91
Persian (wine variety), 325
Pestalozzi (Swiss pedagogue), 120, 457n130
Petit-Vanjeu (farm), 187
Pétry, Consul Jean Baptiste, 90–91, 94, 234–35, 295, 353, 356, 447n20; complaints by about military rank inflation, 194
Pfeuty (Guard artilleryman), travels of, 202
Pfister, from Domingan community, 322; Pfister family, burial of, 337
Philadelphia, 73, 90, 92, 97, 99, 100, 102, 103, 112, 117, 120, 135, 204, 207, 213, 237, 238, 239, 278, 311, 321, 322, 328, 330, 333, 343, 350, 352, 353, 357, 362, 377, 379, 447n20, 448n45, 451n9, 451n20, 453n51, 455n79, 461n1, 462n22, 467n47, 467n49, 471n20, 473n10, 489n14, 495n124; Colonial Society of French Emigrants, base for, 1, 2, 4, 7, 63, 72, 99, 104, 142–56, 176–98, 201, 202, 227, 249, 279, 283, 295, 370, 373, 374, 378; disembarkation point, 7, 53, **61**, 62, 63–64, 68, 69–71, 94, 105, 134, 141, 231, 238, 323, 439n70, 440n23, 441n32, 443n69, 445n93, 452n33, 452n35, 455n96, 457n132, 457n134, 463n31, 489–90n25, 503n138; Domingan refugees in, 9, 105, 145, 270, 336, 341, 477n72; drawing of allotments in, 221, 226–27, 230, 238–39, 249, 300, 306, 329, 336, 374–75; embarkation point, **212**, 213, 215, 236, 278,